T0325512

IONIC LIQUIDS UnCOILED

Critical Expert Overviews

Edited by

Natalia V. Plechkova
The Queen's University of Belfast

Kenneth R. Seddon
The Queen's University of Belfast

A JOHN WILEY & SONS, INC., PUBLICATION

For general information on our other products and services or for technical support, please
contact our Customer Care Department within the United States at (800) 762-2974, outside the
United States at (317) 572-3993 or fax (317) 572-4002.

Wiley also publishes its books in a variety of electronic formats. Some content that appears in
print may not be available in electronic formats. For more information about Wiley products,
visit our web site at www.wiley.com.

Library of Congress Cataloging-in-Publication Data:

Ionic liquids uncoiled : critical expert overviews / edited by Natalia V. Plechkova,
Kenneth R. Seddon.
    p. cm.
    Summary: "The book presents articles on topics at the forefront of ionic liquids research,
ranging from applied to theoretical, from synthetic to analytical, from biotechnology to
electrochemistry, from process engineering to nanotechnology"—Provided by publisher.
    Includes bibliographical references and index.
    ISBN 978-0-470-07470-1 (hardback)
    1. Ionic solutions. 2. Ionic structure. I. Plechkova, Natalia V., 1979– II. Seddon, Kenneth
R., 1950–
    QD541.I56 2012
    541'.3723–dc23
                                                                      2012020575

Printed in the United States of America

10 9 8 7 6 5 4 3 2 1

CONTENTS

COIL CONFERENCES

COIL-1	Salzburg	Austria	2005
COIL-2	Yokohama	Japan	2007
COIL-3	Cairns	Australia	2009
COIL-4	Washington	USA	2011
COIL-5	Vilamoura, Algarve	Portugal	2013
COIL-6	Seoul	Korea	2015
COIL-7	Ottawa[†]	Canada	2017
COIL-8	Belfast[†]	UK	2019

[†] Precise location still to be confirmed

PREFACE

So, why the strange title for this book? Is this just a book of conference proceedings based on lectures at COIL? How is this book different from the series of volumes that were edited with Prof. Robin D. Rogers based on symposia we had organised at American Chemical Society (ACS) meetings? We will attempt here to answer these questions, and others, and hence explain the philosophy behind this book.

COIL (Congress on Ionic Liquids) was a concept that originally arose in discussions with DeChema, following on from a meeting about the Green Solvents series of conferences, which have traditionally been held in Germany. It was noticed that the number of submitted papers on ionic liquids surpassed the total of all the other papers combined. At that time, the only international meeting on ionic liquids had been a NATO Workshop, held in Crete in 2000 (Rogers, R.D., Seddon, K.R., and Volkov, S. (eds.), *Green industrial applications of ionic liquids*, NATO Science Series II: Mathematics, Physics and Chemistry, Vol. 92, Kluwer, Dordrecht, 2002). This was correctly interpreted as a need for a truly dedicated international meeting on ionic liquids, and so COIL was born. With the invaluable support of DeChema, the first meeting was held in Salzburg (literally, and appropriately, "Salt Castle"), and it has since moved around the world biannually, in a carefully planned progression. These meetings have been a resounding success, and their timeline is given on the opposite page. It is, unquestionably, the foremost forum for showcasing and discussing the latest advances in ionic liquids. However, we, and others, resisted the temptation to produce proceedings volumes. Although having a certain value as a time capsule, such volumes date quickly, and individual chapters are poorly cited (as they usually appear around the same time in the primary journals). Nevertheless, taken together, the COIL speakers have been a remarkable group of chemists, chemical engineers, biochemists, and biologists. Surely such a talented assembly of scientists could make a valuable contribution, *en masse*, to the published literature. And so this volume, *Ionic Liquids UnCOILed*, slowly emerged from our collective mind: "UnCOILed" because, although every principal author has presented their work as COIL lectures, the chapter content in this book has never been presented there. We wrote to each of our selected authors (and, amazingly, only one turned us down!), and set them a difficult challenge. The letter we sent out included the following text: "The concept is to select the key speakers from COIL-1, COIL-2 and COIL-3 and invite them to write critical reviews on specific areas of ionic liquid chemistry.

The area we have selected for you is [...]. It is important to emphasise that these are meant to be critical reviews. We are not looking for comprehensive coverage, but insight, appreciation and prospect. We want the type of review which can be read to give a sense of importance and scope of the area, highlighting this by the best published work and looking for the direction in which the field is moving. We would also like the problems with the area highlighting, for example, poor experimental technique, poor selection of liquids, and variability of data. We hope you would like to be involved in this project, as we believe these books will define the field for the next few years." Indeed, we felt rather like Division Seven, contacting the Impossible Missions Force (in the original, more cerebral, TV series, not the recent films!), "Your mission [...] should you decide to accept it ..." However, to the best of our knowledge, the emails did not "self-destruct in five seconds," and the acceptance rate was beyond our best expectations. We also issued guidance as to which abbreviations to use, and so there is concordance between every chapter (unlike a recent book, which contained over 25 different abbreviations for a single ionic liquid!).

The quality and size of the reviews that we received meant that we had to revert to Wiley and ask permission to produce not one, but three books! Wiley generously agreed. Thus, this will be the first of three volumes. The following ones, at six monthly intervals, will be *Ionic Liquids Further UnCOILed* and *Ionic Liquids Completely UnCOILed*. All will contain overviews of the same critical nature.

We look forward to the response of our readership (we can be contacted at quill@qub.ac.uk). It is our view that, in the second decade of the 21st century, reviews that merely regurgitate a list of all papers on a topic, giving a few lines or a paragraph (often the abstract!) to each one, have had their day. Five minutes with an online search engine will provide that information. The value of a review lies in the expertise and insight of the reviewer—and their willingness to share it with the reader. It takes moral courage to say "the work of [...] is irreproducible, or of poor quality, or that the conclusions are not valid"—but in a field expanding at the prestigious rate of ionic liquids, it is essential to have this honest feedback. Otherwise, errors are propagated. Papers still appear using hexafluorophosphate or tetrafluoroborate ionic liquids for synthetic or catalytic chemistry, and calculations on "ion pairs" are still being used to rationalise liquid state properties! We trust that this volume, containing 11 excellently perceptive reviews, will help guide and secure the future of ionic liquids.

NATALIA V. PLECHKOVA
KENNETH R. SEDDON

ACKNOWLEDGEMENTS

This volume is a collaborative effort. We, the editors, have our names emblazoned on the cover, but the book would not exist in its present form without support from many people. First, we thank our authors for producing such splendid, critical chapters, and for their open responses to the reviewers' comments and to editorial suggestions. We are also indebted to our team of expert reviewers, whose comments on the individual chapters were challenging and thought provoking, and to Ian Gibson for his input to the cover design. The backing from the team at Wiley, led by Dr. Arza Seidel, has been fully appreciated—it is always a joy to work with such a professional group of people. Finally, this book would never have been published without the unfailing, enthusiastic support from Deborah Poland, Sinead McCullough, and Maria Diamond, whose patience and endurance never cease to amaze us.

N.V.P.
K.R.S.

CONTRIBUTORS

RIHAB AL SALMAN, Clausthal University of Technology, D-38678 Clausthal-Zellerfeld, Germany

MOHAMMAD AL ZOUBI, Clausthal University of Technology, D-38678 Clausthal-Zellerfeld, Germany

JARED L. ANDERSON, Department of Chemistry, The University of Toledo, Toledo, OH, USA

ROB ATKIN, Centre for Organic Electronics, Chemistry Building, The University of Newcastle, Callaghan, New South Wales 2308, Australia

NATALIA BORISENKO, Clausthal University of Technology, D-38678 Clausthal-Zellerfeld, Germany

TIMO CARSTENS, Clausthal University of Technology, D-38678 Clausthal-Zellerfeld, Germany

ANDRÉ B. DE HAAN, Eindhoven University of Technology/Dept. of Chemical Engineering and Chemistry Process Systems Engineering, P.O. Box 513, 5600 MB Eindhoven, The Netherlands

SHERIF ZEIN EL ABEDIN, Electrochemistry and Corrosion Laboratory, National Research Centre, Dokki, Cairo, Egypt

FRANK ENDRES, Clausthal University of Technology, D-38678 Clausthal-Zellerfeld, Germany

WERNER FREYLAND, Karlsruhe Institute of Technology, Kaiserstrasse 12, D-76128 Karlsruhe, Germany

CHRISTA M. GRAHAM, Department of Chemistry, The University of Toledo, Toledo, OH, USA

ROBERT HAYES, Centre for Organic Electronics, Chemistry Building, The University of Newcastle, Callaghan, New South Wales 2308, Australia

EKATERINA I. IZGORODINA, School of Chemistry, Monash University, Wellington Rd, Clayton, Victoria 3800, Australia

JUNKO KAGIMOTO, Department of Biotechnology, Tokyo University of Agriculture and Technology, Koganei, Tokyo 184-8588, Japan

PETER LICENCE, School of Chemistry, The University of Nottingham, NG7 2RD, UK

KEVIN R.J. LOVELOCK, School of Chemistry, The University of Nottingham, NG7 2RD, UK

WYTZE (G.W.) MEINDERSMA, Eindhoven University of Technology/Department of Chemical Engineering and Chemistry Process Systems Engineering, P.O. Box 513, 5600 MB Eindhoven, The Netherlands

HIROYUKI OHNO, Department of Biotechnology, Tokyo University of Agriculture and Technology, Koganei, Tokyo 184-8588, Japan

MARIJA PETKOVIC, Instituto de Tecnologia Química e Biológica, Universidade Nova de Lisboa, Av. da República, 2780-157, Oeiras, Portugal

JENNIFER M. PRINGLE, ARC Centre of Excellence for Electromaterials Science, Monash University, Wellington Road, Clayton, Victoria 3800, Australia

ALEXANDRA PROWALD, Clausthal University of Technology, D-38678 Clausthal-Zellerfeld, Germany

MARK B. SHIFLETT, DuPont Central Research and Development, Experimental Station, Wilmington, Delaware 19880, USA

CRISTINA SILVA PEREIRA, Instituto de Tecnologia Química e Biológica, Universidade Nova de Lisboa, Av. da República, 2780-157, Oeiras, Portugal, and Instituto de Biologia Experimental e Tecnológica (IBET), Apartado 12, 2781-901, Oeiras, Portugal

DEBORAH WAKEHAM, Centre for Organic Electronics, Chemistry Building, The University of Newcastle, Callaghan, NSW 2308, Australia

AKIMICHI YOKOZEKI, 32 Kingsford Lane, Spencerport, New York 14559, USA

ABBREVIATIONS

IONIC LIQUIDS

GNCS	guanidinium thiocyanate
[HI-AA]	hydrophobic derivatised amino acid
IL	ionic liquid
poly(RTILs)	polymerisable room temperature ionic liquids
[PSpy]$_3$[PW]	[1-(3-sulfonic acid)propylpyridinium]$_3$[PW$_{12}$O$_{40}$]·2H$_2$O

CATIONS

[1-C$_m$-3-C$_n$im]$^+$	1,3-dialkylimidazolium
[C$_n$mim]$^+$	1-alkyl-3-methylimidazolium
[Hmim]$^+$	1-methylimidazolium
[C$_1$mim]$^+$	1,3-dimethylimidazolium
[C$_2$mim]$^+$	1-ethyl-3-methylimidazolium
[C$_3$mim]$^+$	1-propyl-3-methylimidazolium
[C$_4$mim]$^+$	1-butyl-3-methylimidazolium
[C$_5$mim]$^+$	1-pentyl-3-methylimidazolium
[C$_6$mim]$^+$	1-hexyl-3-methylimidazolium
[C$_7$mim]$^+$	1-heptyl-3-methylimidazolium
[C$_8$mim]$^+$	1-octyl-3-methylimidazolium
[C$_9$mim]$^+$	1-nonyl-3-methylimidazolium
[C$_{10}$mim]$^+$	1-decyl-3-methylimidazolium
[C$_{11}$mim]$^+$	1-undecyl-3-methylimidazolium
[C$_{12}$mim]$^+$	1-dodecyl-3-methylimidazolium
[C$_{13}$mim]$^+$	1-tridecyl-3-methylimidazolium
[C$_{14}$mim]$^+$	1-tetradecyl-3-methylimidazolium
[C$_{15}$mim]$^+$	1-pentadecyl-3-methylimidazolium
[C$_{16}$mim]$^+$	1-hexadecyl-3-methylimidazolium
[C$_{17}$mim]$^+$	1-heptadecyl-3-methylimidazolium
[C$_{18}$mim]$^+$	1-octadecyl-3-methylimidazolium
[C$_8$C$_3$im]$^+$	1-octyl-3-propylimidazolium
[C$_{12}$C$_{12}$im]$^+$	1,3-bis(dodecyl)imidazolium
[C$_4$dmim]$^+$	1-butyl-2,3-dimethylimidazolium
[C$_4$C$_1$mim]$^+$	1-butyl-2,3-dimethylimidazolium

$[C_6C_{7O1}im]^+$	1-hexyl-3-(heptyloxymethyl)imidazolium
$[C_4vim]^+$	3-butyl-1-vinylimidazolium
$[D_{mvim}]^+$	1,2-dimethyl-3-(4-vinylbenzyl)imidazolium
$[(allyl)mim]^+$	1-allyl-3-methylimidazolium
$[P_n mim]^+$	polymerisable 1-methylimidazolium
$[C_2mmor]^+$	1-ethyl-1-methylmorpholinium
$[C_4py]^+$	1-butylpyridinium
$[C_4m_\beta py]^+$	1-butyl-3-methylpyridinium
$[C_4m_\gamma py]^+$	1-butyl-4-methylpyridinium
$[C_4mpyr]^+$	1-butyl-1-methylpyrrolidinium
$[C_2C_6pip]^+$	1-ethyl-1-hexylpiperidinium
$[C_8quin]^+$	1-octylquinolinium
$[H_2NC_2H_4py]^+$	1-(1-aminoethyl)-pyridinium
$[H_2NC_3H_6mim]^+$	1-(3-aminopropyl)-3-methylimidazolium
$[Hnmp]^+$	1-methyl-2-pyrrolidonium
$[N_{1\,1\,2\,2OH}]^+$	ethyl(2-hydroxyethyl)dimethylammonium
$[N_{1\,1\,1\,4}]^+$	trimethylbutylammonium
$[N_{4\,4\,4\,4}]^+$	tetrabutylammonium
$[N_{6\,6\,6\,14}]^+$	trihexyl(tetradecyl)ammonium
$[P_{2\,2\,2(1O1)})]$	triethyl(methoxymethyl)phosphonium
$[P_{4\,4\,4\,3a}]^+$	(3-aminopropyl)tributylphosphonium
$[P_{6\,6\,6\,14}]^+$	trihexyl(tetradecyl)phosphonium
$[S_{2\,2\,2}]^+$	triethylsulfonium

ANIONS

$[Ala]^-$	alaninate
$[\beta Ala]^-$	β-alaninate
$[Arg]^-$	arginate
$[Asn]^-$	asparaginate
$[Asp]^-$	aspartate
$[BBB]^-$	bis[1,2-benzenediolato(2-)-$O,O´$]borate
$[C_1CO_2]^-$	ethanoate
$[C_1SO_4]^-$	methylsulfate
$[CTf_3]^-$	tris{(trifluoromethyl)sulfonyl}methanide
$[Cys]^-$	cysteinate
$[FAP]^-$	tris(perfluoroalkyl)trifluorophosphate
$[Gln]^-$	glutaminate
$[Glu]^-$	glutamate
$[Gly]^-$	glycinate
$[His]^-$	histidinate
$[Ile]^-$	isoleucinate
$[lac]^-$	lactate
$[Leu]^-$	leucinate

[Lys]$^-$	lysinate
[Met]$^-$	methionate
[Nle]$^-$	norleucinate
[NPf$_2$]$^-$	bis{(pentafluoroethyl)sulfonyl}amide
[NTf$_2$]$^-$	bis{(trifluoromethyl)sulfonyl}amide
[O$_2$CC$_1$]$^-$	ethanoate
[O$_3$SOC$_2$]$^-$	ethylsulfate
[OMs]$^-$	methanesulfonate (mesylate)
[OTf]$^-$	trifluoromethanesulfonate
[OTs]$^-$	4-toluenesulfonate, [4-CH$_3$C$_6$H$_4$SO$_3$]$^-$ (tosylate)
[Phe]$^-$	phenylalaninate
[Pro]$^-$	prolinate
[Ser]$^-$	serinate
[Suc]$^-$	succinate
[Thr]$^-$	threoninate
[Trp]$^-$	tryphtophanate
[Tyr]$^-$	tyrosinate
[Val]$^-$	valinate

TECHNIQUES

AES	Auger electron spectroscopy
AFM	atomic force microscopy
ANN	associative neural network
ARXPS	angle resolved X-ray photoelectron spectroscopy
ATR-IR	attenuated total reflectance infrared spectroscopy
BPNN	back-propagation neural network
CCC	counter-current chromatography
CE	capillary electrophoresis
CEC	capillary electrochromatography
COSMO-RS	**CO**nductorlike**S**creening**MO**del for Real Solvents
COSY	**CO**rrelation **S**pectroscop**Y**
CPCM	conductor-like polarisable continuum model
CPMD	Car–Parrinello molecular dynamics
DFT	density functional theory
DRS	direct recoil spectroscopy
DSC	differential scanning calorimetry
DSSC	dye-sensitised solar cell
ECSEM	electrochemical scanning electron microscopy
EC-XPS	electrochemical X-ray photoelectron spectroscopy
EFM	effective fragment potential method
EI	electron ionisation
EOF	electro-osmotic flow
EPSR	empirical potential structure refinement

ES	electrospray mass spectrometry
ESI–MS	electrospray ionisation mass spectrometry
EXAFS	extended X-ray absorption fine structure
FAB	fast atom bombardment
FMO	fragment molecular orbital method
GC	gas chromatography
GGA	generalized gradient approximations
GLC	gas–liquid chromatography
GSC	gas–solid chromatography
HM	heuristic method
HPLC	high performance liquid chromatography
HREELS	high resolution electron energy loss spectroscopy
IGC	inverse gas chromatography
IR	infrared spectroscopy
IRAS	infrared reflection absorption spectroscopy
IR-VIS SFG	infrared visible sum frequency generation
ISS	ion scattering spectroscopy
L-SIMS	liquid secondary ion mass spectrometry
MAES	metastable atom electron spectroscopy
MALDI	matrix-assisted laser desorption
MBSS	molecular beam surface scattering
MIES	metastable impact electron spectroscopy
MLR	multi-linear regression
MM	molecular mechanics
MS	mass spectrometry
NMR	nuclear magnetic resonance
NR	neutron reflectivity
PDA	photodiode array detection
PES	photoelectron spectroscopy
PPR	projection pursuit regression
QM	quantum mechanics
QSAR	quantitative structure–activity relationship
QSPR	quantitative structure–property relationship
RAIRS	reflection absorption infrared spectroscopy
RI	refractive index
RNN	recursive neural network
RP-HPLC	reverse phase-high performance liquid chromatography
SANS	small angle neutron scattering
SEM	scanning electron microscopy
SFA	surfaces forces apparatus
SFC	supercritical fluid chromatography
SFG	sum frequency generation
SFM	systematic fragmentation method
SIMS	secondary ion mass spectrometry
STM	scanning tunnelling microscopy

SVN	support vector network
TEM	tunnelling electron microscopy
TGA	thermogravimetric analysis
TLC	thin layer chromatography
TPD	temperature programmed desorption
UHV	ultra-high vacuum
UPLC	ultra-pressure liquid chromatography
UPS	ultraviolet photoelectron spectroscopy
UV	ultraviolet
UV-Vis	ultraviolet-visible
XPS	X-ray photoelectron spectroscopy
XRD	X-ray powder diffraction
XRR	X-ray reflectivity

MISCELLANEOUS

Å	$1 \text{ Ångstrom} = 10^{-10} \text{ m}$
ACS	American Chemical Society
ATPS	aqueous two-phase system
BE	binding energy
BILM	bulk ionic liquid membrane
b.pt.	boiling point
BSA	bovine serum albumin
BT	benzothiophene
calc.	calculated
CB	Cibacron Blue 3GA
CE	crown ether
CLM	charge lever momentum
CMC	critical micelle concentration
CMPO	octyl(phenyl)-N,N-diisobutylcarbamoylmethylphosphine oxide
COIL	Congress on Ionic Liquids
CPU	central processing unit
d	doublet (NMR)
D°_{298}	bond energy at 298 K
2D	two-dimensional
3D	three-dimensional
DBT	dibenzothiophene
DC18C6	dicyclohexyl-18-crown-6
4,6-DMDBT	4,6-dimethyldibenzothiophene
DMF	dimethylmethanamide (dimethylformamide)
2DOM	two-dimensional ordered macroporous
3DOM	three-dimensional ordered macroporous
DOS	density of states

DPC	diphenylcarbonate
DRA	drag-reducing agent
DSSC	dye-sensitised solar cell
E	enrichment
EDC	extractive distillation column
EOR	enhanced oil recovery
EPA	Environmental Protection Agency
EPSR	empirical potential structure refinement
eq.	equivalent
FCC	fluid catalytic cracking
FFT	fast Fourier transform
FIB	focused ion beam
FSE	full-scale error
ft	foot
GDDI	generalised distributed data interface
HDS	hydrodesulfurisation
HEMA	2-(hydroxyethyl) methacrylate
HOMO	highest occupied molecular orbital
HOPG	highly oriented pyrolytic graphite
HV	high vacuum
IgG	immunoglobulin G
IPBE	ion pair binding energy
ITO	indium-tin oxide
IUPAC	International Union of Pure and Applied Chemistry
J	coupling constant (NMR)
LCST	lower critical separation temperature
LLE	liquid–liquid equilibria
LUMO	lowest unoccupied molecular orbital
m	multiplet (NMR)
M	molar concentration
MBI	1-methylbenzimidazole
MCH	methylcyclohexane
MD	molecular dynamics
MDEA	methyl diethanolamine; bis(2-hydroxyethyl)methylamine
MEA	monoethanolamine; 2-aminoethanol
MFC	minimal fungicidal concentrations
MIC	minimal inhibitory concentrations
MNDO	modified neglect of differential overlap
m.pt.	melting point
3-MT	3-methylthiophene
MW	molecular weight
m/z	mass-to-charge ratio
NBB	1-butylbenzimidazole
NCA	N-carboxyamino acid anhydride
NES	New Entrepreneur Scholarship

NFM	N-formylmorpholine
NIP	neutral ion pair
NIT	neutral ion triplet
NMP	N-methylpyrrolidone
NOE	nuclear Overhauser effect
NRTL	non-random two liquid
ocp	open circuit potential
p	pressure
PDMS	polydimethoxysilane
PEDOT	poly(3,4-ethylenedioxythiophene)
PEG	poly(ethyleneglycol)
PEN	poly(ethylene-2,6-naphthalene decarboxylate)
PES	polyethersulfone
pH	$-\log_{10}([H^+])$; a measure of the acidity of a solution
pK_b	$-\log_{10}(K_b)$
PPDD	polypyridylpendant poly(amidoamine) dendritic derivative
PS	polystyrene
psi	1 pound per square inch = 6894.75729 Pa
PTC	phase transfer catalyst
PTFE	poly(tetrafluoroethylene)
PTx	pressure–temperature–composition
r	bond length
RDC	rotating disc contactor
REACH	Registration, Evaluation, Authorisation and restriction of CHemical substances
RMSD	root-mean-square deviation
RT	room temperature
s	singlet (NMR)
S	entropy
$scCO_2$	supercritical carbon dioxide
SDS	sodium dodecyl sulfate
S/F	solvent-to-feed ratio
SILM	supported ionic liquid membrane
SLM	supported liquid membranes
t	triplet (NMR)
TBP	4-(t-butyl)pyridine
TCEP	1,2,3-tris(2-cyanoethoxy)propane
TEGDA	tetra(ethylene glycol) diacrylate
TMB	trimethylborate
TOF	time-of-flight
UCST	upper critical solution temperature
UHV	ultrahigh vacuum
UV	ultraviolet
VLE	vapour–liquid equilibria

VLLE	vapour–liquid–liquid equilibria
VOCs	volatile organic compounds
v/v	volume for volume
w/w	weight for weight
wt%	weight per cent
γ	surface tension
δ	chemical shift in NMR
X	molar fraction

1 Electrodeposition from Ionic Liquids: Interface Processes, Ion Effects, and Macroporous Structures

FRANK ENDRES, NATALIA BORISENKO,
RIHAB AL SALMAN, MOHAMMAD AL ZOUBI,
ALEXANDRA PROWALD, and TIMO CARSTENS

Clausthal University of Technology, Clausthal-Zellerfeld, Germany

SHERIF ZEIN EL ABEDIN

Electrochemistry and Corrosion Laboratory, National Research Centre, Dokki, Cairo, Egypt

ABSTRACT

In this chapter, we discuss the prospects and challenges of ionic liquids for interfacial electrochemistry and electrodeposition processes. In contrast to aqueous or organic solutions, ionic liquids form surprisingly strongly adhering solvation layers that vary with the applied electrode potential and that alter the tunnelling conditions in a scanning tunnelling microscopy (STM) experiment. Different cation–anion combinations can have an impact on the fundamental electrochemical processes, and the purity of ionic liquids is a key factor in interfacial electrochemistry. It is also shown that ionic liquids have a high potential in the making of three-dimensional ordered macroporous structures of semiconductors.

1.1 INTRODUCTION

Until about the year 2000, most papers dealing with electrochemistry or electrodeposition in or from ionic liquids used systems based on aluminium(III)

Ionic Liquids UnCOILed: Critical Expert Overviews, First Edition. Edited by Natalia V. Plechkova and Kenneth R. Seddon.
© 2013 John Wiley & Sons, Inc. Published 2013 by John Wiley & Sons, Inc.

chloride and 1,3-dialkylimidazolium ions, which were first reported in 1982 [1]. Although Walden reported in his paper from 1914 [2] on liquids that we often call today "air and water stable ionic liquids," a community of about 10–20 groups worldwide investigated AlCl₃-based liquids from 1948 onwards, which can only be handled under the conditions of an inert-gas dry box. One can speculate about the reasons, but one explanation might be that these liquids are (still) relatively easy to produce: mix carefully water-free aluminium(III) chloride with a well-dried organic halide (e.g., 1-butyl-3-methylimidazolium chloride) in a glove box and, depending on the ratio of the components, a Lewis acidic or a Lewis basic liquid is obtained. As aluminium(III) chloride adsorbs and reacts with water, even under the conditions of a dry box, some ageing takes place, which produces less defined oxochloroaluminates(III) by hydrolysis. As aluminium can be easily electrodeposited from these liquids, a common purification method is to perform a refining electrolysis with an aluminium anode and a steel cathode, leading to clear and well-defined electrolytes. A major review on these liquids was written by Hussey [3], and his article well summarises the prospects of these liquids, which were mainly used as electrolytes for the electrodeposition of aluminium and its alloys, and a few other metals [4]. One can say that, in 2000, there seemed to be no more surprises with these liquids and that electrochemical processes seemed to be well understood, except for, maybe, a few unusual observations (mainly reported in meetings), for example, that aluminium deposition is rather problematic if tetraalkylammonium ions are used instead of imidazolium ions. Furthermore, there was practically no understanding of the interfacial electrochemical processes. As these liquids can only be handled under the conditions of a dry box, scanning tunnelling or atomic force microscopy experiments (STM/AFM), which are well suited for such purposes, were extremely demanding. Nevertheless, one of the authors of this chapter (FE) and Freyland showed in a pioneering, but hardly cited, paper [5] that STM experiments can be performed in aluminium(III) chloride-based ionic liquids and that the surface of highly oriented pyrolytic graphite (HOPG) can be resolved atomically. In subsequent papers, it was shown that the surface and the initial deposition steps on Au(111) as a well-defined model surface can be probed in these liquids. Underpotential phenomena that can lead to alloying, sub-monolayer island deposition, and Moiré patterns [6–9] were found. Later results showed that the Au(111) surface seemed to be resolved atomically in a limited potential régime [10]. Although the latter result was a good step forward, the question remained as to why atomic resolution is much more difficult in ionic liquids than in aqueous solutions, where even atomic processes in real time were demonstrated [11].

In 1999, we started with what was considered at that time to be more or less air- and water-stable ionic liquids, namely 1-butyl-3-methylimidazolium hexafluorophosphate, which was home-made. The motivation was clear: aluminium(III) chloride-based liquids were well-defined liquids with many prospects, but the deposition of silicon, germanium, titanium, tantalum, and others was impossible. Either there was a co-deposition/alloying with aluminium or there was no deposition at all, presumably due to the complexing of the precursors. It can be said, open and above board, that our first cyclic voltammogram of [C$_4$mim][PF$_6$] on Au(111) was a nightmare: instead of the expected capacitive behaviour (flat line) within an expected electrochemical window of 4 V, a multitude of cathodic and anodic peaks appeared, of which the peaks attributed to chloride (from the synthesis) were still relatively easy to identify. The first STM experiments were also quite disappointing because, in contrast to the nice STM images in the aluminium(III) chloride-based liquids, even the normally easy to probe steps between the gold terraces were hardly seen. Thus, experiments were somehow unexpectedly complicated with these liquids. Without reporting the whole of the story, we found that there are (at least) three requirements for good and reproducible experiments with air- and water-stable ionic liquids: purity, purity, and again purity. With a step-by-step improvement of the quality of the liquid, we could show that the deposition of germanium, silicon, tantalum, Si$_x$Ge$_{1-x}$, and selenium is possible and that the processes can be well probed on the nanoscale with *in situ* STM. There have been attempts in the literature to deposit magnesium [12–14] and titanium [15], but according to our experience, magnesium has not (yet) been obtained as a pure phase [16], and instead of titanium, titanium sub-halides were obtained [17], and maybe Li-Ti alloys. As discussed in Reference [18], one should not forget that transition and rare earth elements can have a rich chemistry with covalent bonds to non-metallic main group elements. Thus, one should not expect that the electrodeposition of titanium from an ionic liquid is as simple as that of silver from an aqueous solution. Apart from these aspects, we have hitherto encountered several other puzzling phenomena, *inter alia*, which we cannot yet answer successfully:

1. We have not yet succeeded with STM in achieving atomic resolution of Au(111) in air- and water-stable ionic liquids.
2. In some ionic liquids, even HOPG can hardly be probed atomically, although this is quite easy under air with the same device.
3. The electrodeposition of aluminium is influenced by the cation. Even slight modifications in cation structure can alter the grain size of the resulting deposit.
4. The electrodeposition of tantalum from [C$_4$mpyr][NTf$_2$] is possible in a narrow potential régime, but much more difficult in [C$_4$mim][NTf$_2$].
5. The composition and thickness of electrochemically made Si$_x$Ge$_{1-x}$ is influenced by the ionic liquid species.

A question that we have to address is the following: What would happen if ionic liquids were strongly adsorbed on electrode surfaces, and if this adsorption was different with different liquids?

1.2   RESULTS AND DISCUSSION

1.2.1   Purity Issues

Before our view on electrodeposition processes in or from ionic liquids is discussed, we would like to draw the reader's attention to purity issues. Ten years ago, almost all ionic liquids were synthesised in individual laboratories. With some efforts, the quality of the liquids, and consequently the quality of the experimental results, was quite reproducible. Typical impurities were alkali metal ions (from the precursors), halides, and water. If the liquids were slightly coloured and if this colour interfered with the envisaged application (e.g., spectroscopy), they were purified over silica or alumina to produce clear liquids. At meetings dealing with ionic liquids, companies advertised their products and coloured liquids were criticised by most people, with a few exceptions. Today, there are many companies worldwide that make and sell ionic liquids. The commercial quality is surely much better than five years ago, but there are still some pitfalls. The supplier normally does not tell the consumer how the liquid was made, and in most cases there is no detailed analysis of the liquid. One finds, in the catalogues, purity levels of between 95% and 99%, and the consumer is normally left to interpret without guidance the nature of the 1–5% of whatever. The most pragmatic approach for an experiment is to test if a reaction or a process gives the desired results or not. From the several thousands of papers published in the field of ionic liquids, we dare concluding that a certain impurity level does not seem to alter tremendously organic, inorganic, or technical chemistry reactions (although halide impurities can poison catalytic processes). Even in the electroplating of aluminium, a certain water level does not seem to disturb the result. Can we now conclude that impurities do not generally disturb? Definitely not! In the last five years, many physicists and physical chemists have started fundamental physical experiments both in the bulk and at the interfaces. Due to their extremely low vapour pressures, ionic liquids are suited to ultra-high vacuum (UHV) experiments; for example, they can be studied with photoelectron spectroscopy (X-ray photoelectron spectroscopy [XPS], or ultraviolet photoelectron spectroscopy [UPS]). At a Bunsen Meeting in 2006 at Clausthal University of Technology, Maier and coworkers [19] presented a paper where they reported on silicon at the ionic liquid/UHV interface detected by XPS. The conclusion was that—somehow—during handling, the silicon grease used for sealing the glass flask contaminated the liquid. The paper was published in the journal *Zeitschrift für Physikalische Chemie* and has been well cited to date [19]. For an overview on the prospects of ionic liquids in surface science, we would like to refer to Reference [20]. We ourselves found, by *in situ* STM, that purification with alumina or silica introduces impurities that segregate on the electrode surface and that strongly disturb the image quality and the nanoscale processes [21]. MacFarlane and coworkers confirmed, by dynamic light scattering experiments, that this "purification" process introduces impurities [22].

From our point of view, for surface science experiments, such a post-treatment should be avoided, if possible. We have already found potassium impurities above the 100 ppm level, where the supplier correctly guaranteed for water, halide, and lithium below 10 ppm. When we found potassium, the supplier admitted, after our insistence, that the production process had been changed. In another (also apparently extremely pure) liquid, we found by XPS 10% of a 1,3-dialkylimidazolium salt in $[C_4mpyr][NTf_2]$. The supplier had no explanation for this observation. From all our experience with the peer-reviewing of papers, we have learnt that the customers often seem to rely "blind" on the suppliers. When we asked the authors to write a few words on the purity, we received comments such as "The liquid was pure within the given limits" or "We know that the impurities mentioned by the referee do not disturb" or "With higher quality liquids, we get exactly the same results," and so on. It is understandable that people want to get their results published in a science system based on international competition, but the question is why there are, for example, contradictory results on the double-layer behaviour of ionic liquids [23–26]. Surely all authors are convinced that they have carried out the best experiments, but what accounts for the different results? We have therefore decided to purchase solely custom-made liquids, where all our experience with the quality of ionic liquids is accounted for. Our recommendation for people interested in fundamental experiments with ionic liquids is the following: start with an extremely pure liquid where all impurities are below 10 ppm. This gives a reliable reference. Then check if the results are the same with lower quality liquids. In surface science in general, and electrochemistry in particular, the approach with low-quality liquids might be dangerous, as the wrong conclusions can be easily drawn. All liquids that we purchase now are subjected to a detailed analytical program. First, we dry them under vacuum at elevated temperature in beakers that were purified with H_2O_2/H_2SO_4 and ultrapure water. Then a cyclic voltammogram is acquired, from which we can already comment on the quality. As the last steps in the analysis, XPS and *in situ* STM experiments are performed. If all these results are convincing, the liquid is cleared for experiments in the laboratory.

1.2.2   Interfacial Layers and Scanning Probe Microscopy Studies

1.2.2.1   AFM Studies   When we realised that atomic resolution of Au(111) does not seem to be simple to achieve in air- and water-stable ionic liquids, we wondered what the reason might be for this. An assumption we mentioned at meetings in 2002 and 2003 was that maybe ionic liquids are strongly adsorbed on electrode surfaces, thus delivering a different tunnelling barrier from the one in aqueous solutions. In 2005 and 2006, we investigated the surface of Au(111) in extremely pure $[C_4mpyr][NTf_2]$ and found both a restructuring of the surface [27] and a surface structure that reminded us of the well-known herringbone superstructure known for Au(111). Furthermore, the grain size of electrodeposited aluminium is influenced by the cation, resulting in nanocrystalline aluminium from [C₄mpyr][NTf₂] instead of microcrystalline aluminium from [C₄mim][NTf₂]. We interpreted this surprising result as due to cation adsorption on the electrode surface and the growing nuclei [28]. We should mention that, in the following two years, we were unable to reproduce the results on the herringbone structure. Eventually, we found that some ionic liquid producers scaled up their processes from laboratory scale to pilot plants and that this obviously had an impact on the quality of ionic liquids, at least for experiments at interfaces. One of us (FE) came in touch with the studies performed by Rob Atkin, who showed by scanning force microscopy that ionic liquids are surprisingly strongly adsorbed on the surface of solid materials [29]. Some more papers appeared proving (with different liquids) Atkin's results. At a meeting at Monash University, Atkin and Endres decided to collaborate on experiments on the adsorption of ionic liquids on Au(111). For this purpose, we selected [C₄mpyr][NTf₂] and [C₄mim][NTf₂] from custom syntheses (with all detectable impurities below 10 ppm) to ensure the best possible experiments. The liquids were sealed into ampoules in Clausthal and sent, together with sealed Au(111) samples, to Atkin's laboratory. After a few weeks, Atkin sent us the curves (already evaluated) shown in Figure 1.1. The result is amazing: both liquids are adsorbed on the surface of Au(111), and 3–5 layers can be identified. The results further show that [C₄mpyr][NTf₂] is roughly four times more strongly adsorbed than [C₄mim][NTf₂]. The adsorption of ionic liquids on metal surfaces is, so far, not really surprising, as water and organic solvents are also adsorbed [30–32]; however, the force to rupture these layers is at least one order of magnitude higher than that for water or organic solvents, and a commercial AFM is sufficient for such experiments. This will have an impact on electrochemical processes and on the image quality in STM experiments. A further interesting result is that the ionic liquid/electrode interfacial layer consists of more than one layer. Since an ionic liquid is solvent and solute at the same time, there does not seem to be a simple electrochemical double layer, as concluded from theoretical considerations performed by Kornyshev and Fedorov [33]. The question now arises as to what the STM "sees" at the ionic liquid/electrode interface.

1.2.2.2   In Situ *STM Studies The *in situ* STM allows direct probing of the processes at the electrode/electrolyte interface at different electrode potentials. Although there is a huge community that applies this technique for electrochemical purposes in aqueous solutions, only two other groups (Freyland at Karlsruhe, and Mao at Xiamen) have hitherto published papers on *in situ* STM studies in ionic liquids. One reason might be that the experimental requirements are one order of magnitude more demanding than for aqueous solutions. From our experience, the quality of an ionic liquid can never be good enough for *in situ* STM studies.

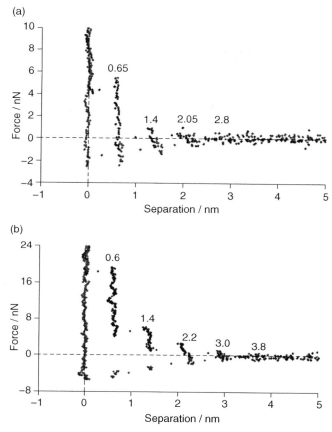

Figure 1.1   (a) Force versus distance profile for an AFM tip (Si_3N_4) approaching and retracting from a Au(111) surface in [C_4mpyr][NTf$_2$]. At least five steps in the force curve can be seen, extending to a separation of 3.8 nm. The force required to rupture the innermost layer, and move into contact with the gold surface, is 20 nN. The separation distances for each layer prior to push-through are given on the plot. (b) Force versus distance profile for an AFM tip approaching and retracting from a Au(111) surface in [C_4mim][NTf$_2$]. At least three steps in the force curve can be seen, extending to a separation of 3 nm. The force required to rupture the innermost layer, and move into contact with the gold surface, is 5 nN. The separation distances for each layer prior to push-through are given on the plot.

1.2.3   HOPG/[C_4mpyr][NTf$_2$]

Let us start with the investigation of the interface between HOPG and [C_4mpyr][NTf$_2$] (ultrapure liquid from Merck KGaA, Darmstadt, Germany). As already mentioned, it is quite difficult (but not impossible) to probe this surface at atomic resolution. Figure 1.2 shows a cyclic voltammogram of the electrode: the surface oxidation of graphite at the rims starts at moderate

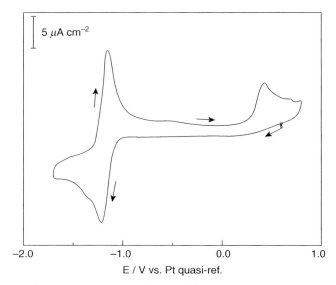

Figure 1.2 Cyclic voltammogram of [C₄mpyr][NTf₂] on HOPG at 25 °C at the scan rate of 10 mV s⁻¹.

potentials of just 0.3–0.4 V. We assume that the peak couple at −1.2 V is due to a surface process. If we now have a look at the interface at the open circuit potential (ocp) (Fig. 1.3a), we identify a flat surface with a defect in the middle. This defect is due to a step consisting of three graphene layers. The surface looks more or less normal, and one would interpret the horizontal stripes as a consequence of a direct tip/surface contact at the step. If we zoom in, however, we clearly see that there is a layer on the surface (Fig. 1.3b), and as a first interpretation one would say this is due to dirt adsorbed to the surface. This interpretation would be tempting and easy, but not understandable, as the liquid is of ultrapure quality. Furthermore, on Au(111), the same liquid does not show such a layer under the same experimental conditions. So, what else can it be? The variation of the electrode potential gives us hints that, at an electrode potential of just 600 mV more in the negative direction, a regular structure appears on the surface which looks hexagonal with a periodicity in the 8- to 10-nm régime (Fig. 1.3c). At slightly more negative electrode potentials, we see a cathodic peak in the cyclic voltammogram. One should not overinterpret such STM images, but from our experience "real" dirt does not give such regular structures upon changing the electrode potential. We can therefore assume that the ionic liquid is subject to a hexagonal superstructure on the hexagonal graphite surface and this reorientation is correlated with an electrochemical response. We can conclude at least that there is a surface layer on top of HOPG that undergoes a reorientation if negative electrode potentials are applied.

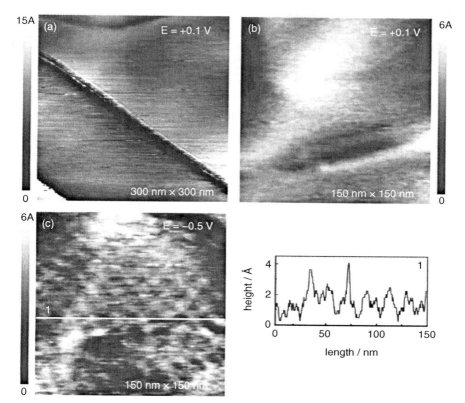

Figure 1.3 A sequence of *in situ* STM images of HOPG in [C$_4$mpyr][NTf$_2$]. (a) At +0.1 V versus Pt quasi-reference, a typical surface structure of HOPG is obtained. (b) A higher resolution shows that an adsorbed layer seems to be present. (c) With decreasing electrode potential, a type of an ordering is observed. The individual islands are about 20 nm in width and less than 0.5 nm in height (height profile 1).

1.2.4   Au(111)/[C$_6$mim][FAP]

Stimulated by a recent paper from Fedorov *et al.* [34], who showed by simulations that the camel shape of the capacitance curve might be due to reorientations at the ionic liquid/electrode interface, we investigated the Au(111) surface under 1-hexyl-3-methylimidazolium tris(pentafluoroethyl)trifluorophosphate ([C$_6$mim][FAP]). We have had good experiences with this liquid, which can be made in very high quality by Merck. The cyclic voltammogram shown in Figure 1.4 shows only slight peaks within the electrochemical window, and from the electrochemical experiment alone one could only speculate on the surface or bulk processes.

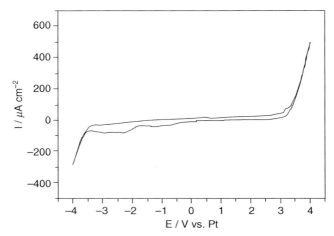

Figure 1.4 Cyclic voltammogram of [C₆mim][FAP] on Au(111) at a scan rate of 10 mV s⁻¹.

The *in situ* STM shows that there are slow processes occurring on the Au(111) surface. Figure 1.5a shows the surface that we obtain in [C₆mim][FAP] under ocp conditions.

The surface shows a wormlike structure with heights of between 200 and 300 pm. We observed a similar surface with [C₄mpyr][NTf₂], whereas with [C₄mim][NTf₂] a rather flat surface was obtained under the same conditions. This result is a hint that the liquid interacts with the surface. If we change the electrode potential to values that are 0.5–1.5 V more negative, a slow reorientation occurs, finally giving a surface with islands on top (Fig. 1.5b). An evaluation of all STM images reveals that the wormlike structures rather merge toform these islands than being the result of a deposition. Interestingly, if we change the electrode potential back to the original value, we do not get the same surface as under ocp conditions. If, on the other hand, we apply initially a potential 0.5–1.5 V more positive than the ocp, we see another restructuring of the surface, as shown in Figure 1.5c. The wormlike surface also transforms slowly, resulting in a differently structured surface. We should mention that all these processes are extremely slow, occurring over hours. This means that the electrochemical experiment alone can be misleading at too high scan rates, and cyclic voltammograms should be acquired with rather slow scan rates, for example, 0.1 mV s⁻¹, if these slow processes are to be probed. However, this will require high-quality instruments to resolve the small currents at low scan rates. Unfortunately, we have so far not yet obtained better quality STM images with this liquid upon zooming in to higher resolution. Furthermore, we have hints that, even with the same liquids, the results are not perfectly reproducible, giving further evidence that even with one and the same liquid, a slightly different outcome cannot be ruled out. In order to shed more light on

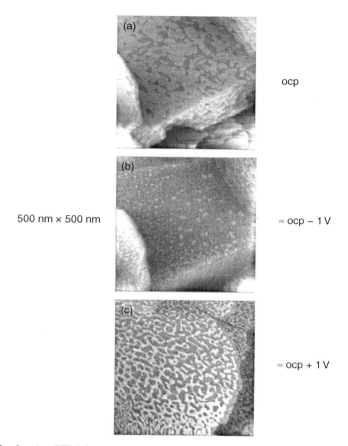

ocp

500 nm × 500 nm

≈ ocp − 1 V

≈ ocp + 1 V

Figure 1.5 *In situ* STM images of Au(111) under [C₆mim][FAP] around the open circuit potential: very slow potential-dependent processes can be observed on the electrode surface.

the STM results with this liquid, we performed tunnelling spectroscopy. This technique is illustrated schematically in Figure 1.6.

In the STM experiment, the distance tip-to-surface is held constant by a feedback circuit after the approach, and in this constant height mode the surface is probed, giving three-dimensional (3D) images of the surface. From the theory of tunnelling, there are now two options to obtain electronic information on the tunnelling process. In one experiment, the feedback circuit is switched off and the tunnelling voltage is varied. The tunnelling current is acquired as a function of the tunnelling voltage, and with the integrated version of Equation (1.1) [35], the tunnelling barrier ϕ can be obtained by fitting the formula to the experimental values:

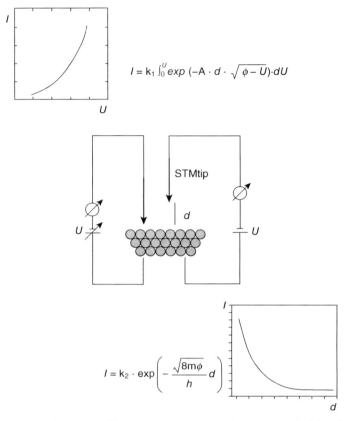

Figure 1.6 A sketch of tunnelling spectroscopy and the formulas behind (see text).

$$I = k_1 \int_0^U \exp\left(-A \cdot d \cdot \sqrt{\phi - U}\right) dU \tag{1.1}$$

k_1, A: constants, d: distance, ϕ: tunnelling barrier, U: voltage.

ϕ includes the work functions of tip and electrode, and of other processes involved in the tunnelling. This evaluation has been successfully applied by Zell and Freyland to STM experiments in aluminium(III) chloride-based ionic liquids, and the apparent tunnelling barrier showed a dependence on the alloy composition [36].

Another approach, based on the same theory, is to vary the distance of tip to sample. This can be done by retracting and approaching the tip with the STM scanner. The formula for evaluation is simpler:

$$I = k_2 \exp\left(-\frac{\sqrt{8m\phi}}{\hbar} d\right)$$

(1.2)

k_2: constant, m: electron mass, d: distance.

Here, the apparent tunnelling barrier can be obtained directly from the plot of $\ln(I)$ versus d. We prefer this evaluation. On the one hand, there is no need to fit a model to the experimental data. Equation (1.1) requires three variables, that is, k_1, d, and ϕ. On the other hand, that experiment requires a variation of the tip potential, in part over ±1 V, to obtain good enough experimental data. We have, together with Rob Atkin, the first results that these interfacial layers vary with the electrode potential, and consequently the tunnelling barrier might be influenced. With the distance variation, these problems do not occur, and from our experience the results are more reproducible. It might depend on the liquid and the electrode, and thus the experimenter is encouraged to perform both tunnelling spectroscopy experiments to decide. Figure 1.7 shows the evaluated results.

From these data we can conclude, that in this experiment, the tunnelling barrier is lowest around the ocp and rises both to the cathodic and to the anodic régime. We might preliminarily conclude, together with the *in situ* STM results, that the restructuring of the interface leads to a higher tunnelling barrier. It has to be expected that the structure of the ionic liquid's ions will have an influence on the surface processes, and that one should refrain from

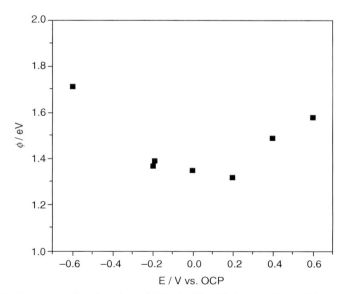

Figure 1.7 The tunnelling barrier of [C$_6$mim][FAP] /Au(111) at different electrode potentials. The tunnelling barrier has a minimum at the open circuit potential.

extrapolating results from experiments with one liquid/electrode system to another one. In the following section, we show that a variation of the cation can lead to totally different STM results.

1.2.5   Au(111)/[C₄mpyr][FAP]

As already mentioned, we had hints in 2006 that the Au(111)/[C₄mpyr][NTf₂] surface shows, in a certain potential régime, a structure similar to the well-known herringbone superstructure, which has been observed by many groups so far, both under electrochemical conditions and in UHV. As one example, we would like to mention the work of Repain et al. [37]. Unfortunately, we were unable to reproduce our results for two years, and in our experience this was due to a lower liquid quality. Furthermore, we also observed that the bistriflamide can be remarkably unstable under mild electrochemical conditions, and the oxidation of copper in such liquids at elevated temperatures can lead quantitatively to the formation of CuF₂ [38]. For this reason, we selected a liquid of the best possible quality and ordered from Merck a custom synthesis of [C₄mpyr][FAP], together with a detailed analytical protocol. The analysis from Merck showed that all detectable impurities were below 10 ppm. Figure 1.8 shows the cyclic voltammogram of this liquid on Au(111) in the potential régime where the STM experiments were performed.

The platinum (quasi-)reference electrode is not perfect but still a good choice for STM experiments as there are only negligible Faraday currents; thus, the electrode potential remains quite stable. If, however, the electrolyte

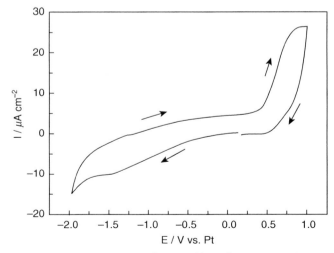

Figure 1.8 Cyclic voltammogram of [C₄mpyr][FAP] on Au(111) in the potential régime from which the STM images of Figure 1.9 were acquired.

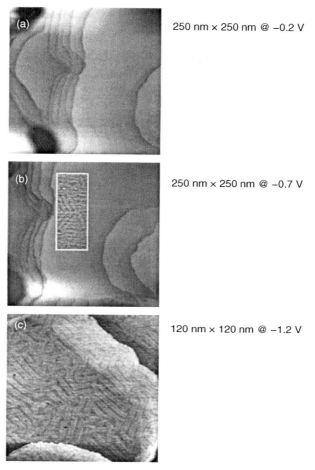

250 nm × 250 nm @ −0.2 V

250 nm × 250 nm @ −0.7 V

120 nm × 120 nm @ −1.2 V

Figure 1.9 *In situ* STM images of Au(111) under [C₄mpyr][FAP]: with decreasing electrode potential, the herringbone superstructure of Au(111) is observed. The inset in (b) shows (with an artificial contrast) the initial herringbone structure.

composition is changed during an experiment, the potential of this electrode will shift by up to 500 mV, depending on the species, in either direction. This quasi-reference electrode can be employed if care is taken, and we employ it since in any case it excludes any side reaction on the electrode surface from reference electrode species. Figure 1.9 shows three *in situ* STM images of Au(111) at different electrode potentials.

At −0.2 V, we see a flat surface. Apart from the typical terraces there is no striking surface structure, even if we zoom in. At −0.7 V, a closer look shows that a structure which is quite similar to the one we observed in 2006 appears [27]. If the electrode potential is further reduced to −1.2 V, and if the resolution is increased to 200 nm × 200 nm, the surface shows a structure that is quite similar to the abovementioned herringbone superstructure. The reported distance between the rims should be 6.4 nm, but one finds slightly varying values in the literature [39, 40]. With a calibrated STM tip, we obtain a value of 5.4 ± 0.5 nm, which is slightly lower than the reported 6.4 nm. The uncertainty arises from the noise in the STM images. If we further zoom in, the STM image becomes a little noisy, but we have hints for the periodic structure of Au(111). Now, the question arises: why in this liquid do we see the herringbone superstructure? We should mention that, in [C$_4$mim][FAP] of the same quality, we have hitherto not yet seen the herringbone superstructure. Experiments performed in Rob Atkin's laboratory reveal that both liquids are differently adsorbed on Au(111), [C$_4$mpyr][FAP] more strongly than [C$_4$mim][FAP], and that the layer thickness varies with the applied electrode potential. The experiments, which will be published separately, show that in the cathodic régime the first layer close to the surface is more compressed compared with the ocp conditions. Preliminary tunnelling spectroscopy experiments performed in our laboratory show that, in the potential régime where the herringbone superstructure is observed, the tunnelling barrier is lowest. All these observations lead us to the preliminary conclusion that, in this limited potential régime, the STM tip is close enough to the surface to allow the electrons to tunnel through the adsorbed ionic liquid layer and make the herringbone structure visible. Maybe it is just a small step now, in this liquid, to probing the surface with atomic resolution. However, this does not mean that this aim can be achieved in all liquids easily. So far, we have not yet seen the herringbone structure with [C$_4$mim][FAP], and at least we can conclude that other electrochemical and tunnelling parameters are required. In contrast to *in situ* STM experiments in water, ionic liquids are much more challenging. Not only the purity plays a rôle, but there is also a dependence on the ions and solutes added (e.g., for the sake of electrodeposition) to ionic liquids, which might again change the interface structure. These studies are in their infancy, but there is an increased awareness now, and more and more people are starting to realise that it is not just the electrochemical window that makes ionic liquids interesting for electrochemical purposes. With a fundamental understanding of the interfacial processes, there is also a great chance of developing tailored technical processes.

1.2.6   Influence of the Cation on Aluminium Deposition

When we started with experiments on the electrodeposition of aluminium from air- and water-stable ionic liquids, we had quite a simple motivation to use [C$_4$mpyr][NTf$_2$]. This liquid can be well dried below 3 ppm H$_2$O. If dry aluminium(III) chloride is added, there is no risk of a reaction with water which, in contrast, is almost unavoidable if aluminium(III) chloride is mixed with, for example, [C$_4$mim]Cl, which is quite difficult to dry. Another reason was that we planned pulsed electrodeposition to make nanocrystalline aluminium deposits, and we expected a wider electrochemical window for [C₄mpyr][NTf₂] compared with [C₄mim][NTf₂]. It was a great surprise when we realised, after the first set of experiments, that even under potentiostatic conditions with not that high overvoltages, reproducibly nanocrystalline aluminium was obtained [41]. Control experiments showed that with [C₄mim][NTf₂] under quite similar experimental conditions, microcrystalline aluminium deposits were obtained [28]. We could exclude viscosity effects. An *in situ* STM study showed that there were more differences between the liquids: we found an underpotential deposition of aluminium and a surface alloy Al/Au in [C₄mim][NTf₂], but not in [C₄mpyr][NTf₂]. Later, we showed that the electroactive species in both liquids are identical [42]. Only one explanation remained: the cation of the liquid influences the deposition process and from Atkin's AFM measurements, it is likely that the [C₄mpyr]⁺ ion behaves like a surface-active species. We found more surprising effects. If the [C₄mim]⁺ ion is electrochemically decomposed prior to deposition experiments, again nanocrystalline aluminium is obtained. From all these results, we can conclude, at least, that the cation (and likely also the anion) of ionic liquids influences the electrodeposition process. Presumably, the interfacial layers play a rôle, but a close insight would require not only STM and AFM experiments, but also other surface-sensitive techniques, such as sum-frequency generation spectroscopy experiments under electrochemical control. In the following, we would like to show that even variations in the cation structure can lead to surprising results. Together with a company, we investigated the aluminium deposition from a mixture of aluminium(III) chloride with 1-(2-methoxyethyl)-3-methylimidazolium chloride [43].

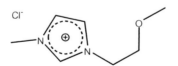

Structural formula of 1-(2-methoxyethyl)-3-methylimidazolium chloride

The cyclic voltammogram was normal, showing a reversible aluminium deposition/stripping (Fig. 1.10). Surprisingly, the deposit was semi-bright from the beginning, and the SEM analysis (as well as an X-ray powder diffraction [XRD] analysis) showed nanocrystals instead of the expected microcrystals (Fig. 1.11).

We tested other organic halides with two ether groups in the imidazolium side chain, and again we obtained different results: in part we only found black deposits, although there was no hint of the decomposition of the liquid.

All our experiences with electrodeposition from ionic liquids lead us to the following conclusion: not only does the electroactive species play a rôle, but even slight variations in the cation and/or anion structure may also have an

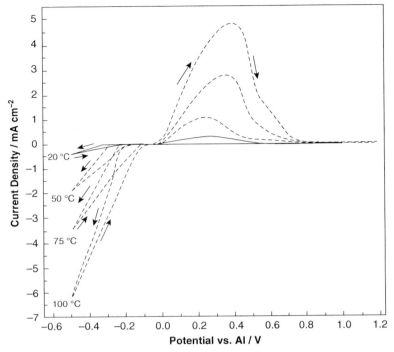

Figure 1.10 Temperature-dependent cyclic voltammograms of aluminium deposition from 1-(2-methoxyethyl)-3-methylimidazolium chloride/aluminium(III) chloride.

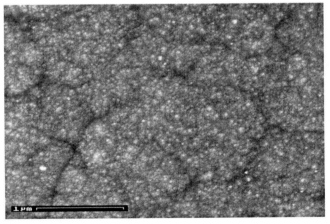

Figure 1.11 Nanocrystalline aluminium deposited from 1-(2-methoxyethyl)-3-methylimidazolium chloride/aluminium(III) chloride.

effect. As ionic liquids are definitely (and different strongly) adsorbed onto electrode surfaces, we assume that the interfacial layers influence the elemental electrochemical processes. Interfacial layers will, according to our experience, be influenced both by the solute and by impurities, and there might be time dependent effects. One should **definitely** not expect that all liquids behave in the same manner.

1.2.7   Challenges in the Making of Macroporous Materials

In the past three years, we have expanded our activities to the template-assisted electrodeposition of nanomaterials. These activities involve the electrodeposition of nanowires (silicon, SiGe, and silver), and of macroporous materials. For nanowires, we employ commercial membranes with a defined pore size in which the deposition is performed; in the case of macroporous materials, we deposit polystyrene spheres on the surface of the electrode and fill the voids by electrodeposition. After dissolution of the polystyrene spheres, we get a replica. Figure 1.12 shows a sketch of the electrodeposition principle.

In contrast to the numerous, and mainly nonreproducible, literature data, we apply the polystyrene spheres by simply dipping the electrode into an alcoholic suspension of commercial polystyrene spheres. Upon retracting from the suspension, the alcohol evaporates and a well-ordered 3D polystyrene opal structure is obtained (Fig. 1.13).

On flat surfaces in particular, we get well-ordered polystyrene layers up to 10-μm thick with ordered islands that have lateral dimensions of up to 20 μm $\times$ 20 μm. We should mention that how well the spheres self-assemble not only depends on the surface roughness, but also depends on the material. So far we have obtained our best quality samples with gold single crystals, but also indium-tin oxide (ITO) can be employed as a substrate for polystyrene precipitation. We could show that photonic crystals of germanium can be made by electrodeposition and that this material exhibits optical band gap behaviour [44]. A typical structure and a typical optical behaviour are shown in Figure 1.14.

Due to the low surface tension, the ionic liquid easily wets the electrode surface underneath the polystyrene spheres, and germanium grows from the bottom to the top quite regularly. We could now say that it is trivial to make such 3D ordered macroporous (3DOM) substrates, and germanium photonic crystals are indeed easy to make. But it is not that simple. Apart from germanium, there are other materials that are interesting, if macroporous. 3DOM silicon, for example, has a high potential for radio frequency applications. Macroporous aluminium and its alloys would be interesting as anode host materials in lithium ion batteries [45]; 3DOM conducting polymers and noble metals have a high potential in catalysis; and so on. However, Figure 1.15 shows that one has to be prepared for surprises when attempting to make 3DOM materials from ionic liquids.

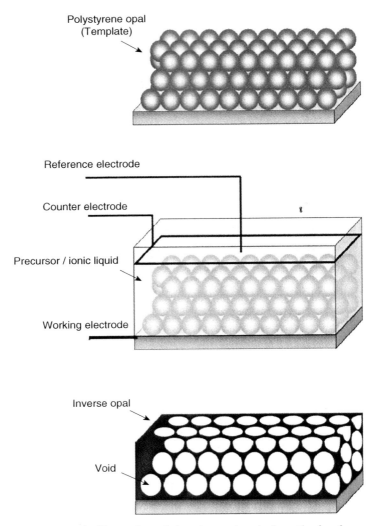

Polystyrene opal
(Template)

Reference electrode

Counter electrode

Precursor / ionic liquid

Working electrode

Inverse opal

Void

Figure 1.12   Schematic illustration of the electrochemical synthesis of macroporous materials from ionic liquids. Top: A template of polystyrene (PS) on a conducting substrate. Middle: Electrochemical cell. Bottom: Inverse opal of the deposited material after removal of the template by tetrahydrofuran.

The first scanning electron microscopy (SEM) pictures show attempts to make 3DOM silicon by potentiostatic electrodeposition, followed by removal of the polystyrene spheres. In contrast to 3DOM–Ge, there is no such long range regular structure and a disintegration of the polystyrene template during electrodeposition has obviously occurred. The growth processes of silicon and germanium are different, and also influenced by the ionic liquid itself. An

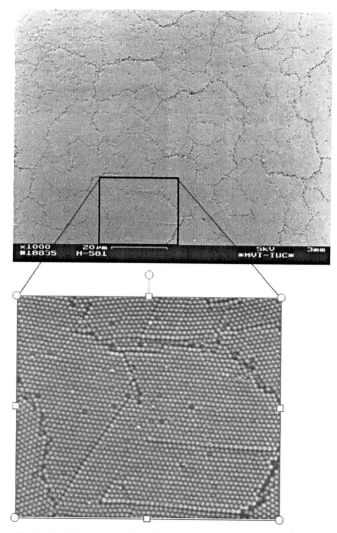

Figure 1.13   Gold thin film (cleaned with limonene) coated with hexagonally ordered polystyrene spheres (diameter 581 nm from Duke Scientific, Palo Alto, CA, USA). For SEM imaging, the template was slightly annealed for 2 hours at 70 °C.

ordered growth of 3DOM–Si will presumably require other parameters. Whereas the deposition of 3DOM–Al on a polished copper electrode from [C$_4$mim]Cl-AlCl$_3$ is relatively easy to control [45], the deposition on ITO under similar conditions is more complicated and can lead to structures such as shown in Figure 1.15. A controlled growth with the aim of making 3DOM–Al on ITO requires well-adapted electrochemical parameters.

(a)

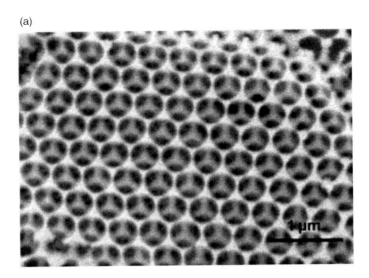

(b)

2DOM Si$_x$Ge$_{1-x}$

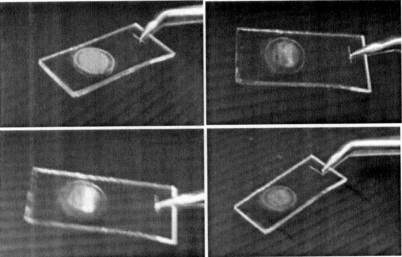

Figure 1.14   (a) SEM image of 3DOM Ge on ITO-glass after removal of the polystyrene template (370 nm), obtained potentiostatically at −2.1 V (vs. Ag quasi-reference electrode) for 30 minutes. 0.1 M GeCl$_4$ in [C$_4$mim][NTf$_2$]. (b) Photographs of 2DOM Si$_x$Ge$_{1-x}$ (pore size ~370 nm) on ITO–glass substrate showing a colour change with changing the angle of the incident visible white light (GeCl$_4$ + SiCl$_4$) (0.1 + 0.1 M) in [C$_4$mim][NTf$_2$]. See colour insert.

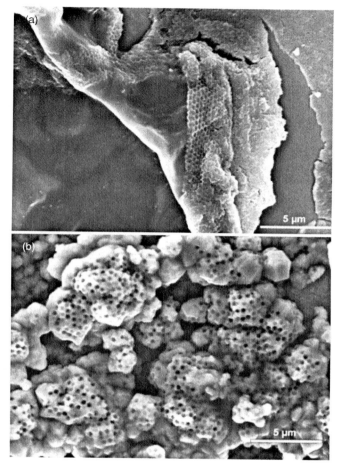

Figure 1.15   (a) Less successful approaches: SEM images of attempted macroporous silicon obtained after overpotential deposition (0.3 M SiCl$_4$ in [C$_4$mpyr][FAP]) for 1 hour, followed by removal of the polystyrene template (400 nm). (b) SEM images of irregular macroporous aluminium from [C$_4$mim]Cl-AlCl$_3$ on ITO–glass substrates: with a too fast growth, the polystyrene template is corrupted and an irregular structure results.

1.3   CONCLUSION

This chapter has shown that electrochemical reactions in ionic liquids are considerably different from those in aqueous solutions. Ionic liquids form remarkably strongly adherent interfacial layers on electrode surfaces, and the adsorption varies with the composition of the liquid and with the applied electrode potential. Furthermore, the surface chemistry is strongly dependent on the quality of the liquid, and from our point of view for fundamental physicochemical and electrochemical experiments, the quality of the liquid cannot be high enough. Our *in situ* STM experiments have shown that the surface chemistry is influenced by the ionic liquid and that even slight changes in cation–anion combination have an effect. In a limited potential régime, the herringbone superstructure of Au(111) is visible in [C_4mpyr][FAP], and maybe it is not a too long way to go to achieve atomic resolution in this liquid. In other liquids, such as [C_4mim][FAP] or [C_6mim][FAP], we have not yet observed this nice superstructure. Results from Atkin's group make it likely that the thickness of the interfacial layers, and maybe their structures, vary with the electrode potential and with the liquid itself, and that in each liquid individual tunnelling parameters have to be found for STM experiments. It is furthermore likely that these interfacial layers influence the electrodeposition process, and even slight modifications such as the introduction of an ether group alter, for example, the grain size of deposited aluminium. Apart from these very fundamental aspects, which are still in their infancy, ionic liquids have the potential to make functional materials with macroporous structures, and photonic crystals. In our opinion, electrochemical processes in ionic liquids are not yet straightforward, and many more fundamental studies will be required.

REFERENCES

1   Wilkes, J.S., Levisky, J.A., Wilson, R.A., and Hussey, C.L., Dialkylimidazolium chloroaluminate melts—a new class of room-temperature ionic liquids for electrochemistry, spectroscopy, and synthesis, *Inorg. Chem.* **21**, 1263–1264 (1982).

2   Walden, P., Über die Molekulargrösse und elektrische Leitfähigkeit einiger geschmolzenen Salze, *Bull. Acad. Impér. Sci. St. Pétersbourg* **8**, 405–422 (1914).

3   Hussey, C.L., Room temperature molten salt systems, in: *Advances in molten salt chemistry*, Vol. 5, ed. G. Mamantov, Elsevier, New York (1983), pp. 185–230.

4   Endres, F., Ionic liquids: solvents for the electrodeposition of metals and semiconductors, *Chemphyschem* **3**, 144–154 (2002).

5   Endres, F., Freyland, W., and Gilbert, B., Electrochemical scanning tunnelling microscopy (EC-STM) study of silver electrodeposition from a room temperature molten salt, *Ber. Bunsenges. Phys. Chem.* **101**, 1075–1077 (1997).

6   Moustafa, E.M., Zein El Abedin, S., Shkurankov, A., Zschippang, E., Saad, A.Y., Bund, A., and Endres, F., Electrodeposition of Al in 1-butyl-1-methylpyrrolidinium bis(trifluoromethylsulfonyl)amide and 1-ethyl-3-methylimidazolium bis(trifluoromethylsulfonyl)amide ionic liquids: in situ STM and EQCM studies, *J. Phys. Chem. B* **111**, 4693–4704 (2007).

7   Freyland, W., Zell, C.A., Zein El Abedin, S., and Endres, F., Nanoscale electrodeposition of metals and semiconductors from ionic liquids, *Electrochim. Acta* **84**, 3053–3061 (2003).

8 Mann, O., and Freyland, W., Mechanism of formation and electronic structure of semiconducting ZnSb nanoclusters electrodeposited from an ionic liquid, *Electrochim. Acta* **53**, 514–524 (2007).

9 Wang, F.-X., Pan, G.-B., Liu, Y.-D., and Xiao, Y., Pb deposition onto Au(1 1 1) from acidic chloroaluminate ionic liquid, *Chem. Phys. Lett.* **488**, 112–115 (2010).

10 Pan, G.-B., and Freyland, W., 2D phase transition of PF_6 adlayers at the electrified ionic liquid/Au(1 1 1) interface, *Chem. Phys. Lett.* **427**, 96–100 (2006).

11 Magnussen, O.M., Zitzler, L., Gleich, B., Vogt, M.R., and Behm, R.J., In-situ atomic-scale studies of the mechanisms and dynamics of metal dissolution by high-speed STM, *Electrochim. Acta* **46**, 3725–3733 (2001).

12 NuLi, Y., Yang, J., and Wang, P., Electrodeposition of magnesium film from $BMIMBF_4$ ionic liquid, *Appl. Surf. Sci.* **252**, 8086–8090 (2006).

13 Wang, P., NuLi, Y., Yang, J., and Feng, Z., Mixed ionic liquids as electrolyte for reversible deposition and dissolution of magnesium, *Surf. Coat. Technol.* **201**, 3783–3787 (2006).

14 Feng, Z., NuLi, Y., Wang, J., and Yang, J., Study of key factors influencing electrochemical reversibility of magnesium deposition and dissolution, *J. Electrochem. Soc.* **135**, C689–C693 (2006).

15 Mukhopadhyay, I., Aravinda, C.L., Borissov, D., and Freyland, W., Electrodeposition of Ti from $TiCl_4$ in the ionic liquid l-methyl-3-butyl-imidazolium bis(trifluoromethylsulfone)imide at room temperature: study on phase formation by in situ electrochemical scanning tunneling microscopy, *Electrochim. Acta* **50**, 1275–1281 (2005).

16 Cheek, G.T., O'Grady, W.E., Zein El Abedin, S., Moustafa, E.M., and Endres, F., Electrochemical studies of magnesium deposition in ionic liquids, *J. Electrochem. Soc.* **155**, D91–D95 (2008).

17 Endres, F., Zein El Abedin, S., Saad, A.Y., Moustafa, E.M., Borisenko, N., Price, W.E., Wallace, G.G., MacFarlane, D.R., Newman, P.J., and Bund, A., On the electrodepsoition of tantalum in ionic liquids, *Phys. Chem. Chem. Phys.* **10**, 2189–2199 (2008).

18 Schubert, T., Zein El Abedin, S., Abbott, A.P., McKenzie, K.J., Ryder, K.S., and Endres, F., Electrodeposition of metals, in: *Electrodeposition from ionic liquids*, eds. F. Endres, D.R. MacFarlane, and A.P. Abbott, Wiley-VCH, Weinheim (2008), pp. 83–123.

19 Gottfried, J.M., Maier, F., Rossa, J., Gerhard, D., Schulz, P.S., Wasserscheid, P., and Steinrück, H.-P., Surface studies on the ionic liquid 1-ethyl-3-methylimidazolium ethylsulfate using X-ray photoelectron spectroscopy (XPS), *Z. Phys. Chem.* **220**, 1439–1453 (2006).

20 Steinrück, H.-P., Surface science goes liquid, *Surf. Sci.* **604**, 481–484 (2010).

21 Endres, F., Zein El Abedin, S., and Borisenko, N., Probing of lithium and alumina impurities in air- and water stable ionic liquids by cyclic voltammetry and in situ scanning tunnelling microscopy, *Z. Phys. Chem.* **220**, 1377–1394 (2006).

22 Clare, B.R., Bayley, P.M., Best, A.S., Forsyth, M., and MacFarlane, D.R., Purification or contamination? The effect of sorbents on ionic liquids, *Chem. Comm.* 23, 2689–2691 (2008).

23   Su, Y.-Z., Fu, Y.-C., Yan, J.-W., Chen, Z.-B., and Mao, B.-W., Double layer of Au(100)/ionic liquid interface and its stability in imidazolium-based ionic liquids, *Angew. Chem. Int. Ed. Engl.* **48**, 5148–5151 (2009).

24   Alam, M.T., Islam, M.M., Okajima, T., and Ohsaka, T., Measurements of differential capacitance in room temperature ionic liquid at mercury, glassy carbon and gold electrode interfaces, *Electrochem. Commun.* **9**, 2370–2374 (2007).

25   Islam, M.M., Alam, M.T., and Ohsaka, T., Electrical double-layer structure in ionic liquids: a corroboration of the theoretical model by experimental results, *J. Phys. Chem. C* **112**, 16568–16574 (2008).

26   Gnahm, M., Pajkossy, T., and Kolb, D.M., The interface between Au(111) and an ionic liquid, *Electrochim. Acta* **55**, 6212–6217 (2010).

27   Borisenko, N., Zein El Abedin, S., and Endres, F., In situ investigation of gold reconstruction and of silicon electrodeposition in the room temperature ionic liquid 1-butyl-1-methylpyrrolidinium bis(trifluoromethylsulfonyl)imide, *J. Phys. Chem. B* **110**, 6250–6256 (2006).

28   Zein El Abedin, S., Moustafa, E.M., Hempelmann, R., Natter, H., and Endres, F., Electrodeposition of nano- and microcrystalline aluminium in some water and air stable ionic liquids, *ChemPhysChem* **7**, 1535–1543 (2006).

29   Atkin, R., and Warr, G.G., Structure in confined room-temperature ionic liquids, *J. Phys. Chem. C* **111**, 5162–5168 (2007).

30   O'Shea, S.J., Welland, M.E., and Pethica, J.B., Atomic force microscopy of local compliance at solid-liquid interfaces, *Chem. Phys. Lett.* **223**, 336–340 (1994).

31   Lim, R., and O'Shea, S.J., Discrete solvation layering in confined binary liquids, *Langmuir* **20**, 4916–4919 (2004).

32   Jeffery, S., Hoffmann, P.M., Pethica, J.B., Ramanujan, C., Ozer, H.O., and Oral, A., Direct measurement of molecular stiffness and damping in confined water layers, *Phys. Rev. B* **70**, 054114 (2004).

33   Fedorov, M.V., and Kornyshev, A.A., Ionic liquid near a charged wall: structure and capacitance of electrical double layer, *J. Phys. Chem. B* **112**, 11868–11872 (2008).

34   Fedorov, M.V., Georgi, N., and Kornyshev, A.A., Double layer in ionic liquids: the nature of the camel shape of capacitance, *Electrochem. Commun.* **12**, 296–299 (2010).

35   Hamers, R.J., Tunneling Spectroscopy, in: *Scanning tunneling microscopy and spectroscopy: theory, techniques and applications*, ed. D.A. Bonnell, Wiley-VCH, New York (1993), pp. 51–103.

36   Zell, C.A., and Freyland, W., In situ STM and STS study of Co and Co-Al alloy electrodeposition from an ionic liquid, *Langmuir* **19**, 7445–7450 (2003).

37   Repain, V., Berroir, J.M., Rousset, S., and Lecoeur, J., Reconstruction, step edges and self-organization on the Au(111) surface, *Appl. Surf. Sci.* **162**, 30–36 (2000).

38   Ismail, A.S., Zein El Abedin, S., Höfft, O., and Endres, F., Unexpected decomposition of the bis(trifluoromethylsulfonyl) amid anion during electrochemical copper oxidation in an ionic liquid, *Electrochem. Commun.* **12**, 909–911 (2010).

39   Hasegawa, Y., and Avouris, P., Manipulation of the reconstruction of the Au(111) surface with the STM, *Science* **258**, 1763–1765 (1992).

40 Oka, H., and Sueoka, K., Connection of herringbone ridges on reconstructed Au(111) surfaces observed by scanning tunneling microscopy, *Jpn. J. Appl. Phys.* **44**, 5430–5433 (2005).

41 Zein El Abedin, S., Moustafa, E.M., Hempelmann, R., Natter, H., and Endres, F., Additive free electrodeposition of nanocrystalline aluminium in a water and air stable ionic liquid, *Electrochem. Commun.* **7**, 1116–1121 (2005).

42 Eiden, P., Liu, Q.X., Zein El Abedin, S., Endres, F., and Krossing, I., An experimental and theoretical study of the aluminium species present in $AlCl_3$/[BMP]Tf_2N and $AlCl_3$/[EMIm]Tf_2N ionic liquids, *Chem. Eur. J.* **15**, 3426–3434 (2009).

43 Zein El Abedin, S., Giridhar, P., Schwab, P., and Endres, F., Electrodeposition of nanocrystalline aluminium from a chloroaluminate ionic liquid, *Electrochem. Commun.* **12**, 1084–1086 (2010).

44 Meng, X., Al-Salman, R., Zhao, J., Borissenko, N., Li, Y., and Endres, F., Electrodeposition of 3D ordered macroporous germanium from ionic liquids: a feasible method to make photonic crystals with a high dielectric constant, *Angew. Chem. Int. Ed. Engl.* **48**, 2703–2707 (2009).

45 Gasparotto, L.H.S., Prowald, A., Borisenko, N., Zein El Abedin, S., and Endres, F., Electrochemical synthesis of macroporous aluminium: a possible host anode material for rechargeable Li-ion batteries, *J. Power Sources* **196**, 2879–2883 (2011).

2 Interfaces of Ionic Liquids (1)

WERNER FREYLAND

Karlsruhe Institute of Technology, Karlsruhe, Germany

ABSTRACT

Interfaces play a key rôle in the research and applications of ionic liquids. This is true for liquid/gas, liquid/liquid, and solid/liquid interfaces. In this chapter, their specific characteristics are described and critically discussed, including the electrified metal/ionic liquid interface and phenomena such as wetting and electrowetting. Several open problems are highlighted.

2.1 INTRODUCTION

In experimental studies of surfaces† and interfaces, a necessary and important requirement is the careful control of impurities, both in the bulk and at the interface. Impurities in the bulk can be strongly enriched at the interfaces and thus affect their properties. This surface or interface excess depends on the interfacial free energies of the components, and is given by the Gibbs adsorption equation. On the other hand, if a surface is not protected by a high-purity inert gas or a good vacuum but is in contact with ambient air, adsorption of O_2 or H_2O will spoil its characteristics. A simple exercise from the theory of gas kinetics will illustrate what is meant by "a good vacuum." Near room temperature, the number of collisions of gas molecules with a wall is given by Equation (2.1):

$$Z/cm^{-2}\ s^{-1} \sim 10^{21} M^{-1/2}\ p/mbar,\tag{2.1}$$

---

† In general, the term *surface* is used for an interface where one of the neighbouring phases is a vapour or gas phase or vacuum.

Ionic Liquids UnCOILed: Critical Expert Overviews, First Edition. Edited by Natalia V. Plechkova and Kenneth R. Seddon.

where p is the gas pressure and M is the relative molecular mass of the gas molecules. Thus, assuming that the surface contains 10^{15} atoms per square centimetre and that the sticking coefficient is 1, it takes about 1 second to adsorb a complete monolayer of O_2 on a solid or liquid surface in a vacuum of 10^{-6} mbar. As most measurements last longer than 1 second, it is desirable to work under ultra-high vacuum (UHV), that is, at pressures $\leq 10^{-9}$ mbar. If possible one should also consider *in situ* surface cleaning, for example, by argon ion sputtering.

Usually, liquids near their melting points have relatively high vapour pressures, which do not allow interfacial measurements under UHV conditions. Exceptions are a few liquid metals such as gallium or bismuth ($p < 10^{-10}$ mbar up to 500 K) and ionic liquids such as [C_2mim][NTf$_2$] or [C_4mim][N(CN)$_2$], which near room temperature have vapour pressures as low as $\sim 10^{-11}$ mbar, according to Heintz and coworkers [1]. So, as with liquid metals, surface-sensitive probes that have been developed in solid-state surface science can also be applied to the study of ionic liquid surfaces. Methods used so far include X-ray photoelectron spectroscopy (XPS) and its angle-resolved modification (for a recent review, see Reference [2]), X-ray reflectivity [3], surface light scattering or capillary wave spectroscopy [4], and various spectroscopies such as sum frequency generation (SFG) [5], and direct recoil spectrometry [6]. These investigations give valuable insights into the composition and structure at the ionic liquid/vacuum and, in some cases, the liquid/gas interface. This will be described and discussed in more detail in Section 2.2, together with results of molecular dynamics (MD) and *ab initio* computer simulation studies [7–9]. The following two sections deal with ionic liquid/immiscible liquid interfaces and the solid/liquid and corresponding electrified interfaces. On the basis of recent X-ray reflectivity measurements [10] and *ab initio*-based simulation studies [11, 12], it has been shown that charge ordering plays a key rôle at these interfaces. This is of particular interest for the electrochemical interface of ionic liquids. Its consequences for the double-layer model in ionic liquids will be discussed in comparison with conventional aqueous electrolytes. Further phenomena include the differential capacitance and potential-induced two-dimensional (2D) interfacial phase transitions of adsorbed ions at the ionic liquid/electrode interface [13]. Finally, the similarities of the thermodynamic interfacial characteristics in different ionic media, both ionic liquids and classical molten salts, are briefly described. In the last section, the phenomena of wetting and electrowetting are introduced and first results of ionic liquids under UHV and at ambient gas or liquid conditions are presented [14, 15].

This review on ionic liquid interfaces is not meant to be comprehensive. Instead, some interfacial aspects are critically analysed and those, where clear conclusions can be drawn, are highlighted. The focus is on basic interfacial phenomena. Applications in different fields, such as biphasic catalysis, liquid–liquid extraction, electrochemistry, or electrowetting, are mentioned only briefly.

2.2 LIQUID/VACUUM AND LIQUID/GAS INTERFACES

The best studied interface in ionic liquids is the liquid/vacuum interface, both theoretically and experimentally. Starting with results from computer simulation studies, these in most cases have been performed with imidazolium-based ionic liquids. The first MD simulation on 1,3-dimethylimidazolium chloride using non-polarisable cations and anions was reported by Lynden-Bell [7]. It was found that there is a region of enhanced density immediately below the interface in which the cations are oriented with their plane perpendicular to the surface and their dipole in the surface plane. Segregation of cations and anions at the uppermost surface layer should be negligible. These calculations have been extended to several ionic liquids containing the $[C_4mim]^+$ cation and anions of different sizes, such as $[PF_6]^-$, $[BF_4]^-$, and Cl^- [16]. Again, a structured surface layer containing oriented cations occurs. The butyl side chains face the vacuum and the methyl side chain the liquid. Traversing the liquid/vacuum interface, a clear electrostatic potential drop is calculated, which in the case of $[C_4mim]Cl$ amounts to 0.75 V at 300 K. This potential change cannot be explained by different distributions of cation and anion densities, but results from the orientational ordering of the cations [16]. Subsequent MD simulations of $[C_4mim][PF_6]$ by Bhargava and Balasubramanian [8] and of $[C_2mim][NO_3]$ by Voth and coworkers [17] essentially confirm the interfacial features described (*vide supra*). This includes, in particular, the density enhancement at the interface, the preferred orientation of the cations, and an orientation of the alkyl chains along the surface normal protruding outward from the surface into the vacuum, thus imparting a hydrophobic character. Voth et al. also compared the effect of a polarisable ion model with that of a rigid ion model and found similar structural results. The polarisable model indicates that the cations more likely segregate at the ionic liquid interface.

The structure and composition at the ionic liquid/vacuum interface has been studied by different experimental methods. Law et al. [6] performed the first direct recoil spectrometry measurements upon $[C_nmim]X$ ($n = 4$, 8, or 12; $X = Cl^-$, Br^-, $[BF_4]^-$, or $[PF_6]^-$) ionic liquids with a pressure in the scattering chamber of ~10^{-6} mbar. On the basis of these measurements, it was concluded that the number of cations and anions at the surface is equal. The data also indicated that the cation ring is perpendicular rather than parallel to the surface. With this technique, the interface can be probed within a depth of 0.2–0.4 nm. Baldelli and coworkers [5, 18, 19] have probed the interfacial structure and composition by SFG vibration spectroscopy for a variety of imidazolium-based ionic liquids and under different environments (vacuum, argon gas, reduced partial pressure of H_2O). Their results suggest that both cations and anions are present in the first layer of the interface. Furthermore, the alkyl chains are extended towards the gas phase. At low water vapour pressure, the plane of the imidazolium ring is found to be parallel to the surface, while at higher water content the ring should reorient parallel to the surface normal for hydrophobic ionic liquids, but should remain parallel to the surface for hydrophilic ionic liquids. [19]. Synchrotron X-ray reflectivity measurements have been employed to elucidate the interfacial structure in ionic liquids of the type [C₄mim]X, with X = I⁻, [BF₄]⁻, or [PF₆]⁻ [3, 20]. This technique usually is applied under UHV conditions; however, measurements with a helium gas flow are also possible. After correction of the surface roughness due to thermally excited capillary waves (see below), this method yields the average electron density profile at the interface, $<d_e(z)>$. It is determined from fits of the measured reflectivities by a "reasonable" model for $<d_e(z)>$. In this way, the density profiles obtained are not necessarily unique. Such a discrepancy is evident between the results reported in References [3] and [20]. In their evaluation of the reflectivity data, Jeon et al. [3] included also the information from their SFG spectra, which they measured separately for the ionic liquids given above. For calibration reasons, they also took SFG spectra on a hexadecanol monolayer. Based on this combined analysis, Jeon et al. found evidence for the following structural model of the liquid/vacuum interface of the ionic liquids studied. The topmost surface layer is occupied by polar-oriented hydrophobic butyl chains having an in-plane density of about one-third of the fully-packed alkyl chains of the hexadecanol monolayer. The reflectivity data are best fitted by a so called two-box model, where the topmost layer describes the butyl chains ($<d_e> \sim 100$ e nm⁻³) and the second layer contains the coexisting cations and anions (thickness ~ 0.5 nm, $<d_e> \sim 500$ e nm⁻³). In the case of [C₄mim]I, the uppermost layer contains the butyl chains and cations, whereas the layer underneath is made up of the anions (thickness ~ 0.3 nm, $<d_e> \sim 800$ e nm⁻³). Sloutskin et al. derive a conflicting model on the surface structure from the analysis of their X-ray reflectivity measurements. From their data, they concluded that only one molecular layer exhibiting a density enhancement of ~18% relative to the bulk exists. In a later publication, these authors discuss in more detail the effect of surface roughening by capillary waves and concluded that additional roughness should be included, although "its origin is a mystery" [21]. This example shows that X-ray reflectivity measurements are not necessarily a straightforward method.

A relatively direct insight into interfacial compositions is obtained by XPS methods and, in particular, its angle-resolved variant. In XPS, surface sensitivity is limited by the inelastic mean free path of photoelectrons, which for organic materials at normal emission and for kinetic energies of about 1 keV can probe the surface region within a depth of ~8 nm. Increasing the electron emission angle ($\Theta = 0°$ at normal emission), the depth resolution—now determined by $\cos(\Theta)$—can be improved significantly, reaching values of about 1 nm. Thus, only the uppermost layers can be probed. Steinrück and coworkers [22, 23] have applied this technique in their systematic investigations of the surface compositions and surface enrichment effects in a number of imidazolium-based ionic liquids measured under UHV conditions ($p \sim 5 \times 10^{-10}$ mbar). Typical results are shown in Figure 2.1.

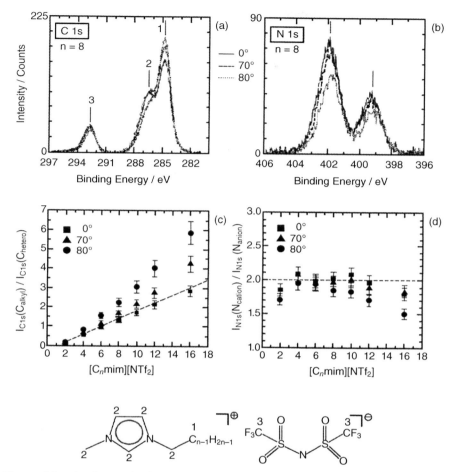

Figure 2.1 Angle resolved XPS study of $[C_n mim][NTf_2]$ ionic liquids ($n = 2$–16) recorded at different emission angles $0 \leq \Theta \leq 80°$ as labelled in the respective panels. (a) C1s spectra corresponding to different carbon positions (i)— see molecular structure at the bottom; (b) N1s spectra; (c) intensity ratio of C1s XPS signals of alkyl chain carbons (1) and hetero carbons (2) plotted versus n at different Θ; the dashed line corresponds to an isotropic distribution of carbon species in the near surface region; (d) intensity ratio of N1s signals of cation and anion. Adapted from Reference [22].

Their main findings are as follows:

1. For $[C_n mim][NTf_2]$ ($n = 1$–16), a clear enhancement of the concentration of alkyl chain carbon atoms occurs at the topmost surface layer (see Fig. 2.1c).

2. Underneath this layer, the number ratio of cations and anions approximately corresponds to a homogeneous distribution of the ions with a

tendency of the anionic nitrogen being nearer to the surface (see Fig. 2.1d).

3. Varying the anion size, the alkyl chain enrichment at the outer surface is more pronounced for smaller anions.

4. In a mixture of [C$_2$mim][NTf$_2$] and [C$_{12}$mim][NTf$_2$], no specific segregation of one of the components is found at the surface.

In conclusion, these observations give a rather detailed picture of the composition profiles at the liquid/vacuum interface of imidazolium-based ionic liquids. They strongly support the conclusions drawn from simulation studies and are in agreement with the SFG observations and the X-ray reflectivity analysis by Jeon et al. [3]. They cannot, however, say much about the structural orientation of cations at the interface.

As indicated above, the surface of a liquid is not plane and smooth but is rough on a molecular scale. This is due to thermal excitations of the molecules vertical to the surface. The vertical displacement, $<z^2>^{1/2}$, increases with temperature and decreases with the restoring force, the surface tension, σ, according to: $<z^2>^{1/2} \propto (T/\sigma)^{1/2}$ (see, for example, Reference [24]). For ionic liquids with $\sigma \sim 30$ mJ m^{-2}, $<z^2>^{1/2}$ is of the order of 1 nm near room temperature. The surface displacement in space and time, $z(r,t)$, defines the surface modes or capillary waves of frequency ω_q and wave vector q. Depending on the bulk viscosity, η, capillary waves at a free surface can either propagate as a damped oscillator or are completely, exponentially damped. In the propagating case the frequency ω_q and the lifetime τ_q are given by Equation (2.2) [24]:

$$\omega_q = (\sigma/\rho)^{1/2} q^{3/2} \text{ and } \tau_q^{-1} = (2\eta/\rho)q^2. \tag{2.2}$$

where ρ is the mass density of the liquid phase, the second phase being a dilute gas or vapour. In the limit of strong damping at high η, $\omega_q = 0$ and $\tau_q^{-1} = \sigma q/2\eta$. The general solution is obtained from the roots of the dispersion relation of surface waves—see, for example, the monograph of Dominique Langevin for details [24]. Experimentally, ω_q and τ_q^{-1} are determined by surface light scattering or capillary wave spectroscopy. Hereby the propagating surface modes act as a moving diffraction grating, where the scattered light is Doppler-shifted by $\pm\omega_q$. The scattered light has two Lorentzian components, which are centred at $\omega_o \pm \omega_m$, with ω_o being the frequency of the incident light (laser) and ω_m that of the Lorentzian at a given q. In the overdamped case, the spectrum has a single Lorentzian centred at ω_o.

The first capillary wave spectra of ionic liquids have been recorded at the liquid/vacuum interface of imidazolium-based liquids [4, 25, 26]. Measurements have been performed at various temperatures ($325 \leq T/K \leq 400$) and wave vectors ($200 \leq q/cm^{-1} \leq 700$) at pressures of $\sim 10^{-8}$ mbar. Results of the experimental dispersion relations, $\omega_m(q;T)$, are shown in Figure 2.2 for [C$_6$mim][PF$_6$] and [C$_6$mim][NTf$_2$], respectively. Obviously, the observed

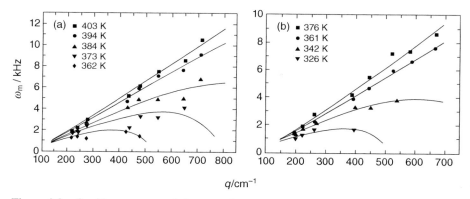

Figure 2.2 Capillary wave peak frequencies ω_m versus wave vector q at different temperatures, recorded at the liquid/vacuum interface of ionic liquids: (a) [C_4mim][PF_6]; (b) [C_6mim][NTf_2]. See also Reference [25].

$\omega_m(q;T)$ dependencies, especially at low temperatures, cannot be reconciled with the behaviour expected for a bare liquid surface according to Equation (2.2). Therefore, in evaluating these data, we have modified the elastic term in the dispersion relation—that due to the surface tension—by an additional contribution, $-\gamma^2 q/\varepsilon_0$, where γ is the dipole moment density at the surface and ε_0 is the vacuum permittivity [4]. This correction takes into account surface polarisation effects, which manifest in a potential drop of $\Delta\varphi \sim 1$ V in the MD simulations (see, e.g., Reference [16]). Using independently measured surface tension data—see below—and densities from the literature, we have fitted the experimental dependencies in Figure 2.2 by the modified dispersion relation (see the full lines in Figure 2.2). The fit results of the viscosities within experimental errors agree with the literature data [25]). Results of the surface dipole moment densities $\gamma(T)$ determined from the fits are plotted in Figure 2.3, as indicated by the full symbols. Also included (open circles) are calculations using the modified dispersion relation and the measured ω_m data, together with the fitted viscosities. As can be seen, for both ionic liquids, the surface dipole moment densities decrease continuously with temperature and reach a nearly constant value of 0.6–0.8 nC m^{-1} at higher temperatures. A very similar behaviour was observed for [C_4mim][PF_6] [26]. According to MD calculations, a strong alignment and ordering of the imidazolium cations is predicted at lower temperatures, which should decrease at higher temperatures. So, it is reasonable to assume that $\gamma(T)$ reflects this change of ordering at the interface. Therefore, we have interpreted the $\gamma(T)$ variation by an order–disorder transition, which in mean field approximation can be described by:

$$\gamma(T)/\gamma_0 = \tanh(T_c\gamma/T\gamma_0). \tag{2.3}$$

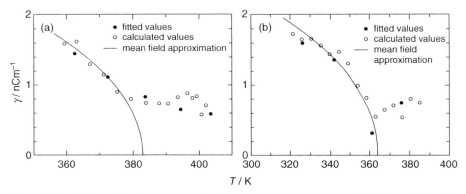

Figure 2.3 Surface dipole moment density γ versus temperature T. Full symbols: Fit results of the experimental $\omega_m(q;T)$ data. Open symbols: Values calculated from $\omega_m(T)$. Full curve: Fit by mean field approximation—see text. (a) [C_6mim][PF$_6$], $\gamma_0 = 3.9$ nC m^{-1}, $T_c = 383$ K; (b) [C_6mim][NTf$_2$], $\gamma_0 = 3.9$ nC m^{-1}, $T_c = 364$ K. See also Reference [25].

The full lines in Figure 2.3 present fits of the $\gamma(T)$ data by Equation (2.3).

Two subsequent studies of the dynamics of capillary waves at the liquid/vacuum interface of imidazolium-based ionic liquids have been reported [27, 28]. Sloutskin et al. [27] used X-ray photocorrelation spectroscopy, while Yao and coworkers [28] employed dynamic light scattering as described above, and analysed their spectra in the time domain (see also Reference [24]). In both studies a transition from overdamped to oscillatory capillary waves is found with increasing temperature, dependent on q. This behaviour is consistent with the strong decrease of the viscosity as a function of raising temperature, which is typical of ionic liquids. It is also not in conflict with surface polarisation as just described. However, Sloutskin et al. criticise the suggestion of a surface dipole layer as discussed above. They argue that a $\gamma \sim 2$ nC m^{-1} at $q = 360$ cm^{-1} should lead to a decrease in ω_m of 40%, which they should have observed with their experimental resolution of ~10% in ω_m. Taking $\gamma = 1.6$ nC m^{-1}—the maximum value in Figure 2.3—and $q = 360$ cm^{-1}, this yields $-\gamma^2 q/\varepsilon_0 = -9$ mJ m^{-2}, that is, a reduction of the surface tension by ~20% for, for example, [C_6mim][PF$_6$] near room temperature. This reduction in σ corresponds to a frequency change of ~10% (see Eq. 2.2), which is difficult to resolve in the experiment of Sloutkin et al., as they state [27]. So, their argument against the surface dipole moment model is not as strong as it was cited in Reference [28] without further justification.

The basic thermodynamic characteristic of an interface is the interfacial free energy or tension, $\sigma_{i,k}$. In Table 2.1, we present a selection of $\sigma_{l,v}$ data for imidazolium-based ionic liquids obtained in our group with the maximum bubble pressure method [4, 25]. Also included in this table are a few values of classical molten salts from the literature [29, 30]. Comparing these numbers, it is striking that the surface tensions $\sigma_{l,v}$ of the ionic liquids are lower by roughly

TABLE 2.1   Interfacial Free Energy, $\sigma_{l,v}$, and Surface Excess Entropy, $-\partial\sigma_{l,v}/\partial T$, at the Liquid/Vapour Interface of Selected Molten Salts and Ionic Liquids Near Their Respective Melting Points or at Room Temperature (RT) Where Indicated

Compound	$\sigma_{l,v}/mJ\ m^{-2}$	Ref.	$-\partial\sigma_{l,v}/\partial T/mJ\ m^{-2}\ K^{-1}$
NaCl	113.8 ± 2	[29]	0.07
KCl	98	[29]	0.07
CsCl	92 ± 3	[30]	0.07
BiCl$_3$	73 ± 1	[30]	0.14
[C$_4$mim]Cl	47 ± 1.5	[26]	0.04
[C$_4$mim][PF$_6$]	43 ± 1.5 (RT)	[4]	0.04
[C$_6$mim][PF$_6$]	40 ± 1.5 (RT)	[25]	0.06
[C$_6$mim][NTf$_2$]	31 ± 1.5 (RT)	[25]	0.08

a factor of 2. Qualitatively, this is due to the smaller ionic radii of molten salts in comparison with ionic liquids. There is also a similar trend in the reduction of surface tensions with increasing the size of the cation or anion. As for the uncertainty of literature data of $\sigma_{l,v}$, a critical remark is timely. In many experimental methods, like the Wilhelmy method, a solid part is wet by the liquid. This wetting strongly depends on the purity of the solid. A second, and an even larger, error source is surface impurities, like adsorbed water or surface-enriched water in ionic liquids. This effect is manifested by a step-like change in the temperature dependence of $\sigma_{l,v}$ at temperatures of around 340 K (see also Reference [4]), where visible evaporation of water commences. Therefore, it is not surprising that surface tension data of ionic liquids in the literature scatter up to 15% (see also Reference [20]).

The surface excess entropy, $-\partial\sigma_{l,v}/\partial T$, gives a phenomenological first insight into surface ordering. Comparing the values in Table 2.1, a trend to higher ordering is indicated in the imidazolium salts in comparison with molten salts like KCl or NaCl. Increasing the length of the alkyl chains, this effect seems to be reduced. However, the surface excess entropy of [C$_6$mim]$^+$ ionic liquids is still lower than that of pure liquid alkanes, which have values of $\sim 9 \times 10^{-5}\ J\ m^{-2}\ K^{-1}$ [31].

2.3   LIQUID/LIQUID INTERFACES

Interfaces of ionic liquids with organic liquids or water are of particular interest in chemical applications, including multiphase catalysis and synthesis [32], liquid/liquid extraction and separation [32], synthesis of nanocrystalline materials [33, 34], and also electrochemistry [35]. Here, we focus on a few fundamental properties of these interfaces.

Generally, the interfacial tension $\sigma_{l\alpha,l\beta}$ between two immiscible liquid phases α and β in contact is given by:

$$\sigma_{1\alpha,1\beta} = \sigma_{1\alpha,v\alpha} + \sigma_{1\beta,v\beta} - w_{\alpha,\beta} \tag{2.4}$$

Here $\sigma_{1,v}$ denotes the respective liquid/vapour interfacial free energies and $w_{\alpha\beta}$ is the free energy of interaction across the liquid/liquid interface. This is determined through the work of adhesion, that is, the work necessary to separate a unit area of the interface α/β into two liquid/vapour interfaces of α and β. Different approximations have been developed to determine $w_{\alpha\beta}$, taking into account the type of intermolecular forces at the interface (see, e.g., Reference [36]). Furthermore, $\sigma_{1\alpha,1\beta}$ depends on the type of the binary bulk phase diagram. If, for example, the phase diagram exhibits a miscibility gap with an upper critical point, then $\sigma_{1\alpha,1\beta}$ approaches zero at this consolute point. Under these conditions, the vertical displacement of the molecules at the interface becomes huge and a diffuse interface results ($<z^2>^{1/2} \propto 1/\sigma$, see above).

Measurements of the interfacial tensions exist both for ionic liquid/organic solvent and for ionic liquid/water interfaces. They have been obtained mainly with the aid of the pendant-drop method (see, e.g., Reference [36]). For transparent fluids of interest here, the contactless method of capillary wave spectroscopy is the method of choice, especially suited for measurements of low $\sigma_{1\alpha,1\beta}$ (see also Reference [24]). Heintz and coworkers [37] studied the temperature dependence of $\sigma_{1\alpha,1\beta}$ in the system ([C$_4$mim][NTf$_2$] + 1,2-hexanediol), which has a critical temperature of $T_c = 330$ K; about 40 K below T_c, and at coexistence they report an interfacial tension of 1 mJ/m^2, which decreases nearly linearly to zero approaching T_c. Very similar observations have been made in Reference [38] for the system ([C$_6$mim][NTf$_2$] + 1-octene) in a temperature interval $283 \le T/K \le 348$; however, the critical temperature in this system is clearly higher and so the lowest $\sigma_{1\alpha,1\beta}$ observed at 348 K is 10 mJ m^{-2}. The liquid/liquid interfacial tensions in the system ([C$_n$mim][NTf$_2$] + 0.1M LiCl aqueous solution) have been measured for different $n = 6, 8, 10,$ or 12 [39]. It is found that with increasing alkyl chain length, the interfacial tension decreases from 12 mJ m^{-2} ($n = 8$) to 10 mJ m^{-2} ($n = 12$); the temperature is not given, but it is presumably room temperature. Ralston and coworkers [15] have studied mixtures of [C$_4$mim][BF$_4$] with water in contact with hexadecane in electrowetting experiments from which they could determine the interfacial tensions.

As was pointed out before, low interfacial tensions imply a diffuse interfacial region; for instance, for a value of 1 mJ m^{-2}, the width of the diffuse interface is about 20–30 nm. For computer simulations, this is a real challenge. Several MD simulation studies of the liquid/liquid interfacial properties have been undertaken where one phase is an ionic liquid [40–42]. Lynden-Bell and coworkers [40] studied the interfaces between [C$_1$mim]Cl and Lennard-Jones fluids and water, where the Lennard-Jones fluid is expected to behave similar to an organic solvent. In the latter case, the orientation of the cations is found to be perpendicular to the interface, comparable with the ionic liquid/vacuum interface. Correspondingly, the ionic liquid has a negative electrostatic potential of $\Delta\varphi \sim -1$ V relative to the Lennard-Jones fluid. The interfacial tension with a weak Lennard-Jones fluid ($\varepsilon = 45$ K) is estimated to be ~70 mJ m^{-2} [40], which seems a little high. In a diffusing interface between the ionic liquid and water, the calculations indicate no orientation of the molecules and the potential change across the interface has the opposite sign [40].

Systems with a broad miscibility gap are solutions of [C$_n$mim][PF$_6$] or [C$_n$mim][NTf$_2$] in water [43]. Phase separation in these mixtures and the properties of the neat liquid/liquid interfaces have been investigated with MD calculation by Sieffert and Wipff [41, 42]. In comparison with classical water–oil mixtures, the randomly mixed water–ionic liquid solutions show a slow phase separation within 20–40 ns. The interface is found to be relatively sharp (1 nm) near room temperature, and it is overall neutral with isotropically oriented molecules, as in the bulk phase [41]. In the case of [C$_8$mim][PF$_6$]–water mixtures, the cations are ordered at the interface, the imidazolium rings pointing towards the aqueous side and the octyl chains towards the ionic liquid. In [C$_4$mim][PF$_6$]–water mixtures, the interface is not well defined and a heterogeneous charge distribution is indicated in the MD results accompanied by a positive potential at the interface [42]. The same authors also investigated the rôle of the liquid/liquid interface in the biphasic rhodium-catalysed hydroformylation of heavy alkenes [44]. Experimental studies on the microscopic structure of ionic liquid/liquid interfaces are rare. Only recently, the interfaces between [C$_n$mim][PF$_6$] and different solvents like CCl$_4$ and D$_2$O have been probed by infrared visible sum frequency generation (IR-VIS SFG) spectroscopy [19]. These measurements indicate that the local structures in the interfacial regions are different from the bulk and also from the ionic liquid/air interface.

During the last decade, several groups have explored the electrochemical properties of the ionic liquid/water interface (see also Reference [35]). Research on this interface is particularly motivated by its use in liquid–liquid extraction applications. Both polarisable and non-polarisable interfaces have been studied by different methods, including measurements of the phase boundary potential, the polarised potential window, the electrocapillary curves, the ion partitioning, and the charge transfer kinetics by scanning electrochemical microscopy. For further details reference is given to a recent review by Samec, Langmaier, and Kakiuchi [45].

2.4 SOLID/LIQUID AND ELECTRIFIED SOLID/IONIC FLUID INTERFACES

As is known from theory, a hard sphere fluid near a rigid wall possesses a layering structure except at low densities (see, e.g., Reference [46]). A very similar observation has been reported by Heyes and Clarke in their first MD simulation of an ionic fluid, molten KCl, contained between two parallel uncharged rigid walls [47]. In the vicinity of the walls, there are clear oscillations in the density $\rho(z)$ (z is aligned perpendicularly to the walls) with a

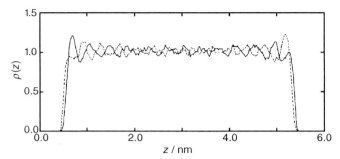

Figure 2.4   Component density profiles $\rho_+(z)$ (——) and $\rho_-(z)$ (-----) determined from MD simulations of a 5-nm-thick liquid film of 504 ions (K^+ and Cl^-) in an electric field of $E = 3.6 \times 10^{-9}$ V m^{-1} at 1050 K. From Reference [47], reproduced with permission of the Royal Society (2010).

periodicity of ~ 0.3 nm, which die away rapidly towards the centre of the fluid film. These oscillations indicate a layering of the ions near the interface caused by hard core repulsions. However, there is no significant difference in the profiles of K^+ and Cl^- ions in the interfacial region. This changes if one applies an electric field across the molten salt film. Now, charge ordering across the film occurs, indicating "multilayers" in the electrified interphase. This is demonstrated in Figure 2.4. In comparison with the unperturbed film, there is a clear splitting of the density profile into a cation, $\rho_+(z)$, and an anion, $\rho_-(z)$, component. This does not mean that the interfacial region is made up of alternating planes of opposite charges, but there is an excess of opposite charges in neighbouring planes.

Qualitatively, a similar charge ordering can be expected at the solid/liquid interface of ionic liquids. Evidence was given in an MD simulation study [48] and in a recent X-ray reflectivity experiment [10]. In their computer simulation, Lynden-Bell and coworkers [48] studied the structural, electrostatic, and dynamic properties of a 4.5-nm-thick liquid [C_1mim]Cl film confined between two parallel walls, both charged and uncharged. In both cases, a layering pattern is found across the liquid film. At the electrified ionic liquid/solid interface, anions segregate significantly towards the positive electrode while both ions coexist in the same layer on the negative side [48]. This different behaviour presumably is due to the different size and shape of the ions. It is also observed that the orientation of the cations is nearly the same for the polarised and non-polarised interface. The electrostatic potential inside the liquid film exhibits oscillatory behaviour in phase with the charge density profiles. The experimental proof is based on high-energy X-ray reflectivity measurements at a charged sapphire (0001-orientation)/ionic liquid interface [10]. The momentum transfer range spanned in this experiment ranged up to $q = 14$ nm^{-1}. Measurements were performed under vacuum ($<10^{-5}$ mbar) in a temperature range between $-15\,°C$ and $110\,°C$ with three ionic liquids having different imidazolium cations and the common anion $[FAP]^{-}$, tris(pentafluoroethyl)trifluorophosphate. One of the main observations is that all reflectivity curves show a clear dip around $q_o \sim 8$ nm^{-1}, which is a strong indication for interfacial layering with a layer spacing of $d = 2\pi/q_o \sim 0.8$ nm. This distance is comparable in magnitude with the thickness of a cation-anion double-layer. A detailed analysis of the Fresnel normalised reflectivity curves yields a layering structure of cation and anion rich strata, which decays into the bulk with a decay length of ~ 2 nm at low temperatures [10]. According to the authors: "This layering is expected to be a generic feature of ionic liquids at charged interfaces."

Charge ordering, as just described, has a strong impact on the electrochemical properties of the electrified ionic liquid/electrode interface—where the electronically conducting electrode material can be a metal or semiconductor. The characteristic differences between a conventional aqueous electrolyte and an ionic fluid—molten salt or ionic liquid—at a charged metal electrode are sketched in Figure 2.5. In the classical double-layer model described by the Gouy–Chapman–Stern theory (see, e.g., Reference [49]), the electrostatic potential $\varphi(z)$ decays continuously into the diffuse layer and the differential capacitance in dilute aqueous solutions exhibits a minimum at the potential of zero charge. In contrast, there is no diffuse layer in ionic fluids and $\varphi(z)$ is characterised by an oscillatory behaviour caused by charge ordering. As for the differential capacitances, the situation is more complex in ionic fluids and seems to be an unsolved problem at present. In the mean-field lattice gas model of Kornyshev [50], maxima and minima are predicted depending on the lattice parameter, Γ. For a densely packed ionic fluid, Γ is near 1 and a maximum is expected in the capacitance-potential curve. Recent experimental investigations of different ionic liquids do not give a clear picture, and are partly conflicting, also due to the fact that the electrode surface structure in many cases is not defined.

In a series of papers, Madden and coworkers have investigated the electrochemical interface with ionic fluids using *ab initio*-based MD simulations. Problems they have dealt with include the electrochemical charge transfer at the metal electrode [51] and potential induced 2D phase transitions of ions adsorbed at the electrode [11, 52]. Here, we briefly discuss the latter phenomenon. Considering the example of an electrified Al(100)/molten LiCl interface, Madden et al. used an interaction potential that includes the polarisation of the ions by interionic interactions and by the interfacial potential and the polarisation of the metal electrode by the ions in the liquid. Varying the applied potential, a strong ordering of ions next to the electrode occurs at a specific potential. This is demonstrated by the in-plane 2D-structure factor, $S_{2D}(q)$, in Figure 2.6. It must be added that this ordering transition is accompanied by a maximum in the differential capacitance at the transition potential [11]. It is also interesting to note that the transition is no longer observed if the electrode or ion polarisation effects are not included.

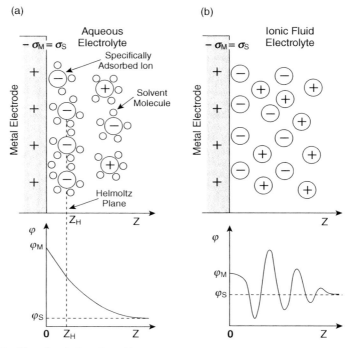

Figure 2.5 Sketch of the electrified metal electrode/electrolyte interface for (a) an aqueous electrolyte and (b) a molten salt or an ionic liquid electrolyte. In part (a), the solution phase is divided into the compact Helmholtz layer and a diffusion layer; the potential $\varphi(z)$ decays exponentially in the diffuse layer provided that the metal potential φ_M is sufficiently low. In part (b), the structure of the ionic fluid near the charged (σ_M) electrode is dominated by charge ordering or layering; this is also reflected in the corresponding potential profile $\varphi(z)$ which decays over several layers towards the bulk solution value φ_S. The comparison shows that the double-layer model in the conventional sense does not apply to the electrified metal/ionic fluid interface.

The experimental verification of such an ordering transition for an ionic liquid was first obtained by scanning tunnelling microscopy (STM) imaging at the electrified Au(111)/[C$_4$mim][PF$_6$] interface [13]. This is shown in Figure 2.7. At potentials positive of -0.4 V (vs. Pt quasi-reference), the [PF$_6$]$^-$ anions on Au(111) form a nearly hexagonal Moiré pattern (Fig. 2.7a), while below -0.4 V the structure transforms into a $\sqrt{3} \times \sqrt{3}$ ordered phase.

2.5 WETTING AND ELECTROWETTING CHARACTERISTICS

Ionic liquids have been considered as lubricants in high friction applications and as electrowetting agents in microfluidic devices—lab-on-a-chip devices—

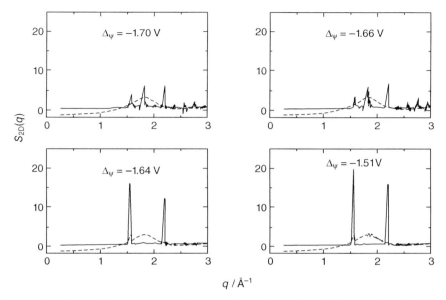

Figure 2.6 Potential induced ordering transition of absorbed Cl⁻ ions at the electrified Al(100)/molten LiCl interface. Shown is the in-plane Cl⁻···Cl⁻ structure factor $S_{2D}(q)$ versus momentum transfer q at different applied potentials $\Delta\psi$. Notice the Bragg peaks at specific q-values. The dashed curves indicate the liquid-like structure factor in adjacent electrolyte layers. From Reference [11], reproduced with permission of the Institute of Physics, London (2010).

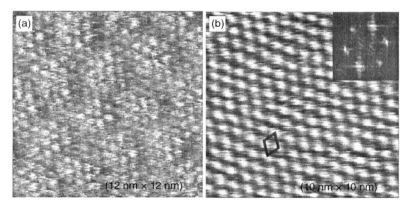

Figure 2.7 High-resolution STM images of [PF₆]⁻ ions adsorbed on Au(111) at different potentials versus Pt quasi-reference. (a) STM image recorded at 0 V showing a Moiré pattern of the ions with a nearest neighbour distance of ~2.5 nm. (b) STM image at −0.4 V with a $\sqrt{3} \times \sqrt{3}$ structure. The inset is the corresponding fast Fourier transform (FFT) of the structure ($V_{bias} = 0.8$ V, $I_t = 0.35$ nA). See also Reference [13].

and in novel display elements [53,54]. All these applications require an in depth knowledge of the wetting and electrowetting characteristics. Yet only a few investigations have been reported so far. A systematic study of the advancing and receding contact angles of several 1-alkyl-3-methylimidazolium ionic liquids on polar and non-polar surfaces has been undertaken [55]. In general, partial wetting is observed under these circumstances, similar to molecular fluids. The wettability of different substrates like soda-lime glass, Teflon, and glassy carbon by the $[C_2mim]Cl-AlCl_3$ system has been studied at room temperature [56]. Complete wetting (contact angle being zero) occurs on glass for $X(AlCl_3) > 0.6$, and on Teflon for $X(AlCl_3) > 0.66$, indicating the key rôle of aluminium(III) chloride content. This type of information is essential in battery development because the wetting behaviour of a battery electrolyte can strongly affect the internal resistance and the charge-discharge behaviour.

In electrowetting, the contact angle Θ of a liquid drop of a conducting fluid on top of a thin dielectric film with capacitance C can be reduced by applying a voltage U across the drop and film. Electrowetting enables the manipulation of tiny amounts of a liquid on a solid surface. The basic equation relating Θ with U is the Young-Lippmann equation (see, e.g., Reference [57]):

$$\cos(\Theta) = \cos(\Theta_o) + (C/2\sigma_{l,v})U^2, \tag{2.5}$$

where $\sigma_{l,v}$ is the liquid/vapour interfacial tension of the liquid provided that the ambient medium is the vapour phase and not a second immiscible and non-conducting liquid. In the latter case, $\sigma_{l,v}$ must be replaced by $\sigma_{l\alpha,l\beta}$. A typical electrowetting curve, $\cos(\Theta)$ versus U, is shown in Figure 2.8 for an ionic liquid. Plotted are the contact angles (left scale) and $\cos(\Theta)$ (right scale) as a function of the applied DC voltage for the example of $[C_6mim][NTf_2]$ on a 100-nm SiO_2-Si substrate [14]. As can be seen, for voltages $|U| < 50$ V, the data follow a parabolic dependence in accordance with Equation (2.5), indicated by the parabolic curve. Clear deviations set in above the saturation potential of $\sim\pm50$ V. Above saturation, an interesting observation is made with ionic liquids studied under UHV conditions: bubbles of different size develop, which we interpreted by a decomposition of the ionic liquid [14]. This explanation assumes an electric breakdown of the 100 nm SiO_2 layer and a corresponding discharging. Recent calculations of the nonlinearities of the electric field at the three phase contact line support this model [58]. However, it must be noted that a generally accepted picture of the phenomenon of contact angle saturation in electrowetting is still missing [59].

For a given electrowetting angle, the voltage in Equation (2.5) can be reduced either by increasing the capacitance of the dielectric film or by decreasing $\sigma_{l,k}$. The latter is possible by changing the ambient medium of the liquid drop, as indicated above.. This was achieved in a recent electrowetting experiment with $([C_4mim][BF_4] + H_2O)$ on a fluoropolymer surface and with hexadecane as ambient phase [15]. In this way, a maximum range of contact

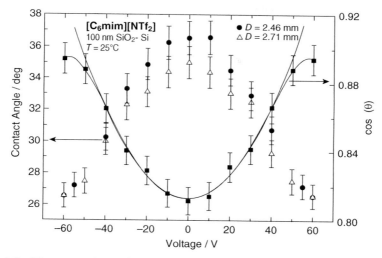

Figure 2.8   Electrowetting of $[C_6mim][NTf_2]$ on a 100-nm-thick SiO_2 film on Si under vacuum at 300 K. Left-hand scale: Contact angle Θ at different DC voltages; different symbols correspond to separate runs with droplets of different contact line diameter D. Right hand scale: Splined electrowetting curve, $\cos(\Theta)$ versus U, determined from averaged Θ-values. The parabolic curve is a fit by Equation (2.5) indicating a saturation potential of $|U| \sim 50$ V. See also Reference [14].

angle variation of ~100° could be reached, which is clearly higher than the typical 10–20°, found in a systematic study of 19 ionic liquids at the liquid/air interface [53]. With regard to potential applications, surely further progress will be made in the near future on the problem of electrowetting of ionic liquids.

2.6   SUMMARY AND CONCLUSIONS

It is amazing how much is known about ionic liquid interfaces, even though this knowledge was gained in less than a decade. The best studied interface is that between liquid and gas or vacuum, where a relatively clear picture of the structure and composition exists. In imidazolium-based ionic liquids, it is generally accepted that the alkyl chains protrude outwards from the liquid, thus imparting a hydrophobic character. The first liquid layer underneath exhibits a slight density enhancement relative to bulk, and the ratio of the cation and anion numbers is approximately the same depending a little on the anion size. In particular, MD simulations indicate an orientational ordering of the imidazolium cations perpendicular to the surface and an electrostatic potential decrease of 0.5–1.0 V across the interface. The corresponding dipole moment densities, as derived from capillary wave spectra, indicate an interfacial order–disorder transition at elevated temperatures. Although the existence of these dipole moments has been questioned, a convincing counter-evidence is lacking. The basic thermodynamic characteristics, the interfacial free energy and the excess entropy, exhibit qualitative similarities with molten salts. The surface tension decreases with increasing size of the cations and anions, while the excess entropy increases.

Interfaces of ionic liquids with organic solvents and aqueous solutions are of particular interest in several chemical applications. Although first results of their interfacial structures and tensions exist, more information about the interfacial electrochemical properties and the bulk phase behaviour of these mixtures is desirable. Comparatively good progress has been made in elucidating the microscopic structure of the solid/liquid interface and its electrified modification. Charge ordering plays an important rôle in ionic liquids and molten salts alike.

The electrified ionic fluid/metal electrode interface is characterised by alternating cation and anion rich layers. This layering decays over several interionic distances into the bulk phase. This structure has far reaching consequences for the electrochemical interface and its properties such as the differential capacitance. The classical double-layer model as developed for aqueous electrolytes does not apply in the conventional sense to the electrode/ionic fluid interface. Further investigations of this problem can be expected. The understanding of the wetting and electrowetting behaviour of ionic liquids on solid substrates is still incomplete. With regard to the potential applications—especially due to their low vapour pressure, thermal stability, and high electrical conductivity, ionic liquids are superior to aqueous media—further progress on electrowetting of ionic liquids should be strived for.

ACKNOWLEDGEMENTS

Financial support of this work by the Network of Excellent Scientists (NES programme of Karlsruhe Institute of Technology) is gratefully acknowledged. J. Szepessy is thanked for his continuous help in editing the figures.

REFERENCES

1 Emel'yanenko, V.N., Verevkin, S.P., and Heintz, A., The gaseous enthalpy of formation of the ionic liquid 1-butyl-3-methylimidazolium dicyanamide from combustion calorimetry, vapour pressure measurements, and ab initio calculations, *J. Am. Chem. Soc.* **129**, 3930–3937 (2007).

2 Steinrück, H.-P., Surface science goes liquid, *Surf. Sci.* **604**, 481–484 (2010).

3 Jeon, Y., Sung, J., Bu, J.W., Vaknin, D., Ouchi, Y., and Kim, D., Interfacial restructuring of ionic liquids determined by sum-frequency generation spectroscopy and X-ray reflectivity, *J. Phys. Chem. C* **112**, 19649–19654 (2008).

4   Halka, V., Tsekov, R., and Freyland, W., Peculiarity of the liquid/vapour interface of ionic liquids: study of surface tension and viscoelasticity of liquid [C_4mim][PF_6] at various temperatures, *Phys. Chem. Chem. Phys.* **7**, 2038–2043 (2005).

5   Rivera-Rubero, S., and Baldelli, S., Surface spectroscopy of room temperature ionic liquids on a platinum electrode: a sum frequency generation study, *J. Phys. Chem. B* **108**, 15133–15140 (2004).

6   Law, G., Whatson, P.R., Carmichael, A.J., and Seddon, K.R., Molecular composition and orientation of the surface of room-temperature ionic liquids: effect of molecular structure, *Phys. Chem. Chem. Phys.* **3**, 2879–2885 (2001).

7   Lynden-Bell, R.M., Gas/liquid interfaces of room-temperature ionic liquids, *Mol. Phys.* **101**, 2625–2633 (2003).

8   Bhargava, B.L., and Balasubramanian, S., Layering at an ionic liquid interface: a molecular dynamics simulation study of [bmim][PF_6], *J. Am. Chem. Soc.* **128**, 10073–10078 (2006).

9   Wang, Y., Feng, S., and Voth, G.H., Transferable coarse grained models for ionic liquids, *J. Chem. Theor. Comput.* **5**, 1091–1098 (2009).

10 Mezger, M., Schröder, H., Reichert, H., Schramm, S., Okasinski, J.S., Schöder, S., Honkimäki, V., Deutsch, M., Ocko, B.M., Ralston, J., Rohwerder, M., Stratmann, M., and Dosch, H., Molecular layering of fluorinated ionic liquids at a charged sapphire (0001) interface, *Science* **322**, 424–428 (2008).

11 Pounds, M., Tazi, S., Salanne, M., and Madden, P.A., Ion adsorption at a metallic electrode: an ab initio based simulation study, *J. Phys. Condens. Matter* **21**, 1–10 (2009).

12 Reed, S.K., Laming, O.J., and Madden, P.A., Electrochemical interphase between an ionic liquid and a model metallic electrode, *J. Chem. Phys.* **126**, 084704/1–13 (2007).

13 Pan, G.-B., and Freyland, W., 2D phase transition of PF_6 adlayers at the electrified ionic liquid/Au(111) interface, *Chem. Phys. Lett.* **427**, 96–100 (2006).

14 Halka, V., and Freyland, W., Electrowetting of ionic liquids under UHV conditions, *Z. Phys. Chem.* **222**, 117–127 (2008).

15 Paneru, M., Priest, R., Sedev, R., and Ralston, J., Electrowetting of aqueous solutions of ionic liquid in solid-liquid-liquid system, *J. Phys. Chem. C* **114**, 8383–8388 (2010).

16 Lynden-Bell, R.M., and Del Popolo, M., Simulation of the surface structure of butyl-methyl imidazolium ionic liquids, *Phys. Chem. Chem. Phys.* **8**, 949–954 (2006).

17 Yan, T., Li, S., Jiang, W., Gao, X., Xiang, B., and Voth, G.A., Structure of the liquid/vacuum interface of room-temperature ionic liquids: a molecular dynamics simulation, *J. Phys. Chem. B* **110**, 1800–1806 (2006).

18 Santos, C.S., and Baldelli, S., Surface orientation of 1-methyl, 1-ethyl, and 1-butyl-3-methylimidazolium methyl sulphate as probed by sum-frequency vibration spectroscopy, *J. Phys. Chem. B* **111**, 4715–4723 (2007).

19 Rivera-Rubero, S., and Baldelli, S., Influence of water on the surface of hydrophilic and hydrophobic room-temperature ionic liquids, *J. Am. Chem. Soc.* **126**, 11788–11789 (2004).

20 Sloutskin, E., Ocko, B.M., Tamam, L., Kuzmenko, L., Gog, T., and Deutsch, M., Surface layering in ionic liquids: an X-ray reflectivity study, *J. Am. Chem. Soc.* **127**, 7796–7804 (2005).

21 Sloutskin, E., Lynden-Bell, R.M., Balasubramanian, S., and Deutsch, M., The surface structure of ionic liquids: comparing simulations with X-ray measurements, *J. Chem. Phys.* **125**, 174715/1–7 (2006).

22 Lovelock, K.R.J., Kolbeck, C., Cremer, T., Paape, N., Schulz, P.S., Wasserscheid, P., Maier, F., and Steinrück, H.-P., Influence of different substituents on the surface composition of ionic liquids studied using ARXPS, *J. Phys. Chem.* **113**, 2854–2864 (2009).

23 Maier, F., Cremer, T., Kolbeck, C., Lovelock, K.R.J., Paape, N., Schulz, P.S., Wasserscheid, P., and Steinrück, H.-P., Insights into the surface composition and enrichment effect of ionic liquids and ionic liquid mixtures, *Phys. Chem. Chem. Phys.* **12**, 1905–1915 (2010).

24 Langevin, D., *Light scattering by liquid surfaces and complementary techniques, surfactant science series*, Vol. 41, Marcel Dekker Inc., New York (1992).

25 Halka, V., Tsekov, R., Mechdiev, I., and Freyland, W., Order-disorder transition at the liquid /vapour interface of imidazolium based ionic lquids: a surface light scattering study, *Z. Phys. Chem.* **221**, 549–558 (2007).

26 Halka, V., Tsekov, R., and Freyland, W., Interfacial phase transitions in imidazolium based ionic liquids, *J. Phys. Condens. Matter* **17**, 3325–3331. (2005).

27 Sloutskin, E., Huber, P., Wolff, M., Ocko, B.M., Madsen, A., Sprung, M., Schön, V., Baumert, J., and Deutsch, M., Dynamics and critical damping of capillary waves in an ionic liquid, *Phys. Rev. E* **77**, 060601/1–3 (2008).

28 Hoshino, T., Ohmasa, Y., Osada, R., and Yao, M., Dispersion relation of capillary waves on ionic liquids: observation of the fast overdamped mode, *Phys. Rev. E* **78**, 061604/1–8 (2008).

29 Evans, R., and Sluckin, T.J., A density functional theory for inhomogeneous charged fluids: application to the surface of molten salts, *Mol. Phys.* **40**, 413–435 (1980).

30 Janz, G.J., *Molten salts*, Academic Press, New York (1967).

31 Earnshaw, J.C., and Hughes, C.J., Surface induced phase transitions in normal alkane fluids, *Phys. Rev. A* **46**, R4494–R4496 (1992).

32 Wasserscheid, P., and Welton, T., eds., *Ionic liquids in synthesis*, 2nd.ed., Wiley-VCH, Weinheim (2008).

33 Kanishka, B., and Rao, C.N.R., Use of ionic liquids , liquid/liquid interfaces and other novel methods for the synthesis of inorganic nanocrystals, in: *Advanced wet-chemical synthesis approaches to inorganic nanostructures (2008)*, ed. P.D. Cozzoli, 69MLM6 Conference, Transworld Research Network Publisher, Trivandrum, India (2010), pp. 79–105.

34 Soejima, T., and Kinizuk, N., Ultrathin gold nanosheets formed by photoreduction at the ionic liquid/water interface, *Chem. Lett.* **34**, 1234–1235 (2005).

35 Kakiuchi, T., Electrochemical aspects of ionic liquid/water two phase systems, *Anal. Chem.* **79**, 6442–6449 (2007).

36 Adamson, A.W., and Gast, A.P., *Physical chemistry of surfaces*, 6th ed., Wiley-Interscience, New York (1997).

37 Wertz, C., Tschersich, A., Lehmann, J.K., and Heintz, A., Liquid-liquid equilibria and liquid/liquid interfacial tension measurements of mixtures containing ionic liquids, *J. Mol. Liq.* **131–132**, 2–6 (2007).

38 Ahosseini, A., Sensenich, B., Weatherley, L.R., and Scurto, A., Phase equilibrium, volumetric, and interfacial properties of the ionic liquid [C_6mim][NTf$_2$] and 1-octene, *J. Chem. Eng. Data* **55**, 1611–1617 (2010).

39 Fitchett, B.D., Rollins, J.B., and Conboy, J.C., Interfacial tension and electrocapillarity measurements of the room-temperature ionic liquid/aqueous interface, *Langmuir* **21**, 12179–12186 (2005).

40 Lynden-Bell, R.M., Kohanoff, J., and Del Popolo, M.G., Simulation of interfaces between room-temperature ionic liquids and other liquids, *Faraday Discuss.* **129**, 57–67 (2005).

41 Sieffert, N., and Wipff, G., The [BMI][Tf$_2$N] ionic liquid/water system: a molecular dynamics simulation of phase separation and of the liquid/liquid interface, *J. Phys. Chem. B* **110**, 13076–13085 (2006).

42 Sieffert, N., and Wipff, G., Aqueous interfaces with hydrophobic ionic liquids: a MD study, *J. Phys. Chem. B* **109**, 18964–18973 (2005).

43 Swatlowski, R.P., Visser, A.E., Reichert, M., Broker, G.A., Farina, L.M., Holbrey, J.D., and Rogers, R.D., On the solubilisation of water with ethanol in hydrophobic ionic liquids, *Green Chem.* **4**, 81–87 (2002).

44 Sieffert, N., and Wipff, G., Adsorption at the liquid/liquid interface in the biphasic rhodium-catalyzed hydroformylation of 1-hexene in ionic liquids. A molecular dynamics study, *J. Phys. Chem. C* **112**, 6450–6461 (2008).

45 Samec, Z., Langmaier, J., and Kakiuchi, T., Charge transfer processes at the interface between hydrophobic ionic liquid and water, *Pure Appl. Chem.* **8**, 1473–1488 (2009).

46 Navasculés, G., and Tarazona, P., Surface tension of a hard sphere fluid near a wall, *Mol. Phys.* **37**, 1077–1087 (1979); and further references therein.

47 Heyes, D.M., and Clarke, J.H.R., Computer simulation of molten salt interfaces: effect of a rigid boundary and an applied electrical field, *J. Chem. Soc. Faraday Trans.* **77**, 1089–1100 (1981).

48 Pinilla, C., Del Popolo, M.G., Kohanoff, J., and Lynden-Bell, R.M., Polarization relaxation in an ionic liquid confined between electrified walls, *J. Phys. Chem. B* **111**, 4877–4884 (2007).

49 Bard, A.J., and Faulkner, L.R., *Electrochemical methods: fundamentals and applications*, John Wiley and Sons, Inc., New York (2001).

50 Kornyshev, A.A., Double-layer in ionic liquids. Paradigm change? *J. Phys. Chem.* **111**, 5545–5557 (2007).

51 Reed, S.K., Madden, P.A., and Papadopoulos, A., Electrochemical charge transfer at a metallic electrode: a simulation study, *J. Chem. Phys.* **128**, 124701/1–10 (2008).

52 Tazi, S., Salanne, M., Simon, C., and Turq, P., Potential induced ordering transition of the adsorbed layer at the ionic liquid/electrified metal interface, *J. Phys. Chem.* **114**, 8453–8459 (2010).

53 Nanyakkara, Y.S., Moon, H., Payagala, T., Wijeratne, A.B., Crank, J.A., Sharma, P.S., and Armstrong, D.W., A fundamental study on electrowetting by traditional and multifunctional ionic liquids: possible use in electrowetting on dielectric microfluidic applications, *Anal. Chem.* **80**, 7690–7698 (2008).

54 Bower, Ch., Rider, C.H., Fyson, J., Simister, E., and Clark, A., Electrowetting display elements, PCT Int. Appl. WO 2008053144, 13pp (2008).

55   Batchelor, T., Cunder, J., and Fadeev, A.Y., Wetting study of imidazolium ionic liquids, *J. Colloid Interface Sci.* **330**, 415–420 (2009).

56   Eberhart, J.G., The wetting behaviour of dialkylimidazolium chloroaluminate, a room temperature molten salt, *J. Electrochem. Soc.* **132**, 1889–1891 (1985).

57   Mugele, F., and Baret, J.C., Electrowetting: from basics to application, *J. Phys. Condens. Matter* **17**, R705–R774 (2005).

58   Drygiannakis, A.I., Papathanasiou, A.G., and Boudouvis, A.G., On the connection between dielectric breakdown strength, trapping of charge, and contact angle saturation in electrowetting, *Langmuir* **25**, 147–152 (2009).

59   Mugele, F., Fundamental challenges in electrowetting: from equilibrium shapes to contact angle saturation and drop dynamics, *Soft Matter* **5**, 3377–3384 (2009).

3 Interfaces of Ionic Liquids (2)

ROBERT HAYES, DEBORAH WAKEHAM, and ROB ATKIN

Centre for Organic Electronics, Chemistry Building, The University of Newcastle, Callaghan, New South Wales, Australia

ABSTRACT

Understanding the properties of ionic liquid interfaces is important for a wide variety of applications and processes. However, ionic liquid interfaces are inherently more complex than traditional solvents, and correlations between ion structural arrangements and physical properties are only beginning to be ascertained. In this chapter, we show that ionic liquids and their interfaces can be divided into three structurally distinct regions: the interfacial layer, the transition zone, and the bulk liquid.

3.1 INTRODUCTION

When liquid molecules come into contact with a second (solid, liquid, or gas) macroscopic phase, an interface forms. The interface is the common physical boundary between the two bulk phases. The study of liquid interfaces probably dates back to Leonardo da Vinci, who demonstrated capillary wetting behaviour in liquids confined within narrow tubes [1]. Our understanding of liquid interfaces has since advanced tremendously, reflecting progress that has been made in both chemistry and physics. Today, liquid interfaces underpin many important chemical, physical, and environmental processes. This research has touched upon our everyday lives in a myriad of ways, including surface coatings, energy storage, and food preparation.

Molecular liquid interfaces are typically diffuse. The density profile and physical properties do not change sharply, but rather vary over a few molecular diameters from the values associated with one bulk phase to those of the

Ionic Liquids UnCOILed: Critical Expert Overviews, First Edition. Edited by Natalia V. Plechkova and Kenneth R. Seddon.
© 2013 John Wiley & Sons, Inc. Published 2013 by John Wiley & Sons, Inc.

other. At air–liquid, liquid–liquid, and most solid substrates, surface roughness greater than the molecular dimensions results in physical properties varying smoothly in a fashion usually best described by an error function [2]. The exception is for atomically smooth solid surfaces, which induce the liquid to pack into solvation layers on account of geometric packing constraints [3]. The level of order in these layers decays into the bulk, such that the molecular density profile oscillates with constant period but decreasing amplitude as the distance from the surface is increased. For all types of interfaces, control of interfacial properties is often achieved through the addition of surfactant molecules that adsorb to the surface and form a structured monolayer, physically separating (at least partially) the two phases [4].

The structure, properties, and dynamics of liquid molecules at interfaces dictate performance in a range of chemical processes, including product yields in heterogeneous catalysis [5], partition coefficients in analytical separations [6], charge-transfer rates in electrochemistry [7], and friction coefficients in lubrication [8], amongst others.

It is in this context that ionic liquid interfaces will be examined. Interestingly, molten salt interfaces were being used to make important chemical discoveries more than 200 years ago, notably in Humphrey Davy's isolation of alkali and alkali earth elements via melt electrolysis [9]. (Note that ionic liquids are distinguished from molten salts by melting point: if a salt's melting point is less than $100\,°C$, it is classed as an ionic liquid; if it is greater than $100\,°C$, the substance is a molten salt.) In spite of this pedigree, ionic liquid interfaces remain largely unexploited in chemical processes because the subtle molecular, bulk, and surface specific effects that govern interfacial behaviour are yet to be elucidated [10].

In this chapter, we describe the current understanding of ionic liquid interfaces and identify avenues for future research. A key theme to emerge in this chapter is that ionic liquids are inherently more complex than molecular solvents and respond to interfaces in markedly different ways. Recent advances in experimental methods have led to unprecedented sensitivity in measuring ionic liquid interfacial phenomena, with the goal of these studies being to develop design rules for tailoring ionic liquid surface properties [10].

Interfacial and bulk ionic liquid structure are intimately related and result from the unusual combination of interionic forces present, which include electrostatic, hydrogen bonding, and solvophobic† interactions [11]. Bulk ionic liquid structure, first predicted in molecular dynamic simulations [12, 13] and later confirmed by X-ray [14, 15] and neutron scattering [16–19] studies (cf. Fig. 3.1), is remarkable for two reasons. First, virtually every other class of molecular solvent is structurally homogeneous, that is, lacking structure beyond a preferred organisation between adjacent molecules. In ionic liquids, the ion

---

† Solvophobicity in a non-aqueous solvent is analogous to the "hydrophobic" effect in water.

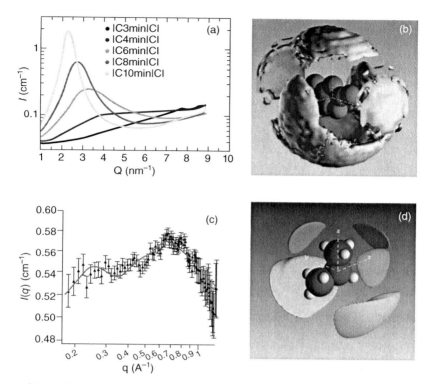

Figure 3.1   (a) X-ray diffraction spectra for a series of [C_nmim]Cl ionic liquids (used with permission from Reference [14]). (b) Empirical potential structure refinement (EPSR) modelling of cation probability distribution around an imidazolium cation in 1,3-dimethylimidazolium chloride (used with permission from Reference [19]). (c) Small angle neutron scattering (SANS) spectra for ethylammonium nitrate (used with permission from Reference [16]). (d) EPSR modelling of cation (bottom lighter two lobes) and anion (top darker three lobes) probability distributions around an ethylammonium cation in ethylammonium nitrate (used with permission from Reference [17]). See colour insert.

arrangements are propagated over much greater distances due to strong clustering of like molecular groups [14, 17]. Second, the internal organisation of polar and apolar domains raises interesting parallels to thermodynamically stable bicontinuous microemulsions, but with length scales at least an order of magnitude smaller [10]. The level of liquid structure is largely dependent on how surfactant-like the ionic liquid is: more pronounced structure has been reported with increasing cation alkyl chain length and, conversely, ionic liquids lacking bulk order have been identified for short-chain ($<C_4$) imidazolium salts [14].

The interionic forces that lead to bulk structure are also expressed at interfaces. However, interactions between the second phase and the ionic liquid ions have an organising effect, such that the interfacial structure is more pronounced than the bulk morphology, and features relating to cation surfactant-like properties are even more prominent. In the same way as revelations of bulk liquid structure have changed the way we think about bulk ionic liquid processes (including solvation [19] and reactivity [20], thermodynamics [21, 22] and kinetic properties [23, 24], transport behaviour [21], etc.), even more pronounced interfacial organisation will change our understanding of interfacial phenomena.

Ionic liquid interfaces can be divided into three regions. The *interfacial layer* consists of the layer of ions in direct contact with the surface of the other phase. This layer is the most organised, and often enriched in the more surface active cation, which interacts with hydrophobic phases via its alkyl chain, and with hydrophilic or anionic surfaces via its charged group. There are few examples of ionic liquid anions adsorbed to cationic surfaces. The region through which the interfacial structure decays to the bulk structure is referred to as the *transition zone*, which typically extends two to five (or more) ion-pair diameters into the bulk. The width of this zone reflects the rate of change between interfacial and bulk structure. The third zone is the *bulk liquid*, which frequently has a bicontinuous structure [17], but can be homogeneous for short chain aprotic ionic liquids [14]. For the purposes of these definitions, it does not matter whether the bulk liquid is structured or not; the key point is that interactions between the surface and the ionic liquid lead to enrichment of one ion species at the interface, which templates an interfacial structure that decays through the transition zone to the bulk morphology.

Obtaining atomic resolution of interfacial ionic arrangements is challenging. Some techniques are better able to resolve structure in either the interfacial layer or the transition zone, but not both. Unsurprisingly, the most detailed knowledge is obtained when two or more techniques are applied to the same interface. In this instance, the diversity in the range of ionic liquids available [25], which provides enormous avenues for scientific innovation, has meant that studies of the same interface using more than one technique have rarely been performed. This complicates analysis, as it is difficult to tease out the relative effects of different investigative techniques versus different ionic liquid species for interfacial structure.

Some types of interfaces are more amenable to investigation than others. Due to their rigidity and fixed interfacial properties, solid surfaces are probably the easiest to investigate. As a result, interfacial structure can be described in greatest detail for solid substrates, especially for atomically smooth surfaces. Studies of gas–liquid interfaces can be more difficult, as the surface is deformable and dynamic, while studies of liquid–liquid interfaces (particularly macroscopic ones) are the most complicated, as scattering and adsorption effects due to both liquid phases can occur.

This chapter aims to present an up-to-date account of ionic liquid interfacial structure, and is organised as follows. Section 3.2 deals with the solid–ionic liquid interfaces, in which there is currently the most interest, and correspondingly the greatest amount of information. Section 3.3 reviews air–ionic liquid interfaces, and Section 3.4 examines liquid–ionic liquid interfaces. In Section 3.5, prospects for future ionic liquid interface research directions are outlined.

3.2   THE SOLID–IONIC LIQUID INTERFACE

3.2.1   Pure Interfaces

Close to a solid surface, ionic liquids exhibit oscillatory density profiles consistent with ion-pair (anion + cation) dimensions. Whilst this behaviour invites comparisons with molecular liquids [3], the arrangement for ionic liquids should not be termed "solvation layering," as there is a preferred ion orientation within surface layers due to clustering of like molecular groups that is absent in molecular liquids. The interfacial structure is better described as "layered" or "lamellar-like". The solid surface plays a key rôle in determining the extent and strength of ionic liquid interfacial structure, as it templates the first layer of ions, which then orients the second layer of ions, and so on. In this section, the structure of solid–ionic liquid interfaces for both protic [26] and aprotic [27] ionic liquids is examined for a range of solid substrates. The similarities and differences between the properties of these surfaces, and how they affect interfacial morphology, are described. Adsorption of surface active molecules at solid–ionic liquid interfaces is then briefly reviewed.

3.2.2   Mica–Ionic Liquid Interfaces

The mica surface is an anionic, atomically smooth crystalline substrate. Negatively charged surface sites are arranged precisely with the surface lattice, leading to a constant surface charge density of one negative lattice charge per $0.48 \, nm^2$ [28]. The atomic smoothness and constant charge of mica make data analysis relatively uncomplicated for this substrate compared with other more commonplace surfaces, whilst the high, well-defined surface charge of mica invites comparison with a charged metal electrode surface.

The first investigation of ionic liquid interfacial structure (of any type) was at a mica interface using the surface forces apparatus (SFA) for the protic ionic liquid ethylammonium nitrate, $[EtNH_3][NO_3]$ in 1988 [29]. Four to five oscillations in the force profile were observed, with the period of 0.5–0.6 nm, consistent with the predicted [29] ion-pair diameter, suggesting cations and anions are present in approximately equal numbers. On the basis of absolute mica–mica separations, up to nine near-surface (anion + cation) layers were inferred. As the force necessary to squeeze out a layer increases as the mica–mica separation decreases, it was suggested that ionic liquid interfacial structuring is more pronounced closer to the surface (and decays out into the bulk). Horn et al. speculated that alternating sublayers of anions and cations may be present, but this could not be resolved from force measurements.

This initial study was not expanded upon for almost a quarter of a century, until the mica–ionic liquid interface was revisited for protic and aprotic ionic liquids using atomic force microscopy (AFM) force curve measurements [30–32]. AFM can probe both ionic liquid interfacial and transition zone structure adjacent to a wide variety of surfaces and was the subject of a recent review [10]. A summary of AFM force curve measurements from six key mica systems are presented in Figure 3.2a–f: ethylammonium nitrate ($[EtNH_3][NO_3]$), propylammonium nitrate ($[PrNH_3][NO_3]$), propylammonium methanoate ($[PrNH_3][HCO_2]$), dimethylethylammonium methanoate ($[Me_2EtNH][HCO_2]$), 1-ethyl-3-methylimidazolium bis{(trifluoromethyl)sulfonyl}amide ($[C_2mim][NTf_2]$), and 1-butyl-1-methylpyrrolidinium bis{(trifluoromethyl)sulfonyl}amide ($[C_4mpyr][NTf_2]$). These systems illustrate the diversity of ionic liquid structuring adjacent to a mica surface and allow mechanisms for controlling surface structure to be revealed. The AFM data are characterised by a series of repeating "push-throughs" at discrete separations on tip approach (dark shade) and (sometimes) retraction (light shade), from the surface. The rupture force increases closer to the surface, indicating that near-surface ionic liquid order is more pronounced closer to the substrate. The contrast between the oscillatory results obtained by Horn et al. [29] and the stepwise AFM data is ascribed to the differences in the experimental methods.

A typical force profile for the $[EtNH_3][NO_3]$–mica system is shown in Figure 3.2a and can be rationalised as follows. Beyond 4 nm, zero force is recorded as the tip experiences negligible resistance moving through the ionic liquid towards the mica. This is significant, as it shows that the AFM is insensitive to the bicontinuous sponge-like structure that exists in the bulk liquid [17]. At 4.1 nm, the tip encounters the first outermost detectable layer and pushes against it. The force increases up to 0.1 nN, then the tip ruptures the layer and "jumps" 0.5 nm before encountering another layer 3.6 nm from the interface, and the process is repeated. The jump interval in all cases is 0.5 nm, in excellent agreement with the diameter of the $[EtNH_3][NO_3]$ ion pair determined from the bulk density and previous SFA results [29]. Attractions between adsorbed $[EtNH_3]^+$ cations on the tip and substrate are responsible for the adhesion force (6 nN) measured upon retraction. Zero force is reached at a separation of 2 nm, which corresponds to the fourth interfacial layer. These forces have a strong temperature dependence, with both the number of detectable layers and rupture force decreasing by half when the temperature is increased from 14 °C to 30 °C [31].

As mica is highly negatively charged, electrostatics dictates that a cation layer is adsorbed to the surface, with the anion largely excluded. As such, a thinner innermost layer should be detected in the force profile. The

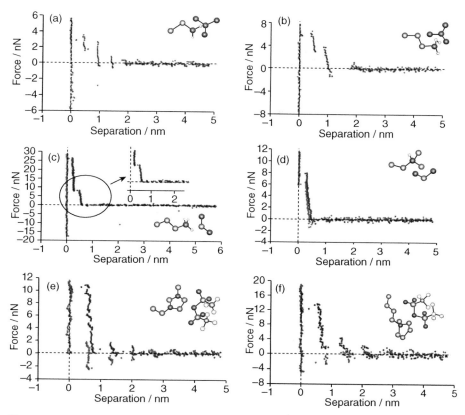

Figure 3.2 Force versus distance profiles for an AFM Si_3N_4 tip approaching (dark shade) and retracting from (light shade) for (a) an [EtNH$_3$][NO$_3$]–mica system at 21 °C (used with permission from Reference [30]), (b) a [PrNH$_3$][NO$_3$]–mica system at 21 °C (used with permission from Reference [30]), (c) a [PrNH$_3$][HCO$_2$]–mica system at 21 °C (used with permission from Reference [31]), (d) a [Me$_2$EtNH][HCO$_2$]–mica system at 21 °C (used with permission from Reference [31]), (e) a [C$_2$mim][NTf$_2$]–mica system at 21 °C (used with permission from Reference [32]), and (f) a [C$_4$mpyr][NTf$_2$]–mica system at 21 °C (used with permission from Reference [32]).

[EtNH$_3$][NO$_3$]–mica system in Figure 3.2a does not show this, and all step sizes are 0.5 nm. It is noted that one cation occupies an area greater than the size of a mica surface charge site (one per 0.48 nm^2) [33], so even at cation saturation coverage, the degree of substrate charge quenching cannot exceed 87% [29]. Thus, the "zero" distance on the force profile actually corresponds to the tip pushing up against, but not penetrating, a strongly-bound, compact cation layer (thin layers due to surface bound cations are detected for less highly charged substrates, described below). The first non-adsorbed layer detected in the transition zone consists of an [EtNH$_3$][NO$_3$] ion pair; the cations are closer to the substrate and oriented with alkyl groups pointing towards the surface to maximise solvophobic interactions with surface-bound cations, similar to the alkyl chain clustering observed in the bulk (cf. Fig. 3.1d). This is followed by a [NO$_3$]$^-$ layer above the ethylammonium cationic headgroup, to quench the electrostatic charge, resulting in an overall substrate A(AB)(BA)(AB) packing arrangement, as in the bulk [17]. The surface serves to align and strengthen the bulk liquid structure, which then decays through the transition zone. This structure is distinct from molecular liquid solvation layers, in which the surface induces structure in an otherwise unstructured fluid, and there is no preferred molecular arrangement.

Fewer and non-vertical steps are detected by AFM when the cation alkyl group is increased from ethyl to propyl in Figure 3.2b [30]. The longer alkyl chain increases rotational freedom and hence [PrNH$_3$][NO$_3$] can pack more efficiently without layering. This demonstrates another important difference between bulk and interfacial ionic liquid structure; the intensity of the small angle neutron scattering (SANS) peak is greater for [PrNH$_3$][NO$_3$] than [EtNH$_3$][NO$_3$] as the longer alkyl chain leads to stronger solvophobic interactions and hence better defined bulk order [16]. However, the converse is sometimes true at a solid surface, where a longer alkyl chain can lead to reduced interfacial order due to molecular flexibility.

Replacing the nitrate anion with methanoate decreases the level of interfacial structure [31], with only two ion-pair layers noted in the force profile (cf. Fig. 3.2c). Ions in layers are less strongly bound together, and the AFM is able to detect anion and cation sublayers, identified by steps thinner than the ion-pair dimension at 0.28, 0.60, 0.83, and 1.17 nm. As zero separation corresponds to a layer of [PrNH$_3$]$^+$ ions electrostatically adsorbed to mica, with propyl chains oriented towards the bulk, the closest measurable layer at 0.28 nm corresponds to a second [PrNH$_3$]$^+$ layer with alkyl tails orientated towards the mica surface, interacting solvophobically with adsorbed cations. The step between 0.28 and 0.60 nm therefore equates to a neutralising layer of [HCO$_2$]$^-$ anions, yielding a total thickness for the anion and cation layers of 0.6 nm, consistent with the ion-pair diameter of [PrNH$_3$][HCO$_2$] [31]. The next two layers have similar thicknesses within the accuracy of the measurement.

The results for [Me$_2$EtNH][HCO$_2$] in Figure 3.2d demonstrate how ionic liquid interfacial layering can be largely prevented at a solid interface. This ionic liquid has only a single, thin layer at 0.45 nm. This is 21% smaller than predicted for ion-pair geometry [31], and provides strong evidence for a single layer of weakly surface-adsorbed cations adjacent to the mica surface, and no subsequent transition zone structure. In contrast to the primary ammonium cations described previously, the [Me$_2$EtNH]$^+$ charge centre is sterically hindered, which prevents close approach of the cation to the substrate. This reduces the strength of electrostatic attractions, which allows the AFM tip to displace the cation layer and move into contact with the mica. The bulk structure of this ionic liquid has not been reported, but we would predict that long range order is absent.

The force profiles for the two aprotic ionic liquids, [C₂mim][NTf₂] and [C₄mpyr][NTf₂], presented (cf. Fig. 3.2e,f) are particularly well defined. At least five steps are noted on approach and retraction. The stepwise retraction indicates that the liquid layers spontaneously reform as the tip moves away from the surface. The step widths are in excellent agreement with the predicted ion-pair diameters (0.75 nm for [C₂mim][NTf₂] and 0.79 nm for [C₄mpyr][NTf₂]) and *d*-spacings from crystallographic data [34]. The non-vertical nature of the "push-throughs" of [C₄mpyr][NTf₂] is due to molecular flexibility imparted by the butyl moiety, as per the [PrNH₃][NO₃] argument developed above. Interfacial forces are stronger for the [C₄mpyr]⁺ cation, due to stronger solvophobic interactions.

3.2.3   Sapphire–Ionic Liquid Interfaces

Like mica, the Al₂O₃(0001) (sapphire) is crystalline and atomically smooth (2–2.5 Å along the basal plane), but its surface charge density is lower. (An ideal sapphire surface at neutral pH in water displays one negative lattice charge per 52.7 nm² [35]; however, this value should be considerably lower due to the ionising effect of the X-ray beam). High-energy X-ray reflectivity experiments for aprotic ionic liquids have revealed an interfacial structure similar to that suggested for mica in most respects [36, 37]. An electron density model for 1-butyl-1-methylpyrrolidinium tris(pentafluoroethyl)trifluorophosphate,

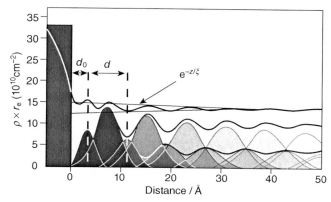

Figure 3.3   Oscillations in cation (smaller amplitude), anion (larger amplitude), and total electron (black line) densities obtained from the best fits to an X-ray reflectivity profile of 1-butyl-1-methylpyrrolidinium tris(pentafluoroethyl)trifluorophosphate ([C₄mpyr][FAP]) at a Al₂O₃(0001) (sapphire) interface (used with permission from Reference [36]).

[C₄mpyr][FAP] is shown in Figure 3.3. Gaussian-type distributions are modelled for both the anions (larger intensity) and cations (smaller intensity) at the interface, and summing these distributions provides the variation in electron density with distance, which has a period of 0.8 nm, corresponding to the size of the ion pair. It was suggested that the cation is in contact with the negatively charged surface, leading to ABAB packing, until the bulk structure is reached after approximately five ion-pair layers. However, given that studies of bulk structure consistently report bilayered (sponge or aggregate) structures due to solvophobic interactions [14–17], it may be that an A(AB)(BA) arrangement is present.

3.2.4 Silica-Ionic Liquid Interfaces

The surface chemistry of silica is quite different from mica and sapphire. Silica is amorphous, not crystalline, and is substantially rougher; typical root mean square (rms) roughness values for silica are 1.3 nm over 5×5 μm [30]. This degree of surface roughness is sufficient to broaden molecular liquid solvation layers [3] (roughness an order of magnitude greater than this will eliminate solvation layers completely [38]). Compared with mica, silica has a greatly reduced and variable surface charge, which arises from an equilibrium between protonated and deprotonated hydroxyl groups. In water, an average nett negative charge of one site per 20 nm² is typical [39]. The corresponding values in ionic liquids are expected to be considerably higher, but still much less than on mica.

The different surface properties of silica compared with mica result in different AFM force profiles. Whilst the basic stepwise form of the data remains, for silica the forces are smeared and non-vertical, due to surface roughness. In Figure 3.4a, for the [EtNH₃][NO₃]–silica system, the thinness of the innermost step (0.25 nm) suggests it consists primarily of an electrostatically bound

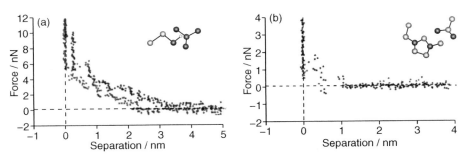

Figure 3.4 Force versus distance profiles for an AFM Si₃N₄ tip approaching (dark shade) and retracting from (light shade) for (a) an [EtNH₃][NO₃]–silica system at 21 °C (used with permission from Reference [30]), and (b) a [C₂mim][O₂CMe]–silica system at 21 °C (used with permission from Reference [30]).

cation layer, inferred but never observed for mica. As the silica is of lower surface charge density than mica, the adsorbed layer can be displaced by the AFM tip at sufficiently high force. The data for 1-ethyl-3-methylimidazolium ethanoate ([C_2mim][O_2CMe] (cf. Fig. 3.4b) are similar to those obtained for this ionic liquid and a mica surface, which suggests an electrostatically bound interfacial layer of [C_2mim]$^+$ cations, followed by two layers. The spacing of the inner cation layer is consistent with the ethyl group being orientated approximately normal to the interface. The increased molecular volume and charge delocalisation on [C_2mim]$^+$ compared with [EtNH$_3$]$^+$ means that adsorption is weaker for the aprotic ionic liquid. The bulk structure for this ionic liquid [40] is not as well defined as for other aprotics [14], and so we should not expect many layers at a solid interface.

Sum frequency generation (SFG) spectroscopy has also been used to investigate the arrangement of aprotic ionic liquids at the silica surface [41, 42]. These experiments reveal that the cation is absorbed with the imidazolium ring slightly tilted towards the silica surface (between 16° and 32° from surface normal), tending to more parallel orientations with decreasing alkyl chain length. The orientation of the cation alkyl chain was nearly normal to the surface, and longer alkyl chains are more ordered and have fewer *gauche* defects.

The structure suggested by these AFM and SFG experiments at a silica interface is consistent with recent all-atom force field modelling performed by Canongia Lopes [43]. Figure 3.5 shows the equilibrium configuration in two simulation boxes: (top) a $5 \times 5 \times 5$ nm cube of 1-(2-hydroxyethyl)-3-methylimidazolium tetrafluoroborate and (bottom) a $5 \times 5 \times 2.5$ nm slab of hydrophilic silica glass. The boundary conditions for both boxes are established in the directions parallel to the glass and are separated in Figure 3.5 for visual interpretation (during the simulation, the ionic liquid box stands directly over the glass slab). The intervening surfaces show the charge distributions across the silica and ionic liquid interfaces. The overall silica surface charge is negative, but positive and negatively charged regions are present. Positive charges are represented by black dots, negative charged by grey dots, and the shaded regions represent partially charged surface domains. Crucially, the simulation shows that anywhere there is a negative charge at the surface, a cation is adsorbed (and vice versa), supporting the argument that the surface templates the composition of the interfacial layer. For very highly charged surfaces like mica, or an electrode under the influence of an applied potential, it is therefore reasonable to suggest that the interfacial layer is essentially entirely composed of oppositely charged ions. On less highly charged or variably charged surfaces (like the one modelled here), it is correct to say that the interfacial layer is enriched in one ion or the other, but in such cases the key effect of the surface in terms of transition zone structure is its smoothness. The smooth substrate flattens the bulk structure into more ordered near-surface layers, which decay to the bulk structure over two molecular layers in Figure 3.5. As experimental techniques (e.g., AFM, SFG, X-ray reflectivity) sample a

Figure 3.5   All-atom force field molecular dynamics simulation of a silica–ionic liquid interface (kindly supplied by Prof. J.N. Canongia Lopes).

large surface area, they are sensitive to the average compositions. For a more complete description of this model, the reader is directed to the chapter by Canongia Lopes in a later volume in this series.

3.2.5   Graphite–Ionic Liquid Interfaces

Graphite is atomically smooth but uncharged, and therefore interacts with ionic liquids in a fundamentally different way from the substrates considered previously. Whereas the surface charge of mica, sapphire, or silica favours electrostatic interactions with ions, this is not the case for graphite, and solvophobic interactions with uncharged molecular regions dominate, similar to those that will occur at air–ionic liquid interfaces. This leads to significant differences in AFM force profiles. "Push-throughs" consistent with the ion-pair dimensions are observed for [EtNH$_3$][NO$_3$]–graphite (Fig. 3.6a) and [C$_2$mim][O$_2$CMe]–graphite (Fig. 3.6b), but they are superimposed on

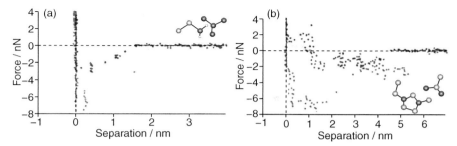

Figure 3.6 Force versus distance profile for an AFM Si_3N_4 tip approaching (dark shade) and retracting from (light shade) for (a) [EtNH$_3$][NO$_3$]–graphite system at 21 °C (used with permission from Reference [30]) and (b) [C$_2$mim][O$_2$CMe]–graphite system at 21 °C (used with permission from Reference [30]).

attractive van der Waals forces. A small final step is noted in both systems (and on retraction for [EtNH$_3$][NO$_3$]), which arises from the cation orientation; alkyl chains lying flat in the case of [EtNH$_3$][NO$_3$], and the imidazolium ring aligned parallel to the surface for [C$_2$mim][O$_2$CMe]–graphite. This favourable cation alignment leads to the formation of six to seven ion-pair layers on graphite for [C$_2$mim][O$_2$CMe], substantially more than on silica or mica for the same ionic liquid, where a perpendicular orientation of interfacial cations was preferred [30].

3.2.6 Gold–Ionic Liquid Interfaces

Au(111) surfaces are excellent metallic electrodes and have been widely used in electrodeposition of elements from ionic liquids [44, 45]. This hexagonal-closed-packed crystalline surface is of comparable roughness to silica (0.1 nm over a 300 nm^2 area) and also negatively charged, although the precise value in the presence of ionic liquids is not known. Figure 3.7a,b shows force profiles for [C$_2$mim][NTf$_2$] and [C$_4$mpyr][NTf$_2$] adjacent to Au(111) surfaces [46].

The force profiles in Figure 3.7 are qualitatively similar to those obtained for the same aprotic ionic liquids on mica (cf. Fig. 3.2e,f), with a series of push-throughs on tip approach and retraction. However, unlike the mica systems, the width of the interfacial layer is much smaller than the ion-pair dimension. Steps are visible at 0.65 nm and 0.60 nm for [C$_2$mim][NTf$_2$] and [C$_4$mpyr][NTf$_2$], respectively. As the surface is negatively charged, this result suggests the interfacial layer is rich in weakly bound cations that the AFM tip is able to penetrate, and then move into contact with the gold substrate, similar to the results for silica. For [C$_2$mim][NTf$_2$], a 13% reduction in layer thickness is measured, suggesting that electrostatic attractions induce the imidazolium ring of the cation to tilt towards the gold surface. The short ethyl group is expected to allow the aromatic ring on [C$_2$mim][NTf$_2$] to be orientated

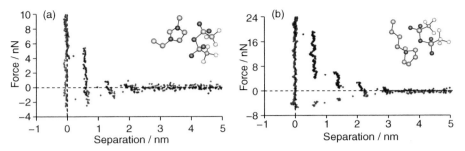

Figure 3.7 Force versus distance profiles for an AFM Si_3N_4 tip approaching (dark shade) and retracting from (light shade) for (a) an [C_2mim][NTf_2]–Au(111) system at 21 °C (used with permission from Reference [46]), and (b) a [C_4mpyr][NTf_2]–Au(111) system at 21 °C (used with permission from Reference [46]).

substantially towards the substrate. A 25% reduction in step size is observed for the [C_4mpyr][NTf_2]-Au(111) system. Strong electrostatic attractions, and the absence of an inflexible aromatic ring, must allow the [C_4mpyr]$^+$ cation to adopt a flatter surface conformation than [C_2mim][NTf_2], thus resulting in the reduced layer thickness. This draws a nice parallel to recent SFG work by Baldelli [47] and modelling by Lynden-Bell et al. [48], who could trace the change in cation conformation in the interfacial layer as a function of applied surface charge. Ionic liquid–surface electrostatic interactions will influence the force required to push-through the layer of ions nearest to the substrate. The significantly higher force required to disrupt [C_4mpyr][NTf_2] compared with [C_2mim][NTf_2] is consistent with increased electrostatic interactions between the surface and the cation, which results from the positive charge being localised on one atom in the case of [C_4mpyr]$^+$ and delocalised across an aromatic ring for [C_2mim]$^+$. Such effects provide a route to tuning interfacial properties [10].

The transition zone extent, and the force required to rupture interfacial layers, is greater for [C_4mpyr][NTf_2] than for [C_2mim][NTf_2], as per results on mica. This is due to differences in cohesive energy within layers. X-ray diffraction experiments for similar ionic liquids [14] have shown the level of order increases with cation alkyl chain length. As [C_4mpyr][NTf_2] possesses a butyl group, and [C_2mim][NTf_2] an ethyl group, stronger solvophobic clustering occurs in the former ionic liquid, and hence it is more organised in the transition region.

3.2.7   Adsorption at the Solid–Ionic Liquid Interface

In molecular solvents, especially water, surface properties are modified through the addition of species that adsorb to the interface. In ionic liquids, added surfactant and ionic liquid ions (usually the cation) compete for surface adsorption sites. As the cation is present at much higher concentrations, and is generally smaller, in most cases the surfactant does not adsorb; only two papers have reported amphiphile adsorption onto solid substrates. Non-ionic surfactants have been shown to adsorb at the graphite-[EtNH₃] [NO₃] interface [49], provided the alkyl chain is sufficiently long to compete with the [EtNH₃]⁺ cation. Micellar aggregates form, similar to those found in aqueous systems [28]. Conventional surfactants do not adsorb at the silica–[EtNH₃][NO₃] interface, as the ionic liquid cation binds more strongly to negative surface adsorption sites. However, Pluronic surfactants are able to adsorb, due to their capacity to have multiple attachment points per molecule [50]. Surface aggregate structures were found to be closely correlated with those present in bulk [EtNH₃][NO₃] [51].

3.3 THE AIR–IONIC LIQUID INTERFACE

3.3.1 Pure Interfaces

The surge in ionic liquid research interest that has occurred in recent times can be traced to a number of factors, but among the most important are their "green characteristics." For the most part, this refers to the fact that many ionic liquids have vanishingly low vapour pressure under standard laboratory conditions [52, 53], which led to ionic liquids being touted as wholesale replacements for volatile organic solvents [54, 55]. Unravelling how ionic liquid ions arrange at air–ionic liquid interfaces is fundamentally important in the context of low vapour pressure, and also for applications ranging from gas capture and storage (e.g., CO_2) [56], to analytical gas separations [57, 58] and functional surface coatings [59].

When a salt is dissolved in water, adsorption of ions at gas interfaces is generally thermodynamically unfavourable due to the absence of solvating water molecules and the formation of image charges [60]. Ionic surfactants, which are salts consisting of a surface-active organic ion and an inorganic counterion, are one exception to this rule, and accumulate at interfaces to form a structured monolayer in response to the entropic hydrophobic force [4]. As ionic liquids consist entirely of ions, charged species must be present at the interface, and enrichment of either the cation or the anion will be a consequence of the balance between entropic and enthalpic contributions. For hydrophilic ionic liquids, it seems likely that the cation, which has some surfactant-like features in most instances, will be enriched at the interface with its hydrocarbon chain protruding into the gas phase. The situation is less clear for hydrophobic ionic liquids. The anions of hydrophobic ionic liquids often include bulky hydrophobic groups, which may be expected to experience a significant attraction to the gas phase. As such, air–liquid interfacial layer composition for hydrophobic ionic liquids will be determined by the relative affinity of the cation and the anion for the interface, which is not simple to predict. For both hydrophobic and hydrophilic ionic liquids, the presence of the macroscopic interface and the likely surface enrichment of one ion or the other has the potential to induce significant transition zone structure.

The air–ionic liquid interface has since been probed over the last decade using a variety of surface sensitive techniques. This includes traditional experimental methods used for free liquid surfaces such as neutron (NR) [61, 62] and X-ray (XRR) reflectivity [62–64], and vibrational SFG [62, 63, 65–71]. The negligible volatility of ionic liquids allows the use of vacuum-based techniques, including direct recoil spectroscopy (DRS) [72, 73], metastable atom electron spectroscopy (MAES) [74, 75], molecular beam surface scattering (MBSS) [76–78], ultraviolet photoemission spectroscopy (UPS) [74], X-ray photoelectron spectroscopy (XPS) [74, 79–82], and angle resolved X-ray photoelectron spectroscopy (ARXPS) [82]. Several molecular dynamic simulations have also been performed [83–87].

3.3.2 Interfacial Layer

The effect of variation in cation and anion species on the composition of the interfacial layer has been investigated using SFG [63, 65–71], MAES [74, 75], MBSS [76–78], and XPS [74, 79–82]. In general, both the cation and the anion are present at the surface. If either species has an alkyl chain, it is oriented towards the gas phase; if either ion has two alkyl groups, the longer alkyl chain will protrude into the air.

SFG has shown that imidazolium cations with alkyl chains longer than four carbons long tend to orientate with their rings parallel to the air interface, regardless of the anion [66]. When the cation alkyl chain is shorter than four carbon atoms, the imidazolium ring orients at some angle normal to the surface, with the exact angle dependent on the anion species [67, 71]. This suggests that, for short alkyl chains, solvophobic interactions are diminished and electrostatic effects become increasingly important, such that the anion is able to influence the orientation of the cation.

Only one study has examined the interfacial layer structure for protic ionic liquids [62]. For ethylammonium nitrate, XRR revealed low electron density immediately adjacent to the gas phase, consistent with cations adsorbed with their ethyl moieties protruding into the air. Complementary SFG experiments confirmed this arrangement; the spectra obtained were dominated by strong sharp peaks originating from CH_3 vibrational modes (cf. Fig. 3.8b). The anion position could not be determined due to the absence of a permanent dipole moment, but reflectivity experiments suggest it is likely positioned directly beneath the cation monolayer, as the small size of the ammonium groups permits lateral interactions between these ethyl groups.

SFG provides information concerning the average orientation of dipolar interfacial ions but cannot resolve the composition of the interfacial layer. MAES [75] and MBSS [76–78] are able to elucidate the ion population density, along with some orientation details. These experiments suggest that when the

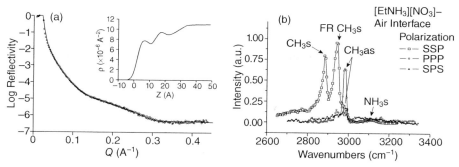

Figure 3.8 (a) X-ray reflectivity profile for the air–$[EtNH_3][NO_3]$ interface at room temperature. The inset shows the scattering length density profile (reproduced with permission from Reference [62]). (b) Vibrational sum frequency spectra of the same interface (reproduced with permission from Reference [62]).

cation alkyl chain has fewer than four carbons, although the alkyl group protrudes into the gas phase, interactions between chains do not occur. When the alkyl chain length is increased to four carbons, the alkyl chains begin to interact with one another, resulting in chain segregation (cf. Fig. 3.9). Further increasing the alkyl chain length results in a fully formed hydrocarbon layer, which coats the anion and cation charged groups. Other studies of aprotic ionic liquids have reported structural differences for short ($<C_4$) and long ($>C_4$) cation alkyl chains [14].

The degree of surface hydrocarbon aggregation is also dependent on the anion species. ARXPS has revealed that smaller, spherical anions, such as chloride, have strong electrostatic interactions with cations, permitting tight packing of charge groups, which results in well-formed hydrocarbon layers. Conversely, larger, bulkier anions reduce hydrocarbon layer formation [82]. XPS studies have shown that the inclusion of ethylene glycol groups in the cation alkyl group decreases chain segregation. This is a consequence of the ethylene glycol groups disrupting solvophobic interactions between alkyl chains [79].

The interfacial layer structures of pyrrolidinium-, guanidinium-, and ammonium-based ionic liquids have received scant attention. The only report that has appeared to date is an SFG study for pyrrolidinium ionic liquids, which showed structural patterns similar to imidazolium ionic liquids with cation alkyl chains oriented towards the gas phase with the ring lying parallel to the surface [69].

3.3.3 Transition Zone Structure

Only a handful of studies have examined the structure of the air–ionic liquid interface transition zone. The first experimental evidence of a transition

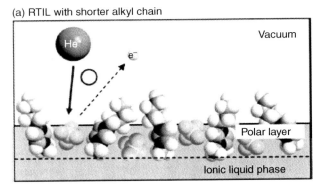

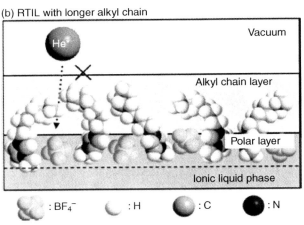

Figure 3.9 Schematic models for the surface structure of ionic liquids with (a) a shorter alkyl chain, and (b) a longer alkyl chain (reproduced with permission from Reference [75]).

zone at air–ionic liquid interfaces was obtained in 2004 [64]. NR experiments using 1-butyl-3-methylimidazolium tetrafluoroborate and 1-octyl-3-methylimidazolium hexafluorophosphate revealed a subsurface region in which the alkyl chains are segregated from the ionic regions in a lamellar-like structure. This arrangement persisted over a distance consistent with two ion pairs. Surprisingly, this subsurface ionic liquid region received no further experimental attention until recently, when the structure of air–[EtNH$_3$][NO$_3$] interface was probed using XRR [62]. The fitted scattering length density profile (cf. Fig. 3.8a) revealed that the bulk liquid structure is not reached until 30 Å, or six ion pairs, from the interface. The oscillations in the density profile indicate the [EtNH$_3$][NO$_3$] transitional zone is composed of alternating polar and non-polar domains, in line with the structure of the solid–ionic liquid transition zone for [EtNH$_3$][NO$_3$].

3.3.4 Relationship between Microscopic Structure of Air–Ionic Liquid Interfaces and Macroscopic Properties

Variation in cation and anion molecular structure also influences the macroscopic physical properties of the air-ionic liquid interface, which may be easily probed via surface tension. Generally, the surface tensions of ionic liquids are lower than water, but higher than pure alkanes [88, 89]. Whilst the relative entropic and enthalpic contributions of surface tension in ionic liquids have not been completely elucidated, a strong entropic driving force is expected. This is based on the observation that longer cation alkyl chains result in reduced surface tensions [88]. In ionic liquids, solvophobic forces result from unfavourable interactions between charged groups and cation alkyl chains, which promote clustering of like groups. Long cation chains produce stronger solvophobic forces, which mean that the propensity of the cation to be enriched at the air–ionic liquid interface will be increased with alkyl chain length, and the interfacial energy will be lowered. This leads to a reduction in surface tension. Currently, the effect of pronounced transition zone structure on surface tension is unclear.

Interfacial ionic liquid properties can, to some extent, be manipulated by varying the morphology of the interface by changing the structure of the ions. However, this will also change bulk structure and properties in a way that may be undesirable. For example, addition of an alcohol group to the cation alkyl chain disrupts solvophobic interactions and decreases interfacial order, but can also increase bulk viscosity. An alternative means of controlling interfacial properties is the use of surfactant molecules, which adsorb to liquid interfaces at low bulk concentrations.

3.3.5 Surfactant Adsorption at Air–Ionic Liquid Interfaces

Surfactant adsorption at air–ionic liquid interfaces was first reported in the 1980s when Evans et al. demonstrated that cationic and non-ionic surfactants adsorb at the air–[EtNH$_3$][NO$_3$] interface [90, 91]. The surface tension of [EtNH$_3$][NO$_3$] decreased with increasing surfactant concentration, from a pure solvent value of 46.6 mN m^{-1} to a limiting value of between 27.5 and 35.6 mN m^{-1} at the critical micelle concentration (CMC), which was dependent on surfactant type. Since then, surfactant adsorption at air–ionic liquid interfaces has been reported for a variety of protic [92–96] and aprotic [97–100] ionic liquids. Critically, surfactant CMC values in ionic liquids can be orders of magnitude higher than water, suggesting that surfactant molecular architectures optimised for water are inappropriate for ionic liquids; there is a need for the development of new ionic liquid-specific surfactant molecules.

To date, there are no dynamic surface tension measurements reported for surfactants in ionic liquids. When a new air liquid interface is created in a surfactant solution, the equilibrium surface tension is not achieved instantly, as surfactant molecules must diffuse to the interface, then adsorb, with diffusion typically the rate-limiting step. As a result, for applications employing an interface, it will be necessary to determine whether the characteristic process time is less than the time required for equilibrium to be reached, as reduced efficiencies could otherwise result.

Only one study has probed adsorbed surfactant interfacial structure at an air-ionic liquid interface. Non-ionic oligo-oxyethylene alkyl ether surfactant ($C_{12}E_4$, $C_{14}E_4$, $C_{16}E_4$) adsorption at the [EtNH$_3$][NO$_3$]-air interface was investigated using NR and SFG, with both similarities and differences to aqueous systems noted [101]. The surfactants adsorb as an oriented monolayer at the [EtNH$_3$][NO$_3$]–air interface and form a saturated film above the CMCs. The ethylene oxide headgroups form a tightly packed layer that is only marginally penetrated by solvent—much less so than similar aqueous systems. The surfactant tails face the gas phase, forming a monolayer slightly thinner than the length of a fully extended tail group.

As stated previously, traditional surfactants are relatively inefficient in ionic liquids. Therefore it is imperative that other surface active species, and their effect on the interface, are investigated. One such species could be water. Although often present as an impurity, water sometimes resides at the air–ionic liquid interface and thereby changes interfacial properties [102, 103]. In hydrophilic ionic liquids, the water is located within the bulk liquid, while in hydrophobic ionic liquids, water phase separates from the bulk liquid and resides at the interface. This has consequences for the interfacial ion arrangement.

The effect of water on cation orientation in the interfacial layer for imidazolium ionic liquids has been studied by SFG [103]. At low water concentrations, the imidazolium ring lies parallel to the interface, with the long alkyl chain protruding into the gas phase (similar to the structure in the absence of water) for both water miscible and immiscible ionic liquids. As the water concentration is increased, C–H stretching modes of the imidazolium ring appear in the SFG spectra for hydrophobic ionic liquids, indicating that the ring is tilting toward the surface normal, with the alkyl chain orientation unaffected [103]. Conversely, the spectra for hydrophilic ionic liquids did not alter in the presence of water, as water molecules are stabilised by favourable interactions within the bulk liquid, and do not migrate to the surface.

These SFG studies did not reveal the effect of water on anions in the interfacial layer. However, a combined XRR and molecular dynamic simulation study of [C$_4$mpyr][NTf$_2$] suggests that the layer adjacent to the gas phase consists of only cations and water molecules [102], with the hydrophobic bistriflamide anion excluded. Through the transition zone, the number of anions increases and water molecules decreases with distance, until the bulk composition is reached (Fig. 3.10).

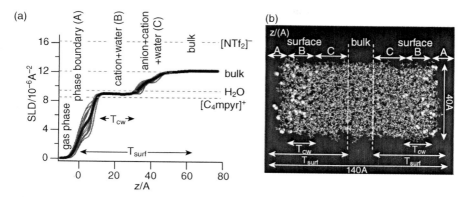

Figure 3.10 (a) Scattering length density profile, and (b) molecular dynamics simulations for [C$_4$mpyr][NTf$_2$] and water at the air interface (used with permission from Reference [102]).

3.4 LIQUID–IONIC LIQUID INTERFACES

3.4.1 Pure Interfaces

Liquid–ionic liquid interfaces are important in many applications, particularly those in which the ionic liquid replaces one or more bulk components (e.g., heterogeneous catalysis, liquid–liquid extractions, and mineral separations). Only a handful of studies [104–109] have investigated the structure at macroscopic liquid–ionic liquid interfaces, and of these only one paper is based on experimental (SFG) measurements [104], with the remainder being molecular dynamics simulations [105–109]. Based on these studies, and the structure of solid–ionic liquid and air–ionic liquid interfaces discussed previously, inferences can be made with respect to the liquid–liquid structure in the interfacial layer and transition zone.

At the interface with a hydrophobic solvent, it may be expected that cation alkyl groups will penetrate into the apolar phase in a surfactant-like fashion, provided their alkyl groups are sufficiently long, as this will minimise the interfacial energy. Recent work by Ouchi et al. confirms this orientation [104] via SFG measurements of air–[C$_n$mim][PF$_6$], CCl$_4$–[C$_n$mim][PF$_6$], and butanol–[C$_n$mim][PF$_6$] interfaces (where $n = 4$ or 8). Almost identical spectra were obtained for the air–[C$_n$mim][PF$_6$] and CCl$_4$–[C$_n$mim][PF$_6$] systems (cf. Fig. 3.11), and analysis of the spectra suggested that cations are preferentially adsorbed in the interfacial layer, forming a monolayer with alkyl chains pointing towards the hydrophobic CCl$_4$ or air phases. For the butanol–[C$_n$mim][PF$_6$] systems, the cation chain is still oriented towards the other phase, but peak broadness indicates that the cation monolayer is incomplete with butanol penetrating the cation alkyl groups.

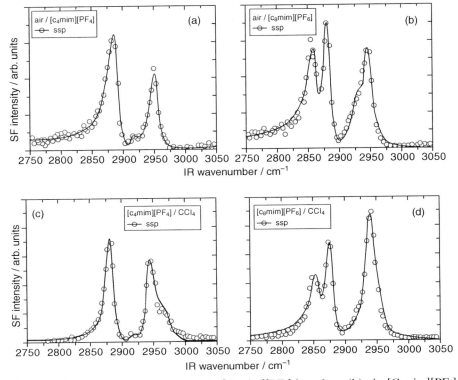

Figure 3.11   SFG spectra of the (a) air–[C$_4$mim][PF$_6$] interface, (b) air–[C$_8$mim][PF$_6$] interface, (c) CCl$_4$–[C$_4$mim][PF$_6$] interface, and (d) CCl$_4$–[C$_8$mim][PF$_6$] interface (used with permission from Reference [104]).

The interface between a hydrophobic ionic liquid and an immiscible hydrophilic solvent, such as water or methanol, has received even less attention. Ouchi et al. reported no characteristic functional group peaks in the D$_2$O–[C$_n$mim][PF$_6$] interfaces (where n = 4 or 8), which is consistent with either random cation orientation or the chains pointing towards the bulk ionic liquid [104]. In the absence of clear experimental data, predictions of interfacial layer composition and transition zone structure can be made based on the properties of the ions and the structure present at other similar interface types. The anions of hydrophobic ionic liquids usually include bulky, hydrophobic groups. These anions are unlikely to be attracted to the interface with a hydrophilic liquid. The cation, on the other hand, is generally less hydrophobic and more surfactant-like, and is thus likely to be enriched in the interfacial layer, oriented with charged groups facing the hydrophilic phase. Based on results for other systems and the bulk structure, the first layer in the transition zone will consist mostly of cations oriented with alkyl groups facing towards the hydrophilic phase, interacting solvophobically with alkyl groups of cations in the adsorbed layer. It is not possible to predict for how many ion-pair layers the transition zone will extend, but given the roughness of liquid–liquid and air–liquid interfaces are typically similar, results at the air–liquid interface for similar ionic liquids could serve as a guide. The interfacial molecules in the hydrophilic molecular solvent will be oriented with their partial negative charge, on average, towards the ionic liquid phase. Given the attractive interaction that will exist between the cation and the hydrophilic solvent, it is reasonable to ask why dissolution does not occur. Insolubility results from the attractive cohesive interactions between cation and anion being sufficient to overcome the free energy gain of mixing.

A fascinating extension to this field would be the study of interfacial arrangements of two immiscible ionic liquids [110–113]. It is unclear what types of interactions will occur between the interfacial layers of the two liquids: does the cation of one liquid interact with the anion of the other, or do solvophobic interactions between cation alkyl groups dominate? If the ionic liquids consist of the same cation and differ only in the anion, it is expected that the cation can exchange between phases, which suggests strong cohesive interaction between anions. This could lead to pronounced interfacial order. The converse question, relating to the structure of the interface between two ionic liquids that share a common anion but different cations, is equally appealing.

3.4.2   Adsorption at Liquid–Ionic Liquid Interfaces and Microemulsions

Many ionic liquid cations are surface active. Ionic liquids are versatile materials for colloidal systems because the ionic liquid can function as a hydrophobic or hydrophilic continuous phase [19, 100, 114], as the surfactant [115, 116], or even as a co-surfactant [114].

The first report of an ionic liquid-stabilised microemulsion was by Gao et al. [117], using $[C_4mim][BF_4]$ as the polar phase, cyclohexane as the hydrophobic phase, and Triton X-100 as the surfactant. Light scattering data revealed domain sizes of 0.1 μm. This system was later revisited by Eastoe et al. [118] using SANS, which demonstrated that surfactant-stabilised nanodroplets with ionic liquid cores were present. The droplet size increased as the ionic liquid volume fraction was raised.

Subsequent studies examined a range of ionic liquid-in-oil and water-in-ionic liquid systems, sometimes with conflicting results. For example, Han et al. [119] and Sarkar et al. [120] reported microemulsion formation for the $[C_4mim][PF_6]$/Triton X-100/water system but Pandey et al. [121] suggested that the ionic liquid partitions into the TX-100 aqueous phase without forming a structured system.

Protic ionic liquid microemulsions [94, 115, 122–124] are generally more similar to corresponding aqueous systems in terms of phase behaviour and patterns of self-assembly than aprotic ionic liquid microemulsions. The phase

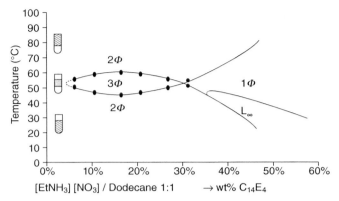

Figure 3.12 Vertical section through the phase prism of $C_{14}E_4$–$[EtNH_3][NO_3]$–dodecane (used with permission from Reference [122]).

diagram for the $C_{14}E_4$–$[EtNH_3][NO_3]$–dodecane system, presented in Figure 3.12, is strikingly similar to those found in corresponding aqueous systems. The key difference between protic ionic liquid and aqueous microemulsions is that amphiphiles with longer hydrocarbon chains are required to achieve similar behaviour, due to increased surfactant tail group solubility in the ionic liquid compared with water.

Future work could use an ionic liquid as an ionic surfactant to stabilise oil/water or ionic liquid-based systems. Recent publications by Kunz et al. [115, 123–125] have shown this is feasible, insofar as mixtures of alkanes and $[EtNH_3][NO_3]$ are concerned.

3.5 FUTURE DIRECTIONS

Ionic liquid interfaces are more complex than those of molecular solvents, and in several respects are more structurally similar to absorbed surfactant layers [126]. This may be expected, as many ionic liquids are amphiphilic [14, 16, 17], with the bulk and interfacial structures present due to solvophobic interactions between alkyl groups. The key concept developed in this chapter is that ionic liquid interfaces consist of three separate but related zones: the *interfacial layer*, which refers to ions in direct contact with the other phase; the *bulk phase*, which refers to the bulk liquid region, which may be structured or unstructured depending mostly on the length of the cation alkyl group; and the *transition zone*, which refers to the region over which the more pronounced interfacial layer structure decays to the bulk morphology. The properties, and therefore usefulness of ionic liquids in interface-dependent chemical applications, will depend on the structure of the ions in all three regions.

The interfacial layer consists of ions in direct contact with the second phase, and displays the greatest degree of organisation. The preferred ion orientation depends principally on the nature of the interface. For solid charged interfaces, ions are adsorbed electrostatically and therefore will be organised to maximise the interaction between charged moieties and surface sites. High surface charge density will produce a monolayer of counter ions, while lower surface charges will produce an interfacial ionic liquid layer that is enriched in one ion or the other. As studies of charged surfaces have so far been limited to anionic substrates, cation monolayers have only ever been observed. Conversely, for uncharged and/or hydrophobic solid interfaces, neutral segments (typically the hydrocarbon chain) arrange facing towards, or lying along, the interface to maximise solvophobic interactions. This means that, at hydrophobic interfaces, the interfacial layer is composed of both cations and anions, with the relative population primarily determined by hydrophobicity. For short alkyl chain cations, Coulombic forces dominate and a mixed layer results. As the alkyl chains become longer, solvophobic interactions become more important than electrostatic ones. A dense hydrocarbon layer results when the alkyl chains are sufficiently long to overcome steric distances.

The transition zone refers to the region between the interfacial layer and the bulk liquid, through which the well-defined interfacial structure decays to the less ordered morphology of the bulk liquid. For ionic liquids with sponge-like bulk order [16, 17], the transition that occurs is from a surface-induced lamellar structure to the bulk sponge, analogous to the surface-induced phase transitions that occur in aqueous surfactant systems [126]. If the bulk order is not present, the interfacial structure decays to an amorphous fluid, but as this change still occurs over a discrete distance, the transition zone definition holds. The distance over which the transition zone extends is a function of the level of order induced by the surface and the strength of the bulk structure. Atomically smooth surfaces that interact strongly with the ionic liquid (either electrostatically or solvophobically) will lead to the most pronounced interfacial structure. The level of structure is decreased by surface roughness or reduced surface–ionic liquid interaction strength. In general, the transition zone is widest for strongly structured ionic liquids with surfaces that are highly charged and atomically smooth. However, increasing the cation alkyl chain length also leads to increased molecular flexibility, which can decrease the extent of the transition zone. Experiments are required to determine which effect is of greater consequence.

In conventional molecular solvents, surfactants and other amphiphilic species are used to control interfacial forces and hence interfacial properties [4]. Amphiphiles can serve the same purpose in ionic liquids [49, 50], but often the ionic liquid itself is surface active. As interfacial arrangements can be altered by modifying ion structure, the properties of ionic liquid–solid, liquid and gas interfaces can be designed. The ability to fix a robust ionic liquid interfacial nanostructure that can endure static and dynamic operating conditions is attractive, especially for situations where surfactant or polymer adsorption is undesirable.

The intermolecular forces that generate bulk nanostructure are still present at the interface; hence the tendency to self-assemble is not diminished. The bulk network of polar and non-polar domains is oriented and aligned by the presence of the interface. The scale and neatness to which this occurs depends on factors such as surface roughness, temperature, and deformability. Revelations of multiple "protective ionic liquid layers" around metal nanoparticles [127–129] show that even small surface areas can induce interfacial structure, which in this case prevents particle aggregation. Recent work with a protic ionic liquid has shown that interfacial ionic liquid structure can also stabilise colloidal-sized particles [127].

Whilst our understanding of the complexity of ionic liquid interfaces has advanced considerably in recent years, there is still much fundamental work to be done. As it is impractical to screen every ionic liquid structure [25], a systematic approach to the study of ionic liquid interfaces in conjunction with bulk structure research is desirable. This could include exploring different molecular functional groups for given cation/anion/surface systems, the effect of surface charge variation, or of surface pressure on interfacial morphology.

The kinetics associated with ionic liquid interfaces remains largely unexplored. Experiments could examine the rates at which ionic liquid ions (and other dissolved solutes) can diffuse through the layered ionic liquid structure towards, and away from, an interface, or on conformational changes due to the presence of dissolved species within the layers. Although some papers have dealt with this for bulk ionic liquids [23, 24], these studies have not yet been extended to interfaces.

An improved understanding of ionic liquid interfaces will facilitate optimisation of established ionic liquid systems, such as charge-transfer processes in electrochemistry [130], current generation in dye-sensitised solar cells (DSSCs) [131], nucleation rates and deposit morphology in electrodeposition [44], as well as reaction rates and selectivity in heterogeneous catalysis [132]. It will also drive new technologies and improve understanding in other areas, including ionic liquid flow behaviour under confinement [133, 134], the mechanism of particle stability in ionic liquids [127], and the integration of ionic liquids into analytical technologies [135, 136].

In order to gain further insight into fundamental aspects of ionic liquid interfaces, some consideration needs to be made of the techniques used. Ionic liquids cannot be treated as "just another solvent" interface because of the complicated nature of their interfacial structure. For some techniques, it is unclear whether the effects of different regions can be completely deconvoluted. Ideally, researchers should use more than one technique to investigate the same interface, as this affords confidence in the structural model developed. Where this is not possible, consideration should be given to results obtained at other interfaces, as structural correlations across different interface types can certainly be made.

ACKNOWLEDGEMENTS

The authors wish to thank Prof. José Nuno Canongia Lopes for supplying the simulation presented in Figure 3.5, and the Australian Research Council for funding support. RH thanks AINSE for a PGRA.

REFERENCES

1   Bikerman, J.J., *Physical surfaces*, Academic, New York (1970).

2   Ocko, B.M., Wu, X.Z., Sirota, E.B., Sinha, S.K., and Deutsch, M., X-ray reflectivity study of thermal capillary waves on liquid surfaces, *Phys. Rev. Lett.* **72**, 242–245 (1994).

3   Israelachvili, J.N., *Intermolecular and surface forces*, Academic Press, London (1992).

4   Hunter, R.J., *Foundations of colloid science*, Oxford University Press, Oxford, UK (1985).

5   Chakrabarty, D.K., and Viswanathan, B., *Heterogeneous catalysis*, New Age Science, New Delhi (2009).

6   Harris, D.C., *Quantitative chemical analysis*, 5th ed., W.H. Freeman & Company, New York (1998).

7   Bockris, J.O.M., and Khan, S.U.M., *Surface electrochemistry: a molecular level approach*, 1st ed., Plenum Press, New York (1993).

8   Rudnick, L.R., *Lubricant additives: chemistry and applications*, 2nd ed., CRC Press, New York (2009).

9   Davy, H., *Researches, chemical and philosophical*, Biggs and Cottle, Bristol (1800).

10   Hayes, R., Warr, G.G., and Atkin, R., At the interface: solvation and designing ionic liquids, *Phys. Chem. Chem. Phys.* **12**, 1709–1723 (2010).

11   Ray, A., Solvophobic interactions and micelle formation in structure forming nonaqueous solvents, *Nature* **231**, 313–315 (1971).

12   Wang, Y., and Voth, G.A., Unique spatial heterogeneity in ionic liquids, *J. Am. Chem. Soc.* **127**, 12192–12193 (2005).

13   Canongia Lopes, J.N.A., and Padua, A.A.H., Nanostructural organization in ionic liquids, *J. Phys. Chem. B* **110**, 3330–3335 (2006).

14   Triolo, A., Russina, O., Bleif, H.J., and DiCola, E., Nanoscale segregation in room temperature ionic liquids, *J. Phys. Chem. B* **111**, 4641–4644 (2007).

15   Greaves, T.L., Kennedy, D.K., Mudie, S.T., and Drummond, C.J., Diversity observed in the nanostructure of protic ionic liquids, *J. Phys. Chem. B* **114**, 10022–10031 (2010).

16   Atkin, R., and Warr, G.G., The smallest amphiphiles: nanostructure in protic room-temperature ionic liquids with short alkyl groups, *J. Phys. Chem. B* **112**, 4164–4166 (2008).

17   Hayes, R., Imberti, S., Warr, G.G., and Atkin, R., Amphiphilicity determines nanostructure in protic ionic liquids, *Phys. Chem. Chem. Phys.* **13**, 3237–3247 (2011).

18   Hardacre, C., Holbrey, J.D., McMath, J.S.E., Bowron, D.T., and Soper, A.K., Structure of molten 1,3-dimethylimidazolium chloride using neutron diffraction, *J. Chem. Phys.* **118**, 273–278 (2003).

19   Hardacre, C., Holbrey, J.D., Nieuwenhuyzen, M., and Youngs, T.G.A., Structure and solvation in ionic liquids, *Acc. Chem. Res.* **40**, 1146–1155 (2007).

20   Chowdhury, S., Mohan, R.S., and Scott, J.L., Reactivity of ionic liquids, *Tetrahedron* **63**, 2363–2389 (2007).

21   Xu, W., Cooper, E.I., and Angell, C.A., Ionic liquids: ion mobilities, glass temperatures, and fragilities, *J. Phys. Chem. B* **107**, 6170–6178 (2003).

22   Heintz, A., Recent developments in thermodynamics and thermophysics of nonaqueous mixtures containing ionic liquids. A review, *J. Chem. Thermodyn.* **37**, 525–535 (2005).

23   Triolo, A., Russina, O., Arrighi, V., Juranyi, F., Janssen, S., and Gordon, C.M., Quasielastic neutron scattering characterization of the relaxation processes in a room temperature ionic liquid, *J. Chem. Phys.* **119**, 8549–8557 (2003).

24   Rivera, A., Brodin, A., Pugachev, A., and Rössler, E.A., Orientational and translational dynamics in room temperature ionic liquids, *J. Chem. Phys.* **126**, 114503 (2007).

25   Earle, M.J., and Seddon, K.R., Ionic liquids. green solvents for the future, *Pure Appl. Chem.* **72**, 1391–1398 (2000).

26   Greaves, T.L., and Drummond, C.J., Protic ionic liquids: properties and applications, *Chem. Rev.* **108**, 206–237 (2008).

27   Angell, C.A., Byrne, N., and Belieres, J.-P., Parallel developments in aprotic and protic ionic liquids: physical chemistry and applications, *Acc. Chem. Res.* **40**, 1228–1236 (2007).

28   Mao, G., Tsao, Y., Tirrel, M.H., Davis, H.T., Hessel, V., and Ringsdorf, H., Self-assembly of photopolymerizable bolaform amphiphile mono- and multilayer, *Langmuir* **9**, 3461–3470 (1993).

29   Horn, R.G., Evans, D.F., and Ninham, B.W., Double-layer and solvation forces measured in a molten salt and its mixtures with water, *J. Phys. Chem.* **92**, 3531–3537 (1988).

30   Atkin, R., and Warr, G.G., Structure in confined room-temperature ionic liquids, *J. Phys. Chem. C* **111**, 5162–5168 (2007).

31   Wakeham, D., Hayes, R., Warr, G.G., and Atkin, R., Influence of temperature and molecular structure on ionic liquid solvation layers, *J. Phys. Chem. B* **113**, 5961–5166 (2009).

32   Hayes, R., El Abedin, S.Z., and Atkin, R., Pronounced structure in confined aprotic room-temperature ionic liquids, *J. Phys. Chem. B* **113**, 7049–7052 (2009).

33   Patrick, H.N., Warr, G.G., Manne, S., and Aksay, I.A., Self-assembly structures of nonionic surfactants at graphite/solution interfaces, *Langmuir* **13**, 4349–4356 (1997).

34   Choudhury, A.R., Winterton, N., Steiner, A., Cooper, A.I., and Johnson, K.A., In-situ crystallization of low-melting ionic liquids, *J. Am. Chem. Soc.* **127**, 16792–16793 (2005).

35   Xu, Z., Ducker, W., and Israelachvili, J.N., Forces between crystalline alumina (sapphire) surfaces in aqueous sodium dodecylsulfate solutions, *Langmuir* **12**, 2263–2270 (1996).

36   Mezger, M., Schröder, H., Reichert, H., Schramm, S., Okasinski, J.S., Schöder, S., Honkimäki, V., Deutsch, M., Ocko, B.M., Ralston, J., Rohwerder, M., Stratmann, M., and Dosch, H., Molecular layering of fluorinated ionic liquids at a charged sapphire (0001) surface, *Science* **322**, 424–428 (2008).

37   Mezger, M., Schramm, S., Schröder, H., Reichert, H., Deutsch, M., Souza, E.J.D., Okasinski, J.S., Ocko, B.M., Honkimäki, V., and Dosch, H., Layering of [BMIM]+-based ionic liquids at a charged sapphire interface, *J. Chem. Phys.* **131**, 094701 (2009).

38   Horn, R.G., and Israelachvili, J.N., Direct measurement of structural· forces between two surfaces in a nonpolar liquid, *J. Chem. Phys.* **75**, 1400–1411 (1981).

39   Iler, R.K., *The chemistry of silica*, Wiley-Interscience Publishers, New York (1979).

40   Bowron, D.T., D'Agostino, C., Gladden, L.F., Hardacre, C., Holbrey, J.D., Lagunas, M.C., McGregor, J., Mantle, M.D., Mullan, C.L., and Youngs, T.G.A., Structure and dynamics of 1-ethyl-3-methylimidazolium acetate via molecular dynamics and neutron diffraction, *J. Phys. Chem. B* **114**, 7760–7768 (2010).

41   Fitchett, B.D., and Conboy, J.C., Structure of the room-temperature ionic liquid/ SiO_2 interface studied by sum-frequency vibrational spectroscopy, *J. Phys. Chem. B* **108**, 20255–20262 (2004).

42   Rollins, J.B., Fitchett, B.D., and Conboy, J.C., Structure and orientation of the imidazolium cation at the room-temperature ionic liquid/SiO_2 interface measured by sum-frequency vibrational spectroscopy, *J. Phys. Chem. B* **111**, 4990–4999 (2007).

43   Canongia Lopes, J.N.A., *Personal Communications* (2010).

44   Endres, F., Ionic liquids: solvents for the electrodeposition of metals and semi-conductors, *ChemPhysChem* **3**, 144–154 (2002).

45   Endres, F., Hofft, O., Borisenko, N., Gasparotto, L.H., Prowald, A., Al-Salman, R., Carstens, T., Atkin, R., Bund, A., and Zein El Abedin, S., Do solvation layers of ionic liquids influence electrochemical reactions? *Phys. Chem. Chem. Phys.* **12**, 1724–1732 (2010).

46   Atkin, R., Abedin, S.Z.E., Hayes, R., Gasparotto, L.H.S., Borisenko, N., and Endres, F., AFM and STM studies on the surface interaction of [BMP]TFSA and (EMIm]TFSA ionic liquids with Au(111), *J. Phys. Chem. C* **113**, 13266–13272 (2009).

47   Baldelli, S., Surface structure at the ionic liquid–electrified metal interface, *Acc. Chem. Res.* **41**, 421–431 (2008).

48   Pinilla, C., Del Pópolo, M.G., Kohanoff, J., and Lynden-Bell, R.M., Polarization relaxation in an ionic liquid confined between electrified walls, *J. Phys. Chem. B* **111**, 4877–4844 (2007).

49   Atkin, R., and Warr, G.G., Self-assembly of a nonionic surfactant at the graphite/ ionic liquid interface, *J. Am. Chem. Soc.* **127**, 11940–11941 (2005).

50   Atkin, R., De Fina, L.-M., Kiederling, U., and Warr, G.G., Structure and self assembly of pluronic amphiphiles in ethylammonium nitrate and at the silica surface, *J. Phys. Chem. B* **113**, 12201–12213 (2009).

51   Araos, M.U., and Warr, G.G., A nonaqueous liquid crystal emulsion: fluorocarbon oil in a hexagonal phase in an ionic liquid, *Langmuir* **24**, 9354–9360 (2008).

52   Seddon, K.R., Ionic liquids for clean technology, *J. Chem. Technol. Biotechnol.* **68**, 351–356 (1997).

53   Earle, M.J., Esperanca, J.M.S.S., Gilea, M.A., Canongia Lopes, J.N., Rebelo, L.P.N., Magee, J.W., Seddon, K.R., and Widegren, J.A., The distillation and volatility of ionic liquids, *Nature* **439**, 831–834 (2006).

54   Forsyth, S.A., Pringle, J.M., and Macfarlane, D.R., Ionic liquids—an overview, *Aust. J. Chem.* **57**, 113–119 (2004).

55   Cull, S.G., Holbrey, J.D., Vargas-Mora, V., Seddon, K.R., and Lye, G.J., Room-temperature ionic liquids as replacements for organic solvents in multiphase bioprocess operations, *Biotechnol. Bioeng.* **69**, 227–233 (2000).

56   Bates, E.D., Mayton, R.D., Ntai, I., and Davis, J.H., CO_2 capture by a task-specific ionic liquid, *J. Am. Chem. Soc.* **124**, 926–927 (2002).

57   Anthony, J.L., Maginn, E.J., and Brennecke, J.F., Solubilities and thermodynamic properties of gases in the ionic liquid 1-*n*-butyl-3-methylimidazolium hexafluorophosphate, *J. Phys. Chem. B* **106**, 7315–7320 (2002).

58   Armstrong, D.W., He, L., and Liu, Y.-S., Examination of ionic liquids and their interaction with molecules, when used as stationary phases in gas chromatography, *Anal. Chem.* **71**, 3873–3876 (1999).

59   Weyershausen, B., and Lehmann, K., Industrial application of ionic liquids as performance additives, *Green Chem.* **7**, 15–19 (2005).

60   Onsager, L., and Samaras, N.N.T., The surface tension of Debye-Hückel electrolytes, *J. Chem. Phys.* **2**, 528–536 (1934).

61   Bowers, J., Vergara-Gutierrez, M.C., and Webster, J.R.P., Surface ordering of amphiphilic ionic liquids, *Langmuir* **20**, 309–312 (2004).

62   Niga, P., Wakeham, D., Nelson, A., Warr, G.G., Rutland, M., and Atkin, R., Structure of the ethylammonium nitrate surface: an X-ray reflectivity and vibrational sum frequency spectroscopy study, *Langmuir* **26**, 8282–8288 (2010).

63   Jeon, Y., Sung, J., Bu, W., Vaknin, D., Ouchi, Y., and Kim, D., Interfacial restructuring of ionic liquids determined by sum-frequency generation spectroscopy and X-ray reflectivity, *J. Phys. Chem. C* **112**, 19649–19654 (2008).

64   Sloutskin, E., Ocko, B.M., Tamam, L., Kuzmenko, I., Gog, T., and Deutsch, M., Surface layering in ionic liquids: an x-ray reflectivity study, *J. Am. Chem. Soc.* **127**, 7796–7804 (2005).

65   Iimori, T., Iwahashi, T., Ishii, H., Seki, K., Ouchi, Y., Ozawa, R., Hamaguchi, H.-o., and Kim, D., Orientational ordering of alkyl chain at the air/liquid interface of ionic liquids studied by sum frequency vibrational spectroscopy, *Chem. Phys. Lett.* **389**, 321–326 (2004).

66   Rivera-Rubero, S., and Baldelli, S., Surface characterization of 1-butyl-3-methylimidazollum Br-, I-, PF_6-, BF_4-, $(CF_3SO_2)_2N$-, SCN-, CH_3SO_3-, CH_3SO_4-, and $(CN)_2N$- ionic liquids by sum frequency generation, *J. Phys. Chem. B* **110**, 4756–4765 (2006).

67   Santos, C., and Baldelli, S., Surface orientation of 1-methyl-, 1-ethyl-, and 1-butyl-3-methylimidazolium methyl sulfate as probed by sum-frequency generation vibrational spectroscopy, *J. Phys. Chem. B* **111**, 4715–4723 (2007).

68   Iimori, T., Iwahashi, T., Kanai, K., Seki, K., Sung, J., Kim, D., Hamaguchi, H.-o., and Ouchi, Y., Local structure at the air/liquid interface of room-temperature

ionic liquids probed by infrared-visible sum frequency generation vibrational spectroscopy: 1-alkyl-3-methylimidazolium tetrafluoroborates, *J. Phys. Chem. B* **111**, 4860–4866 (2007).

69 Aliaga, C., Baker, G.A., and Baldelli, S., Sum frequency generation studies of ammonium and pyrrolidinium ionic liquids based on the bis-trifluoromethanesulfonimide anion, *J. Phys. Chem. B* **112**, 1676–1684 (2008).

70 Iwahashi, T., Miyamae, T., Kanai, K., Seki, K., Kim, D., and Ouchi, Y., Anion configuration at the air/liquid interface of ionic liquid [bmim]OTf studied by sum-frequency generation spectroscopy, *J. Phys. Chem. B* **112**, 11936–11941 (2008).

71 Santos, C., and Baldelli, S., Alkyl chain interaction at the surface of room temperature ionic liquids: systematic variation of alkyl chain length (R = C_1-C_4, C_8) in both cation and anion of [RMIM][R-OSO_3] by sum frequency generation and surface tension, *J. Phys. Chem. B* **113**, 923–933 (2009).

72 Gannon, T., Law, G., and Watson, P., First observation of molecular composition and orientation at the surface of a room-temperature ionic liquid, *Langmuir* **15**, 8429–8434 (1999).

73 Law, G., Watson, R., Carmichael, A., and Seddon, K.R., Molecular composition and orientation at the surface of room-temperature ionic liquids: effect of molecular structure, *Phys. Chem. Chem. Phys.* **3**, 2879–2885 (2001).

74 Höfft, O., Bahr, S., Himmerlich, M., Krischok, S., Schaefer, J.A., and Kempter, V., Electronic structure of the surface of the ionic liquid [EMIM][Tf_2N] studied by metastable impact electron spectroscopy (MIES), UPS, and XPS, *Langmuir* **22**, 7120–7123 (2006).

75 Iwahashi, T., Nishi, T., Yamane, H., Miyamae, T., Kanai, K., Seki, K., Kim, D., and Ouchi, Y., Surface structural study on ionic liquids using metastable atom electron spectroscopy, *J. Phys. Chem. C* **113**, 19237–19243 (2009).

76 Waring, C., Bagot, P.A.J., Slattery, J.M., Costen, M.L., and McKendrick, K.G., O(3P) atoms as a probe of surface ordering in 1-alkyl-3-methylimidazolium-based ionic liquids, *J. Phys. Chem. Lett.* **1**, 429–433 (2010).

77 Wu, B., Zhang, J., Minton, T.K., McKendrick, K.G., Slattery, J.M., Yockel, S., and Schatz, G.C., Scattering dynamics of hyperthermal oxygen atoms on ionic liquid surfaces: [emim][NTf_2] and [C_{12}mim][NTf_2], *J. Phys. Chem. C* **114**, 4015–4027 (2010).

78 Waring, C., Bagot, P.A.J., Slattery, J.M., Costen, M.L., and McKendrick, K.G., O(3P) atoms as a chemical probe of surface ordering in ionic liquids, *J. Phys. Chem. A* **114**, 4896–4904 (2010).

79 Kolbeck, C., Killian, M., Maier, F., Paape, N., Wassercheid, P., and Steinrück, H.-P., Surface characterization of functionalized imidazolium-based ionic liquids, *Langmuir* **24**, 9500–9507 (2008).

80 Lockett, V., Sedev, R., Bassell, C., and Ralston, J., Angle-resolved X-ray photoelectron spectroscopy of the surface of imidazolium ionic liquids, *Phys. Chem. Chem. Phys.* **10**, 1330–1335 (2008).

81 Maier, F., Cremer, T., Kolbeck, C., Lovelock, K.R.J., Paape, N., Schulz, P.S., Wassercheid, P., and Steinrück, H.-P., Insights into the surface composition and enrichment effects of ionic liquids and ionic liquid mixtures, *Phys. Chem. Chem. Phys.* **12**, 1905–1915 (2010).

82   Kolbeck, C., Cremer, T., Lovelock, K.R.J., Paape, N., Schulz, P.S., Wassercheid, P., Maier, F., and Steinrück, H.-P., Influence of different anions on the surface composition of ionic liquids studied using ARXPS, *J. Phys. Chem. B* **113**, 8682–8688 (2009).

83   Lynden-Bell, R.M., Gas-liquid interfaces of room temperature ionic liquids, *Mol. Phys.* **101**, 2625–2633 (2003).

84   Lynden-Bell, R.M., and Del Pópolo, M.G., Simulation of the surface structure of butylmethylimidazolium ionic liquids, *Phys. Chem. Chem. Phys.* **8**, 949–954 (2006).

85   Yan, T., Li, S., Jiang, W., Gao, X., Xiang, B., and Voth, G.A., Structure of the liquid-vacuum interface of room-temperature ionic liquids: a molecular dynamics study, *J. Phys. Chem. B* **110**, 1800–1806 (2006).

86   Bhargava, B.L., and Balasubramanian, S., Layering at an ionic liquid-vapor interface: a molecular dynamics simulation study of [bmim][PF$_6$], *J. Am. Chem. Soc.* **128**, 10073–10078 (2006).

87   Chang, T., and Dang, L.X., Computational studies of structures and dynamics of 1,3-dimethylimidazolim salt liquids and their interfaces using polarizable potential models, *J. Phys. Chem. A* **113**, 2127–2135 (2009).

88   Law, G., and Watson, P.R., Surface tension measurements of N-alkylimidazolium ionic liquids, *Langmuir* **17**, 6138–6141 (2001).

89   Restolho, J., Mata, J.L., and Saramago, B., On the interfacial behavior of ionic liquids: surface tensions and contact angles, *J. Colloid Interface Sci.* **340**, 82–86 (2009).

90   Evans, D.F., Yamauchi, A., Roman, R., and Casassa, E.Z., Micelle formation in ethylammonium nitrate, a low-melting fused salt, *J. Colloid Interface Sci.* **88**, 89–96 (1982).

91   Evans, D.F., Yamauchi, A., Wei, G.J., and Bloomfield, V.A., Micelle size in ethylammonium nitrate as determined by classical and quasi-elastic light-scattering, *J. Phys. Chem.* **87**, 3537–3541 (1983).

92   Araos, M.U., and Warr, G.G., Self-assembly of nonionic surfactants into lyotropic liquid crystals in ethylammonium nitrate, a room-temperature ionic liquid, *J. Phys. Chem. B* **109**, 14275–14277 (2005).

93   Araos, M.U., and Warr, G.G., Structure of nonionic surfactant micelles in the ionic liquid ethylammonium nitrate, *Langmuir* **24**, 9354–9360 (2008).

94   Atkin, R., Bobillier, S.M.C., and Warr, G.G., Propylammonium nitrate as a solvent for amphiphile self-assembly into micelles, lyotropic liquid crystals, and microemulsions, *J. Phys. Chem. B* **114**, 1350–1360 (2010).

95   Greaves, T.L., Weerawardena, A., Fong, C., and Drummond, C.J., Formation of amphiphile self-assembly phases in protic ionic liquids, *J. Phys. Chem. B* **111**, 4082–4088 (2007).

96   Greaves, T.L., Weerawardena, A., Krodkiewska, I., and Drummond, C.J., Protic ionic liquids: physicochemical properties and behavior as amphiphile self-assembly solvents, *J. Phys. Chem. B* **112**, 896–905 (2008).

97   Fletcher, K.A., and Pandey, S., Surfactant aggregation within room-temperature ionic liquid 1-ethyl-3-methylimidazolium bis(trifluoromethylsulfonyl)imide, *Langmuir* **20**, 33–36 (2004).

98   Patrascu, C., Gauffre, F., Nallet, F., Bordes, R., Oberdisse, J., de Lauth-Viguerie, N., and Mingotaud, C., Micelles in ionic liquids: aggregation behavior of alkyl

poly(ethyleneglycol)-ethers in 1-butyl-3-methyl-imidazolium type ionic liquids, *ChemPhysChem* **7**, 99–101 (2006).

99 Gao, Y., Li, N., Li, X., Zhang, S., Zheng, L., Bai, X., and Yu, L., Microstructures of micellar aggregations formed within 1-butyl-3-methylimidazolium type ionic liquids, *J. Phys. Chem. B* **113**, 123–130 (2009).

100 Anderson, J.L., Pino, V., Hagberg, E.C., Sheares, V.V., and Armstrong, D.W., Surfactant solvation effects and micelle formation in ionic liquids, *Chem. Commun.* **19**, 2444–2445 (2003).

101 Wakeham, D., Niga, P., Warr, G.G., Rutland, M., and Atkin, R., Nonionic surfactant adsorption at the ethylammonium nitrate surface: a neutron reflectivity and vibrational sum frequency spectroscopy study, *Langmuir* **26**, 8313–8318 (2010).

102 Lauw, Y., Horne, M.D., Rodopoulos, T., Webster, N.A.S., Minofar, B., and Nelson, A., X-Ray reflectometry studies on the effect of water on the surface structure of [C_4mpyr][NTf$_2$] ionic liquid, *Phys. Chem. Chem. Phys.* **11**, 11507–11514 (2009).

103 Rivera-Rubero, S., and Baldelli, S., Influence of water on the surface of hydrophilic and hydrophobic room-temperature ionic liquids, *J. Am. Chem. Soc.* **126**, 11788–11789 (2004).

104 Iwahashi, T., Miyamae, T., Kanai, K., Seki, K., Kim, D., and Ouchi, Y., Interfacial structure at ionic-liquid/molecular-liquid interfaces probed by sum-frequency generation vibrational spectroscopy, in: *Ionic liquids: from knowledge to application*, Vol. 1030, ed. N.V. Plechkova, R.D. Rogers, and K.R. Seddon, ACS Symposium Series, Vol. 1030, American Chemical Society, Washington, DC (2009), p. 305.

105 Lynden-Bell, R.M., Kohanoff, J., and Del Pópolo, M.G., Simulation of interfaces between room temperature ionic liquids and other liquids, *Faraday Discuss.* **129**, 57–67 (2005).

106 Chaumont, A., Schurhammer, R., and Wipff, G., Aqueous interfaces with hydrophobic room-temperature ionic liquids: a molecular dynamics study, *J. Phys. Chem. B* **109**, 18964–18973 (2005).

107 Sieffert, N., and Wipff, G., The [BMI][Tf$_2$N] ionic liquid/water binary system: a molecular dynamics study of phase separation and of the liquid-liquid interface, *J. Phys. Chem. B* **110**, 13076–13085 (2006).

108 Sieffert, N., and Wipff, G., Comparing an ionic liquid to a molecular solvent in the cesium cation extraction by a calixarene: a molecular dynamics study of the aqueous interfaces, *J. Phys. Chem. B* **110**, 19497–19506 (2006).

109 Sieffert, N., and Wipff, G., Rhodium-catalyzed hydroformylation of 1-hexene in an ionic liquid: a molecular dynamics study of the hexene/[BMI][PF$_6$] interface, *J. Phys. Chem. B* **111**, 4951–4962 (2007).

110 Arce, A., Earle, M.J., Katdare, S.P., Rodriguez, H., and Seddon, K.R., Phase equilibria of mixtures of mutually immiscible ionic liquids, *Fluid Phase Equilib.* **261**, 427–433 (2007).

111 Arce, A., Earle, M.J., Katdare, S.P., Rodriguez, H., and Seddon, K.R., Mutually immiscible ionic liquids, *Chem. Commun.* **24**, 2548–2550 (2006).

112 Canongia Lopes, J.N., Cordeiro, T.C., Esperança, J.M.S.S., Guedes, H.J.R., Huq, S., Rebelo, L.P.N., and Seddon, K.R., Deviations from ideality in mixtures of two ionic liquids containing a common ion, *J. Phys. Chem. B* **109**, 3519–3525 (2005).

113 Navia, P., Troncoso, J., and Romani, L., Excess magnitudes for ionic liquid binary mixtures with a common ion, *J. Chem. Eng. Data* **52**, 1369–1374 (2007).

114   Greaves, T.L., and Drummond, C.J., Ionic liquids as amphiphile self-assembly media, *Chem. Soc. Rev.* **37**, 1709–1726 (2008).

115   Zech, O., Thomaier, S., Bauduin, P., Rück, T., Touraud, D., and Kunz, W., Microemulsions with an ionic liquid surfactant and room temperature ionic liquids AS polar pseudo-phase, *J. Phys. Chem. B* **113**, 465–473 (2009).

116   Kang, W., Dong, B., Gao, Y., and Zheng, L., Aggregation behavior of long-chain imidazolium ionic liquids in ethylammonium nitrate, *Colloid Polym. Sci.* **288**, 1225–1232 (2010).

117   Gao, H., Li, J., Han, B., Chen, W., Zhang, J., Zhang, R., and Yan, D., Microemulsions with ionic liquid polar domains, *Phys. Chem. Chem. Phys.* **6**, 2914–2916 (2004).

118   Eastoe, J., Gold, S., Rogers, S.E., Paul, A., Welton, T., Heenan, R.K., and Grillo, I., Ionic liquid-in-oil microemulsions, *J. Am. Chem. Soc.* **127**, 7302–7303 (2005).

119   Gao, Y., Han, S., Han, B., Li, G., Shen, D., Li, Z., Du, J., Hou, W., and Zhang, G., TX-100/water/1-butyl-3-methylimidazolium hexafluorophosphate microemulsions, *Langmuir* **21**, 5681–5684 (2005).

120   Seth, D., Chakraborty, A., Setua, P., and Sarkar, N., Interaction of ionic liquid with water in ternary microemulsions (triton X-100/water/1-butyl-3-methylimidazolium hexafluorophosphate) probed by solvent and rotational relaxation of coumarin 153 and coumarin 151, *Langmuir* **22**, 7768–7775 (2006).

121   Behera, K., Dahiya, P., and Pandey, S., Effect of added ionic liquid on aqueous triton X-100 micelles, *J. Colloid Interface Sci.* **307**, 235–245 (2007).

122   Atkin, R., and Warr, G.G., Phase behavior and microstructure of microemulsions with a room-temperature ionic liquid as the polar phase, *J. Phys. Chem. B* **111**, 9309–9316 (2007).

123   Zech, O., Thomaier, S., Kolodziejski, A., Touraud, D., Grillo, I., and Kunz, W., Ionic liquids in microemulsions—a concept to extend the conventional thermal stability range of microemulsions, *Chem. Eur. J.* **16**, 783–786 (2010).

124   Zech, O., Thomaier, S., Kolodziejski, A., Touraud, D., Grillo, I., and Kunz, W., Ethylammonium nitrate in high temperature stable microemulsions, *J. Colloid Interface Sci.* **347**, 227–232 (2010).

125   Thomaier, S., and Kunz, W., Aggregates in mixtures of ionic liquids, *J. Mol. Liq.* **130**, 104–107 (2007).

126   Hamilton, W.A., Porcar, L., Butler, P.D., and Warr, G.G., Local membrane ordering of sponge phases at a solid-solution interface, *J. Chem. Phys.* **116**, 8533–8546 (2002).

127   Smith, J.A., Werzer, O., Webber, G.B., Warr, G.G., and Atkin, R., Surprising particle stability and rapid sedimentation rates in an ionic liquid, *J. Phys. Chem. Lett.* **1**, 64–68 (2010).

128   Machado, G., Scholten, J.D., de Vargas, T., Teixeira, S.R., Ronchi, L.H., and Dupont, J., Structural aspects of transition-metal nanoparticles in imidazolium ionic liquids, *Int. J. Nanotechnol.* **4**, 541–563 (2007).

129   Scheeren, C.W., Machado, G., Teixeira, S.R., Morais, J., Domingos, J.B., and Dupont, J., Synthesis and characterization of Pt(0) nanoparticles in imidazolium ionic liquids, *J. Phys. Chem. B* **110**, 13011–13020 (2006).

130 McFarlane, D.R., Pringle, J.M., Howlett, P.C., and Forsyth, M., Ionic liquids and reactions at the electrochemical interface, *Phys. Chem. Chem. Phys.* **12**, 1659–1669 (2010).

131 Gorlov, M., and Kloo, L., Ionic liquid electrolytes for dye-sensitized solar cells, *Dalton Trans.* **20**, 2655–2666 (2008).

132 Parvulescu, V.I., and Hardacre, C., Catalysis in ionic liquids, *Chem. Rev.* **107**, 2615–2665 (2007).

133 Ueno, K., Imaizumi, S., Hata, K., and Watanabe, M., Colloidal interaction in ionic liquids: effects of ionic structures and surface chemistry on rheology of silica colloidal dispersions, *Langmuir* **25**, 825–831 (2009).

134 Perkin, S., Albrecht, T., and Klein, J., Layering and shear properties of an ionic liquid, 1-ethyl-3-methylimidazolium ethylsulfate, confined to nano-films between mica surfaces, *Phys. Chem. Chem. Phys.* **12**, 1243–1247 (2010).

135 Pandey, S., Analytical applications of room-temperature ionic liquids: a review of recent efforts, *Anal. Chim. Acta* **556**, 38–45 (2006).

136 Han, X., and Armstrong, D.W., Ionic liquids in separations, *Acc. Chem. Res.* **40**, 1079–1086 (2007).

4 Ionic Liquids in Separation Science

CHRISTA M. GRAHAM and JARED L. ANDERSON

Department of Chemistry, The University of Toledo, Toledo, Ohio, USA

ABSTRACT

Ionic liquids possess a multitude of unique properties that make them particularly useful in separation science. This chapter discusses the use of ionic liquids and polymeric ionic liquids in chromatographic and electrophoretic separations. Ionic liquids have been used as mobile phase modifiers and as selective stationary phases in high performance liquid chromatography, planar chromatography, and supercritical fluid chromatography. By exploiting the density, viscosity, and water-immiscibility of some classes of ionic liquids, they have been applied as selective phases in counter-current chromatography. The unique physicochemical properties of ionic liquids and polymeric ionic liquids have made them a new and useful class of gas chromatographic stationary phases that exhibit broad selectivity while also possessing high column bleed temperatures. New separation methods in which ionic liquids are used as run buffer additives, and as wall-coated stationary phases, have significantly improved electrophoretic separations in capillary electrophoresis and capillary electrochromatography.

4.1 BRIEF HISTORY OF THE DEVELOPMENT OF IONIC LIQUIDS AND POLYMERIC IONIC LIQUIDS IN SEPARATION SCIENCE

A challenge for separation science today is the development of green or environmentally benign processes and analyses. With the global population growth and accompanying industrial, pharmacological, and energy developments,

Ionic Liquids UnCOILed: Critical Expert Overviews, First Edition. Edited by Natalia V. Plechkova and Kenneth R. Seddon.

there are a growing number of environmental toxins to monitor. The scientific community is now pressed to move away from the use of toxic compounds and towards safer and more environmentally friendly reagents. Many under-developed countries have experienced recent industrial expansion resulting in increased carbon dioxide emissions. Growing emissions are of concern as increasing evidence indicates that they may have a negative impact on the global climate by trapping solar energy, as do all greenhouse gases. This is believed to cause a rise in the global average temperature, and it is feared that this rise will bring about great changes in the global climate [1]. As the world population grows and ages, the demand for pharmaceutical products also increases. These products—such as antibiotics, anti-inflammatory medications, and even common caffeine—are released into the environment after passing through the digestive tracts of patients [2]. Chemical impurities in pharmaceuticals can pose additional health risks, and efforts to improve the testing and control of such impurities are underway [3]. As the environmental levels of contaminants increase, they have the potential to change the ecological balance of the ecosphere. Because of these concerns, there are ongoing efforts to detect, quantify, and clean up environmental toxins. The processes developed to meet these goals often utilise volatile organic compounds (VOCs), which may be toxic, forcing scientists and engineers to find alternative solvent systems. Ionic liquids are one such class of alternative solvents that exhibit great potential to replace some VOCs, or replace some components in more complicated industrial processes. Ionic liquids offer unique physical and chemical characteristics, such as low vapour pressure at ambient temperature and variable solvation interactions. These properties allow ionic liquids to be used in the development of technologies that are not amenable to solvents that possess high vapour pressures, such as VOCs.

Ionic liquids have been known for quite some time, and Wilkes, in his personal account of the development of ionic liquids, mentioned that ionic liquids were first observed as the "red oil" formed during Friedel–Crafts reactions catalysed by aluminium(III) chloride [4]. The first reported ionic liquid, ethyl-ammonium nitrate, with a melting point of 12 °C, was synthesised by Walden in the early part of the 20th century, although the term "ionic liquid" had yet to be coined. While there were other ionic liquids synthesised intentionally (often referred to as fused salts or melts), they were reactive, especially towards water, and had to be handled carefully [5–8]. It was the discovery of water- and air-stable ionic liquids [9, 10] that opened the door to the use of ionic liquids as a versatile class of solvents in the field of separation science.

Many of the challenges associated with the use of organic solvents in separation methods can be partially or completely circumvented by replacing those solvents with ionic liquids. The low volatility, high thermal stability, and variable viscosity of ionic liquids make them an interesting solvent system to separation scientists. Despite their positive attributes, ionic liquids, as neat solvents or as mixtures, are not the "magic bullet" of separation science, at times offering little to no improvement over current methodologies or even proving detrimental to the separation at hand. Therefore, it is important to note that ionic liquids are capable of providing "niche applications" in separation science, but their direct use does not always lead to favourable results or offer advantages over more conventional solvents. In this chapter, we highlight the major niches that ionic liquids have found in chromatography, and briefly discuss their utility, limitations, and future directions.

4.2   IONIC LIQUIDS IN CHROMATOGRAPHIC AND ELECTROPHORETIC SEPARATIONS

All chromatographic techniques have some basic similarities. They all possess a mobile phase, stationary phase, some means of containing the stationary phase, and a means of detection. Analytes are introduced at the beginning of a chromatographic column or bed as a sample plug, and undergo various modes of interaction with the mobile and stationary phases prompting them to separate into bands. As these bands elute, they are detected and produce a chromatogram. All chromatographic techniques seek to limit the amount of band broadening to maximise the efficiency of the separation. Band broadening can arise from factors outside of the chromatographic column, such as during injection of the sample and variation of detector response. Band broadening that occurs in the chromatographic column, in which a pressure driven flow is used, can be described by the van Deemter equation, Equation (4.1):

$$H = A + B\mu^{-1} + C\mu \tag{4.1}$$

The term H is the height equivalent to a theoretical plate, a measure of the amount of band broadening; A describes eddy diffusion; B describes longitudinal molecular diffusion; C describes the resistance to mass transfer of the analyte; and μ is the mobile phase linear velocity. Eddy diffusion results from flow inequities throughout the stationary phase, particularly in a packed column. Therefore, chromatographic techniques that employ a packed column will have larger A terms than those that do not. The longitudinal diffusion arises from the diffusion of analytes from areas of high concentration to those of low concentration. Thus, the B term is much larger for chromatographic techniques that use carrier gases than for those with liquid mobile phases. The resistance to mass transfer, C, is composed of two terms—the resistance to mass transfer in the mobile phase, C_m, and the resistance to mass transfer in the stationary phase, C_s. For gaseous mobile phases, C_m is negligible. The optimum μ for a particular chromatographic technique can be determined by plotting H versus μ. The parameter μ, which produces the lowest H is the optimum flow rate, and will generate chromatograms with the sharpest peaks. Of the three van Deemter terms, C is the most dependent on the interactions of the analytes with the stationary and mobile phases. These interactions are governed by the solvation interactions available based on the polarity and structure of both the analyte and stationary/mobile phases. The interactions between analytes and stationary phases are often studied in order to characterise new stationary phases, or to determine the partitioning behaviour of analytes under various chromatographic conditions.

In order to characterise the solvation interactions between solutes and stationary phases in gas chromatography, Rohrschneider and McReynolds developed the Rohrschnieder–McReynolds system, which is still one of the most widely used methods of stationary phase classification [11]. This system is based on the interaction of five solutes (benzene, 2-pentanone, nitropropane, pyridine, and 1-butanol) chosen to measure specific solvation interactions with the stationary phase. The Rohrschneider–McReynolds constants can be related to the overall difference in the Kovats retention index (ΔI) by Equation (4.2):

$$\Delta I = aX' + bY' + cZ' + dU' + eS' \tag{4.2}$$

The Rohrschneider–McReynolds phase constants for benzene (X'), 1-butanol (Y'), 2-pentanone (Z'), nitropropane (U'), and pyridine (S') are determined by the difference between the retention index for that probe on a squalane stationary phase and the retention index for the same probe on the stationary phase under investigation. The overall polarity of the stationary phase can be determined by averaging the phase constants. This has been done for two imidazolium-based ionic liquid stationary phases by Armstrong et al., and the values were compared with commercial gas chromatographic stationary phases [12]. Using this method, the polarities of the ionic liquid phases were found to be similar to each other, and to polydimethoxysilane (PDMS) stationary phases. However, the fact that the separation selectivities between the ionic liquid and PDMS stationary phases were found to be distinct has highlighted the shortcomings of the method for classifying ionic liquid stationary phases. The limited number of probe molecules used in the Rohrschneider–McReynolds system inhibits its capability of accurately measuring individual solvation interactions.

The Abraham solvation parameter model is a linear solvation free energy model that can be used to characterise the interactions between solute molecules in the liquid or gas phase with liquid stationary phases [13]. The model assumes a three-step process for solvation: the formation of a cavity in the solvent, reorganisation of solvent molecules around the formed cavity, and the incorporation of the solute molecule into the cavity with the accompanying introduction of solute–solvent interactions. In contrast to the Rohrschneider–McReynolds system, this model utilises a large number of probe molecules in which each interacts with the stationary phase through various solvation interactions resulting in varied retention times for the analytes. The Abraham model allows the calculation of individual solvation interactions from the retention time data and has been used successfully to quantify these interactions for ionic liquids and polymeric ionic liquids [13–17]. The model is described by Equation (4.3):

$$\log_{10}(k) = c + eE + sS + aA + bB + lL \tag{4.3}$$

In Equation (4.3), $\log_{10}(k)$ is the solute retention factor calculated by measuring the retention time of the analyte and dead volume of the chromatographic column. The solute descriptors E, S, A, B, L are specific for the probe molecules used to characterise the stationary phase, and have been previously determined for many molecules [13]. They are defined as E, an excess molar refraction calculated from the solute's refractive index; S, the solute dipolarity/polarisability; A, the solute hydrogen bond acidity; B, the solute hydrogen bond basicity; and L, the solute–gas hexadecane partition coefficient determined at 298 K. The system constants (e, s, a, b, l) are defined as e, the ability of the ionic liquid to interact with π- and nonbonding electrons of the solute; s, a measure of the dipolarity/polarisability of the ionic liquid; a, a measure of the ionic liquid hydrogen bond basicity; b, a measure of the hydrogen bond acidity; and l, describing dispersion forces.

4.3 HIGH PERFORMANCE LIQUID CHROMATOGRAPHY

High performance liquid chromatography (HPLC) is one of the most widely used chromatographic techniques, enjoying great popularity in the pharmaceutical industry. HPLC utilises a solid support with coated or bound stationary phase. Analytes injected at the head of the column are separated based on the differences in their mobile phase–stationary phase partition coefficients. The stationary phase can be coated on the solid support (liquid–liquid chromatography), or it can be chemically bound to the support, which offers the advantage of being a more robust class of HPLC stationary phase.

HPLC can be operated in "normal phase" mode, with a polar stationary phase and a non-polar mobile phase or, and more commonly, in "reverse phase" mode with a non-polar stationary phase, such as the common "C_{18}" columns, and a polar mobile phase such as water with some amount of organic modifier, often methanol, added. Whether the stationary phase is coated or chemically bound, free or isolated silanol groups on the stationary phase can interfere with the efficient separation of compounds, especially basic analytes, resulting in poor reproducibility and peak tailing [18, 19]. This is a challenge for all chromatographic techniques that employ silica-based stationary phases, such as HPLC and thin layer chromatography (TLC). To mask these interfering silanol groups, suppressors are often added to the mobile phase in HPLC.

The ability to synthesise ionic liquids with varied solubility in water and organic solvents makes their use in liquid chromatography appealing. The majority of ionic liquids possess viscosities that preclude their use as neat mobile phases for HPLC, due to the extremely high pressures required to force them through the column [20]. However, ionic liquids are often miscible with common HPLC solvents, such as methanol or ethanenitrile, and when mixed with these solvents their viscosity decreases [21]. Poole and coworkers investigated the use of various ionic liquids, including ethylammonium nitrate and propylammonium nitrate [22], and Danielson and Waichigo explored the use of ethylammonium methanoate [23] as mobile phases for HPLC. Poole and coworkers found that ionic liquids paired with thiocyanate were corrosive to the column and therefore unsuitable [22]. The viscosities of the ionic liquids ranged from 32.1 to 636.9 cP, but when used in a binary mixture with water (up to 60% ionic liquid), problems with pumping were alleviated. Danielson used 40–80% ethylammonium methanoate in phosphate buffer for the separation of both aromatic carboxylic acids and amines [23]. While it was noted that the ionic liquid exhibited higher viscosity than methanol, no mention was made of excessive backpressure due to the increased viscosity. Although many ionic liquids may be unsuitable for use as neat mobile phases due to their high viscosities, they can be used as additives in fairly high ratios with aqueous- or organic-based phases.

Ionic liquids have also been studied as mobile phase additives for reverse phase HPLC (RP-HPLC) [24–31]. When used as mobile phase additives, imidazolium-based ionic liquids have been shown to form liquid clathrates with aromatic hydrocarbons [32], effectively creating an extended structure that can modify the separation. Preliminary studies by Marszall et al. using alkylimidazolium-based ionic liquids demonstrated that ionic liquids reduced the retention of basic analytes to a greater extent than alkylammonium-based mobile phase additives [28]. In addition, the shielding of silanol groups by ionic liquids improves the peak shape and resolution of the separation [26, 28, 31]. The improvement of peak shape and modification of the separation of aromatic amines is highlighted in Figure 4.1. All of the ionic liquid additives showed improved peak shape over that obtained by the use of the common ammonium ethanoate mobile phase additive. When $[C_4mim]^+$ and $[C_2mim]^+$ were paired with the methyl sulfate ($[C_1SO_4]^-$) anion, good resolution was achieved for the separation of 2-chlorophenol and 2,4,6-trichlorophenol in RP-HPLC using 80% MeOH:H$_2$O mobile phase concentration [24]. However, it was found that when paired with the $[BF_4]^-$ anion, there was little to no improvement in resolution.

In a study by Zhang et al., it was found that, for the separation of ephedrines by liquid chromatography, the addition of ionic liquids to the mobile phase produced peaks with less tailing and lower retention volumes [27]. This observation was attributed to the shielding of residual silanol groups by the imidazolium-based ionic liquid. The effects were more pronounced when ionic liquids with longer alkyl chains were used, presumably due to the longer alkyl chains shielding the silanols better than the shorter chained alkyl groups. Zheng and Row were able to separate previously unresolved guanine and hypoxanthine via RP-HPLC by utilisation of 1-octyl-3-methylimidazolium methyl sulfate ($[C_8mim][C_1SO_4]$) as a mobile phase additive [33]. The cations of ionic liquids modify the interactions between the analytes and stationary phase in liquid chromatography by the adsorption of the ionic liquid cations to the octadecyl group of the stationary phase [25, 33]. The interactions of the cations

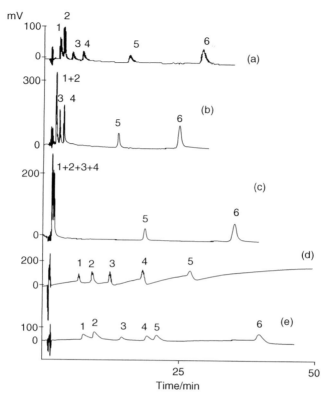

Figure 4.1 Chromatograms of all amines in a Nova-Pak stationary phase using (a) 10 mM [C$_4$mim][BF$_4$], (b) 10 mM [C$_6$mim][BF$_4$], (c) 10 mM [C$_8$mim][BF$_4$], (d) 5 mM ammonium ethanoate, and (e) without additives in the mobile phase. Heterocyclic aromatic amines and concentrations: (1) 9*H*-pyrido[4,3-*b*]indole (1.45 g L^{-1}), (2) 1-methyl-9*H*-pyrido[4,3-*b*]indole (0.490 g L^{-1}), (3) 3-amino-1-methyl-5*H*-pyrido[4,3-*b*] indole (0.443 g L^{-1}), (4) 3-amino-1,4-dimethyl-5*H*-pyrido[4,3-*b*]indole (1.40 g L^{-1}), (5) 2-amino-9*H*-pyrido[2,3-*b*]indole (4.92 g L^{-1}), and (6) 2-amino-3-methyl-9*H*-pyrido[2,3-*b*]indole (0.583 g L^{-1}). The retention time for 2-amino-3-methyl-9*H*-pyrido[2,3-*b*] indole in (d) is higher than 50 minutes. The ethanenitrile content was 18% (v/v). Reproduced with permission from Reference [31].

with the silanol groups of the stationary phase surface compete with the polar groups of the analytes and shield the residual silanol groups improving the peak shape and reducing the retention times for the separation. It would also be expected that the addition of ionic liquids to the mobile phase would modify the interactions and partitioning of the analytes to the mobile phase, in addition to shielding the silanol groups on the stationary phase. In addition, the cation–anion pair of an ionic liquid can act as an ion-pairing agent for ionised analytes. Xiao et al. studied the effect that 1,3-dialkylimidazolium-based ionic liquids have on the RP-HPLC separation of amines, and compared the performance of ionic liquids with that of the more commonly used tetrabutylammonium bromide ($[N_{4444}]$Br) when used as additives [25]. As the concentration of $[C_4mim][BF_4]$ was increased, the retention of amines decreased. It was determined that the hydrogen-bonding ability of the C-2 proton on the imidazolium ring greatly affected the separation of amines, and the retention of the basic amine analytes decreased as the alkyl chain substituent on the imidazolium cation increased in length. When compared to $[N_{4444}]$Br, ionic liquids as RP-HPLC additives reduced the retention of o-, m-, and p-phthalic acids to a lesser extent. This was attributed to the delocalisation of charge on the imidazolium ring, which is thought to lead to a weaker association between analyte and ionic liquid than between analyte and $[N_{4444}]$Br. Additionally, the repulsion between the cationic imidazolium ring and the amines in their ionised state also contributes to the reduced retention of those analytes.

Ruiz-Angel et al. compared the effects that triethylamine and $[C_2mim][BF_4]$ had on peak shape, retention factor, and resolution for the HPLC separation of a set of β-blockers [34]. The imidazolium cation was found to associate with the exposed silanol groups of the stationary phase, thereby improving peak shapes. On the other hand, the chaotropic tetrafluoroborate anions associated with the cationic analytes, and the resulting ion pair was more attracted to the stationary phase, resulting in reduced retention. These two effects compete with each other, resulting in less of an increase in retention factor with the ionic liquid additive than with the triethylamine additive. This property means that it is possible to vary which of the effects will dominate by modification of the cation and/or anion. Separations optimised with $[C_3mim][BF_4]$, ammonium ethanoate, and triethylamine demonstrated that the best resolution for all analytes was achieved with the ionic liquid, while the other additives had peak overlap or peak tailing. By increasing the hydrophobicity of the cation and pairing it with a kosmotropic anion, ion adsorption will be favoured with a reduction in analysis time and improved peak shape. On the other hand, if the cation is made to be less hydrophobic with a shorter alkyl chain, and if a chaotropic anion is used, then ion pairing will dominate, and basic analytes will be more strongly retained with improved peak shapes.

Imidazolium tetrafluoroborates are useful as mobile phase additives for separation of basic analytes using silica-based stationary phases. Previous studies have found that the added ionic liquids did not harm the column, as evidenced by similar column performance before and after exposure to ionic liquid modifiers. It is apparent that the silanol shielding ability of the ionic liquid is due to the cation, while the retention factor is probably due to the hydrophobic nature of the anion [34]. It has been demonstrated that chaotropic anions have a positive effect on the solvation of cationic analytes [35]. Chaotropic anions only decrease retention; they do not improve peak shape [35, 36].

Ionic liquids are not restricted to use as mobile phase modifiers, but can also be used as chemically bound stationary phases for liquid chromatography [37–41]. Immobilised 1-butyl-3-methylimidazolium bromide ($[C_4mim]$Br) has been used as a stationary phase to separate a group of 28 small aromatic test compounds in RP-HPLC [37]. The [C$_4$mim]Br was bound to silica particles and packed into HPLC columns. The solvation interactions of the stationary phase were evaluated by applying multiple linear regression analysis to the retention volumes of the test compounds. It was found that the immobilised stationary phase retained solutes more like aromatic phases than aliphatic ones. In addition, the dipole/induced-dipole interactions for the ionic liquid stationary phase increased with the percentage of organic modifier in the mobile phase, in contrast to the apparent organic modifier insensitivity of such interactions in alkyl-based systems. When halogen-substituted benzenes and benzene alcohols were separated using the [C$_4$mim]Br-bound column, the efficiency (measured by the number of theoretical plates) was much higher than that typically found for ion exchange columns. This indicates increased interaction kinetics for the reversed phase interactions for the ionic liquid-bound phase. In subsequent work, Stalcup and coworkers further character-ised this stationary phase and compared it with a commercially available quaternary amine silica-based stationary phase column [38]. It was found that the immobilised [C$_4$mim]Br exhibited higher hydrophobic interactions than the commercial stationary phase at all mobile phase compositions. The parti-tion coefficients for analytes in [C$_4$mim]Br determined by the shake flask method were found to correlate with the retention of compounds studied, indicating the applicability of HPLC to determination of partition coefficients of analytes with ionic liquids. Stalcup and coworkers also demonstrated that the ionic liquid stationary phase was able to undergo ion-exchange interac-tions as the ionic strength of the mobile phase was modified. This, along with further work by other groups [39–41], shows that the ionic liquid stationary phase can separate in both the reversed phase and ion exchange modes for HPLC, demonstrating once again the versatility and tuneability of ionic liquids in separation science.

Ultraviolet (UV) detection is most commonly used with HPLC [24, 25, 27, 42], but other detection techniques include mass spectrometric (MS) detection [43, 44], photodiode array detection (PDA) [45, 46], and refractive index detec-tion (RI) [44, 47], depending on the needs of the analysis and the nature of the analytes. While the use of ionic liquids in RP-HPLC has some great ben-efits, one challenge is in detection of analyte peaks. Many ionic liquids, such as the imidazolium-based ionic liquids, absorb in the UV region [48], which can cause difficulties in detection. In order to avoid interference, the detection of analytes should not be close to 212 nm, the absorption maximum for the alkylimidazolium cations [20, 29]. The same care in choice of detection wave-length must be taken when using a PDA detector, so that the absorbance of the ionic liquid does not interfere with the detection of the sample. MS detec-tion can be difficult for HPLC separations that use ionic liquids in the mobile phase because the negligible vapour pressure of the ionic liquids makes it dif-ficult to remove them from the sample by evaporation.

4.4   COUNTER-CURRENT CHROMATOGRAPHY

Counter-current chromatography (CCC) is a separation technique similar to traditional liquid chromatography, but differs in that it does not employ a solid stationary phase support [49]. The elimination of the solid support in CCC translates into greatly increased stationary phase volume, and a simplified retention mechanism as compared to chromatographic techniques that rely upon the use of solid stationary phase supports. CCC is particularly useful in preparative separations, due to the increased volume of liquid stationary phase which can accommodate analyte loading as large as 1 g. CCC is useful for the analysis of pharmaceutically interesting compounds of natural products, which are generally polar and/or labile materials, and as a method for cleaning up dirty samples prior to HPLC analysis.

The use of CCC also avoids the problems inherent in other chromatographic techniques that employ solid supports, such as the elimination of contaminants leached from adsorbents, destruction of the column by incompatible solutes, and irreversible adsorption of solutes onto the stationary phase. In addition, either phase can be chosen to be the stationary or mobile phase by modifying the system, termed "dual-mode" when performed within the same separation, while still producing resolutions of 350 to 1000 plates. The nature of CCC allows it to be extended even further than dual-mode, with the possibility of "elution–extrusion" CCC where the mobile phase is eluted and then the stationary phase is extruded for analytes that partition strongly into the stationary phase, with the additional benefit of higher resolution than normal mode CCC [50].

In the selection of solvent systems for CCC, the settling time should be less than 30 seconds, the partition coefficient of the analyte should be between 0.5 and 2.0, and the separation factor between any two components should be greater than 1.5 [51]. The retention of the stationary phase is determined by the flow rate of mobile phase and the centrifugal forces applied to the system. The relative retention (S_f) can be linearly related to the square root of the flow rate (F) by Equation (4.4), with $S_f = V_S/V_C$ and V_S and V_C equal to the stationary and column volumes, respectively [52].

$$S_f = A - BF^{1/2} \tag{4.4}$$

The A and B values are system dependent, with A corresponding to the solvent composition of the two-phase system and B corresponding to the volume ratio of the solvents. From Equation (4.4), it would be expected that a larger volume of stationary phase would result in better retention of that phase by the column.

The use of ionic liquids as stationary phases for CCC has been discussed recently by Berthod et al. [53]. Ionic liquids are interesting solvents for use in CCC because of their low to negligible volatility and tuneable solvent interactions. These properties of ionic liquids mean that there is less loss of solvent due to evaporation compared with organic solvents, and the polarity of the solvent system often does not need to be modified by a mix of multiple solvents, simplifying the solvent selection for CCC. In addition, CCC can be used to determine partition coefficients for ionic liquids with less interfering interactions than other chromatographic techniques, due to the lack of solid support material for the stationary phase. Berthod and Carda-Broch determined the ionic liquid–water partition coefficients ($K_{il/w}$) of various compounds when they investigated [C_4mim][PF_6] as a stationary phase for CCC [54]. They found that they needed to reduce the viscosity of the ionic liquid by the addition of ethanenitrile in order to execute the CCC experiment at reasonable pressures. Continuous UV detection was not possible because of the UV absorption of the ionic liquid. Evaporative light scattering detection could not be used because of the non-volatility of the ionic liquid phase. While they were able to measure partition coefficients directly, the range of values that could be measured was low due to the long retention times and the reduced detection sensitivity.

Ionic liquids have also been investigated as replacements for trichloromethane as modifiers in CCC for the isolation of components of traditional Chinese medicines [55, 56]. The ionic liquids investigated in both works were [$C_n$mim][PF_6] with n = 4, 6, and 8. Liu et al. tested these ionic liquids as modifiers for the ethyl ethanoate–water CCC separation of five flavonoid compounds from the traditional Chinese herbal medicine *Oroxylum indicum* [55]. They were able to successfully separate and purify the flavonoids using the ionic liquids studied. It was found that the separation using the ionic liquids with shorter alkyl chains exhibited better partitioning, due to its higher polarity as compared to the longer chained analogues. The ionic liquids with the shorter chains also had lower viscosity, making them more amenable to CCC. The ionic liquid was then removed from the separated fractions by running the fractions through a macroporous resin column and eluted with 20% MeOH in water, and the flavonoids were eluted with 55% MeOH in water. Xu et al. used the same set of ionic liquids as modifiers for the CCC separation of nomangiferin and mangiferin from *Rhizoma anemarrhenae* [56]. Their procedure closely mirrored that of their previous work with flavonoids, except that the separation of the ionic liquid from the fractions via the macroporous resin column was accomplished with 20% and 50% ethanol for the elution of the ionic liquid and sample, respectively. Xu et al. found that if the concentration of the ionic liquid modifier was too high, emulsification occurred that caused greater loss of the stationary phase. The higher polarity, lower viscosity, and reduction of emulsification led both groups to use [C_4mim][PF_6] as the modifier instead of the longer alkyl chain analogues.

In addition to the benefits that ionic liquids have in terms of solvents and/or solvent additives for CCC, they also carry with them considerable challenges that have reduced the investigation of these solvents in CCC. One significant drawback of ionic liquids is their large viscosity as compared with traditional organic solvents. Common pumps for chromatographic columns demand that the liquids being pumped have viscosities less than approximately 5–10 cP in order to avoid excessive pressure buildup. Ionic liquids tend to have viscosities higher than common pumps can handle and must be modified by the addition of another solvent, such as ethanenitrile, before use as mobile phases [20].

In addition, the separation of the ionic liquid from the sample is complicated by the same properties of ionic liquids that make them attractive as CCC solvents. Ionic liquids cannot be easily distilled (because of their low to negligible vapour pressure), and they are often able to dissolve both polar and non-polar analytes. Methods of separating ionic liquids from the sample include extraction by supercritical CO_2 [57–60], membrane extraction [61], distillation [62], aqueous extraction, and additional chromatographic techniques [55, 56]. Each of these methods adds to the time required for analysis, and often requires the use of additional solvents.

Popular detection methods for CCC include UV detection [50, 54–56, 63–65], evaporative light scattering [66–71], and mass spectrometry [72, 73]. When UV detection is used, there is always the possibility of interference by droplets of stationary phase that are flushed into the absorption flow cell. Many ionic liquids absorb strongly in the UV range, especially the most common imidazolium-based ionic liquids, and this can reduce the sensitivity of detection. Evaporative light scattering and mass spectrometry are not useful as detection methods for CCC utilising ionic liquids since both detection techniques require the evaporation of the solvent prior to detection, and the low vapour pressure of the ionic liquids precludes such evaporation [54].

4.5   IONIC LIQUIDS IN GAS CHROMATOGRAPHY

Gas chromatography (GC) is an analytical technique in which a gaseous mobile phase is coupled with a solid stationary phase termed, gas–solid chromatography (GSC), or a liquid, termed gas–liquid chromatography (GLC). GLC is the most commonly encountered type of GC. Ionic liquids are attractive stationary phases for GC because they have viscosities that can be varied over a wide range, can possess high thermal stability [74], possess unique solvation properties [12, 74], can be coated on fused-silica capillaries with high efficiency, and can be immobilised and cross-linked [75–79].

The use of ionic liquids as GC stationary phases began with the separation of metal halides [80] and organic analytes [81] by ionic liquids composed of inorganic ions. In the 1980s, Poole and coworkers used the organic ionic liquid, ethylammonium nitrate, as a GC stationary phase to separate organic analytes [82]. This ionic liquid stationary phase was found to be more polar than the commercially available Carbowax 20 M but was limited to an operational temperature range of 40–120 °C. The discovery of additional ionic liquids containing organic cations that possess increased stability [5, 9] has led to multiple

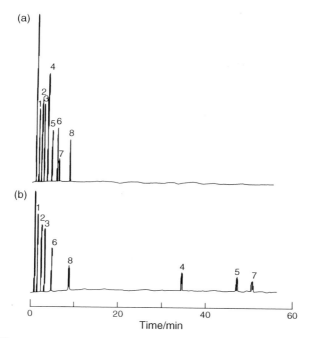

Figure 4.2 Chromatograms comparing the retention and separation of eight compounds on the same size GC columns (15 m, 0.25 mm i.d.) and under identical conditions (isothermal at 100 °C and a pressure of 9 psi). Column A is a commercial DB-5 column, and Column B utilises the ionic liquid [C₄mim][PF₆] as the stationary phase. The test compounds are: (1), butyl ethanoate; (2), 1-heptanol; (3), 1,4-dichlorobenzene; (4), 1,2-cresol; (5), 2,5-dimethylphenol; (6), 1-dodecane; (7), 4-chloroaniline; and (8), 1-tridecane. Reproduced with permission from Reference [12].

studies on the use of ionic liquids as GC stationary phases [12, 74, 77–79, 83–88]. Armstrong et al. studied the utility of [C₄mim][PF₆] and [C₄mim]Cl as GC stationary phases and characterised their solvation interactions with solutes using inverse GC (IGC) [12]. They found that ionic liquid stationary phases were not only able to separate polar analytes from non-polar analytes but were also able to separate the analytes within each group (polar and non-polar) from each other. Figure 4.2 shows the chromatograms of eight compounds generated by using a commercial DB-5 GC column and a column coated with [C₄mim][PF₆]. The ionic liquid coated GC column is shown to be able to separate the polar from the non-polar compounds, reversing the elution order of some pairs of compounds. In subsequent work, a series of ionic liquid GC stationary phases was evaluated and the importance of the anion on the thermal stability of ionic liquid stationary phases for GC was demonstrated by the change in the bleed temperature of the columns [74]. Ionic liquids

containing [OTf]⁻ anions demonstrated greater thermal stability than those
with Cl⁻, [NTf₂]⁻, or [PF₆]⁻ anions. The anion of the ionic liquid stationary phase
was also found to greatly affect the efficiency of separating hydrogen-bond
basic compounds, such as alcohols, as demonstrated by the improved peak
symmetry for the triflate-based ionic liquids. It was also noted that compounds
with high hydrogen-bond acidity were tenaciously retained by stationary
phases incorporating the chloride anion, but were not retained as strongly
when the same ionic liquid cation was paired with the less hydrogen-bond
basic hexafluorophosphate anion. The ionic liquid GC stationary phases possessed solvation interactions that were distinct from the commercially available squalane and DB-5 GC stationary phases, and offered a new and unique
option for chromatographic stationary phases.

In an effort to generate more thermally robust ionic liquid GC stationary
phases, cross-linked ionic liquid stationary phases based on a variety of monomeric ionic liquids with vinylimidazolium cations were developed [75]. The
efficiency of the cross-linked ionic liquid stationary phases was improved over
that of monomeric ionic liquid stationary phases when the GC column was
conditioned at high temperatures. The efficiencies of the cross-linked ionic
liquid stationary phases were much less dependent on the length of the alkyl
chain attached to the imidazolium cation than were the efficiencies of the corresponding monomeric ionic liquid stationary phases. In addition, when the
ionic liquid was highly cross-linked, the efficiency increased with conditioning
at elevated temperatures (300–350 °C), demonstrating the usefulness of these
types of ionic liquid stationary phases for high-temperature GC applications.
While the more highly cross-linked ionic liquid stationary phase was under-
standably more ordered than the monomeric ionic liquid stationary phase, the
selectivity and chromatographic performance remained fairly consistent, indicating that polymerisation and cross-linking does not greatly change the solvation interactions of the ionic liquid.

Armstrong et al. recently summarised the application of ionic liquids as
stationary phases for GC [89]. The two main anions used for ionic liquids are
[NTf₂]⁻ and [OTf]⁻. Ionic liquids with [NTf₂]⁻ anions tend to possess lower
melting points and polarities than those with [OTf]⁻. They also produce good
peak shapes for non-hydrogen-bonding analytes, but tailing peaks for hydrogen-
bonding analytes such as alcohols, carboxylic acids, and amines. The [NTf₂]⁻
anion has a more delocalised charge than [OTf]⁻, while the [NTf₂]⁻ is more
hydrophobic. The tailing associated with the [NTf₂]⁻ ionic liquids can be miti-
gated by pairing it with a cation containing an imide functional group. When
stationary phases containing the [OTf]⁻ anion are employed as GC stationary
phases, acidic and basic analytes have good peak shapes.

The cations of the ionic liquids can be chosen and/or modified to tune the
solvation properties of the ionic liquid. By modifying the cationic structure
with various substituent groups, the ionic liquid can be made to take on a more
or less polar character. By choosing phosphonium cations, polar ionic liquid

stationary phases can exhibit higher thermal stability [78], with lower bleed than the traditional PEG wax-type stationary phases for GC [89]. This low bleeding translates into lower limits of detection because of the reduced background interference from the ionic liquid stationary phase. In comparison with the most polar commercially available GC stationary phases, the moisture- and oxygen-sensitive 1,2,3-tris(2-cyanoethoxy)propane (TCEP), a divinylimidazolium bistriflamide ionic liquid from Supelco (SBIL-100), has been engineered to offer improved oxygen and moisture stability in addition to unique selectivity. In addition, the SBIL-100-coated column demonstrates faster elution times compared with the TCEP-coated column, while maintaining good resolution of analytes [89].

Ionic liquid stationary phases are also useful in conjunction with traditional GC stationary phases to produce highly successful separations for $GC \times GC$ applications [85–87]. The saturated hydrocarbons, monoaromatic, and diaromatic compounds from diesel fuel were successfully separated using a trihexyl(tetradecyl)phosphonium bistriflamide ($[P_{66614}][NTf_2]$) ionic liquid column followed by a non-polar HP-5 column [85]. The $[P_{66614}][NTf_2]$ stationary phase demonstrated stable performance even with continuous use at temperatures reaching 290 °C. Ionic liquids have also been used in $GC \times GC$ in the analysis of phosphorus–oxygen containing analytes [86], and the headspace analysis of U.S. currency [87]. Reid et al. demonstrated that the triflate ionic liquid coated column used as the second column in the $GC \times GC$ separation of 32 distinct analytes demonstrated higher selectivity than the poly (ethyleneglycol) (PEG)-coated column for phosphorus–oxygen containing compounds [86]. Siegler et al. recently described the use of the triflate ionic liquid, 1,9-di(3-vinylimidazolium)nonane triflate, as one stationary phase in a three-dimensional gas chromatograph (GC^3) instrument [88]. The GC^3 instrumentation achieved larger peak capacity than standard $GC \times GC$ instrumentation and the use of the triflate ionic liquid imparted high selectivity for phosphonated compounds in diesel fuel samples.

The physicochemical properties of ionic liquids can be tuned extensively by choosing various modifications of the constituent ions. This tuning changes the selectivity characteristics of the ionic liquid, complementing the ability of ionic liquid stationary phases to separate polar analytes as if they were polar stationary phases and non-polar analytes as if they were non-polar stationary phases [12, 90–92]. In addition, work done by Anderson and coworkers has shown that binary mixtures of ionic liquids and polymeric ionic liquids possessing different anions can be used to fine tune the separation selectivity for hydrogen bonding analytes when used as stationary phases in GC [93, 94]. The hydrogen-bond basicity of the stationary phase was found to increase linearly with the percentage of chloride-based polymeric ionic liquid incorporated into the mixture. This difference in the hydrogen-bond basicity was shown by the improved separation selectivity for alcohols, including the reversal of elution order for some pairs of analytes.

4.6   IONIC LIQUIDS IN SUPERCRITICAL FLUID CHROMATOGRAPHY

The use of supercritical fluids in separation science has been investigated due to the "green" potential of many such fluids. Supercritical fluid chromatography (SFC) is a chromatographic technique in which a supercritical fluid is used as the mobile phase. A supercritical fluid is a material formed when the temperature and pressure is above its critical temperature and critical pressure. Supercritical fluids are useful in chromatography because very small changes in pressure or temperature can greatly affect the density of the fluid and tune separations. SFC uses the same columns as HPLC, and the entire system operates under pressure to maintain the supercritical fluid state of the mobile phase. The most commonly used supercritical fluid is supercritical carbon dioxide ($scCO_2$) because it is affordable, readily available, and environmentally benign. Despite the high solubility of CO_2 in ionic liquids [95, 96], the majority of ionic liquids exhibit little to no solubility in CO_2 regardless of the pressure [97, 98]. The inability of some ionic liquids to dissolve in CO_2 means that $scCO_2$ can be used to extract analytes from ionic liquids and that ionic liquids can be used as stationary phases for SFC. In fact, when imidazolium-based ionic liquids are utilised, the addition of a co-solvent such as ethanol or propanone is required for the ionic liquid to dissolve into the $scCO_2$ [57]. It was also noted that ionic liquids saturated with water had even lower solubility in CO_2 than dried ionic liquids [97], allowing the analytes from aqueous samples to be extracted by ionic liquids and removed by $scCO_2$ [59, 99]. The solubility of CO_2 and $scCO_2$ in ionic liquids has been discussed recently by Roth in a review of the biphasic ionic liquid–$scCO_2$ systems with respect to organic compound partitioning [57]. It was noted that ionic liquids were attractive for separations utilising supercritical fluids, as the solvent properties of the system can be modified by changing the temperature and pressure of the system as well as the identity or structure of the constituent ions of the ionic liquid.

Miyawaki and Tatsuno investigated the use of $scCO_2$ to selectively extract butyl oleate from the ionic liquid methyltrioctylammonium trifluoroethanoate [99]. The use of an ionic liquid as a reaction solvent for the lipase-catalysed butanolysis of triolein was employed in a continuing effort to characterise ionic liquids for biochemical reactions. The unique solvation properties offered by the ionic liquid and their response to $scCO_2$ allowed for the selective extraction of butyl oleate over glycerides. Machida and et al. measured the partition coefficients of a group of aromatic compounds in [C_4mim][PF_6]–$scCO_2$ systems [100]. They found that naphthalene had lower partition coefficients than the other organic molecules they studied, and attributed that to differences in the π–π interactions of the compounds. They also noted that the solubilities of the compounds varied with the density of the CO_2, and depended mostly on vapour pressure at low density, and solubility in CO_2 at high density. Planeta et al. probed the partitioning of compounds in a trihexyl(tetradecyl) phosphonium chloride ionic liquid–scCO$_2$ system [101]. They used Abraham's solvation parameter model and SFC with the ionic liquid as the stationary phase to evaluate solute partitioning. It was found that the chloride-based ionic liquid, much like that which has been determined in inverse gas chromatography (IGC) studies, possessed an elevated hydrogen-bond acidity descriptor compared with [C$_2$mim][BF$_4$] [102], and correspondingly long retention times for acidic solutes.

Chou et al. used a covalently bound 1-octyl-3-propylimidazolium chloride ([C$_8$C$_3$im]Cl) stationary phase for the SFC separation of a group of acidic, basic, and neutral compounds [103]. They were able to successfully separate the group of analytes using optimised separation conditions of 20% MeOH in scCO$_2$ at a pressure of 110 bar and temperature of 35 °C. They determined that the ionic liquid stationary phase interacted with the analytes via electrostatic and hydrogen-bonding interactions, producing a separation unattainable via commercial C$_{18}$ columns. As in other types of chromatography, the unique solvation properties that can be imparted to ionic liquids make their use in SFC appealing.

4.7   CAPILLARY ELECTROPHORESIS AND CAPILLARY ELECTROCHROMATOGRAPHY

Capillary electrophoresis (CE) is a separation technique that relies on differences in electrophoretic mobility (μ) of analytes through a running buffer. The basic setup for CE consists of a reservoir that supplies running buffer at the inlet of a fused silica capillary and a reservoir at the end of the capillary that collects the spent running buffer. When a voltage is applied between the inlet and outlet buffer, an electro-osmotic flow (EOF) is induced as the result of the flow of hydrated cations in the capillary and analytes will migrate down the length of the column, separate based on the differences in their mobility, and be detected before exiting into the outlet buffer [104]. The advantage of separation by EOF versus hydrodynamic flow is the flow profile. EOF exhibits a plug-flow profile, which greatly reduces the band broadening of analytes as they move through the capillary. Hydrodynamic flow profiles are induced when mobile phases are forced through a capillary or column, with higher mobile phase velocities at the centre of the flow and lower velocities at the walls of the column. In CE, when a voltage is applied to the capillary, positively charged ions from the running buffer adsorb to the negatively charged silanol groups on the surface of the capillary, and negatively charged ions from the running buffer are attracted to the previously mentioned positively charged buffer ions. Just past this thin Stern Layer (ca. 0.1 nm thick) exists the mobile layer, composed of hydrated cations, which exhibits positive EOF toward the cathode. The migration of these hydrated cations produces the bulk flow along the capillary. Separation in CE is based on the charge and size of the hydrated analytes. Positively charged analytes will migrate faster and elute before neutral analytes, which also elute before negatively charged analytes. Small positively charged analytes elute before large positively charged analytes while small negatively charged analytes elute after large negatively charged analytes. A modification of CE is capillary electrochromatography (CEC), in which a stationary phase is added. In CEC, separation of analytes is accomplished by both differences in mobility and differences in their partition coefficients to the stationary phase.

The general requirements for CE and CEC running buffers and additives include wide liquid range, relatively high permittivity for the analytes of interest, low viscosity and small molar volume, good solvating power for cations and anions as well as analytes, low vapour pressure, high chemical stability, and non-toxicity. The typical solvents that meet these criteria include ethanenitrile, methanamide, N,N-dimethylmethanamide, methanol, propylene carbonate, dimethyl sulfoxide, nitromethane, and N,N-dimethylethanamide [105].

Because CE and CEC are operated most often using fused silica capillaries, they face similar challenges due to exposed surface silanols as do other separation techniques, namely protein adsorption, the tailing of amine peaks, and hydrolytic instability. The strategies employed in other separation techniques to deal with the challenges presented by the use of fused silica supports have been adapted for application to CE and CEC. Ionic liquids have been applied in CE as run buffer additives, as well as dynamic or covalently attached coatings on the capillary walls for CEC. When ionic liquids are used as dynamic coatings and as electrolytes in the run buffer, they facilitate the separation of neutral compounds that would otherwise elute with the EOF [106, 107]. Separation is accomplished through the enhanced solute interactions with the capillary walls, as well as in the bulk running buffer. This is shown schematically in Figure 4.3, where the ionic liquid both coats the walls of the CE capillary and acts as a running buffer additive.

Ionic liquids have been employed as background electrolytes for CE as they are miscible with common solvents and have been found to modify, and sometimes reverse, the EOF. Yanes et al. utilised ionic liquids incorporating alkylimidazolium cations paired with a variety of anions as both EOF modifiers

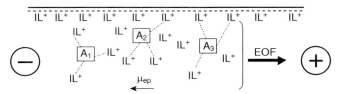

Figure 4.3 The double role of ionic liquids in separations. Separation could be achieved depending on the degree of the interaction between the analyte and the cation of the ionic liquid. IL$^+$: cation of ionic liquid. A$_n$: analyte to be separated. μ_{ep}: electrophoretic mobility. Copyright Wiley-VCH Verlag GmbH & Co. KGaA. Reproduced with permission from Reference [112].

and running electrolytes for the CE separation of polyphenolic compounds from grape seed extract [106]. They found that when ionic liquids were used as running electrolytes, the pH of the system was sufficiently acidic to ensure that the analytes were in their neutral form. In addition, the ionic liquids coated the walls of the capillary, creating an anodic EOF for the separation. This was similar to the results from work performed previously by Yanes et al. using $[N_{2222}][BF_4]$ as the running buffer [107], except that the ionic liquid running electrolytes produced higher EOFs for the same concentrations [106]. It was noted that the delocalised charge of the $[C_2mim][BF_4]$ ionic liquid compared with the $[N_{2222}][BF_4]$ ionic liquid produced lower association constants for the ionic liquid ions, allowing higher migration of ions. The association of polyphenolic compounds with the free cations of the ionic liquids appeared more dominant compared with the cations adsorbed to the walls of the capillary, as the effective migration of the compounds increased with the increase in the concentration of $[C_2mim][BF_4]$. When the alkyl chain on the ionic liquid cation was varied from ethyl to butyl, the resolution of the polyphenolic compounds improved, but it was not evident if the change was due to higher association of the compound with the imidazolium cation or the decreased EOF from the longer alkyl chain on the ionic liquid. With respect to the anions studied, not all ionic liquid pairs were able to produce separations of the analytes and the ionic liquids containing the nitrate and trifluoromethanesulfonate were found to be entirely unsuitable for CE separation. Cations appear to exert the greatest effect on the CE separation.

Mwongela et al. studied the use of ionic liquids as modifiers in micellar electrokinetic chromatography in place of more commonly employed organic solvents [108]. They found that the EOF was modified by the concentration of ionic liquid in the buffer, and that the separation of analytes was dependent on the ionic liquid chosen. When higher concentrations of ionic liquid were used, they noted longer migration times for more hydrophobic analytes. In a subsequent study, Yue and Shi explored the interactions between $[C_nmim]^+$ ionic liquid additives and flavonoids extracted from Chinese herbal medicines in CE separations [109]. It was observed that hydrogen bonding between the C-2 proton of the imidazolium cation and the phenolic compounds was largely responsible for the resolution of the separation, as the resolution of the compounds was severely compromised when the C-2 proton was replaced with a methyl group. They also found that the resolution improved as the length of the alkyl chain on the imidazolium cation was increased from C_2 to C_4, but greatly decreased for the C_5 substituent. When the effect of the anion was considered, the resolution of the flavonoids was improved as the melting point of the ionic liquids decreased. This led to the conclusion that the hydrogen-bonding interaction between the ionic liquids and the flavonoids is the main contributing factor in the resolution of the phenolic flavonoid compounds by CE.

François et al. studied the effects of ionic liquid concentration, the nature and percentage of alcohol in the alcohol-ethanenitrile solvent mixture, and the buffer salt concentration on the CE separation of 2-arylpropanoic acids [110]. The achiral $[C_4mim][NTf_2]$ was used in the separation of four chiral anionic compounds. It was found that with an initial increase in ionic liquid concentration, the electro-osmotic mobility of the running buffer and the electrophoretic mobility of the 2-arylpropanoic acids greatly decreased. This was attributed to the adsorption of the ionic liquid cation to the capillary wall resulting in a dynamic coating of ionic liquid. It was noted that the effect of changing ionic liquid concentration on electrophoretic mobility was not linear, indicating that at low concentrations of ionic liquid, the ionic liquid adsorbed to the walls of the capillary seemed to act as a stationary phase, causing an increase in migration times. As the ionic liquid concentration was increased beyond 7 mM, it was suggested that the capillary wall became fully coated and ion-pair interactions between the 2-arylpropanoic acids and free ionic liquid cations dominated.

Tian et al. used $[C_4mim][BF_4]$ to dynamically coat a silica monolith as a stationary phase for electrochromatographic separation of phenols and nucleoside monophosphates [111]. By dynamically coating the monolith with $[C_4mim][BF_4]$, the authors were able to generate anodic EOF, which, coupled with the chromatographic interactions of the stationary phase, provided improved separation of acidic compounds. The EOF was also found to be sensitive to the concentration of ethanenitrile, presumably due to decreased amount of dynamically coated $[C_4mim][BF_4]$ with increasing ethanenitrile concentration. At high concentrations of $[C_4mim][BF_4]$, the retention of analytes increased, even though the EOF remained constant. This was attributed to the increased association of analytes with the ionic liquids in the running buffer, in addition to the coated ionic liquid on the walls of the solid support. Overall, the addition of $[C_4mim][BF_4]$ improved the separation of phenols and nucleoside monophosphates.

While this overview of electrophoretic techniques is by no means exhaustive, a common theme is evident. The addition of ionic liquids to electrophoretic techniques increases the interactions available to separate analytes. As background electrolytes, ionic liquids increase the number of solute interactions and can modify the running buffer pH and ionic strength. As the concentration of ionic liquids is increased, dynamically coated stationary phases are produced through the interaction of the cation of the ionic liquid and the silanol groups of the capillary wall. At even higher concentrations of ionic liquid, solutes are introduced to both chromatographic and ion-pairing interactions, in addition to the tremendous separating power differences in electrophoretic mobility.

However, one must keep in mind the integrity of ionic liquid additives in CE systems. Work done by López-Pastor et al. with $[C_4mim][PF_6]$, and reported in their 2008 review on ionic liquids and CE, demonstrated that it is possible to separate and detect the individual ions of an ionic liquid [112]. The low integrity of ionic liquids in an electric field limits the possibility of recycling of ionic liquid from CE applications, and a system using 100% ionic liquid for the running buffer would suffer the loss of the ionic liquid as the constituent ions were separated. Furthermore, the study by López-Pastor et al. indicates that ionic liquids, as generally defined, cannot exist in a CE application, but their constituent ions do interact and play an important role in the modification of electrophoretic separations.

An additional challenge to using ionic liquids is the issue of detection. Detection techniques often coupled with electrophoretic techniques include ultraviolet-visible (UV-Vis), fluorescence, electrochemical, and mass spectrophotometric (MS) detectors. Due to their low vapour pressure, ionic liquids hinder the use of techniques that require the evaporation of the solvent (such as MS detection). Also, many ionic liquids absorb in the UV region and create large background noise, reducing the sensitivity of the detection in absorbance and fluorescence techniques. The use of electrochemical detection with ionic liquids is promising due to their large electrochemical window.

4.8   PLANAR CHROMATOGRAPHY

In planar chromatography, the stationary phase is present as a flat bed on some type of support, frequently glass or plastic. Planar chromatography utilising a cellulose stationary phase, commonly termed paper chromatography, has been used since the mid-1800s [113]. The introduction of paper chromatography eventually led to the development of TLC, which is arguably the most well-known and developed form of planar chromatography. In TLC, the stationary phase is deposited on a glass or plastic surface and the analytes are placed as a small, concentrated spot at one end of the slide or surface. Separation of the various analytes is achieved as the mobile phase passes over the stationary phase, with different analytes partitioning to various extents between the mobile and stationary phases. TLC makes use of various stationary phase materials with silica, alumina, and cellulose being some of the more common ones. The distance a particular compound moves divided by the distance the solvent front travelled is known as the retention factor, or R_f, of the compound. Compounds with R_f values close to 1 are not retained by the stationary phase and compounds with R_f values close to zero do not dissolve in and move with the mobile phase. Stationary phases used for TLC often contain a fluorescing moiety to allow detection of many analytes at 254 nm, but it does not interfere with the separation or subsequent detection techniques [113]. As was discussed in Section 4.3 (HPLC), when silica is used as the stationary phase, the free silanol groups of the stationary phase must be suppressed to be able to best separate basic analytes. As stated previously, the ability of ionic liquids to shield these silanol groups has been studied by Marszall et al., and they found that ionic liquids were superior to alkylamine additives in their shielding ability [28]. This property of ionic liquids makes the study of their application to TLC interesting.

In Sherma's recent review of current research efforts using planar chromatography, it was noted that stationary phases for TLC are more varied than for all other chromatographic techniques [113]. This should come as no surprise considering how much longer TLC has been in use as compared with most other types of chromatographic techniques. However, ionic liquid stationary phases for planar chromatography were not mentioned. Ionic liquids have the potential to be an interesting stationary phase because of the previously noted solvation characteristics; however, they can be more expensive to manufacture than other TLC stationary phases. Since TLC plates are only used once and disposed of, the cost associated with manufacturing single-use plates could be inhibitory with respect to ionic liquid stationary phases. However, ionic liquids can be used as mobile phase additives for planar chromatography without the economic limitation.

Kaliszan et al. studied [C_2mim][BF_4], [C_6mim][BF_4], and 1-hexyl-3-(heptyloxymethyl)imidazolium ([$C_6C_7O_1$im][BF_4]) as mobile phase additives for the TLC separation of eight basic drugs [26]. It was shown that the imidazolium tetrafluoroborate-based ionic liquids are able to suppress the surface silanols on the octadecyl bound stationary phase and produce R_M values for amines that can be linearly correlated to the organic modifier concentration of the mobile phase where $R_M = \log_{10}(1/[R_f - 1])$. The use of [$C_2$mim][$BF_4$] as a mobile phase additive produced analyte spots that were free of tailing and well separated. This was impressive, as the ionic liquid additives were found to be more effective at suppressing free silanols than the widely used suppressors triethylamine, dimethyloctylamine, or ammonium hydroxide.

Bączek et al. investigated the effect of the addition of [C_2mim][BF_4] to the mobile phase on the normal mode TLC separation of peptides using a silica gel stationary phase [114]. They used computer simulations to model the effect of the addition of ionic liquid on the change in R_f values as the ethanenitrile concentration of the mobile phase changed. It was found that the change in R_f values with change in ethanenitrile concentration changed from a quadratic response without ionic liquid to a third-order polynomial response with the ionic liquid. This difference was more pronounced for peptides containing a larger number of amino groups, and demonstrates the potential for ionic liquids to finely tune TLC separation of peptides. It also provides further evidence that the ionic liquid modifiers themselves interact with analytes, as has been demonstrated in other chromatographic and electrophoretic techniques mentioned previously.

Santos et al. exploited the ability of ionic liquids to absorb UV in the development of a TLC-MS method to analyse low-molecular-weight alkaloids [115]. The ionic liquid was applied directly to the TLC spots in order to facilitate the use of matrix-assisted laser desorption (MALDI) analysis of low-molecular-weight analytes. While the ionic liquid in this study was not used in the application of TLC directly, it was used to bridge two complementary techniques for an application that was previously not possible, and suffered from high background noise.

4.9   SUMMARY AND FUTURE DIRECTIONS

Ionic liquids are a class of compounds that have proven to be very useful in separation science. When applied to liquid phase techniques, ionic liquids impart both modified chromatographic interactions and ion-pair interactions with analytes in the mobile phase, depending on their concentration. In addition, the charged nature of ionic liquids leads to good suppression of isolated silanols on the stationary phase surface, resulting in improved peak shapes or TLC spots. For CE separations, ionic liquids promote a reverse EOF, which greatly enhances the separation of analytes. Ionic liquids can be modified to be task specific, as in CO_2 capture by both taurinate–ionic liquids and even reversible ionic liquids. The low to negligible volatility of ionic liquids makes them attractive as matrices for GC stationary phases, and can separate polar analytes like a polar stationary phase and non-polar analytes like a non-polar stationary phase.

While the benefits of ionic liquids as applied to chromatography are great, they are not without challenge. The high viscosity of many ionic liquids preclude their use as mobile phases in HPLC and CCC due to the high pressures required to force them through the column. Detection for separations utilising ionic liquids can be complicated by the fact that many ionic liquids absorb in the UV region, are not volatile, and cannot be removed by evaporation prior to detection. To remove ionic liquids prior to detection, an additional separation or extraction must be performed. In addition, when ionic liquids are used in CE separations, the ions migrate independently, effectively destroying the ionic liquid during the electrophoretic run. This greatly reduces the ability to recycle ionic liquids for future runs. Currently, ionic liquids are fairly expensive to manufacture, and so their use in large volumes is not economically feasible unless they can be recycled efficiently.

While the challenges associated with the use of ionic liquids in separation science are real, the benefits far outweigh the problems. Work is underway in many laboratories to address the aforementioned issues, and to further characterise ionic liquids and investigate their roles in separation science. The viscosity of ionic liquids can be modified by the addition of small amounts of organic solvent to make them more amenable to application as liquid mobile phases. The thermal stability of ionic liquids composed of anions that generally impart low thermal stability is being improved through work on various immobilisation and mixing techniques. To address the tendency of ionic liquids to absorb UV radiation, electrochemical detection is being investigated for HPLC separations using ionic liquid mobile phase additives [31]. The retention mechanisms of ionic liquids in liquid chromatography are also under investigation [116, 117]. As the number of synthesised ionic liquids increases, each with unique attributes, the properties and characteristics of those ionic liquids with respect to separation science will need to be determined. As technology continues to improve, the use of ionic liquids for the mentioned separation methods will only become more beneficial and new methods will continue to be developed. In the field of separation science, ionic liquids are truly "COILed" for action.

REFERENCES

1 Schellnhuber, H.J., Global warming: stop worrying, start panicking? *Proc. Natl. Acad. Sci. U.S.A.* **105**(38), 14239–14240 (2008).

2 Rodríguez-Gil, J.L., Catalá, M., Alonso, S.G., Maroto, R.R., Valcárcel, Y., Segura, Y., Molina, R., Melero, J.A., and Martínez, F., Heterogeneous photo-fenton treatment for the reduction of pharmaceutical contamination in Madrid rivers and ecotoxicological evaluation by a miniaturized fern spores bioassay, *Chemosphere* **80**(4), 381–388 (2010).

3 Müller, L., Mauthe, R.J., Riley, C.M., Andino, M.M., De Antonis, D., Beels, C., DeGeorge, J., De Knaep, A.G.M., Ellison, D., Fagerland, J.A., Frank, R., Fritschel, B., Galloway, S., Harpur, E., Humfrey, C.D.N., Jacks, A.S., Jagota, N., Mackinnon, J., Mohan, G., Ness, D.K., O'Donovan, M.R., Smith, M.D., Vudathala, G., and Yotti, L., A rationale for determining, testing, and controlling specific impurities in pharmaceuticals that possess potential for genotoxicity, *Regul. Toxicol. Pharmacol.* **44**(3), 198–211 (2006).

4 Wilkes, J.S., A short history of ionic liquids—from molten salts to neoteric solvents, *Green Chem.* **4**(2), 73–80 (2002).

5 Wilkes, J.S., Levisky, J.A., Wilson, R.A., and Hussey, C.L., Dialkylimidazolium chloroaluminate melts—a new class of room-temperature ionic liquids for electrochemistry, spectroscopy, and synthesis, *Inorg. Chem.* **21**(3), 1263–1264 (1982).

6 Gale, R.J., and Osteryoung, R.A., Electrochemical reduction of pyridinium ions in ionic aluminum-chloride—alkylpyridinium halide ambient-temperature liquids, *J. Electrochem. Soc.* **127**(10), 2167–2172 (1980).

7 Gale, R.J., and Osteryoung, R.A., Potentiometric investigation of dialuminum heptachloride formation in aluminum chloride 1-butylpyridinium chloride mixtures, *Inorg. Chem.* **18**(6), 1603–1605 (1979).

8 Hurley, F.H., and Wier, T.P., Electrodeposition of metals from fused quaternary ammonium salts, *J. Electrochem. Soc.* **98**(5), 203–206 (1951).

9 Wilkes, J.S., and Zaworotko, M.J., Air and water stable 1-ethyl-3-methylimidazolium based ionic liquids, *J. Chem. Soc. Chem. Commun.* **13**, 965–967 (1992).

10 Fuller, J., Carlin, R.T., Delong, H.C., and Haworth, D., Structure of 1-ethyl-3-methylimidazolium hexafluorophosphate—model for room-temperature molten-salts, *J. Chem. Soc. Chem. Commun.* **3**, 299–300 (1994).

11 Poole, C.F., *The essence of chromatography*, Elsevier Science B. V., Amsterdam (2003), p. 138.

12 Armstrong, D.W., He, L.F., and Liu, Y.S., Examination of ionic liquids and their interaction with molecules, when used as stationary phases in gas chromatography, *Anal. Chem.* **71**(17), 3873–3876 (1999).

13 Abraham, M.H., Scales of solute hydrogen-bonding—their construction and application to physicochemical and biochemical processes, *Chem. Soc. Rev.* **22**(2), 73–83 (1993).

14   Revelli, A.L., Mutelet, F., and Jaubert, J.N., Partition coefficients of organic compounds in new imidazolium based ionic liquids using inverse gas chromatography,
J. Chromatogr. A **1216**(23), 4775–4786 (2009).

15   Sprunger, L.M., Gibbs, J., Proctor, A., Acree, W.E., Abraham, M.H., Meng, Y., Yao,
C., and Anderson, J.L., Linear free energy relationship correlations for room
temperature ionic liquids: revised cation-specific and anion-specific equation
coefficients for predictive applications covering a much larger area of chemical
space, *Ind. Eng. Chem. Res.* **48**(8), 4145–4154 (2009).

16   Mutelet, F., Revelli, A.L., Jaubert, J.N., Sprunger, L.M., Acree, W.E., Baker, G.A.
Partition coefficients of organic compounds in new imidazolium and tetralkylam-
monium based ionic liquids using inverse gas chromatography, *J. Chem. Eng. Data*
55(1), 234–242 (2010).

17   Yao, C., and Anderson, J.L., Retention characteristics of organic compounds on
molten salt and ionic liquid-based gas chromatography stationary phases, *J. Chromatogr. A* **1216**(10), 1658–1712 (2009).

18   Vervoort, R.J.M., Debets, A.J.J., Claessens, H.A., Cramers, C.A., and de Jong, G.J.,
Optimisation and characterisation of silica-based reversed-phase liquid chromatographic systems for the analysis of basic pharmaceuticals, *J. Chromatogr. A*
897(1–2), 1–22 (2000).

19   Nawrocki, J., The silanol group and its role in liquid chromatography, *J. Chromatogr. A* **779**(1–2), 29–71 (1997).

20   Berthod, A., and Carda-Broch, S., A new class of solvents for CCC: the room
temperature ionic liquids, *J. Liq. Chromatogr. Relat. Technol.* **26**(9–10), 1493–1508
(2003).

21   Marszall, M.P., and Kaliszan, R., Application of ionic liquids in liquid chromatography, *Crit. Rev. Anal. Chem.* **37**(2), 127–140 (2007).

22   Shetty, P.H., Youngberg, P.J., Kersten, B.R., and Poole, C.F., Solvent properties of
liquid organic salts used as mobile phases in microcolumn reversed-phase liquid-
chromatography, *J. Chromatogr.* **411**, 61–79(1987).

23   Waichigo, M.M., and Danielson, N.D., Ethylammonium formate as an organic
solvent replacement for ion-pair reversed-phase liquid chromatography, *J. Chromatogr. Sci.* **44**(10), 607–614 (2006).

24   Zheng, J., and Row, K.H., Effects of ionic liquid on the separation of 2-chlorophenol
and 2,4,6-trichlorophenol in RP-HPLC, *J. Chromatogr. Sci.* **47**(5), 392–395 (2009).

25   Xiao, X.H., Zhao, L., Liu, X., and Jiang, S.X., Ionic liquids as additives in high
performance liquid chromatography—analysis of amines and the interaction
mechanism of ionic liquids, *Anal. Chim. Acta* **519**(2), 207–211 (2004).

26   Kaliszan, R., Marszall, M.P., Markuszewski, M.J., Baczek, T., and Pernak, J., Suppression of deleterious effects of free silanols in liquid chromatography by imidazolium tetrafluoroborate ionic liquids, *J. Chromatogr. A* **1030**(1–2), 263–271
(2004).

27   He, L.J., Zhang, W.Z., Zhao, L., Liu, X., and Jiang, S.X., Effect of 1-alkyl-3-
methylimidazolium-based ionic liquids as the eluent on the separation of ephedrines by liquid chromatography, *J. Chromatogr. A* **1007**(1–2), 39–45 (2003).

28   Marszall, M.P., Baczek, T., and Kaliszan, R., Reduction of silanophilic interactions
in liquid chromatography with the use of ionic liquids, *Anal. Chim. Acta* **547**(2),
172–178 (2005).

29 Zhang, W.Z., He, L.J., Gu, Y.L., Liu, X., and Jiang, S.X., Effect of ionic liquids as mobile phase additives on retention of catecholamines in reversed-phase high-performance liquid chromatography, *Anal. Lett.* **36**(4), 827–838 (2003).

30 Waichigo, M.M., and Danielson, N.D., Comparison of ethylammonium formate to methanol as a mobile-phase modifier for reversed-phase liquid chromatography, *J. Sep. Sci.* **29**(5), 599–606 (2006).

31 Martín-Calero, A., Pino, V., Ayala, J.H., González, V., and Afonso, A.M., Ionic liquids as mobile phase additives in high-performance liquid chromatography with electrochemical detection: application to the determination of heterocyclic aromatic amines in meat-based infant foods, *Talanta* **79**(3), 590–597 (2009).

32 Holbrey, J.D., Reichert, W.M., Nieuwenhuyzen, M., Sheppard, O., Hardacre, C., and Rogers, R.D., Liquid clathrate formation in ionic liquid-aromatic mixtures, *Chem. Commun.* **4**, 476–477 (2003).

33 Zheng, J., and Row, K.H., Separation of guanine and hypoxanthine with some ionic liquids in RP-HPLC, *Am. J. Appl. Sci.* **3**(12), 2160–2166 (2006).

34 Ruiz-Angel, M.J., Carda-Broch, S., and Berthod, A., Ionic liquids versus triethylamine as mobile phase additives in the analysis of beta-blockers, *J. Chromatogr. A* **1119**(1–2), 202–208 (2006).

35 Pan, L., LoBrutto, R., Kazakevich, Y.V., and Thompson, R., Influence of inorganic mobile phase additives on the retention, efficiency and peak symmetry of protonated basic compounds in reversed-phase liquid chromatography, *J. Chromatogr. A* **1049**(1–2), 63–73 (2004).

36 Berthod, A., Ruiz-Angel, M.J., and Huguet, S., Nonmolecular solvents in separation methods: dual nature of room temperature ionic liquids, *Anal. Chem.* **77**(13), 4071–4080 (2005).

37 Sun, Y., Cabovska, B., Evans, C.E., Ridgway, T.H., and Stalcup, A.M., Retention characteristics of a new butylimidazolium-based stationary phase, *Anal. Bioanal. Chem.* **382**(3), 728–734 (2005).

38 Van Meter, D.S., Sun, Y.Q., Parker, K.M., and Stalcup, A.M., Retention characteristics of a new butylimidazolium-based stationary phase. Part II: anion exchange and partitioning, *Anal. Bioanal. Chem.* **390**(3), 897–905 (2008).

39 Qiu, H.D., Jiang, S.X., and Liu, X., N-methylimidazolium anion-exchange stationary phase for high-performance liquid chromatography, *J. Chromatogr. A* **1103**(2), 265–270 (2006).

40 Wang, Q., Baker, G.A., Baker, S.N., and Colon, L.A., Surface confined ionic liquid as a stationary phase for HPLC, *Analyst* **131**(9), 1000–1005 (2006).

41 Qiu, H.D., Jiang, S.X., Xia, L., and Zhao, L., Novel imidazolium stationary phase for high-performance liquid chromatography, *J. Chromatogr. A* **1116**(1–2), 46–50 (2006).

42 Chaves, A.R., Júnior, G.C., and Queiroz, M.E.C., Solid-phase microextraction using poly(pyrrole) film and liquid chromatography with UV detection for analysis of antidepressants in plasma samples, *J. Chromatogr. B* **877**(7), 587–593 (2009).

43 Brix, R., Bahi, N., de Alda, M.J.L., Farré, M., Fernandez, J.M., and Barceló, D., Identification of disinfection by-products of selected triazines in drinking water by LC-Q-ToF-MS/MS and evaluation of their toxicity, *J. Mass Spectrom.* **44**(3), 330–337 (2009).

44 Endo, Y., Tagiri-Endo, M., Seo, H.S., and Fujimoto, K., Identification and quantification of molecular species of diacyl glyceryl ether by reversed-phase high-performance liquid chromatography with refractive index detection and mass spectrometry, *J. Chromatogr. A* **911**(1), 39–45 (2001).

45 Savaliya, A.A., Shah, R.P., Prasad, B., and Singh, S., Screening of Indian aphrodisiac ayurvedic/herbal healthcare products for adulteration with sildenafil, tadalafil and/or vardenafil using LC/PDA and extracted ion LC-MS/TOF, *J. Pharm. Biomed. Anal.* **52**(3), 406–409 (2010).

46 Silva, M.M., Bergamasco, J., Lira, S.P., Lopes, N.P., Hajdu, E., Peixinho, S., and Berlinck, R.G.S., Dereplication of bromotyrosine-derived metabolites by LC-PDA-MS and analysis of the chemical profile of 14 aplysina sponge specimens from the Brazilian coastline, *Aust. J. Chem.* **63**(6), 886–894 (2010).

47 Varandas, S., Teixeira, M.J., Marques, J.C., Aguiar, A., Alves, A., and Bastos, M.M.S.M., Glucose and fructose levels on grape skin: interference in Lobesia botrana behaviour, *Anal. Chim. Acta* **513**(1), 351–355 (2004).

48 Koel, M., Physical and chemical properties of ionic liquids based on the dialkyl-imidazolium cation, *Proc. Estonian Acad. Sci. Chem.* **49**(3), 145–155 (2000).

49 Conway, W.D., *Counter-current chromatography: apparatus, theory, and applications*, VCH Publishers, Inc., New York (1990), pp. 1–4.

50 Berthod, A., Ruiz-Angel, M.J., and Carda-Broch, S., Elution-extrusion counter-current chromatography. Use of the liquid nature of the stationary phase to extend the hydrophobicity window, *Anal. Chem.* **75**(21), 5886–5894 (2003).

51 Costa, F.D., and Leitao, G.G., Strategies of solvent system selection for the isolation of flavonoids by counter-current chromatography, *J. Sep. Sci.* **33**(3), 336–347 (2010).

52 Du, Q.Z., Wu, C.J., Qian, G.J., Wu, P.D., and Ito, Y., Relationship between the flow-rate of the mobile phase and retention of the stationary phase in counter-current chromatography, *J. Chromatogr. A* **835**(1–2), 231–235 (1999).

53 Berthod, A., Ruiz-Angel, M.-J., and Carda-Broch, S., Ionic liquids as stationary phases in counter-current chromatography, in: *Ionic liquids in chemical analysis*, ed. M. Koel, CRC Press, Boca Raton (2009), pp. 211–228.

54 Berthod, A., and Carda-Broch, S., Use of the ionic liquid 1-butyl-3-methylimidazolium hexafluorophosphate in counter-current chromatography, *Anal. Bioanal. Chem.* **380**(1), 168–177 (2004).

55 Liu, R.M., Xu, L.L., Li, A.F., and Sun, A.L., Preparative isolation of flavonoid compounds from *Oroxylum indicum* by high-speed counter-current chromatography by using ionic liquids as the modifier of two-phase solvent system, *J. Sep. Sci.* **33**(8), 1058–1063 (2010).

56 Xu, L.L., Li, A.F., Sun, A.L., and Liu, R.M., Preparative isolation of neomangiferin and mangiferin from *Rhizoma anemarrhenae* by high-speed counter-current chromatography using ionic liquids as a two-phase solvent system modifier, *J. Sep. Sci.* **33**(1), 31–36 (2010).

57 Roth, M., Partitioning behaviour of organic compounds between ionic liquids and supercritical fluids, *J. Chromatogr. A* **1216**(10), 1861–1880 (2009).

58 Mellein, B.R., and Brennecke, J.F., Characterization of the ability of CO_2 to act as an antisolvent for ionic liquid/organic mixtures, *J. Phys. Chem. B* **111**(18), 4837–4843 (2007).

59 Mekki, S., Wai, C.M., Billard, I., Moutiers, G., Yen, C.H., Wang, J.S., Ouadi, A., Gaillard, C., and Hesemann, P., Cu(II) extraction by supercritical fluid carbon dioxide from a room temperature ionic liquid using fluorinated beta-diketones, *Green Chem.* **7**(6), 421–423 (2005).

60 Blanchard, L.A., and Brennecke, J.F., Recovery of organic products from ionic liquids using supercritical carbon dioxide, *Ind. Eng. Chem. Res.* **40**(1), 287–292 (2001).

61 Schäfer, T., Rodrigues, C.M., Afonso, C.A.M., and Crespo, J.G., Selective recovery of solutes from ionic liquids by pervaporation—a novel approach for purification and green processing, *Chem. Commun.* **17**, 1622–1623 (2001).

62 Earle, M.J., Esperança, J.M.S.S., Gilea, M.A., Lopes, J.N.C., Rebelo, L.P.N., Magee, J.W., Seddon, K.R., and Widegren, J.A., The distillation and volatility of ionic liquids, *Nature* **439**(7078), 831–834 (2006).

63 Hu, R.L., Dai, X.J., Lu, Y.B., and Pan, Y.J., Preparative separation of isoquinoline alkaloids from Stephania yunnanensis by pH-zone-refining counter-current chromatography, *J. Chromatogr. B* **878**(21), 1881–1884 (2010).

64 Duanmu, Q.P., Li, A.F., Sun, A.L., Liu, R.M., and Li, X.P., Semi-preparative high-speed counter-current chromatography separation of alkaloids from embryo of the seed of Nelumbo nucifera Gaertn by pH-gradient elution, *J. Sep. Sci.* **33**(12), 1746–1751 (2010).

65 Lu, Y., Hu, R., Dai, Z., and Pan, Y., Preparative separation of anti-oxidative constituents from Rubia cordifolia by column-switching counter-current chromatography, *J. Sep. Sci.* **33**, 2200–2205 (2010).

66 Cicek, S.S., Schwaiger, S., Ellmerer, E.P., and Stuppner, H., Development of a fast and convenient method for the isolation of triterpene saponins from Actaea racemosa by high-speed counter-current chromatography coupled with evaporative light scattering detection, *Planta Med.* **76**(5), 467–473 (2010).

67 Yoon, K.D., and Kim, J., Application of centrifugal partition chromatography coupled with evaporative light scattering detection for the isolation of saikosaponins-a and -c from Bupleurum falcatum roots, *J. Sep. Sci.* **32**(1), 74–78 (2009).

68 Cheng, Y.J., Liang, Q.L., Hu, P., Wang, Y.M., Jun, F.W., and Luo, G.A., Combination of normal-phase medium-pressure liquid chromatography and high-performance counter-current chromatography for preparation of ginsenoside-Ro from panax ginseng with high recovery and efficiency, *Sep. Purif. Methods* **73**, 397–402 (2010).

69 Yao, S., Liu, R.M., Huang, X.F., and Kong, L.Y., Preparative isolation and purification of chemical constituents from the root of Adenophora tetraphlla by high-speed counter-current chromatography with evaporative light scattering detection, *J. Chromatogr. A* **1139**(2), 254–262 (2007).

70 Liu, Z.L., Jin, Y., Shen, P.N., Wang, J., and Shen, Y.J., Separation and purification of verticine and verticinone from Bulbus Fritillariae Thunbergii by high-speed counter-current chromatography coupled with evaporative light scattering detection, *Talanta* **71**(5), 1873–1876 (2007).

71 Qi, X.C., Ignatova, S., Luo, G.A., Liang, Q.L., Jun, F.W., Wang, Y.M., and Sutherland, I., Preparative isolation and purification of ginsenosides Rf, Re, Rd and Rb1 from the roots of Panax ginseng with a salt/containing solvent system and flow step-gradient by high performance counter-current chromatography coupled

with an evaporative light scattering detector, *J. Chromatogr. A* **1217**(13), 1995–2001 (2010).

72   Inoue, K., Hattori, Y., Hino, T., and Oka, H., An approach to on-line electrospray mass spectrometric detection of polypeptide antibiotics of enramycin for high-speed counter-current chromatographic separation, *J. Pharm. Biomed. Anal.* **51**(5), 1154–1160 (2010).

73   Wybraniec, S., Jerz, G., Gebers, N., and Winterhalter, P., Ion-pair high-speed counter-current chromatography in fractionation of a high-molecular weight variation of acyl-oligosaccharide linked betacyanins from purple bracts of *Bougainvillea glabra*, *J. Chromatogr. B* **878**(5–6), 538–550 (2010).

74   Anderson, J.L., and Armstrong, D.W., High-stability ionic liquids. A new class of stationary phases for gas chromatography, *Anal. Chem.* **75**(18), 4851–4858 (2003).

75   Anderson, J.L., and Armstrong, D.W., Immobilized ionic liquids as high-selectivity/high-temperature/high-stability gas chromatography stationary phases, *Anal. Chem.* **77**(19), 6453–6462 (2005).

76   Han, X., and Armstrong, D.W., Ionic liquids in separations, *Acc. Chem. Res.* **40**(11), 1079–1086 (2007).

77   Huang, K., Han, X., Zhang, X., and Armstrong, D.W., PEG-linked geminal dicationic ionic liquids as selective, high-stability gas chromatographic stationary phases, *Anal. Bioanal. Chem.* **389**(7–8), 2265–2275 (2007).

78   Breitbach, Z.S., and Armstrong, D.W., Characterization of phosphonium ionic liquids through a linear solvation energy relationship and their use as GLC stationary phases, *Anal. Bioanal. Chem.* **390**(6), 1605–1617 (2008).

79   Payagala, T., Zhang, Y., Wanigasekara, E., Huang, K., Breitbach, Z.S., Sharma, P.S., Sidisky, L.M., and Armstrong, D.W., Trigonal tricationic ionic liquids: a generation of gas chromatographic stationary phases, *Anal. Chem.* **81**(1), 160–173 (2009).

80   Juvet, R.S., and Wachi, F.M., Gas chromatographic separation of metal halides by inorganic fused salt substrates, *Anal. Chem.* **32**, 290–291 (1960).

81   Hanneman, W.W., Spencer, C.F., and Johnson, J.F., Molten salt mixtures as liquid phases in gas chromatography, *Anal. Chem.* **32**(11), 1386–1388 (1960).

82   Pacholec, F., Butler, H.T., and Poole, C.F., Molten organic salt phase for gas-liquid-chromatography, *Anal. Chem.* **54**(12), 1938–1941 (1982).

83   Qiu, H.D., Wang, L.C., Liu, X., and Jiang, S.X., Preparation and characterization of silica confined ionic liquids as chromatographic stationary phases through surface radical chain-transfer reaction, *Analyst* **134**(3), 460–465 (2009).

84   Zhao, Q.C., and Anderson, J.L., Highly selective GC stationary phases consisting of binary mixtures of polymeric ionic liquids, *J. Sep. Sci.* **33**(1), 79–87 (2010).

85   Seeley, J.V., Seeley, S.K., Libby, E.K., Breitbach, Z.S., and Armstrong, D.W., Comprehensive two-dimensional gas chromatography using a high-temperature phosphonium ionic liquid column, *Anal. Bioanal. Chem.* **390**(1), 323–332 (2008).

86   Reid, V.R., Crank, J.A., Armstrong, D.W., and Synovec, R.E., Characterization and utilization of a novel triflate ionic liquid stationary phase for use in comprehensive two-dimensional gas chromatography, *J. Sep. Sci.* **31**(19), 3429–3436 (2008).

87   Lambertus, G.R., Crank, J.A., McGuigan, M.E., Kendler, S., Armstrong, D.W., and Sachs, R.D., Rapid determination of complex mixtures by dual-column gas

chromatography with a novel stationary phase combination and spectrometric detection, *J. Chromatogr. A* **1135**(2), 230–240 (2006).

88   Siegler, W.C., Crank, J.A., Armstrong, D.W., and Synovec, R.E., Increasing selectivity in comprehensive three-dimensional gas chromatography via an ionic liquid stationary phase column in one dimension, *J. Chromatogr. A* **1217**(18), 3144–3149 (2010).

89   Armstrong, D.W., Payagala, T., and Sidisky, L.M., The advent and potential impact of ionic liquid stationary phases in GC and GCxGC, *LC GC North Am.* **27**(8), 596–605 (2009).

90   Anderson, J.L., Ding, J., Welton, T., and Armstrong, D.W., Characterizing ionic liquids on the basis of multiple solvation interactions, *J. Am. Chem. Soc.* **124**(47), 14247–14254 (2002).

91   Anderson, J.L., Ding, R.F., Ellern, A., and Armstrong, D.W., Structure and properties of high stability geminal dicationic ionic liquids, *J. Am. Chem. Soc.* **127**(2), 593–604 (2005).

92   Payagala, T., Huang, J., Breitbach, Z.S., Sharma, P.S., and Armstrong, D.W., Unsymmetrical dicationic ionic liquids: manipulation of physicochemical properties using specific structural architectures, *Chem. Mater.* **19**(24), 5848–5850 (2007).

93   Zhao, Q., and Anderson, J.L., Highly selective GC stationary phases consisting of binary mixtures of polymeric ionic liquids, *J. Sep. Sci.* **33**(1), 79–87 (2010).

94   Baltazar, Q.Q., Leininger, S.K., and Anderson, J.L., Binary ionic liquid mixtures as gas chromatography stationary phases for improving the separation selectivity of alcohols and aromatic compounds, *J. Chromatogr. A* **1182**(1), 119–127 (2008).

95   Kim, Y.S., Choi, W.Y., Jang, J.H., Yoo, K.P., and Lee, C.S., Solubility measurement and prediction of carbon dioxide in ionic liquids, *Fluid Phase Equilib.* **228**, 439–445 (2005).

96   Cadena, C., Anthony, J.L., Shah, J.K., Morrow, T.I., Brennecke, J.F., and Maginn, E.J., Why is CO_2 so soluble in imidazolium-based ionic liquids? *J. Am. Chem. Soc.* **126**(16), 5300–5308 (2004).

97   Blanchard, L.A., Gu, Z.Y., and Brennecke, J.F., High-pressure phase behavior of ionic liquid/CO_2 systems, *J. Phys. Chem. B* **105**(12), 2437–2444 (2001).

98   Blanchard, L.A., Hancu, D., Beckman, E.J., and Brennecke, J.F., Green processing using ionic liquids and CO_2, *Nature* **399**(6731), 28–29 (1999).

99   Miyawaki, O., and Tatsuno, M., Lipase-catalyzed butanolysis of triolein in ionic liquid and selective extraction of product using supercritical carbon dioxide, *J. Biosci. Bioeng.* **105**(1), 61–64 (2008).

100   Machida, H., Sato, Y., and Smith, R.L., Measurement and correlation of infinite dilution partition coefficients of aromatic compounds in the ionic liquid 1-butyl-3-methyl-imidazolium hexafluorophosphate ([bmim][PF_6])-CO_2 system at temperatures from 313 to 353 K and at pressures up to 16 MPa, *J. Supercrit. Fluids* **43**(3), 430–437 (2008).

101   Planeta, J., Karasek, P., and Roth, M., Limiting partition coefficients of solutes in biphasic trihexyltetradecylphosphonium chloride ionic liquid-supercritical CO_2 system: measurement and LSER-based correlation, *J. Phys. Chem. B* **111**(26), 7620–7625 (2007).

102   Planeta, J., and Roth, M., Solute partitioning between the ionic liquid 1-n-butyl-3-methylimidazolium tetrafluoroborate and supercrifical CO2 from capillary-column chromatography, *J. Phys. Chem. B* **109**(31), 15165–15171 (2005).

103   Chou, F.M., Wang, W.T., and Wei, G.T., Using subcritical/supercritical fluid chromatography to separate acidic, basic, and neutral compounds over an ionic liquid-functionalized stationary phase, *J. Chromatogr. A* **1216**(16), 3594–3599 (2009).

104   Oda, R.P., and Landers, J.P., Introduction to capillary electrophoresis, in: *Handbook of capillary electrophoresis*, 2nd ed., ed. J.P. Landers, CRC Press, Boca Raton (1997), pp. 2–42.

105   Vaher, M., and Kaljurand, M., Ionic liquids as background electrolyte additives in capillary electrophoresis, in: *Ionic liquids in chemical analysis*, ed. M. Koel, CRC Press, Boca Raton (2009), p. 190.

106   Yanes, E.G., Gratz, S.R., Baldwin, M.J., Robison, S.E., and Stalcup, A.M., Capillary electrophoretic application of 1-alkyl-3-methylimidazolium-based ionic liquids, *Anal. Chem.* **73**(16), 3838–3844 (2001).

107   Yanes, E.G., Gratz, S.R., and Stalcup, A.M., Tetraethylammonium tetrafluoroborate: a novel electrolyte with a unique role in the capillary electrophoretic separation of polyphenols found in grape seed extracts, *Analyst* **125**(11), 1919–1923 (2000).

108   Mwongela, S.M., Numan, A., Gill, N.L., Agbaria, R.A., and Warner, I.M., Separation of achiral and chiral analytes using polymeric surfactants with ionic liquids as modifiers in micellar electrokinetic chromatography, *Anal. Chem.* **75**(22), 6089–6096 (2003).

109   Yue, M.E., and Shi, Y.P., Application of 1-alkyl-3-methylimidazolium-based ionic liquids in separation of bioactive flavonoids by capillary zone electrophoresis, *J. Sep. Sci.* **29**(2), 272–276 (2006).

110   François, Y., Varenne, A., Juillerat, E., Servais, A.C., Chiap, P., and Gareil, P., Non-aqueous capillary electrophoretic behavior of 2-aryl propionic acids in the presence of an achiral ionic liquid—a chemometric approach, *J. Chromatogr. A* **1138**(1–2), 268–275 (2007).

111   Tian, Y., Feng, R., Liao, L.P., Liu, H.L., Chen, H., and Zeng, Z.R., Dynamically coated silica monolith with ionic liquids for capillary electrochromatography, *Electrophoresis* **29**(15), 3153–3159 (2008).

112   López-Pastor, M., Simonet, B.M., Lendl, B., and Valcárcel, M., Ionic liquids and CE combination, *Electrophoresis* **29**(1), 94–107 (2008).

113   Sherma, J., Planar chromatography, *Anal. Chem.* **82**(12), 4895–4910 (2010).

114   Bączek, T., Marszall, M.P., Kaliszan, R., Walijewski, L., Makowiecka, W., Sparzak, B., Grzonka, Z., Wiśniewska, K., and Juszczyk, P., Behavior of peptides and computer-assisted optimization of peptides separations in a normal-phase thin-layer chromatography system with and without the addition of ionic liquid in the eluent, *Biomed. Chromatogr.* **19**(1), 1–8 (2005).

115   Santos, L.S., Haddad, R., Höehr, N.F., Pilli, R.A., and Eberlin, M.N., Fast screening of low molecular weight compounds by thin-layer chromatography and "on-spot" MALDI-TOF mass spectrometry, *Anal. Chem.* **76**(7), 2144–2147 (2004).

116   Nichthauser, J., and Stepnowski, P., Retention mechanism of selected ionic liquids on a pentafluorophenylpropyl polar phase: investigation using RP-HPLC, *J. Chromatogr. Sci.* **47**(3), 247–253 (2009).

117   Buszewski, B., Welerowicz, T., and Kowalkowski, T., Effect of chemically bonded stationary phases and mobile phase composition on beta-blockers retention in RP-HPLC, *Biomed. Chromatogr.* **23**(3), 324–333 (2009).

5 Separation Processes with Ionic Liquids

WYTZE (G. W.) MEINDERSMA and ANDRÉ B. DE HAAN
Department of Chemical Engineering and Chemistry Process Systems
Engineering, Eindhoven University of Technology, Eindhoven,
The Netherlands

ABSTRACT

Ionic liquids can be used to replace conventional solvents in liquid–liquid extraction. In order to replace a conventional separation process by a process that utilises ionic liquids, this process must be more economic, meaning a lower energy demand, lower use of feedstock, less waste, and so on. In general, it is difficult to change existing processes, except when the savings in the investments and running costs are substantial (at least 20%), and/or if the novel ionic liquid process can be carried out in existing equipment, with no or small changes. To apply a novel ionic liquid separation process is, in principle, easier than to retrofit, but the technology must be proven on a reasonable scale and during a prolonged period. Ionic liquids can be suitable extractants for several compounds, as is proven in batch equilibrium experiments. It is shown that a continuous extraction using ionic liquids in a pilot plant produces good results. There are no industrial applications for separations using ionic liquids yet. In order to introduce separation with ionic liquids in industry, more applied research is required, especially on pilot plant scale, to determine optimal process conditions.

The main challenges for applying ionic liquids in separations, and other processes for that matter, are the stability of the ionic liquid in the end, especially at higher temperatures, and regeneration or reuse of the ionic liquid in the process. However, many authors do not report the recovery of the product from the ionic liquid solution and/or the regeneration of the ionic liquids. These steps are essential parts of the whole separation process. Moreover, for

Ionic Liquids UnCOILed: Critical Expert Overviews, First Edition. Edited by Natalia V. Plechkova and Kenneth R. Seddon.
© 2013 John Wiley & Sons, Inc. Published 2013 by John Wiley & Sons, Inc.

some candidates that may perform well in the separation task under investigation, the reduction of their toxicity in the aquatic environment remains an important issue. In addition, the physical properties of ionic liquids must be determined in order to better predict their properties in different applications and for modelling of (separation) processes.

Task-specific ionic liquids are generally more selective than standard ionic liquids and, therefore, more focus must be on the development of these ionic liquids.

5.1 INTRODUCTION

Since ionic liquids have the reputation of being "green," they are of interest in many fields, including separations. A low or negligible vapour pressure, non-flammability and non-toxicity are often equated with greenness. However, some ionic liquids can be distilled, some can be flammable, or at least their decomposition products are, and some are certainly toxic. Therefore, the application of ionic liquids in environmentally benign separation processes is not straightforward and is assessed in the following.

The designer character of ionic liquids is claimed to be an advantage, which is certainly true, but conventional solvents or their mixtures can also be designed, and affinity extractants are molecularly tailored. Therefore, the designer character is not unique for ionic liquids and certainly not easier to understand.

However, ionic liquids do possess unique separation properties, enabling novel separation concepts and new products, and offer the opportunity for novel intensified processes. Ionic liquids can be used in separations because of their stability, non-volatility, adjustable miscibility, and polarity. Ionic liquids can be hydrophilic or hydrophobic, depending on the structure of their cations and anions. The anion seems more important in determining the water miscibility of ionic liquids [1]. For example, common 1,3-dialkylimidazolium ($[1\text{-}C_m\text{-}3\text{-}C_n\text{im}]^+$) salts with halide, ethanoate, nitrate, and trifluoroethanoate anions are fully miscible with water, while hexafluorophosphate ($[PF_6]^-$) or bis{(trifluoromethyl)sulfonyl}amide ($[NTf_2]^-$)-based 1,3-dialkylimidazolium salts are water immiscible, and tetrafluoroborate ($[BF_4]^-$) or trifluoromethane-sulfonate ($[OTf]^-$)-based 1,3-dialkylimidazolium salts can be totally miscible or immiscible, depending on the substituents on the cation [2]. Since the ionic liquids with $[PF_6]^-$ and $[NTf_2]^-$ anions are normally water immiscible, they are the ionic liquids of choice for a large number of investigations involving the formation of biphasic systems required for most ionic liquid extraction applications. However, $[PF_6]^-$ containing ionic liquids cannot be considered green: they are water-sensitive and will form hydrogen fluoride (HF) at higher temperatures [3], and ionic liquids with the $[NTf_2]^-$ anion are usually expensive.

Novel separation processes should possess one or more of the following properties, compared with conventional separation processes:

- More efficient production/less separation steps: higher selectivity
- Higher capacity
- Higher energy efficiency
- Broader operational windows
- Lower investments in the separation process
- Lower use of feedstock
- Reuse of the ionic liquids in the process
- Reduced amount of waste/emissions, direct or indirect, into the environment (atmosphere/water)

The advantages of ionic liquids in separation processes are:

- Tunable physicochemical properties.
- Wide temperature range.
- Easy regeneration by evaporating or stripping of the compounds separated.
- Negligible vapour pressure: no emissions into the atmosphere.

The disadvantages are:

- Often high viscosity.
- Higher price compared with standard organic solvents.
- Some ionic liquids have been found to be toxic in aquatic environments.

In order to replace a conventional separation process by a process that utilises ionic liquids, this process must be more economic, meaning a lower energy demand, lower use of feedstock, less waste, and so on. In general, it is difficult to change existing processes, except when the savings in the investments and running costs are substantial (at least 20%) and/or if the novel ionic liquid process can be carried out in existing equipment, with no (or small) changes. To apply a novel ionic liquid separation process is, in principle, easier than to retrofit, but the technology must be proven on a reasonable scale and during a prolonged period.

The main challenges for applying ionic liquids in separations, and other processes for that matter, are the stability of the ionic liquid in the end, especially at higher temperatures, and regeneration or reuse of the ionic liquid in the process. Moreover, for some candidates that may perform well in the separation task under investigation, the reduction of their toxicity in the aquatic environment remains an important issue. In addition, the physical properties of ionic liquids must be determined in order to better predict their properties in different applications and for modelling of (separation) processes.

5.2 LIQUID SEPARATIONS

5.2.1 Liquid–Liquid Extraction

Liquid–liquid extraction is based on the partial miscibility of liquids, and is used to separate a dissolved component from its solvent by its transfer into a second solvent. The equilibrium in liquid–liquid extraction may be characterised by the distribution coefficient, or partition coefficient, D_i, which is defined by the ratio of the concentrations, C_i, of solute i in the extract phase (E) and in the raffinate† phase (R), according to:

$$D_1 = C_1^E / C_1^R$$
$$D_2 = C_2^E / C_2^R \tag{5.1}$$

The selectivity, $S_{1/2}$, of species 1 over species 2, is defined as the ratio of the distribution coefficients of species 1 and species 2:

$$S_{1/2} = D_1 / D_2 = (C_1^E / C_1^R) / (C_2^E / C_2^R) \tag{5.2}$$

The definition of the distribution coefficient at infinite dilution, D_i^∞, is given by:

$$D_1^\infty = 1/\gamma_1^\infty \text{ and } D_2^\infty = 1/\gamma_2^\infty, \tag{5.3}$$

where γ_i^∞ is the activity coefficient at infinite dilution of species i.

The selectivity at infinite dilution, $S_{1/2}^\infty$, of species 1 over species 2 is defined as the ratio of the distribution coefficients at infinite dilution of species 1 and species 2:

$$S_{1/2}^\infty = D_1^\infty / D_2^\infty = \gamma_2^\infty / \gamma_1^\infty \tag{5.4}$$

5.2.2 Metal Extraction

5.2.2.1 Conventional Process Conventional processes to recover metal oxides are based upon dissolution in mineral acids and bases, followed by extraction with a wide variety of organic solvents or their mixtures. In traditional solvent extraction technologies, adding extractants that reside quantitatively in the extracting phase increases the metal ion partitioning to the more hydrophobic phase. The added extractant molecules dehydrate the metal ions and provide a more hydrophobic environment enabling their transport to the extracting phase. The use of solvent combinations with specific extractants, or chelating agents or ligands, improves their combinatorial extractant performance on separation factors of a particular species. Commercial extractants,

† The portion of an original liquid that remains after other components have been dissolved by a solvent.

such as Alamine 336 and Aliquat 336, whose major components are ternary aliphatic amines and quaternary ammonium salts, respectively, have been used in solvent extraction of metals to a large extent [4]. In addition, crown ethers (cyclic polyether molecules) have been intensively investigated for metal extraction in liquid–liquid separations [5].

5.2.2.2 Liquid–Liquid Extraction with Ionic Liquids

Due to the hydrophobic character of some ionic liquids, it is possible to extract hydrophobic compounds in biphasic separations. In order to separate metal ions from the aqueous phase into hydrophobic ionic liquids, again extractants are required to form complexes to increase the hydrophobicity of the metal compounds [6–8]. The extractants used are neutral compounds, such as crown ethers [6, 7, 9–11], calix[4]arenes [11–16], or tributylphosphate [17]; acidic or anionic extractants, such as organophosphorus acids and pseudohalides. Furthermore, task-specific ionic liquids, where the ionic liquid is both solvent and extractant, have been recently developed [18].

Dai et al. [6] stated that, from a thermodynamic perspective, the solvation of ionic species, such as crown ether complexes, in ionic liquids should be much more favoured than in conventional solvent extractions. This is one of the key advantages of using ionic liquids in separations involving ionic species. Dai et al. used the crown ether dicyclohexyl-18-crown-6 (DC18C6) for extraction of Sr^{2+} from an aqueous solution because DC18C6 is known to form strong complexes with Sr^{2+}. Without the use of an extractant, the D_{Sr} values are below 1 mole per mole, varying from not measurable with toluene or trichloromethane to 0.89 mole per mole for the ionic liquid 1-butyl-3-methylimidazolium hexafluorophosphate ([C_4mim][PF_6]). The distribution coefficients of Sr^{2+} with 1,3-dialkylimidazolium ionic liquids with [PF_6]$^-$ and [NTf_2]$^-$ as anions with the crown ether were in the range of 24 to 11,000 mole per mole, which is exceptionally high compared with the distribution coefficients in organic solvents with this crown ether, such as trichloromethane ($D = 0.77$ mole per mole) and toluene ($D = 0.76$ mole per mole).

To achieve optimal extraction efficiency, there are several controlling factors [19]:

- Varying the structure of an ionic liquid (especially the side chain of its cation) to change its hydrophobicity can improve the distribution coefficients of metal ions [7, 20–22].
- The type of extractant (such as crown ethers) can be modified to achieve optimal selectivity for a specific application [23].
- The extraction efficiency of metal complexes can also be controlled by the pH of the system [24].

There is a variety of extraction mechanisms possible in metal extraction with ionic liquids, that is, solvent ion-pair extraction, ion exchange, and combinations of these [25, 26].

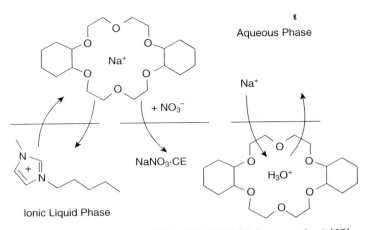

Figure 5.1   Extraction of Na$^+$ with DC18C6 (crown ether) [27].

Partitioning of sodium ions between aqueous nitrate media and 1-alkyl-3-methylimidazolium bis{(trifluoromethyl)sulfonyl}amide ([C$_n$mim][NTf$_2$]) in the presence of DC18C6 takes place via as many as three pathways: conventional nitrato complex extraction and/or one or two ion exchange processes, the relative importance of which is determined by the acidity of the aqueous phase and the hydrophobicity of the ionic liquid cation. Increasing the alkyl chain length of the ionic liquid cation is not always sufficient to eliminate the possibility of ion exchange as a mode of metal ion partitioning between the two phases, with negative implications for the utility of ionic liquids as environmentally benign extraction solvents [27]. Figure 5.1 shows the extraction of Na$^+$ from an aqueous phase.

The extraction of radioactive metals (lanthanides and actinides) has particular industrial significance in metal extraction for the handling of nuclear materials [19].

Octyl(phenyl)-*N,N*-diisobutylcarbamoylmethyl phosphine oxide (CMPO) dissolved in [C$_4$mim][PF$_6$] showed extremely high extraction ability and selectivity of metal ions, as compared with the ordinary diluents, dodecane. Partitioning of the metal cations into the ionic liquid with octyl(phenyl)-*N,N*-diisobutylcarbamoylmethylphosphine oxide (CMPO) appears to proceed via a cation-exchange mechanism, which is quite different from that of the conventional solvent extraction system [28]. The metal was successfully stripped from the ionic liquid.

Selective extraction–separation of yttrium(III) from heavy lanthanides into 1-octyl-3-methylimidazolium hexafluorophosphate ([C$_8$mim][PF$_6$]) containing Cyanex 923† was achieved by adding a water-soluble complexing agent

---

† CYANEX 923 is a mixture of four trialkylphosphine oxides: R$_3$P = O, R$_2$R′P = O, RR$_2'$P = O and R$_3'$P = O, where R = octyl and R′ = hexyl.

(ethylenediaminetetraacetic acid) to the aqueous phase. The selectivity of the $[C_n mim][PF_6]/[NTf_2]$-based extraction system was enhanced to about 4.3 without increasing the loss of $[C_n mim]^+$ [29].

Dietz et al. have shown that the mode of metal ion extraction by crown ethers can be shifted from cation exchange to extraction of a neutral ion-pair complex by increasing the lipophilicity of the ionic liquid cation [22, 23, 30, 31]. Formation of such neutral metal–crown ether–anion complexes is the conventional transport mechanism with crown ether extractants in organic solvents. In order to minimise the loss of ionic liquid, the ionic liquid cation can be added to the initial metal-containing solutions before extraction. Moreover, solutions with the ionic liquid anion are used for stripping. Several recycles were carried out for the extraction of cerium(IV) [32, 33].

The mutual solubility of ionic liquids and water (and/or aqueous acid solutions) determines the suitability of ionic liquids as replacements for conventional molecular solvents in applications for the removal of metal ions. The solubility of water in ionic liquids can be increased by the presence of a neutral ligand, such as a crown ether, capable of extracting significant quantities of acid. The primary problem with cation and anion exchange equilibria in ionic liquid-based systems is that ionic liquid components are transferred to the aqueous phase. This process pollutes the water phase and this can make it difficult to subsequently recover the extracted metal ion [34]. Both cations and anions of the ionic liquids can be exchanged into the aqueous phase [35]. An extraction system in which metal ion partitioning is accompanied by increased dissolution of the ionic liquid in the aqueous phase cannot be regarded as "green" [36]. The process involved in the transfer of the metal ion into an ionic liquid in the presence of a crown ether is surprisingly complex. In principle, such complexity can provide a large number of opportunities for the design of ionic liquid-based extraction systems with improved efficiency and selectivity. However, the full potential of ionic liquids as environmentally benign extraction solvents will clearly not be realised until methods are devised to reduce or eliminate the aqueous dissolution of the ionic liquid that can accompany metal ion partitioning [27]. These are currently major limitations for using ionic liquids in metal ion separation.

5.2.2.3 Liquid–Liquid Extraction with Functionalised (Task-Specific) Ionic Liquids

Functionalised (or task-specific) ionic liquids are designed to have targeted functionality by attaching a metal ion-coordinating group directly onto the ionic liquid cation, which makes the extractant an integral part of the hydrophobic phase and thereby reduces the chance of ionic liquid loss to the aqueous phase. Therefore, functionalised ionic liquids act as hydrophobic solvents and extractants at the same time. However, the cost of functionalised ionic liquids is generally high, as their preparation usually requires multi-step organic syntheses. In order to eliminate this drawback, functionalised ionic liquids may be added to mixtures of less expensive ionic liquids.

Visser et al. [37] designed and synthesised several ionic liquids by appending different functional groups (namely thioether, urea, and thiourea) to imidazolium cations to remove mercury(II) and cadmium(II) from contaminated water. The functionalised ionic liquid cations were combined with the $[PF_6]^-$ anion and used alone, or in a mixture with $[C_4mim][PF_6]$. The distribution coefficients for mercury(II) and cadmium(II) in liquid–liquid separations with the thioether-functionalised ionic liquid were high: 220 mole per mole for mercury(II) and 330–375 mole per mole for cadmium(II), depending on the pH value of the aqueous phase. With the thiourea-appended functionalised ionic liquid, distribution coefficients of 345 mole per mole for mercury(II) and 20 mole per mole for cadmium(II) were obtained. The separation of the metal ions from the functionalised ionic liquids was not reported.

Ammonium bistriflamide salts with carboxylic acid functionalities show the ability to dissolve large quantities of metal oxides. This metal-solubilising power is selective. The oxides of the trivalent rare earths, as well as uranium(VI) oxide, zinc(II) oxide, cadmium(II) oxide, mercury(II) oxide, nickel(II) oxide, copper(II) oxide, palladium(II) oxide, lead(II) oxide, manganese(II) oxide, and silver(I) oxide, are soluble in the ionic liquid. Iron(III), manganese(IV) and cobalt oxides, as well as aluminium(III) oxide and silicon(IV) oxide, are insoluble or very poorly soluble. The metals can be stripped from the ionic liquid by treatment of the ionic liquid with an acidic aqueous solution. After transfer of the metal ions to the aqueous phase, the ionic liquid can be recycled for re-use [38].

Chelate extraction can be used for separating several metal ions from aqueous solutions [39–41]. 1,3-Dialkylimidazolium salts with appended ethylaminodiacetic acid moieties (employed as a di-*tert*-butyl ester) can be used for the formation of metal chelates with copper(II), nickel(II), and cobalt(II) in aqueous solutions [41]. Figure 5.2 shows that an increase in the length of the N-alkyl chain improves the hydrophobicity: complex (d) ($R = C_{10}H_{21}$) is insoluble in water and precipitates after the chelate formation is completed. The efficiency of the extraction is 93%. However, the recovery of the metal from the complex and the regeneration of the complex were not reported.

The separation of the noble organometallic Wilkinson's (W) and Jacobsen's (J) catalysts from a homogeneous organic phase is possible by extraction with functionalised ionic liquids [42], while with non-functionalised 1-ethyl-3-methylimidazolium tetrafluoroborate ($[C_2mim][BF_4]$), no extraction of the catalysts was observed. The best functionalised ionic liquids were those with an amino acid-based anion, glycinate (Gly) or methionate (Met): $[C_4mim]$[Gly] ($D_W = 1.7$ kg per kg, $D_J = 71$ kg per kg), 1-hexyl-3-methylimidazolium glycinate ($[C_6mim][Gly]$) ($D_W = 8.8$ kg per kg, $D_J = 889$ kg per kg), and [(allyl)mim][Met] ($D_W = 76$ kg per kg, $D_J = 66.4$ kg per kg). The recovery of the catalysts from the ionic liquid solution is still under investigation.

5.2.2.4 Conclusions The removal of metal ions from aqueous solutions with ionic liquids is not straightforward. Ion exchange is not a suitable process,

Figure 5.2   Formation of copper(II) complexes with a functionalised ionic liquid [41].

as either the ionic liquid cation or the ionic liquid anion will end up in the aqueous phase. The aqueous dissolution of the ionic liquid that can accompany metal ion partitioning must be avoided or largely eliminated. Therefore, further improvements of the ionic liquid process for the separation of metals are required. One option is to design functionalised ionic liquids. Additionally, many authors did not report the recovery of the metals from the ionic liquid solution and/or the regeneration of the ionic liquids. However, these steps are essential parts of the whole separation process.

5.2.3   Extraction of Aromatic Hydrocarbons

5.2.3.1   *Conventional Process for Aromatic/Aliphatic Separation*   The separation of aromatic hydrocarbons (benzene, toluene, ethyl benzene, and xylenes) from C_4–C_{10} aliphatic hydrocarbon mixtures is challenging since these hydrocarbons have boiling points in a close range, and several combinations form azeotropes. Conventional processes for the separation of these aromatic and aliphatic hydrocarbon mixtures are solvent extraction, suitable for the range of 20–65 wt% aromatic content, extractive distillation for the range of 65–90 wt% aromatics, and azeotropic distillation for high aromatic content, >90 wt% [43]. Typical solvents used are polar components such as sulfolane

[44–51], 1-methylpyrrolidone (NMP) [47, 52], *N*-formylmorpholine (NFM) [50, 53], ethylene glycol [48, 54, 55], or propylene carbonate [56]. This implies additional distillation steps to separate the extraction solvent from both the extract and raffinate phases and to purify the solvent, which leads consequently to additional investments and higher energy consumption. The costs of the regeneration of sulfolane are high since sulfolane, which has a boiling point of 287.3 °C, is in the current aromatic/aliphatic separation process taken overhead from the regenerator and returned to the bottom of the aromatics stripper as a vapour [57]. Overviews of the use of extraction and extractive distillation for the separation of aromatic hydrocarbons from aliphatic hydrocarbons can be found elsewhere [58–61].

5.2.3.2 *Extraction with Ionic Liquids*

The application of ionic liquids for extraction processes is promising because of their negligibly low vapour pressure [62, 63]. This facilitates solvent recovery using techniques as simple as flash distillation or stripping. Thus, extraction of aromatics from mixed aromatic/aliphatic streams with ionic liquids is expected to require fewer process steps and a lower energy consumption than extraction with conventional solvents, provided that the mass-based aromatic distribution coefficient and/or the aromatic/aliphatic selectivity are higher than those of the current state-of-the-art solvents such as sulfolane [64, 65]. The solvent sulfolane is used as a benchmark for the separation of aromatic and aliphatic hydrocarbons because it is one of the most widely used solvents for this separation in industry. A large number (121) of ionic liquids was evaluated for the aromatic/aliphatic extraction [66]. Only 32 ionic liquids showed a higher mol-based aromatic distribution coefficient and a higher aromatic/aliphatic selectivity for at least one aromatic/aliphatic separation (benzene/hexane, toluene/heptane, or *p*-xylene/octane) than sulfolane, and 13 ionic liquids, in boldface in Table 5.1, showed also a higher mass-based distribution coefficient than sulfolane. The results of the extraction with these ionic liquids for these separations are shown in Table 5.1. However, mass-based distribution coefficients are used in industry and not on mole basis because the weight amount of a solvent determines the solvent-to-feed ratio in an extraction column and, hence, the costs of an extraction. If the mass-based distribution coefficient or the selectivity of the ionic liquids is lower than the ones for sulfolane, replacement of this solvent is not considered feasible. Several of the 13 ionic liquids in Table 5.1 that show a higher mass-based aromatic distribution coefficient and a higher aromatic/aliphatic selectivity are not suitable because of their halogen-containing anions (reaction with water: $[AlCl_4]^-$ or corrosion: $[I_3]^-$). Therefore, of the systems summarised in Table 5.1, only four ionic liquids are suitable for the separation of the three aromatic and aliphatic hydrocarbon systems considered (benzene/hexane, toluene/heptane, and *p*-xylene/octane): 1-butyl-3-methylimidazolium tricyanomethanide ($[C_4mim][C(CN)_3]$), 1-butyl-3-methylpyridinium dicyanamide ($[C_4m_\beta py][N(CN)_2]$), 1-butyl-3-methylpyridinium tricyanomethanide ($[C_4m_\beta py][C(CN)_3]$), and 1-butyl-3-methylpyridinium tetracyanoborate ($[C_4m_\beta py][B(CN)_4]$).

TABLE 5.1 Results of Extraction of Aromatic Hydrocarbons with Ionic Liquids

Solvent	Separation	T/°C	% arom	D_{arom}/mol per mol	D_{arom}/kg per kg	$S_{arom/aliph}$	Ref.
Sulfolane	Benzene/hexane	50	7.9	0.58	0.43	28.5	[50]
	Toluene/heptane	40	5.9	0.31	0.26	30.9	[67, 68]
[C$_2$mim][C$_2$H$_5$SO$_4$]	p-Xylene/octane	25	14.3	0.27	0.26	24.9	[51]
	Benzene/hexane	40	6.5	0.59	0.22	66.4	[69]
	Toluene/heptane	40	10.7	0.22	0.10	50.5	[67, 68]
[C$_2$mim][NTf$_2$]	Benzene/hexane	40	12.3	1.20	0.31	27.8	[70]
	Toluene/heptane	25	6.4	0.83	0.22	27.7	[71]
	p-Xylene/octane	25	6.3	0.48	0.15	32.3	[72]
[C$_2$mim][I$_3$]	Toluene/heptane	45	7.5	0.84	0.17	48.6	[73]
[C$_2$mim][AlCl$_4$]	**Toluene/heptane**	**40**	**4.5**	**1.80**	**0.69**	**69.6**	**[74]**
[C$_4$mim][BF$_4$]	Benzene/hexane	40	–	0.57	0.23	32.7	[75]
	Toluene/heptane	40	7.8	0.34	0.15	46.1	[67]
[C$_4$mim][CH$_3$SO$_4$]	Benzene/hexane	40	5.8	0.81	0.30	46.4	[76]
	Toluene/heptane	40	8.6	0.33	0.14	31.2	[68]
[C$_4$mim][I$_3$]	**Toluene/heptane**	**35**	**17.0**	**2.32**	**0.71**	**30.1**	**[73]**
[C$_4$mim][N(CN)$_2$]	Benzene/hexane	30	2.0	0.75	0.32	30.5	[66]
	Toluene/heptane	**40**	**10.0**	**0.63**	**0.32**	**59.0**	**[77]**
[C$_4$mim][C(CN)$_3$]	**Benzene/hexane**	**30**	**2.0**	**1.54**	**0.61**	**32.3**	**[66]**
	Toluene/heptane	**40**	**10.0**	**0.85**	**0.39**	**48.6**	**[66]**
	p-Xylene/octane	**40**	**10.0**	**0.60**	**0.31**	**23.1**	**[66]**
[C$_4$mim][SCN]	Benzene/hexane	30	2.0	0.70	0.31	55.4	[66]
	Toluene/heptane	**40**	**10.0**	**0.50**	**0.26**	**65.8**	**[77]**
[C$_4$mim][AlCl$_4$]	**Toluene/heptane**	**40**	**6.7**	**1.57**	**0.57**	**35.7**	**[74]**
[C$_4$mim][Al$_2$Cl$_7$]	**Benzene/heptane**	**20**	**10.0**	**1.96**	**0.53**	**80.0**	**[78]**

(*Continued*)

TABLE 5.1 (Continued)

Solvent	Separation	T/°C	% arom	D_{arom}/mol per mol	D_{arom}/kg per kg	$S_{arom/aliph}$	Ref.
[C$_2$py][NTf$_2$]	Benzene/hexane	40	–	1.24	0.30	29.7	[79]
	Toluene/heptane	40	–	0.52	0.14	24.7	[80]
	p-Xylene/octane	40	–	0.38	0.11	27.7	[80]
[C$_4$py][BF$_4$]	Toluene/heptane	40	7.7	0.43	0.20	74.4	[81]
[C$_4$m$_\alpha$py][BF$_4$]	Toluene/heptane	40	5.6	0.52	0.23	60.0	[82]
[C$_4$m$_\beta$py][BF$_4$]	Benzene/hexane	30	2.0	0.67	0.25	36.2	[66]
	Toluene/heptane	40	10.0	0.54	0.24	51.5	[66]
[C$_4$m$_\beta$py][N(CN)$_2$]	**Benzene/hexane**	**30**	**6.6**	**1.45**	**0.62**	**54.3**	[77]
	Toluene/heptane	**40**	**10.0**	**0.86**	**0.41**	**44.7**	[77]
	p-Xylene/octane	**40**	**10.0**	**0.54**	**0.30**	**25.2**	[66]
[C$_4$m$_\beta$py][C(CN)$_3$]	**Benzene/hexane**	**30**	**2.0**	**1.84**	**0.70**	**34.8**	[66]
	Toluene/heptane	**40**	**10.0**	**1.12**	**0.51**	**34.8**	[66]
	p-Xylene/octane	**40**	**10.0**	**0.83**	**0.42**	**25.8**	[66]
[C$_4$m$_\beta$py][B(CN)$_4$]	**Benzene/hexane**	**30**	**2.0**	**1.92**	**0.74**	**27.0**	[66]
	Toluene/heptane	**30**	**10.0**	**1.48**	**0.64**	**38.6**	[66]
	p-Xylene/octane	**40**	**10.0**	**1.09**	**0.51**	**22.0**	[66]
[C$_4$m$_\gamma$py][BF$_4$]	Benzene/hexane	40	10.6	0.95	0.39	55.9	[68, 83]
	Toluene/heptane	40	10.0	0.43	0.21	50.0	[68, 83]
	m-Xylene/octane	40	8.1	0.36	0.18	42.7	[68, 83]
[C$_4$m$_\gamma$py][CH$_3$SO$_4$]	Toluene/heptane	40	8.1	0.61	0.24	42.3	[67]
[C$_4$m$_\gamma$py][N(CN)$_2$]	Benzene/hexane	30	2.0	0.83	0.34	34.3	[66]
	Toluene/heptane	**40**	**10.0**	**0.63**	**0.31**	**35.1**	[66]
[C$_4$m$_\gamma$py][SCN]	Benzene/hexane	30	2.0	0.67	0.28	45.2	[66]
	Toluene/heptane	40	10.0	0.51	0.25	45.0	[66]

[C₃OHpy][(C₂F₅)₃PF₃]	Benzene/hexane	45	—	1.09	0.18	36.5	[84]
	Toluene/heptane	45	—	0.75	0.14	37.5	[84]
	p-Xylene/octane	45	—	0.50	0.10	36.9	[84]
[C₄mpyr][CF₃SO₃]	Benzene/hexane	45	—	0.68	0.21	31.8	[85]
	Toluene/heptane	45	—	0.45	0.16	26.7	[85]
	p-Xylene/octane	45	—	0.30	0.12	23.7	[85]
[C₄mpyr][N(CN)₂]	Benzene/hexane	30	2.0	0.71	0.30	15.4	[66]
	Toluene/heptane	**40**	**10.0**	**0.55**	**0.27**	**42.0**	[66]
[C₄mpyr][SCN]	Benzene/hexane	30	2.0	0.73	0.32	17.3	[66]
	Toluene/heptane	**40**	**10.0**	**0.53**	**0.27**	**47.7**	[66]
[N₁₁₁₂OH][NTf₂]ᵃ	Benzene/hexane	25	17.0	1.24	0.33	20.5	[86]
	m-Xylene/octane	25	10.0	0.30	0.09	27.0	[86]
[Me₃NH] [Cl-2.0AlCl₃]	**Benzene/heptane**	**20**	**10.0**	**1.54**	**0.70**	**35.0**	[78]
[Et₃NH][Cl-2.0AlCl]	**Benzene/heptane**	**20**	**10.0**	**1.63**	**0.47**	**35.0**	[78]
[N₁₁₁₂OH][NTf₂]	Benzene/hexane	40	12.0	0.67	0.16	41.9	[79]
[S₂₂₂][NTf₂]	Benzene/hexane	30	2.0	1.23	0.28	32.2	[66]
	Toluene/heptane	45	—	0.63	0.17	21.2	[87]
	p-Xylene/octane	45	—	0.43	0.13	20.8	[87]

ᵃ ethyl(2-hydroxyethyl)dimethylammonium bistriflamide.

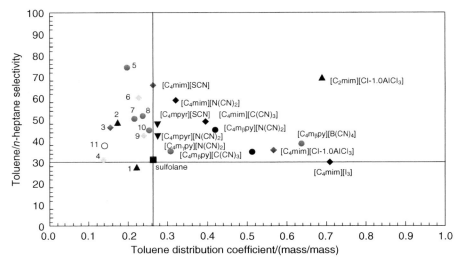

Figure 5.3 Toluene/heptane separation with ionic liquids, ~10 wt% toluene, T ~40°C. (1) $[C_2mim][NTf_2]$, (2) $[C_2mim]I_3$, (3) $[C_4mim][BF_4]$, (4) $[C_4mim][CH_3SO_4]$, (5) $[C_4py]$ $[BF_4]$, (6) $[C_4m_\alpha py][BF_4]$, (7) $[C_4m_\beta py][BF_4]$, (8) $[C_4m_\gamma py][BF_4]$, (9) $[C_4m_\gamma py][CH_3SO_4]$, (10) $[C_4m_\gamma py][SCN]$, (11) $[N-C_3OHpy][(C_2F_5)_3PF_3]$.

Figure 5.3 shows the ionic liquids for the toluene/heptane separation with a higher mol-based toluene distribution coefficient and a higher toluene/heptane selectivity. In this figure, the aromatic/aliphatic selectivity is shown as function of the mass-based distribution coefficient of the aromatic compound.

There are considerable differences in values of the distribution coefficients and the selectivities reported for same separation obtained through activity coefficients at infinite dilution and actual liquid–liquid equilibrium (LLE) experiments. The mol-based aromatic distribution coefficients measured with LLE experiments are generally higher than those obtained through activity coefficients at infinite dilution. The average difference is a factor of 1.5 with a standard deviation of 0.6. The selectivities measured with LLE experiments are generally higher than those calculated with activity coefficients at infinite dilution. The average difference between the two values is a factor of 1.6, but the standard deviation (s.d.) is higher than for the difference in distribution coefficient, s.d. = 1.6, which is to be expected as the selectivity is a ratio of the aromatic and aliphatic distribution coefficients [66].

5.2.3.3 Economic Evaluation An economic evaluation was made for the separation of aromatic compounds from the feed of a naphtha cracker with several ionic liquids ($[C_2mim][C_2H_5SO_4]$, 1-butyl-4-methylpyridinium tetrafluoroborate ([C$_4$m$_\gamma$py][BF$_4$]), [C$_4$m$_\gamma$py][CH$_3$SO$_4$], [C$_4$mim][AlCl$_4$], [C$_2$mim][AlCl$_4$] and later [C$_4$m$_\beta$py][N(CN)$_2$]) and compared to that with sulfolane [88, 89]. The separation of toluene from a mixed toluene/heptane stream was used to model the aromatic/aliphatic separation. Most ethylene cracker feeds contain 10–25% of aromatic components, depending on the source of the feed (naphtha or gas condensate). The aromatic compounds are not converted to olefins and, in fact, small amounts are formed during the cracking process in the cracker furnaces [90]. Therefore, they occupy a part of the capacity of the furnaces and put an extra load on the separation section of the stream containing C$_5$–C$_{10}$ aliphatic compounds. If a major part of the aromatic compounds present in the feed to the crackers could be separated upstream of the furnaces, it would offer several advantages: higher capacity, higher thermal efficiency, and less fouling. The improved margin will be around €20 per ton of feed or €48,000,000 per year for a naphtha cracker with a feed capacity of 300 ton h^{-1}, due to lower operational costs. Figure 5.4 shows a process scheme for the extraction of aromatic hydrocarbons from a naphtha feed.

For a naphtha feed of 300 ton h^{-1} containing about 10% aromatic hydrocarbons, the total investment costs in the sulfolane extraction were estimated to be about €86,000,000 and with [C$_4$m$_\gamma$py][BF$_4$], about €56,000,000, including an ionic liquid inventory of €20,000,000. In the calculations, an ionic liquid price of €20 per kg was used since German chemical company BASF has indicated that it is indeed possible to reach a level of €10–25 per kg of ionic

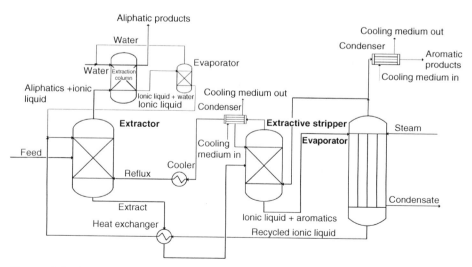

Figure 5.4 Conceptual flow scheme for the separation of aromatic and aliphatic hydrocarbons.

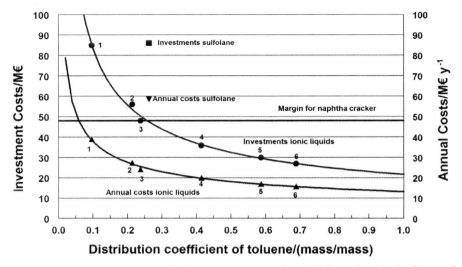

Figure 5.5 Investment and variable costs for extraction with ionic liquids. (1) $[C_2mim]$ $[C_2H_5SO_4]$, (2) $[C_4m_\gamma py][BF_4]$, (3) $[C_4m_\gamma py][CH_3SO_4]$, (4) $[C_4m_\beta py][N(CN)_2]$, (5) $[C_4mim][AlCl_4]$, (6) $[C_2mim][AlCl_4]$ [88, 89].

liquid with production on a large scale [91–93]. After this calculation was carried out, a more suitable ionic liquid was found by using **CO**nductorlike-**S**creening**MO**del for Real Solvents (COSMO-RS): $[C_4m_\beta py][N(CN)_2]$ with $D_{tol} = 0.41$ kg per kg and $S_{tol/hept} = 44.7$.

The lower investment in the ionic liquid process results mainly from the fact that the regeneration of the ionic liquid is much simpler than that of sulfolane. Since also the energy costs are lower, the total annual costs with the $[C_4m_\gamma py][BF_4]$ process are estimated to be €27,400,000, compared with €58,400,000 for sulfolane. The investment and annual costs for the separation of 10% aromatics from a cracker feed with sulfolane and ionic liquids are shown in Figure 5.5. The loss of ionic liquid to the raffinate phase is minimal, estimated to be 0.006%. This seems to be a very small amount, but for a cracker with a capacity of 300 ton h^{-1}, this amounts to 140 ton of ionic liquid per annum. With the cost price of €20 per kg, this amounts to a loss of €2,800,000 per annum. Furthermore, it is unknown what will happen in the cracker with the ionic liquid and where the ionic liquid, or its decomposition products, will end up. However, the ionic liquid can be recovered from the raffinate with a simple one-stage extraction with water.

5.2.3.4 Conclusions Ionic liquids can replace conventional solvents in liquid–liquid extraction of aromatic hydrocarbons, provided the aromatic distribution coefficient and/or the aromatic/aliphatic selectivity are higher than those with sulfolane. The main conclusion of the process evaluation is that ionic liquids showing a high aromatic distribution coefficient with a reasonable aromatic/aliphatic selectivity could reduce the investment costs of the aromatic/aliphatic separation by a factor of two. The best ionic liquids for the separation of aromatic and aliphatic hydrocarbons are $[C_4mim][C(CN)_3]$, $[C_4m_\beta py][N(CN)_2]$, $[C_4m_\beta py][C(CN)_3]$, and $[C_4m_\beta py][B(CN)_4]$. The aromatic distribution coefficients of these ionic liquids are a factor of 1.2 to 2.5 higher and the aromatic/aliphatic selectivities are up to a factor of 1.9 higher than with sulfolane.

5.2.4 Desulfurisation of Fuels

5.2.4.1 Conventional Desulfurisation Conventional desulfurisation of fuels is achieved by catalytic hydroprocessing. However, further or deep hydrodesulfurisation (HDS) leads to a high consumption of energy and hydrogen. The sulfur compounds are converted into hydrogen sulfide, which is easy to remove. Ultra-low sulfur gasoline and diesel oil (<15 ppm S in the United States, and <10 ppm in the European Union) are needed for new engines and catalysts, and for further reduction of CO- and NO_x-emissions. The HDS process is normally only effective for the removal of aliphatic and alicyclic organosulfur compounds. The aromatic sulfur compounds such as thiophenes, benzothiophenes (BTs), dibenzothiophenes (DBTs), and their alkylated derivatives (e.g., 3-methylthiophene, 3-MT) are very difficult to convert to H_2S through HDS catalysts and require high operating and investment costs. Therefore, alternative processes for deep desulfurisation are desirable. However, in a review about novel processes for removing sulfur compounds from refinery streams by Ito and Van Veen of Shell Research, it was concluded that the classical hydrotreating options and their offshoots are competitive in transport-fuel desulfurisation, considering the ongoing development of these processes [94]. Only if the sulfur levels have to be below 1 ppmv will polishing processes become interesting.

5.2.4.2 Desulfurisation with Ionic Liquids Liquid–liquid extraction with chloroaluminate-containing ionic liquids shows promising results, but their industrial use is not desirable with respect to corrosion, environmental concerns, hydrolytic stability, and regeneration aspects. With $[C_4mim]Cl\text{-}AlCl_3$ ($\chi = 0.65$), a reduction in sulfur content in real diesel of 80% was obtained in a five-stage extraction at 60 °C [95].

The ionic liquid $[C_4mim][AlCl_4]$ is partly effective in removing sulfur compounds from real fuels, both diesel and gasoline [96]. Nitrogen levels were significantly reduced, mainly due to a pretreatment with a molecular sieve. The fuels had to be dried before the extraction with activated 13X molecular sieve in order to prevent decomposition of the ionic liquid to form HCl. The ionic liquid/fuel ratio was 1/6 and the extraction was carried out in four stages. After the extraction of partially desulfurised gasoline, almost no thiophenes (C_6-thiophenes: 0.71 ppm) and no sulfides were detected, and the amount of other

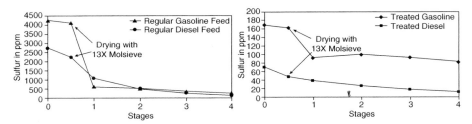

Figure 5.6   Four-stage desulfurisation extraction of diesel and gasoline. Ionic liquid/ fuel = 1/6, room temperature extraction [96].

sulfur species was 3.84 ppm. However, C_1-C_6 benzothiophenes, which were not detectable in the feed, were present in the final product, most likely due to a Lewis-acid catalysed Diels-Alder reaction. Also, with untreated gasoline, C_4-BT and C_{3+}–DBT were formed during the extraction. Diesels did not show any formation of new sulfur compounds. The level of sulfur compounds after the four-stage extraction is still high: 129.15 ppm for regular diesel and 239.08 ppm for regular gasoline. The levels are lower for the fuels pretreated with a molecular sieve: 11.56 ppm for treated diesel and 82.33 ppm for treated gasoline (see Figure 5.6). Regeneration of the ionic liquid was not carried out.

The ionic liquid $[C_4mim][Cu_2Cl_3]$ showed a higher sulfur removal rate from a model oil (DBT and piperidine in dodecane, 500–764 ppm S) than $[C_4mim][AlCl_4]$ or $[C_4mim][BF_4]$: 23.4%, 16%, and 11%, respectively. The desulfurisation rate with gasoline with different sulfur contents with $[C_4mim][Cu_2Cl_3]$ varied from 16.2% (950 ppm S in the feed) to 37.4% (196 ppm S in the feed). The sulfur removal rate is lower with gasoline (21.6%) than with the model oil (23.4%) with the same sulfur content of 680 ppm in the feed [97].

The ionic liquid $[C_4mim]Cl$ can remove 50% of nitrogen compounds from straight-run diesel (13,240 ppm S and 105 ppm N), but only 5% of the sulfur compounds [98]. Nitrogen compounds are inhibitors for the HDS reaction and removal of these compounds can improve the reaction. Regeneration of the ionic liquid was carried out by back extraction with water (50 wt%) or metha-nol (10 wt%).

Effective removal of DBT can be achieved by addition of iron(III) chloride to $[C_4mim]Cl$. The best results were obtained with $[C_4mim][Fe_2Cl_7]$: DBT was completely removed from a model oil containing 5000 ppm DBT. This combi-nation was also effective in removing 1180 ppm sulfur compounds from diesel oil [99]. However, removal of other sulfur compounds and the regeneration of the ionic liquid were not mentioned.

Zhang and Zhang used the ionic liquids $[C_2mim][BF_4]$, $[C_4mim][BF_4]$, $[C_4mim][PF_6]$ [100], $[C_6mim][PF_6]$, $[C_8mim][BF_4]$, $[Me_3NH]Cl$-$AlCl_3$ ($\chi = 0.60$), and $[Me_3NH][Al_2Cl_7]$ [101] to lower the sulfur content. With a low (240 ppm S) or a high sulfur (820 ppm S) gasoline, only 10–30% of the sulfur compounds were removed [100]. After 10 extraction cycles with [C_6mim][PF_6] and [C_8mim][BF_4], the sulfur content in gasoline was lowered from 820 ppm to 400 ppm and 320 ppm, respectively. Besides the sulfur compounds, aromatics were also removed, about 8% after 10 cycles [101]. In a three-stage extraction of a model oil with [C_8mim][BF_4], the thiophene and DBT content was reduced by 79% and 87%, respectively [102]. With the ionic liquids [Me_3NH][Al_2Cl_7] and [Me_3NH]Cl-$AlCl_3$ ($\chi = 0.60$), the removal rate of sulfur compounds was only 15–20%. The regeneration of the 1,3-dialkyimidazolium-based ionic liquids was carried out by evaporating the absorbed compounds and the removal rate was retained. However, the [Me_3NH]Cl-$AlCl_3$ ionic liquids could not be regenerated. After extraction with 1-(4-sulfonic acid)butyl-3-methylimidazolium 4-toluenesulfonate, 250 ppm DBT was removed from a model oil (500 ppmw DBT in tetradecane) at 60 °C and a mass ratio oil/ionic liquid = 4/1. After a five-stage extraction with this ionic liquid of a pre-desulfurised oil with a sulfur content of 438 ppmw, the sulfur content was 45 ppmw. The process conditions were: $T = 80\,°C$, mass ratio diesel/ionic liquid = 4/1, and extraction time 25 minutes [103].

1,3-Dialkylimidazolium dialkylphosphate ionic liquids are useful for extractive desulfurisation of fuels [104–107]. The best results were obtained with [C_4mim][(MeO)$_2PO_2$], although the solubility of gasoline in this ionic liquid is 35.3 mg (g ionic liquid)$^{-1}$ at 298.15 K, which may lead to increased separation costs. The ionic liquid is not soluble in the gasoline. The mass-based distribution coefficients of sulfur compounds in straight-run gasoline with this ionic liquid at 298.15 K are: $D_{3-MT} = 0.59$ kg per kg, $D_{BT} = 1.37$ kg per kg, and $D_{DBT} = 1.59$ kg per kg [107]. The removal of sulfur compounds from gasoline is more difficult than for straight-run gasoline, which is likely due to co-extraction of aromatic compounds from the gasoline. The regeneration of these ionic liquids still needs more investigation.

With pyridinium-containing ionic liquids, the sulfur removal rate of a model oil (thiophene in heptane) is 45.5% with 1-butylpyridinium tetrafluoroborate ([C_4py][BF_4]) and lower for other pyridinium-containing ionic liquids [108]. The mass ratio ionic liquid/model oil was 1:1 and the extraction time was 30–40 minutes. The sulfur removal rate increased somewhat with temperature: 48.3% at 60 °C. After six extraction cycles, the sulfur content was decreased from 498 ppm to 18 ppm. Regeneration of the ionic liquids was carried out by evaporation of thiophene at 100 °C, or by re-extraction with CCl_4.

The distribution coefficient of DBT in its extraction from dodecane showed a clear variation in cation types:

1,4-dimethylpyridinium > methylpyridinium > pyridinium
≈ imidazolium ≈ pyrrolidinium

and much less significant variation with the anion type (see Fig. 5.7 [109]). Polyaromatic quinolinium-based ionic liquids showed even greater sulfur removal (90% at 60 °C), but those ionic liquids have higher melting points.

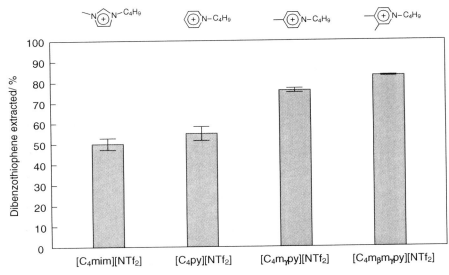

Figure 5.7 Sulfur removal with several ionic liquids [109].

[C₄mim][C₈H₁₇SO₄] and [C₂mim][C₂H₅SO₄] are other promising ionic liquids, with a DBT distribution coefficient of 1.9 and 0.8 kg per kg, respectively (500 ppm DBT in dodecane, ionic liquid : dodecane 1:1) [110]. With diesel and fluid catalytically cracked (FCC) gasoline, the distribution coefficient of sulfur compounds was lower for [C₄mim][C₈H₁₇SO₄]: 0.3–0.8 kg per kg (200–400 ppm S in the feed) and 0.5 kg per kg (300 ppm in the feed). The process conditions were: mass ratio 1:1, mixing time 15 minutes at room temperature. Regeneration of [C₂mim][C₂H₅SO₄] containing 20 mg tetrahydrothiophene per kg and 2.5 wt% cyclohexane was carried out by stripping with air at 100 °C for 30 minutes. However, regeneration of sulfur-loaded ionic liquids from diesel oil was not possible by stripping only, and additional re-extraction and distillation steps will be required.

A major drawback of these ionic liquids is the cross-solubility of hydrocarbons in the ionic liquids, as the co-extracted hydrocarbons must be separated from the ionic liquid together with the sulfur compounds. The cross-solubility is lower for [C₂mim][C₂H₅SO₄] than for [C₄mim][C₈H₁₇SO₄]. In Figure 5.8, a possible process scheme of an ionic liquid extraction implemented into an existing refinery for the desulfurisation of diesel is depicted [110].

The Institut Français du Pétrole has obtained a patent in which sulfur and nitrogen compounds can be removed from fuels with ionic liquids containing an alkylating agent, allowing formation of ionic sulfur- and nitrogen-derivatives that are soluble in the ionic liquid [111]. The ionic liquid is separated from the oil by decantation.

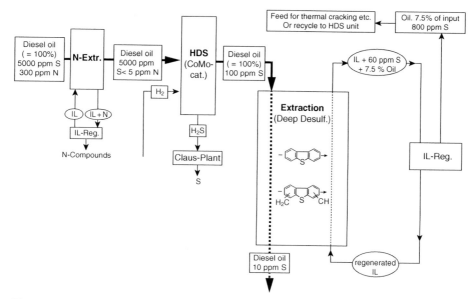

Figure 5.8 Extractive desulfurisation of diesel with $[C_4mim][C_8H_{17}SO_4]$, cross-solubility of oil 5%, mass ratio ionic liquid/oil 1.5, $D_S = 0.8$ kg kg^{-1}, six stages [110].

5.2.4.3 Oxidative Desulfurisation

A low removal rate of sulfur compounds of 7–8% was found by Lo et al., but the rate was increased after oxidation of the sulfur compounds by H_2O_2-ethanoic acid to form sulfones, to 55% with $[C_4mim][BF_4]$ and to 85% with $[C_4mim][PF_6]$ [112]. Zhao et al. found that DBT was completely removed after 50 minutes at 60 °C with the ionic liquid $[C_4m_7py][BF_4]$ (Fig. 5.9) in combination with oxidation by H_2O_2 ($H_2O_2/S = 3$) [113]. The ionic liquid can be used seven times without significant decrease in activity (99.6% sulfur removal). The sulfur content of an actual diesel fuel with 3240 ppm S was reduced to 20 ppm after extraction with the ionic liquid, oxidation and extraction with DMF (99.4% removal). The process conditions were: $T = 60$ °C, $V_{oil}/V_{IL} = 1$, $H_2O_2/S = 6$, reaction time 2 hours. The group of Zhao also investigated a deep eutectic solvent, $[N_{4444}]Br·2C_6H_{11}NO$, as an effective catalyst for the oxidative desulfurisation of thiophene with $H_2O_2/$ethanoic acid [114]. Most of the oxidation products were transferred to the aqueous phase due to their higher polarity. The desulfurisation rates of thiophene-containing model oil and actual FCC gasoline were 98.8% and 95.3%, respectively.

The ionic liquid $[C_4py][HSO_4]$ was used as a phase transfer catalyst (PTC) at 60 °C, 60 minutes, and $H_2O_2/$sulfur molar ratio (O/S) of 4. The desulfurisation rate of a model oil reaches the maximum (93.3%), and the desulfurisation of the real gasoline is also investigated; 87.7% of sulfur contents are removed under optimal reaction conditions. $[C_4py][HSO_4]$ could be recycled five times

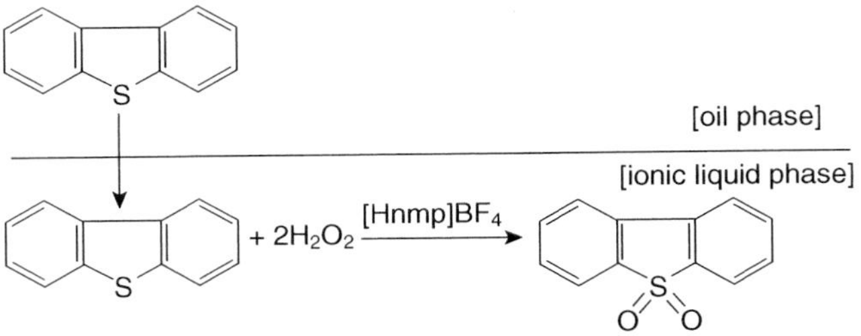

Figure 5.9 Extraction and oxidation reaction of DBT ([Hnmp][BF$_4$] is a protonated ionic liquid, 1-methyl-2-pyrrolidonium tetrafluoroborate) [113].

without significant decrease in activity [115]. A series of functionalised ionic liquids, immiscible with oil, halogen-free, and containing a –COOH group in the cation, were used for oxidative desulfurisation as both the catalyst and extractant. The removal of DBT and 4,6-dimethyldibenzothiophene (4,6-DMDBT) from model diesel at 298 K could reach 96.7% and 95.1%, respectively. The functionalised ionic liquid could be recycled five times without any apparent loss of the catalytic activity [116].

An interesting new heteropolyanion-based functionalised ionic liquid, [1-(3-sulfonic acid)propylpyridinium]$_3$[PW$_{12}$O$_{40}$·2H$_2$O], [PSpy]$_3$[PW$_{12}$O$_{40}$·2H$_2$O] (abbreviated [PSpy]$_3$[PW]), was used as an effective catalyst for desulfurisation of fuels in [C$_8$mim][PF$_6$] by using aqueous H$_2$O$_2$ as oxidant. The removal of 4,6-DMDBT could be up to 98.8%, and the system could be recycled at least nine times without significant decrease in activity. The sulfur level of FCC gasoline could be reduced from 360 to 70 ppm with this extraction and catalytic oxidation system [117]. In a related paper, a simple liquid–liquid extraction and catalytic desulfurisation system, composed of 30% H$_2$O$_2$ and [C$_4$mim][BF$_4$] with added molybdate(VI) or [PMo$_{12}$O$_{40}$]$^{3-}$ salts, was effective in removing 99% of sulfur compounds (BT, DBT and 4,6-DMDBT) from a model oil. The desulfurisation system was recyclable for five times with very little decrease in activity [118]. Ultraclean Fuel in Australia has claimed an oxidative desulfurisation process, which is capable of reducing the sulfur content from 500 ppm to 30 ppm [119].

5.2.4.4 Conclusions Desulfurisation of fuels by extraction with ionic liquids as the only process step does not lead to sulfur levels of 10 ppm and lower at the present state of research. Only if pre-desulfurised fuels are used, or if extraction is combined with other processes, such as oxidation, can the required sulfur levels be reached. Regeneration of the ionic liquid still needs attention, although a number of authors report that the ionic liquid can be recycled several times. Some authors report a decrease in sulfur removal rate after several extraction cycles. Next to experiments with a model oil, measurements with real fuels must be carried out because the sulfur removal rates will differ due to the presence of other compounds in the fuel.

5.2.5 Proteins

5.2.5.1 Conventional Process Efficient bioseparation is very important in modern biotechnology since the purification costs are between 20% and 60% of the total production costs, and in special cases, this can even amount to 90%. Industrially relevant separation techniques include membrane separation, adsorption, and extraction [120–123]. Aqueous two-phase system (ATPS) extraction can result in a high extractability and good retention of the bioactivity [124, 125]. However, high viscosities of one or two phases complicate the scale-up of the process [126].

5.2.5.2 Liquid–Liquid Extraction with Ionic Liquids Extraction of myoglobin, human serum albumin and immunoglobulin G (IgG) with the ionic liquid trihexyl(tetradecyl)phosphonium bis{(trifluoromethyl)sulfonyl}amide ($[P_{66614}][NTf_2]$) showed almost 100% retention. Also, $[P_{66614}][N(CN)_2]$ is a good solvent for myoglobin [127]. However, IgG is not stable enough to be used in extraction with the ionic liquids investigated. The extraction of bovine serum albumin (BSA), trypsin, cytochrome c, and γ-globulins in ionic liquid-based aqueous two-phase systems ($[C_4mim]Br$, $[C_6mim]Br$ or $[C_8mim]Br$ and $K_2[HPO_4]$) was effective. About 75–100% of the proteins were extracted into the ionic liquid phase in a single step extraction [128]. Extraction efficiencies of the proteins were found to increase with increasing temperature and increasing alkyl chain length of cation of the ionic liquids. Compared with the traditional poly(ethyleneglycol) (PEG) aqueous two-phase system (ATPS), the ionic liquid-based ATPS have the advantages of lower viscosity, little emulsion formation, quick phase separation, and others [128]. The ATPS ($[C_4mim][N(CN)_2] + K_2[HPO_4]$) was demonstrated to be effective for the selective separation of BSA from aqueous saccharides. It is found that about 82.7–100% BSA was enriched into the top phase, while almost quantitative saccharides were preferentially extracted into the bottom phase of the ATPS. The extraction efficiency of BSA from the aqueous saccharide solutions increases with increasing number of hydroxyl groups of the saccharides (see Fig. 5.10) [129].

Penicillin G can be extracted from its fermentation broth with an ionic liquid ATPS consisting of the hydrophilic ionic liquid $[C_4mim][BF_4]$ and NaH_2PO_4, while leaving miscellaneous proteins in the ionic liquid-poor phase (top phase in Fig. 5.11A) [130]. The hydrophilic $[C_4mim][BF_4]$ can be recovered by transfer into an hydrophobic ionic liquid-rich phase of $[C_4mim][PF_6]$ (bottom phase in Fig. 5.11C), leaving most of the penicillin in the conjugated water phase (top phase in Fig. 5.11D). In comparison with the butyl ethanoate/water system or the polymer-ATPS, the integrated ionic liquid system shows several advantages: penicillin is efficiently extracted into the ionic liquid-rich phase at neutral pH and hydrophobic ionic liquids can separate hydrophilic ionic liquids from the penicillin-containing aqueous phase. Protein emulsification occurring in the organic solvent system is avoided and the

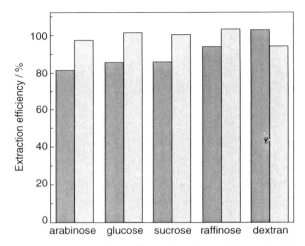

Figure 5.10 The extraction efficiencies of BSA and saccharides in ([C₄mim] [N(CN)₂] + K₂[HPO₄]) ATPS at 298.15 K: ▉ BSA; ☐ saccharides. The test concentrations of BSA and saccharides were all at 4 mg/cm³; 0.6 g ionic liquid, 0.6 g K₂[HPO₄], and 1.5 cm³ H₂O were added for the formation of the ATPS [129].

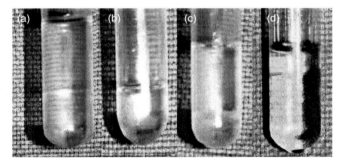

Figure 5.11 Separation of penicillin from the fermentation broth containing miscellaneous protein [130].

trouble for recovering the phase-forming material in the polymer-ATPS is overcome [130].

An ionic liquid/ATPS based on the hydrophilic ionic liquid [C₄mim]Cl and K₂[HPO₄] has been employed for direct extraction of proteins from human body fluids. Proteins present at low concentration levels were quantitatively extracted into the [C₄mim]Cl-rich upper phase, with a distribution ratio of about 10 kg per kg between the upper and lower phase, and an enrichment factor of 5. Addition of an appropriate amount of K₂[HPO₄] to the separated upper phase results in a further phase separation, giving rise to an improved enrichment factor of 20. The use of an ionic liquid as a green solvent offers clear advantages over traditional liquid–liquid extractions, in which the use of toxic organic solvents is unavoidable [131].

The commercial ionic liquid Ammoeng™ 110 (Solvent Innovation, Merck, Darmstadt, Germany), which contains cations with oligoethylene glycol units in combination with an inorganic salt mixture ($K_2[HPO_4]/K[H_2PO_4]$), was found to be highly effective for the formation of ATPS that can be used for the biocompatible purification of active enzymes, such as alcohol dehydrogenases (ADH), from *Lactobacillus brevis* and from a thermophilic bacterium [132]. Both enzymes were enriched in the ionic liquid phase, resulting in an increase of specific activity by a factor of 2 and 4, respectively. Furthermore, the presence of ionic liquid within the system provided the opportunity to combine the extraction process with the performance of enzyme-catalysed reactions [132].

The transfer of lysozyme in the liquid–liquid extraction from an aqueous solution to a solution consisting of $[C_4mim]_3[CB]$ (CB = an affinity-dye, Ciba-cron Blue 3GA) in $[C_4mim][PF_6]$ decreased as the pH of the aqueous phase increased. An extraction higher than 90% was observed at pH 4. At a high ionic strength, the affinity of lysozyme is higher for the aqueous phase. Lyso-zyme molecules were thus almost quantitatively recovered from the ionic liquid phase by contacting with aqueous solutions of 1 M KCl at pH 9–11. The resultant recoveries of lysozyme using back-extraction at pH 11 after each complete forward and backward cycle, for all eight extractions, ranged between 87% and 93%. Similarly, the recoveries of lysozyme at pH 8 stayed above 85% for all eight cycles. The extraction was specific for lysozyme in contrast to cytochrome c, ovalbumin, and BSA [133].

5.2.5.3 Conclusions Ionic liquids can be successfully employed to extract proteins from aqueous solutions with an ATPS. The viscosity of the ionic liquid-based system is generally lower than that of traditional PEG systems. Furthermore, there is little emulsion formation, a quick phase separation, and the ionic liquids can be recycled.

5.3 EXTRACTIVE DISTILLATION

5.3.1 Conventional Process

Conventional distillation is the most applied separation process in the chemical industry. However, mixtures with close-boiling compounds and mixtures with a low relative volatility are difficult, impossible, or economically unat-tractive to separate by ordinary distillation. One of the most useful ways to separate chemicals that cannot be distilled is to employ selective solvents. These solvents take advantage of the non-ideality of a mixture of components having different chemical structures. Liquid–liquid extraction, azeotropic distillation, and extractive distillation are examples of the use of solvents in a separation process. Volatile entrainers are used in azeotropic distillation and non-volatile solvents in extractive distillation. Of course, the added selective solvent has to be isolated from the process mixtures in order to allow its reuse. This implicates additional investments and energy costs. In extractive distillation, selective solvents, such as sulfolane, 1-methylpyrrolidone (NMP), dimethylformamide (DMF), or ethanenitrile are added at the top of the distillation column to increase the relative volatility [134]. Therefore, extractive distillation is, in principle, more attractive than azeotropic distillation because the solvent is not evaporated and the energy requirements are much lower.

Compounds with no or little interaction with the solvent leave the extraction column via the top and the compounds with a strong interaction with the solvent leave the column via the bottom. In the next step, the solvent is separated from the compounds and is recycled to the top of the extraction column. Figure 5.12 shows an extractive distillation process for the separation and purification of benzene and toluene.

The solvent-to-feed ratio (S/F) in extractive distillation is usually 5 to 8, which implicates high additional costs due to larger equipment and a high energy requirement. Sometimes, a salt is added to increase the relative volatility of the compounds through the salting-out effect [134, 136–142]. The advantages

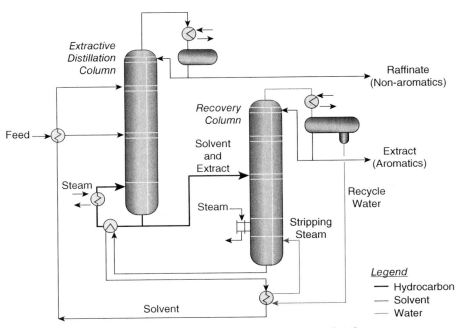

Figure 5.12   Extractive distillation process [135].

of adding a salt are that a lower solvent-to-feed ratio is required. Additionally, the salt possesses no vapour pressure and is therefore not present in the vapour phase, unless it is entrained. Furthermore, the salt can easily be separated from the bottom product. However, the disadvantages are that salts are often corrosive; precipitation of the salt can occur; the salt can cause decomposition of components at higher temperatures; and the salt can only be applied in aqueous systems, such as in the separation of alcohols and water.

5.3.2   Ionic Liquids in Extractive Distillation

Ionic liquids as solvents combine the advantages of both organic solvents and salts: increasing the relative volatility of one of the components and reducing the solvent-to-feed ratio by the salting-out effect without the disadvantages of a solid salt [134, 143–153]. Ionic liquids are suitable as entrainers for a whole range of azeotropic systems [144, 146, 149].

The selected solvent for an extractive distillation process should have the following characteristics:

- High boiling point to avoid solvent losses.
- High selectivity and solvency to obtain high-purity products and to reduce the quantity of solvent used. Easy recovery to increase the efficiency in the solvent recovery process and to reduce the recovery costs.

Figure 5.13 shows a process scheme with ionic liquids as solvents. The best option for the regeneration column will probably be a multi-effect evaporator, which requires a low amount of energy, possibly followed by a strip column for the removal of last traces of products.

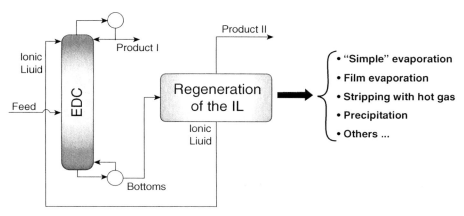

Figure 5.13   Process scheme of extractive distillation with ionic liquids. EDC, extractive distillation column.

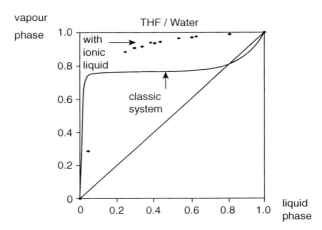

Figure 5.14   Equilibrium phase diagram for the system THF/water. The solid line shows the classic azeotropic mixture. The data points indicate how the azeotrope has been broken after addition of the ionic liquid. The amount of THF in the vapour phase is always larger than in the liquid phase [161].

5.3.2.1 Olefin/Paraffin (Alkene/Alkane) Separation   Since the boiling points of olefins and paraffins lie within a close range, separation by conventional distillation is difficult and expensive. The separation of 1-hexene and hexane is taken as an example for the separation of olefins and paraffins. The highest selectivity of hexane to 1-hexene with a solvent concentration of 0.3 mole fraction in the liquid phase is with the conventional solvent NMP ($S = 1.1613$ at 313.15 K, and $S = 1.632$ at 333.15 K), measured by headspace gas chromatography (GC). From a series of ionic liquids tested, the best ionic liquid at 313.15 K was [C_2mim][NTf_2] ($S = 1.0576$), and at 333.15 K was 1-octylquinolinium bistriflamide ([C_8quin][NTf_2]) ($S = 1.0390$) [152]. A suitable solvent should possess both a high selectivity and a high capacity for the components to be separated. The selectivity and capacity at infinite dilution were calculated with a COSMO-RS model. NMP showed a selectivity of $S^\infty = 1.50$ and a capacity of $1/\gamma^\infty = 0.41$. The best ionic liquid was [C_8quin][NTf_2] with $S^\infty = 1.93$ and $1/\gamma^\infty = 0.16$, although it exhibited the lowest selectivity of the ionic liquids tested, but it had the highest capacity [154]. It is clear that there is a large difference in the selectivity at finite and at infinite dilution, especially for the use of ionic liquids. At a finite dilution, $x_{solvent} = 0.3$, the best solvent is NMP, but at infinite dilution, an ionic liquid exhibits the highest selectivity.

Since both the selectivity and the capacity at finite dilution of the conventional solvent NMP are considerably higher than those of the ionic liquids tested, replacement of NMP for this separation would only be advantageous if the thermal and chemical stability of NMP would cause a problem at the process conditions of the extractive distillation.

5.3.2.2 Aromatic/Aliphatic Hydrocarbon Separation   The separation of aromatic from non-aromatic hydrocarbons is challenging because of close-boiling points and azeotrope formation. The mixture of toluene and methyl-cyclohexane (MCH) has been used as representative of the compounds found in an industrial stream. Promising ionic liquids for the separation of this mixture are $[C_6mim][NTf_2]$ ($S_{MCH/tol}^{\infty} = 6.97$ and $1/\gamma_{tol}^{\infty} = 1.06$ at 313.15 K) [155] and $[C_8quin][NTf_2]$ ($S_{MCH/tol}^{\infty} = 3.24$ and $1/\gamma_{tol}^{\infty} = 1.52$ at 373.15 K) [156]. The selectivities and capacities with conventional solvents at approximately 333 K for this separation are: sulfolane $S_{MCH/tol}^{\infty} = 7.43$ and $1/\gamma_{tol}^{\infty} = 0.28$ [157]; NFM $S_{MCH/tol}^{\infty} = 6.98$ and $1/\gamma_{tol}^{\infty} = 0.38$; NMP $S_{MCH/tol}^{\infty} = 5.47$ and $1/\gamma_{tol}^{\infty} = 0.60$ [158]; DMF $S_{MCH/tol}^{\infty} = 3.12$ and $1/\gamma_{tol}^{\infty} = 0.37$; and furfural $S_{MCH/tol}^{\infty} = 3.37$ and $1/\gamma_{tol}^{\infty} = 0.23$ [159]. Ionic liquids exhibit comparable selectivities to the conventional solvents, but their capacities are higher than those of the conventional solvents.

5.3.2.3 Organic Compounds/Water Separation   Especially when water is part of the azeotropic mixture, very high separation factors can be achieved because many ionic liquids are hygroscopic materials with a strong affinity to water. Figure 5.14 displays the classic vapour liquid diagram for tetrahydrofuran (THF)/water [160]. The data points indicate the change after the addition of the ionic liquid. Suitable ionic liquids for this separation are: $[C_4mim]Cl$, $[C_4mim][BF_4]$, $[C_2mim][BF_4]$, and $[C_8mim][BF_4]$. To afford a sufficient separation, the amount of ionic liquid added has to be in the range of 30–50 wt%.

As the ionic liquid has no relevant vapour pressure, a second separation column for entrainer distillation is not required and energy can be saved. For the separation of ethanol from water, a process with $[C_2mim][BF_4]$ as solvent was compared with a conventional process with 1,2-ethanediol as solvent. The overall heat duty for the conventional process is 2213 kJ (kg ethanol)$^{-1}$ and for the ionic liquid process, 1656 kJ (kg ethanol)$^{-1}$, a reduction of 25% [147].

5.3.2.4 Separation of Trimethylborate (TMB) and Methanol   BASF has carried out an extractive distillation in a mini-plant with an column diameter of 30 mm with either DMF, $[C_2mim][CH_3SO_3]$ or $[C_2mim][OTs]$ (OTs = tosylate = 4-toluenesulfonate, $[4\text{-}CH_3C_6H_4SO_3]^-$) as solvent. Either a stripper or an evaporator replaced the distillation column for the regeneration of the DMF in the ionic liquid process. The S/F (entrainer/feed) ratio was for $[C_2mim][CH_3SO_3]/[C_2mim][OTs]/DMF$ 1.0/1.7/4.7 kg S/kg F: almost a factor 5 lower for the ionic liquid $[C_2mim][CH_3SO_3]$ than for DMF. The extractive distillation column with DMF as solvent required more than twice the number of theoretical stages compared with the ionic liquid process. A benchmark calculation for a capacity of 5000 tons per year has revealed that a saving potential of 37–59% on energy cost and 22–35% on the investment can be achieved, depending on the ionic liquid used and the regeneration process of the ionic liquid. The total savings are in the range of 25–35%.

BASF has been conducting this extractive distillation process in the mini-plant continuously for three months. Although the ionic liquid faced a severe thermal treatment of about 250 °C in the recycling step, its performance was retained completely without a purge [150].

5.3.2.5 Conclusions Extractive distillation with ionic liquids can be attractive because of the potential savings in investments, fewer stages in the extraction column, lower energy costs, lower solvent-to-feed ratio, and an easier regeneration, but in some cases conventional solvents perform better. The use of activity coefficients at infinite dilution sometimes provides results that differ from the distribution coefficients and selectivities at finite solutions. Therefore, for the design of an extractive distillation, experimental values of the distribution coefficients and selectivities will give more reliable results than activity coefficients at infinite dilution.

5.4 COMBINATION OF SEPARATIONS IN THE LIQUID PHASE WITH MEMBRANES

The separation of several compounds with ionic liquids can also be combined with membranes, as a bulk ionic liquid membrane (BILM) [161, 162], as a support for the ionic liquid (SILMs, supported ionic liquid membrane) [163–167], or as a separating barrier between the ionic liquid and product phase [168–172].

The selectivity of toluene/heptane in a BILM with [C$_8$mim]Cl at 25 °C was around 4 to 9, depending on the initial toluene concentration [161]. The toluene/heptane selectivity with liquid–liquid extraction, calculated from activity coefficients at infinite dilution, is around 8 at 35 °C [173]. The separation factor at $T = 25$ °C with a BILM with [C$_2$mim][C$_2$H$_5$SO$_4$], is 2.75 after 40 hours, but decreases to 1.5 at 80 hours. With liquid–liquid extraction using the same ionic liquid, the selectivity is around 50 at 40 °C [67].

SILMs combine extraction and stripping, and the amount of solvent in the SILM process is much less than in a solvent extraction process. [C$_4$mim][PF$_6$] in a porous polyvinylidene fluoride film showed a benzene/heptane selectivity of 67, a toluene/heptane selectivity of 11, and the mass transfer coefficients of benzene and toluene were 6.2×10^{-4} m h^{-1} and 9.9×10^{-4} m h^{-1}, respectively [166]. The receiving phase was hexadecane. The toluene/heptane selectivity is lower than for several other ionic liquids (see Table 5.1) and certainly lower than for sulfolane. Moreover, the product has to be separated from the receiving phase. Ionic liquids with silver salts impregnated on porous membranes facilitate the transport of alkenes through the membrane [167, 174, 175]. In these processes, the combination of the silver salt and the ionic liquid determines the selectivity of the alkene.

Nanofiltration is a suitable process for the separation of non-volatile products from the ionic liquid [169, 170]. However, the feed phase must be diluted in order to decrease the viscosity, and this implicates an extra energy requirement to concentrate the product afterwards.

The selectivity in both BILM and SILM processes is determined by the nature of the ionic liquids used. In general, the capacity or mass transfer rates in the BILM and SILM processes are relatively low. The advantages claimed by several authors are that the solvent inventory in BILM and SILM processes is lower than for extraction processes, and that extraction and stripping, including continuous regeneration of the ionic liquid, are combined in one process. However, it remains to be seen if these applications are useful, as a membrane poses an extra barrier for transport of the desired compounds from the feed phase to the ionic liquid, resulting in a lower capacity or lower transfer rates. Moreover, the product has to be separated from the receiving phase by another separation process and the feed phase has to be diluted to lower the viscosity in case of separating the products from the ionic liquids. Only in the case of separating a heat-sensitive or non-volatile product from the ionic liquid phase could a separation process with a membrane be useful.

5.5 GAS SEPARATIONS

5.5.1 Conventional Processes

Chemical and physical absorption processes are extensively used in the natural gas, petroleum, and chemical industries for the separation of CO_2 [176]. Physical absorption is preferred when acid gases (e.g., H_2S, or CO_2) are present at elevated concentrations in the gas stream. Physical solvents are non-reactive polar organic compounds with an acid gas affinity. Chemical absorption is typically used for the removal of remaining acidic impurities, and when gas purity is a down-stream constraint. For chemical CO_2 removal, aqueous solutions of primary, secondary, tertiary, sterically hindered amines and formulated amine mixtures are the most widely used solvents. About 75–90% of the CO_2 is captured using a MEA-based (monoethanolamine) technology producing a gas stream of high CO_2 content (>99%) after desorption [177].

The major drawbacks of the traditional gas absorption separation processes are mainly caused by the nature of the solvent and by the type of interactions given between the solute and the solvent. In an industrial gas absorption process, it is desirable to achieve high absorption rates and high solute capacities of a solvent that is easily regenerated and for which volume make-up is minimised.

5.5.2 CO_2 Separation with Standard Ionic Liquids

Ionic liquids can be used for gas separations, and the removal of CO_2 from several gases [178–190] has mainly been investigated. The CO_2 solubility is larger in ionic liquids with anions containing fluoroalkyl groups, such as $[NTf_2]^-$ and $[CTf_3]^-$, regardless of whether the cation is 1,3-dialkylimidazolium, 1,1-dialkylpyrrolidinium, or tetraalkylammonium. These results suggest that

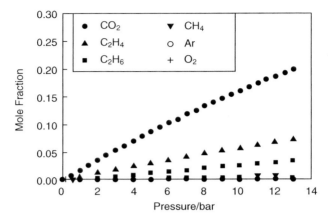

Figure 5.15   Solubilities of gases in [C$_4$mim][PF$_6$] at 298 K [178].

the nature of the anion has the most significant influence on the gas solubilities [180, 181, 183, 185, 190].

An increase in the alkyl chain length on the cation increases the CO_2 solubility marginally. Oxygen is scarcely soluble in several ionic liquids, and H_2, N_2, and CO have a solubility below the detection limit in several studies [178, 185]. Figure 5.15 shows the solubility of a number of gases in [C$_4$mim][PF$_6$] at 298 K [178]. The CO_2/C_2H_4 selectivity is around 3 and the CO_2/CH_4 selectivity is about 30.

5.5.3   CO$_2$ Separation with Functionalised Ionic Liquids

In order to increase CO_2 solubility in ionic liquids, functionalised task-specific ionic liquids can be used. Since the conventional solvents for CO_2 absorption are amine based, the functionalisation with amine groups seemed obvious [189, 191–194]. Figure 5.16 shows the CO_2 absorption in MEA, Sulfinol, methyldiethanolamine (MDEA), Selexol, and several NH$_2$-functionalised ionic liquids: [H$_2$NC$_2$H$_4$py][BF$_4$], [H$_2$NC$_3$H$_6$mim][BF$_4$], and [H$_2$NC$_3$H$_6$mim][NTf$_2$].

The ionic liquid [H$_2$NC$_3$H$_6$mim][BF$_4$] has about the same performance as a 30% MEA solution [191]. The functionalised ionic liquids show both physical and chemical absorption behaviour.

5.5.4   CO$_2$ Separations with Ionic Liquid (Supported) Membranes

Copolymers of polymerisable RTILs—poly(RTILs)—and different lengths of PEG polymers were found to a have CO_2/N_2 separation performance exceeding the upper bound of the "Robeson Plot" [195, 196]. The ideal CO_2/N_2 selectivities with [C$_n$mim][NTf$_2$] ionic liquids are in the order of 20, while the selectivity with the poly(RTIL) [P$_n$mim][NTf$_2$] is around 30.

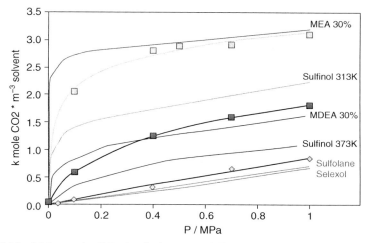

Figure 5.16   Volumetric CO_2 loads in several liquids at 333 K [191], where ■ is $[H_2NC_2H_4py][BF_4]$, □ is $[H_2NC_3H_6mim][BF_4]$, and ◇ is $[H_2NC_3H_6mim][NTf_2]$.

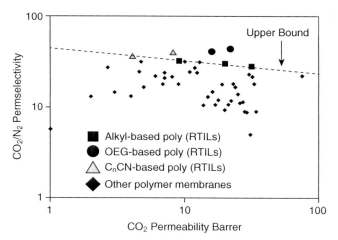

Figure   5.17   Robeson   plot   for   the CO_2/N_2 separation   [196].   OEG, oligoethyleneglycol.

The CO_2/CH_4 selectivity is only slightly higher with the poly(RTILs) than with the corresponding 1,3-dialkylimidazolium-based ionic liquids, that is, 12 *vs.* 9. Alkyl-terminated nitrile groups are also of interest for gas separations in poly(RTILs) [196]. Both the CO_2/N_2 and the CO_2/CH_4 selectivities are higher with these poly(RTILs), 40 and 32, respectively (see Fig. 5.17). Supported liquid membranes (SLM) based on functionalised ionic liquids, such

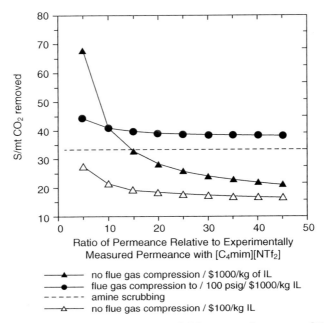

Figure 5.18   Comparison of costs of CO_2 removal processes [197].

as $[H_2NC_3H_6mim][NTf_2]$ in a poly(tetrafluoroethylene) (PTFE) membrane, showed a high CO_2/CH_4 selectivity of the order of 65, and a high stability for more than 260 days. The CO_2 permeability decreased gradually from 690 to 560 Barrer (517.5 to 420×10^{-12} m^3 m^{-2} m s^{-1} bar) during that period [198]. $[C_2mim][N(CN)_2]$ in a polyethersulfone (PES) membrane showed a CO_2/CH_4 selectivity of 20 and a CO_2/N_2 selectivity of 60. The CO_2 permeability was 610 Barrer (457.5×10^{-12} m^3 m^{-2} m s^{-1} bar) [199].

Porous alumina membranes saturated with $[C_2mim][NTf_2]$ showed a CO_2/N_2 selectivity of 127. A cost comparison with conventional amine scrubbing was carried out by Baltus et al. [197]. The measured CO_2 permeance in $[C_4mim][NTf_2]$ was 4×10^{-5} mol $bar^{-1}m^{-2}$ s^{-1}. From Figure 5.18, it is clear that a price for the ionic liquid of the order of $1000 per kg is required at a permeance of at least a factor 15 times higher than the permeance of $[C_4mim][NTf_2]$.

5.5.5   Olefin/Paraffin Separations with Ionic Liquids

Apart for the removal of CO_2 from other gases, ionic liquids can also be used for ethylene/ethane or propylene/propane separations [188, 200–212]. For the separation of hydrocarbon gases, it was found that the solubility increased as the number of carbons increased, and also when the number of carbon–carbon double bonds increased [202]. The separation factors with standard ionic liquids for the separation of ethylene/ethane or propylene/propane are relatively small, as the solubilities of these gases in these ionic liquids are very low (see Fig. 5.15). Therefore, ionic liquids with functional groups or dissolved carrier molecules are more effective, such as the use of Ag-salts, for example, $Ag[BF_4]$ in an ionic liquid with the $[BF_4]^-$ anion [200, 201, 207, 209, 212]. The propylene/propane selectivity with an 0.25 M $Ag[BF_4]$ in $[C_4mim][BF_4]$ is 20 at a low partial C_3H_x pressure of 1 bar, but decreases at higher pressures [212]. The flux rates with liquid functionalised ionic liquid membranes are in the order of 2×10^{-2} L m^{-2} s^{-1}, and the selectivity can vary between 100 and 540 for 1-hexene/hexane or 1-pentene/pentane [207].

5.5.6 Conclusions

Standard ionic liquids do often not perform better in gas separations than conventional solvents. However, ionic liquids do have the advantage of a negligible volatility, they are less corrosive than conventional solvents, and they can operate at high temperatures. Moreover, functionalised ionic liquids show a higher solubility and a higher selectivity than conventional solvents. Functionalised ionic liquids combine the physical and chemical absorption of gases.

5.6 ENGINEERING ASPECTS

5.6.1 Equipment

Successful introduction of ionic liquids into extraction operations also requires knowledge of their hydrodynamic and mass transfer characteristics because the viscosity and the density of ionic liquids are usually higher than those of conventional solvents. Common extraction contactors may not be suitable for separation processes with ionic liquids as extractants. Therefore, mechanical energy has to be used to enhance mass transfer into ionic liquids.

A centrifugal extractor was used for the separation of ethylbenzene from octane with $[C_4mim][PF_6]$ [213]. The centrifugal extraction system contained four 50-mm diameter annular space extractors. The density of the ionic liquid is 1030 kg m^{-3}, and its viscosity is 450 mPa s at 298.15 K and 80 mPa s at 313.15 K. At optimum process conditions (rotation speed of 3500 rpm for this extractor), a single phase efficiency of 90% was obtained.

Extraction of aromatics from aliphatics in a pilot plant rotating disc contactor (RDC) with the ionic liquids $[C_4m_\gamma py][BF_4]$ [214] and $[C_4m_\beta py][N(CN)_2]$, and sulfolane as the solvent, was investigated. In the extraction process, hydrodynamics (drop size, hold-up, and operational window) and mass transfer efficiency determine the column performance [215]. The investigated parameters concern the total flux, the rotating speed of the RDC [216, 217], and the concentration of toluene in the organic (heptane) phase.

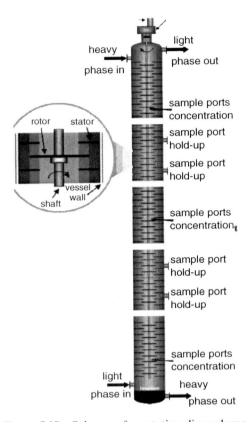

Figure 5.19   Scheme of a rotating disc column.

The pilot RDC extraction column is schematically shown in Figure 5.19. The column consists of nine jacketed glass segments, each 360 mm in length, and an inside diameter of 60 mm, with eight stirred compartments each. Settlers of 500 mm (top and bottom) with an internal diameter of 90 mm enclose the stirred segments. The solvent (ionic liquid or sulfolane) is the dispersed phase, which is fed at the top of the column, and the extract phase is collected from the bottom settler. The heptane/toluene phase is fed from the bottom and the raffinate phase is collected from the top settler. Regenerated ionic liquid was used in the extraction experiments.

5.6.2 Hydrodynamics

Figure 5.20 shows that the drop sizes for the extraction of 10 wt% toluene from heptane decreases with increasing rotor speed for all three solvents, as expected. At a flux of 8 m^3 m^{-2} h^{-1}, with constant flow rates for both phases,

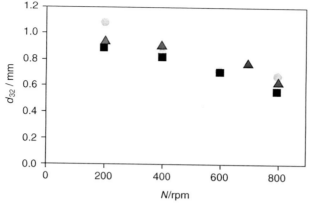

Figure 5.20   Drop sizes for extraction of 10 wt% toluene from heptane with sulfolane (○), [C₄mγpy][BF₄] (■) [214] and [C₄mβpy][N(CN)₂] (▲).

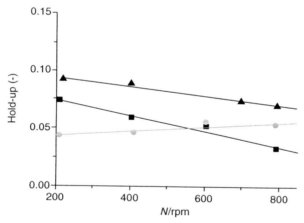

Figure 5.21   Hold-up for sulfolane (○, S/F = 7.5), [C₄mγpy][BF₄] (■, S/F = 6) and [C₄mβpy][N(CN)₂] (▲, S/F = 4), extraction 10 wt% toluene, flux 8 m³ m⁻² h⁻¹.

the hold-up increases with increasing rotor speed with sulfolane as the solvent, but decreases when the ionic liquids [C₄mγpy][BF₄] and [C₄mβpy][N(CN)₂] are used as solvents (Figs. 5.21 and 5.22). This last phenomenon is not expected since an increase in rotor speed should result in a greater torroidal motion of the dispersed phase, which results in a greater residence time of the dispersed phase. The explanation for this is that cohesion of droplets at the stator rings causes an increase in the hold-up at lower fluxes. Moreover, the decrease in

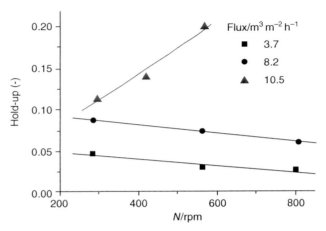

Figure 5.22   Hold-up for $[C_4m_\gamma py][BF_4]$ extraction 75 wt% toluene (S/F = 7.5) [214].

hold-up with increasing rotor speed was also found by others [218]. At low fluxes and low rotor speeds, there is cohesion of the droplets at the stator ring and when the rotor speed is increased, the droplets at the stator ring tend to fall down through the column. Because there is an almost free fall of droplets of the dispersed phase, drops just fall faster if the rotor speed increases, causing a decrease in the hold-up. There is less cohesion at higher fluxes, which is confirmed by the decrease in drop size at increasing flux. Because sulfolane has a much lower viscosity (8 mPa s at 313 K) than $[C_4m_\beta py][N(CN)_2]$ (46 mPa s at 298 K) or $[C_4m_\gamma py][BF_4]$ (80 mPa s at 313 K), there is less cohesion when sulfolane is used and, therefore, the sulfolane hold-up increases with increasing rotor speed, as expected.

For sulfolane, an S/F ratio of 7.5 (10 wt%) or 6 (50 wt%), for $[C_4m_\gamma py][BF_4]$ an S/F ratio of 6 (10 wt%) or 7.5 (50 and 75 wt%), and for $[C_4m_\beta py][N(CN)_2]$ an S/F ratio of 4 (10 wt% toluene) was used for the extractive removal of toluene from a mixture of 10–75 wt% toluene in heptane at 40 °C. It can be concluded from Figures 5.23 and 5.24 that, for extraction of 10 wt% toluene as well for extraction of 50 wt% toluene in the RDC, the maximum achievable flux is higher for $[C_4m_\gamma py][BF_4]$ than for sulfolane.

The operational window of sulfolane is larger than of $[C_4m_\beta py][N(CN)_2]$, which can be explained by the differences in density, viscosity, and interfacial tension between the solvent and feed phase. Sulfolane has a density of 1261 kg m^{-3} and a viscosity of 8 mPa s, $[C_4m_\beta py][N(CN)_2]$ has a density of 1050 kg m^{-3} and a viscosity of 20 mPa s, and $[C_4m_\gamma py][BF_4]$ has a density of 1184 kg m^{-3} and a viscosity of 80 mPa s (all at 313 K). The interfacial tension between sulfolane and the toluene/heptane phase is lower than for both ionic liquids.

Extraction of more than 65 wt% toluene from heptane is not possible with sulfolane, but extraction with an ionic liquid is still possible. Figure 5.25 shows

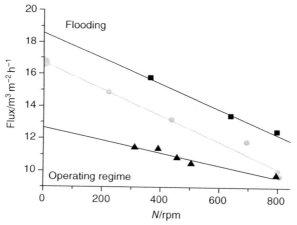

Figure 5.23 Operational region for extraction of 10 wt % toluene from *n*-heptane in the RDC at T = 313 K with sulfolane (●), [C₄m₇py][BF₄] (■) [214] and [C₄mβpy][N(CN)₂] (▲).

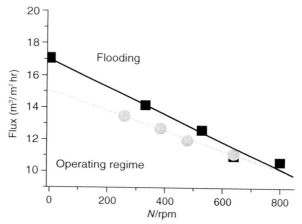

Figure 5.24 Operational region for extraction of 50 wt% toluene from *n*-heptane in the RDC at $T = 313$ K with sulfolane (●) and [C₄m₇py][BF₄] (■) [214].

the operational region for extraction of 75 wt% toluene from heptane in the RDC.

5.6.3 Mass Transfer

Figures 5.26 and 5.27 show the concentration profiles of toluene over the column length. From Figure 5.26, it can be concluded that with sulfolane as

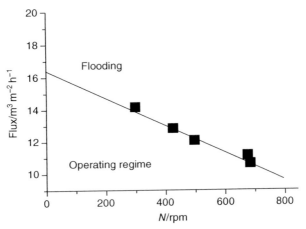

Figure 5.25 Operational region for extraction of 75 wt% toluene from *n*-heptane in the RDC at $T = 313$ K with $[C_4m_\gamma py][BF_4]$ [214].

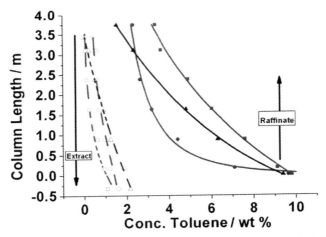

Figure 5.26 Concentration profile of toluene over the column, $T = 313$ K, $N = 640$ rpm, raffinate (closed symbols), extract (open symbols), sulfolane (●, ○), $[C_4m_\gamma py][BF_4]$ (■,), and $[C_4m_\beta py][N(CN)_2]$ (▲,) as solvents.

solvent more toluene is extracted than with $[C_4m_\gamma py][BF_4]$, although the distribution ratios of toluene in sulfolane and $[C_4m_\gamma py][BF_4]$ are comparable on a weight basis (0.22 kg per kg for $[C_4m_\gamma py][BF_4]$ vs. 0.26 kg per kg for sulfolane). The ionic liquid $[C_4m_\beta py][N(CN)_2]$ shows a slightly higher mass transfer of toluene than sulfolane. The raffinate phase with this ionic liquid contains

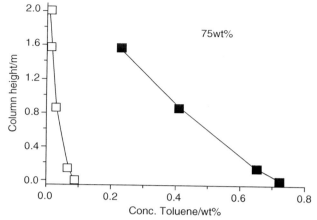

Figure 5.27 Concentration profile of toluene over the column, $T = 313$ K, $N = 640$ rpm, closed symbols for $[C_4m_ypy][BF_4]$ as solvent, raffinate (■), extract (□).

less toluene than with sulfolane, and the extract phase with the ionic liquid has a higher toluene content than with sulfolane. The difference in maximum flux can be explained by the difference in drop size. Due to the higher viscosity of $[C_4m_ypy][BF_4]$ compared with that of sulfolane, the ionic liquid has larger drop sizes than sulfolane and, consequently, a higher gravitational force and less resistance. The $[C_4m_ypy][BF_4]$ droplets fall faster through the heptane and higher counteracting forces can be overcome.

Extraction of more than 65 wt% toluene is not possible with sulfolane since the compositions approach the plait point in the ternary diagram, so separation is not possible anymore [67, 68], but, from Figure 5.27, it can be concluded that with $[C_4m_ypy][BF_4]$, a rapid decrease of toluene concentration in the raffinate phase occurred. Ionic liquids such as $[C_4m_ypy][BF_4]$ offer the advantage of extracting feeds containing high (>65%) aromatics, which is not possible with the currently used solvents.

As well as the toluene/heptane separation, experiments with model FCC, light catalytically cracked spirit (LCCS), and diesel feeds were also carried out. In addition, refinery streams from the BP refinery in Rotterdam were used. The performance with the actual refinery stream as feed is comparable with that of the model feeds. The removal of benzene and toluene from the actual refinery FCC stream was 81% and 71% respectively, and from the model feed, 89% and 75%, respectively.

5.6.4 Conclusions

Ionic liquids can be suitable extractants for several compounds, as is proven in batch equilibrium experiments. Currently, it is also shown that a continuous extraction using ionic liquids in a pilot plant produces good results. The extraction with ionic liquids was compared to an extraction with a known solvent in industry, sulfolane, and the results are that the ionic liquid $[C_4m_\gamma py][BF_4]$ has a larger operational window, but that sulfolane has better mass transfer than the more viscous ionic liquid $[C_4m_\gamma py][BF_4]$. A less viscous ionic liquid, such as $[C_4m_\beta py][N(CN)_2]$, has a smaller operational window than sulfolane, due to a smaller density difference with the feed phase, but has a better performance in mass transfer than sulfolane.

5.7   DESIGN OF A SEPARATION PROCESS

5.7.1   Introduction

Since the properties of an ionic liquid are defined by the combination of the cation and anion, so-called tailoring offers the possibility of creating a special solvent for a specific task. Because of the large number of combinations ($>10^{12}$), it is impossible to synthesise all ionic liquids and measure their properties [219]. Thus, to determine suitable ionic liquids for a certain problem, simulation tools will be very useful. The use of group contribution methods such as UNIFAC is difficult because the specific interaction energy parameters of ionic liquids are not always available yet. Therefore, a dielectric continuum model (COSMO-RS) has been chosen by a large number of authors. COSMO-RS is a quantum chemical approach, recently proposed by Klamt and Eckert [220–222] for the *a priori* prediction of activity coefficients and other thermophysical data using only structural information of the molecules. This method enables screening of ionic liquids based on the surface charge, the polarity. In addition, it is possible to calculate activity coefficients at infinite dilution.

5.7.2   Application of COSMO-RS

Several authors have used COSMO-RS for the prediction of activity coefficients at infinite dilution with several ionic liquids [223–225]. In aqueous systems, COSMO-RS provides good results for systems with haloalkanes or aromatics as solutes in water, but COSMO-RS was less successful for non-aqueous systems. The predictions were evaluated with experimental data and the deviations varied up to 16% [224].

Liquid–liquid and vapour–liquid equilibria were evaluated with COSMO-RS [226–233] and also for the use of ionic liquids solvents in extractive distillation [152, 154, 156]. COSMO-RS calculations for a non-polar mixture of MCH and toluene showed that the selection of the solvent is complicated.

Figure 5.28 shows the COSMO-RS predictions and the experimental values for the separation factors of MCH and toluene with a molar fraction of 0.30 of the selected ionic liquids in the liquid phase at 373.15 K [156]. COSMO-RS over-predicts the separation factors values for this system by about 20%.

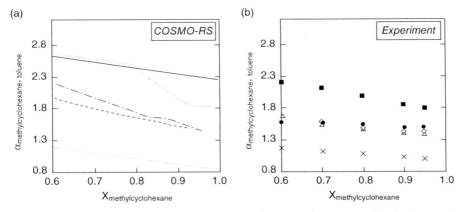

Figure 5.28   (a) COSMO-RS predictions and (b) experiments. $T = 373.15$ K. ($\blacksquare$ —)
[C$_8$quin][NTf$_2$], ($\triangle$---) [C$_8$quin][BBB], ($\diamond$---) [C$_4$mim][NTf$_2$], ($\bullet$---) ECOENG^{TM} 500
(Solvent Innovation, Merck, Darmstadt, Germany), ($\times$-----) binary [156].

COSMO-RS is also used for screening of ionic liquids as green solvents in
denitrification and desulfurisation of fuels [234–236]. The ionic liquid 4-ethyl-
4-methylmorpholinium thiocyanate ([C$_2$mmor][SCN]) proved to be the most
viable ionic liquid for the removal of thiophene, BT, and DBT [234]. The
COSMO-RS screening of ionic liquids for the denitrification process showed
that, in general, cations without aromatic rings (such as [C$_2$mpyr]$^+$, [C$_2$mpip]$^+$,
and [C$_2$mmor]$^+$) gave higher selectivity and capacity for both five- and six-
membered ring nitrogen species. The anions [SCN]$^-$ and [O$_2$CMe]$^-$ showed the
highest selectivity [235]. BT was extracted from hexane with [C$_2$mim][C$_2$H$_5$SO$_4$]
and [C$_2$mim][O$_2$CMe]. COSMO-RS was used to predict the performance of
single, as well as mixed, ionic liquids. Root mean square deviation (RMSD)
values of 4.36% and 7.87% were achieved for [C$_2$mim][C$_2$H$_5$SO$_4$] and [C$_2$mim]
[O$_2$CMe], respectively [236].

Not only activity coefficients, but also other thermophysical data, can be
predicted by COSMO-RS, such as vapour pressure and vaporisation enthalpy
[237], densities and molar volumes [238], and water solubilities [239, 240]. The
calculated enthalpies are in good agreement with the experimental data, but
the vapour pressure is underestimated.

Comparison of COSMO-RS with other methods was carried out by several
authors, for example, COSMOSPACE [241], UNIFAC [242], or UNIQUAC
[243]. COSMOSPACE yielded better results than UNIQUAC and COSMO-
RS. On the other hand, calculations with COSMO-RS of the miscibility gap
of [C$_4$mim][PF$_6$] with an alcohol strongly differed from the experimental
results, and a far better agreement with experimental data was found with a
UNIQUAC-based correlation [243].

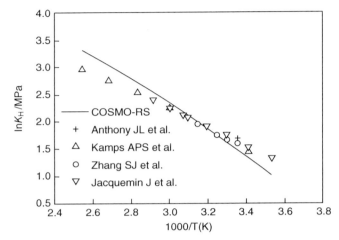

Figure 5.29   Henry's constants for CO_2 in [C_4mim][PF_6] [245]. — COSMO-RS, + Anthony et al. [178], △ Pérez-Salado Kamps et al. [179], ○ Zhang et al. [246], ▽ Jacquemin et al. [247].

Provided that the appropriate interaction parameters are available, the modified UNIFAC (Dortmund) can successfully be applied for systems with ionic liquids [242, 244]. COSMO-RS can also be used as a tool to screen suitable ionic liquids for the absorption of CO_2 [245]. Figure 5.29 shows the comparison of the Henry's constants of CO_2 in [C_4mim][PF_6].

5.7.3   Conclusions

The general conclusion is that γ^∞-values determined with COSMO-RS can be used for a first screening. However, quantitative predictions are still inaccurate and COSMO-RS does not always give sound predictions of γ-values for finite concentrations. Therefore, it cannot be used for calculation of distribution coefficients and selectivities in real solutions yet. COSMO-RS appears to be a very promising tool to support the design of suitable ionic liquids for specific (separation) problems.

5.8   CONCLUSIONS

There are no industrial applications for separations using ionic liquids yet. In order to introduce separation with ionic liquids in industry, more applied research is required, especially on pilot plant scale, to determine optimal process conditions.

Development of a complete separation process with ionic liquids, that is, including the primary separation, recovery of the products, and the regeneration of the ionic liquid for reuse, must be carried out.

Recovery of the separated products is rarely investigated, but it is a required aspect for a complete separation process. Regeneration of ionic liquids is rarely reported but is also a very important issue for industrial use as only ionic liquids that can be regenerated will be used.

Activity coefficients at infinite dilution of ionic liquids are useful for screening purposes, but for separations with ionic liquids, real distribution coefficient and selectivity values at finite dilutions will have to be obtained, as these are concentration dependent.

Ionic liquids are not always better than conventional solvents, but they can show advantages in some specific cases. Unfortunately, no benchmarks with conventional solvents are mentioned in a large number of publications.

Too much emphasis still exists on ionic liquids with hexafluorophosphate as the anion because these are versatile hydrophobic ionic liquids. In industry, these ionic liquids will never be used, due to their instability towards water with concomitant HF formation.

Task-specific ionic liquids are generally more selective than standard ionic liquids and, therefore, more focus must be on the development of these ionic liquids.

REFERENCES

1 Huddleston, J.G., Visser, A.E., Reichert, W.M., Willauer, H.D., Broker, G.A., and Rogers, R.D., Characterization and comparison of hydrophilic and hydrophobic room temperature ionic liquids incorporating the imidazolium cation, *Green Chem.* **3**(4), 156–164 (2001).

2 Seddon, K.R., Stark, A., and Torres, M.-J., Influence of chloride, water, and organic solvents on the physical properties of ionic liquids, *Pure Appl. Chem.* **72**(12), 2275–2287 (2000).

3 Swatloski, R.P., Holbrey, J.D., and Rogers, R.D., Ionic liquids are not always green: hydrolysis of 1-butyl-3-methylimidazolium hexafluorophosphate, *Green Chem.* **5**(4), 361–363 (2003).

4 Filiz, M., Sayar, N.A., and Sayar, A.A., Extraction of cobalt(II) from aqueous hydrochloric acid solutions into alamine 336-m-xylene mixtures, *Hydrometallurgy* **81**(3–4), 167–173 (2006).

5 Bradshaw, J.S., and Izatt, R.M., Crown ethers: the search for selective ion ligating agents, *Acc. Chem. Res.* **30**(8), 338–345 (1997).

6 Dai, S., Ju, Y.H., and Barnes, C.E., Solvent extraction of strontium nitrate by a crown ether using room-temperature ionic liquids, *J. Chem. Soc. Dalton Trans.* **8**, 1201–1202 (1999).

7 Visser, A.E., Swatloski, R.P., Reichert, W.M., Griffin, S.T., and Rogers, R.D., Traditional extractants in nontraditional solvents: groups 1 and 2 extraction by crown ethers in room-temperature ionic liquids, *Ind. Eng. Chem. Res.* **39**(10), 3596–3604 (2000).

8   Visser, A.E., Swatloski, R.P., Griffin, S.T., Hartman, D.H., and Rogers, R.D., Liquid/liquid extraction of metal ions in room temperature ionic liquids, *Sep. Sci. Technol.* **36**(5&6), 785–804 (2001).

9   Bartsch, R.A., Chun, S., and Dzyuba, S.V., Ionic liquids as novel diluents for solvent extraction of metal salts by crown ethers, in: *Ionic liquids industrial applications to green chemistry*, Vol. 818, eds. R.D. Rogers, and K.R. Seddon, American Chemical Society, Washington, DC (2002), pp. 58–68.

10 Hirayama, N., Okamura, H., Kidani, K., and Imura, H., Ionic liquid synergistic cation-exchange system for the selective extraction of lanthanum(III) using 2-thenoyltrifluoroacetone and 18-crown-6, *Anal. Sci.* **24**(6), 697–699 (2008).

11 Luo, H., Yu, M., and Dai, S., Solvent extraction of Sr^{2+} and Cs^+ based on hydrophobic protic ionic liquids, *Z. Naturforsch., A, Phys. Sci.* **62**(5/6), 281–291 (2007).

12 Shimojo, K., and Goto, M., First application of calixarenes as extractants in room-temperature ionic liquids, *Chem. Lett.* **33**(3), 320–321 (2004).

13 Visser, A.E., Swatloski, R.P., Hartman, D.H., Huddleston, J.G., and Rogers, R.D., Calixarenes as ligands in environmentally-benign liquid-liquid extraction media, aquous biphasic systems and room temperature ionic liquid, *ACS Symp. Ser.* **757**, Calixarenes for Separations, 223–236 (2000).

14 Shimojo, K., and Goto, M., Solvent extraction and stripping of silver ions in room-temperature ionic liquids containing calixarenes, *Anal. Chem.* **76**(17), 5039–5044 (2004).

15 Sieffert, N., and Wipff, G., Alkali cation extraction by calix[4]crown-6 to room-temperature ionic liquids. The effect of solvent anion and humidity investigated by molecular dynamics simulations, *J. Phys. Chem. A* **110**(3), 1106–1117 (2006).

16 Sieffert, N., and Wipff, G., Comparing an ionic liquid to a molecular solvent in the cesium cation extraction by a calixarene: a molecular dynamics study of the aqueous interfaces, *J. Phys. Chem. B* **110**(39), 19497–19506 (2006).

17 Stepinski, D.C., Jensen, M.P., Dzielawa, J.A., and Dietz, M.L., Synergistic effects in the facilitated transfer of metal ions into room-temperature ionic liquids, *Green Chem.* **7**(3), 151–158 (2005).

18 Davis, J.H.J., Task-specific ionic liquids, *Chem. Lett.* **33**(9), 1072–1077 (2004).

19 Zhao, H., Xia, S., and Ma, P., Use of ionic liquids as "green" solvents for extractions, *J. Chem. Technol. Biotechnol.* **80**(10), 1089–1096 (2005).

20 Chun, S., Dzyuba, S.V., and Bartsch, R.A., Influence of structural variation in room-temperature ionic liquids on the selectivity and efficiency of competitive alkali metal salt extraction by a crown ether, *Anal. Chem.* **73**(15), 3737–3741 (2001).

21 Keskin, S., Kayrak-Talay, D., Akman, U., and Hortacsu, O., A review of ionic liquids towards supercritical fluid applications, *J. Supercrit. Fluids* **43**(1), 150–180 (2007).

22 Dietz, M.L., Dzielawa, J.A., Laszak, I., Young, B.A., and Jensen, M.P., Influence of solvent structural variations on the mechanism of facilitated ion transfer into room-temperature ionic liquids, *Green Chem.* **5**(6), 682–685 (2003).

23 Dietz, M.L., Jakab, S., Yamato, K., and Bartsch, R.A., Stereochemical effects on the mode of facilitated ion transfer into room-temperature ionic liquids, *Green Chem.* **10**(2), 174–176 (2008).

24   Wei, G.-T., Yang, Z., and Chen, C.-J., Room temperature ionic liquid as a novel medium for liquid/liquid extraction of metal ions, *Anal. Chim. Acta* **488**(2), 183–192 (2003).

25   Cocalia, V.A., Jensen, M.P., Holbrey, J.D., Spear, S.K., Stepinski, D.C., and Rogers, R.D., Identical extraction behavior and coordination of trivalent or hexavalent f-element cations using ionic liquid and molecular solvents, *Dalton Trans.* **11**, 1966–1971 (2005).

26   Cocalia, V.A., Holbrey, J.D., Gutowski, K.E., Bridges, N.J., and Rogers, R.D., Separations of metal ions using ionic liquids: the challenges of multiple mechanisms, *Tsinghua Sci. Technol.* **11**(2), 188–193 (2006).

27   Dietz, M.L., and Stepinski, D.C., A ternary mechanism for the facilitated transfer of metal ions into room-temperature ionic liquids (RTILs): implications for the greenness of RTILs as extraction solvents, *Green Chem.* **7**(10), 747–750 (2005).

28   Nakashima, K., Kubota, F., Maruyama, T., and Goto, M., Feasibility of ionic liquids as alternative separation media for industrial solvent extraction processes, *Ind. Eng. Chem. Res.* **44**(12), 4368–4372 (2005).

29   Sun, X., Peng, B., Chen, J., Li, D., and Luo, F., An effective method for enhancing metal-ions' selectivity of ionic liquid-based extraction system: adding water-soluble complexing agent, *Talanta* **74**(4), 1071–1074 (2008).

30   Dietz, M.L., Ionic liquids as extraction solvents: where do we stand? *Sep. Sci. Technol.* **41**(10), 2047–2063 (2006).

31   Dietz, M.L., and Stepinski, D.C., Anion concentration-dependent partitioning mechanism in the extraction of uranium into room-temperature ionic liquids, *Talanta* **75**(2), 598–603 (2008).

32   Zuo, Y., Liu, Y., Chen, J., and Li, D.Q., The separation of cerium(IV) from nitric acid solutions containing thorium(IV) and lanthanides(III) using pure [C_8mim] PF_6 as extracting phase, *Ind. Eng. Chem. Res.* **47**(7), 2349–2355 (2008).

33   Zuo, Y., Liu, Y., Chen, J., and Li, D.Q., Extraction and recovery of cerium(IV) along with fluorine(I) from bastnasite leaching liquor by DEHEHP in [C_8mim] PF_6, *J. Chem. Technol. Biotechnol.* **84**(7), 949–956 (2009).

34   Kozonoi, N., and Ikeda, Y., Extraction mechanism of metal ion from aqueous solution to the hydrophobic ionic liquid, 1-butyl-3-methylimidazolium nonafluorobutanesulfonate, *Monatsh. Chem.* **138**(11), 1145–1151 (2007).

35   Jensen, M.P., Neuefeind, J., Beitz, J.V., Skanthakumar, S., and Soderholm, L., Mechanisms of metal ion transfer into room-temperature ionic liquids: the role of anion exchange, *J. Am. Chem. Soc.* **125**(50), 15466–15473 (2003).

36   Dietz, M.L., and Dzielawa, J.A., Ion-exchange as a mode of cation transfer into room-temperature ionic liquids containing crown ethers: implications for the "greenness" of ionic liquids as diluents in liquid-liquid extraction, *Chem. Commun.* **20**, 2124–2125 (2001).

37   Visser, A.E., Swatloski, R.P., Reichert, W.M., Mayton, R., Sheff, S., Wierzbicki, A., Davis, J.H., Jr, and Rogers, R.D., Task-specific ionic liquids incorporating novel cations for the coordination and extraction of Hg^{2+} and Cd^{2+}: synthesis, characterization, and extraction studies, *Environ. Sci. Technol.* **36**(11), 2523–2529 (2002).

38 Nockemann, P., Thijs, B., Pittois, S., Thoen, J., Glorieux, C., VanHecke, K., Van-Meervelt, L., Kirchner, B., and Binnemans, K., Task-specific ionic liquid for solubilizing metal oxides, *J. Phys. Chem. B* **110**(42), 20978–20992 (2006).

39 Ajioka, T., Oshima, S., and Hirayama, N., Use of 8-sulfonamidoquinoline derivatives as chelate extraction reagents in ionic liquid extraction system, *Talanta* **74**(4), 903–908 (2008).

40 Harjani, J.R., Friscic, T., MacGillivray, L.R., and Singer, R.D., Metal chelate formation using a task-specific ionic liquid, *Inorg. Chem.* **45**(25), 10025–10027 (2006).

41 Harjani, J.R., Friscic, T., MacGillivray, L.R., and Singer, R.D., Removal of metal ions from aqueous solutions using chelating task-specific ionic liquids, *Dalton Trans.* **34**, 4595–4601 (2008).

42 Li, M., Wang, T., Pham, P.J., Pittman, C.U., Jr, and Li, T., Liquid phase extraction and separation of noble organometallic catalysts by functionalized ionic liquids, *Sep. Sci. Technol.* **43**(4), 828–841 (2008).

43 Weissermel, K., and Arpe, H.-J., *Aromatics—production and conversion, in industrial organic chemistry, 4*th *completely revised edition*, Wiley-VCH, Weinheim (2003), pp. 313–336.

44 Chen, J., Duan, L.-P., Mi, J.-G., Fei, W.-Y., and Li, Z.-C., Liquid-liquid equilibria of multi-component systems including n-hexane, n-octane, benzene, toluene, xylene and sulfolane at 298.15 K and atmospheric pressure, *Fluid Phase Equilib.* **173**(1), 109–119 (2000).

45 Chen, J., Li, Z., and Duan, L., Liquid-liquid equilibria of ternary and quaternary systems including cyclohexane, 1-heptene, benzene, toluene, and sulfolane at 298.15 K, *J. Chem. Eng. Data* **45**(4), 689–692 (2000).

46 Choi, Y.J., Cho, K.W., Cho, B.W., and Yeo, Y.K., Optimization of the sulfolane extraction plant based on modeling and simulation, *Ind. Eng. Chem. Res.* **41**(22), 5504–5509 (2002).

47 Krishna, R., Goswami, A.N., Nanoti, S.M., Rawat, B.S., Khanna, M.K., and Dobhal, J., Extraction of aromatics from 63–69°C naphtha fraction for food-grade hexane production using sulfolane and NMP as solvents, *Indian J. Technol.* **25**(12), 602–606 (1987).

48 Yorulmaz, Y., and Karpuzcu, F., Sulfolane versus diethylene glycol in recovery of aromatics, *Chem. Eng. Res. Des.* **63**(3), 184–190 (1985).

49 De Fre, R.M., and Verhoeye, L.A., Phase equilibria in systems composed of an aliphatic and an aromatic hydrocarbon and sulfolane, *J. Appl. Chem. Biotechnol.* **26**(9), 469–487 (1976).

50 Mahmoudi, J., and Lotfollahi, M.N., (Liquid + liquid) equilibria of (sulfolane + benzene + n-hexane), (N-formylmorpholine + benzene + n-hexane), and (sulfolane + N-formylmorpholine + benzene + n-hexane) at temperatures ranging from (298.15 to 318.15) K: experimental results and correlation, *J. Chem. Thermodyn.* **42**(4), 466–471 (2010).

51 Lee, S., and Kim, H., Liquid-liquid equilibria for the ternary systems sulfolane + octane + benzene, sulfolane + octane + toluene and sulfolane + octane + p-xylene, *J. Chem. Eng. Data* **40**(2), 499–503 (1995).

52 Al-Jimaz, A.S., Fandary, M.S., Alkhaldi, K.H.A.E., Al-Kandary, J.A., and Fahim, M.A., Extraction of aromatics from middle distillate using N-methyl-2-pyrrolidone:

experiment, modeling, and optimization, *Ind. Eng. Chem. Res.* **46**(17), 5686–5696 (2007).

53 Chen, D.C., Ye, H.Q., and Wu, H., Measurement and correlation of liquid-liquid equilibria of methylcyclohexane + toluene + N-formylmorpholine at (293, 303, 313, and 323) K, *J. Chem. Eng. Data* **52**(4), 1297–1301 (2007).

54 Wang, W., Gou, Z., and Zhu, S., Liquid-liquid equilibria for aromatics extraction systems with tetraethylene glycol, *J. Chem. Eng. Data* **43**(1), 81–83 (1998).

55 Al-Sahhaf, T.A., and Kapetanovic, E., Measurement and prediction of phase equilibria in the extraction of aromatics from naphtha reformate by tetraethylene glycol, *Fluid Phase Equilib.* **118**(2), 271–285 (1996).

56 Ali, S.H., Lababidi, H.M.S., Merchant, S.Q., and Fahim, M.A., Extraction of aromatics from naphtha reformate using propylene carbonate, *Fluid Phase Equilib.* **214**(1), 25–38 (2003).

57 Schneider, D.F., Avoid sulfolane regeneration problems, *Chem. Eng. Progr.* **100**(7), 34–39 (2004).

58 Firnhaber, B., Emmrich, G., Ennenbach, F., and Ranke, U., Separation process for the recovery of pure aromatics, *Erdoel Erdgas Kohle* **116**(5), 254–260 (2000).

59 Hombourger, T., Gouzien, L., Mikitenko, P., and Bonfils, P., Solvent extraction in the oil industry, in: *Petroleum refining, 2. separation processes*, Vol. 2, ed. J.P. Wauquier, Editions Technip, Paris (2000), pp. 359–456.

60 Hamid, S.H., and Ali, M.A., Comparative study of solvents for the extraction of aromatics from naphtha, *Energy Sources* **18**(1), 65–84 (1996).

61 Rawat, B.S., and Gulati, I.B., Liquid-liquid equilibrium studies for separation of aromatics, *J. Appl. Chem. Biotechnol.* **26**(8), 425–435 (1976).

62 Huddleston, J.G., and Rogers, R.D., Room temperature ionic liquids as novel media for "clean" liquid-liquid extraction, *Chem. Commun.* **16**, 1765–1766 (1998).

63 Heintz, A., Recent developments in thermodynamics and thermophysics of non-aqueous mixtures containing ionic liquids. A review, *J. Chem. Thermodyn.* **37**(6), 525–535 (2005).

64 Anjan, S.T., Ionic liquids for aromatic extraction: are they ready? *Chem. Eng. Progr.* **102**(12), 30–39 (2006).

65 Abu-Eishah, S.I., and Dowaidar, A.M., Liquid-liquid equilibrium of ternary systems of cyclohexane + (benzene, + toluene, + ethylbenzene, or + o-xylene) + 4-methyl-*N*-butylpyridinium tetrafluoroborate ionic liquid at 303.15 K, *J. Chem. Eng. Data* **53**(8), 1708–1712 (2008).

66 Meindersma, G.W., Hansmeier, A.R., and de Haan, A.B., Ionic liquids for aromatics extraction. present status and future outlook, *Ind. Eng. Chem. Res.* **49**(16), 7530–7540 (2010).

67 Meindersma, G.W., Podt, A., and de Haan, A.B., Selection of ionic liquids for the extraction of aromatic hydrocarbons from aromatic/aliphatic mixtures, *Fuel Process. Technol.* **87**(1), 59–70 (2005).

68 Meindersma, G.W., Podt, A.J.G., and de Haan, A.B., Ternary liquid-liquid equilibria for mixtures of toluene + *n*-heptane + an ionic liquid, *Fluid Phase Equilib.* **247**(1–2), 158–168 (2006).

69 García, J., Fernández, A., Torrecilla, J.S., Oliet, M., and Rodríguez, F., Liquid-liquid equilibria for {hexane + benzene + 1-ethyl-3-methylimidazolium ethylsulfate} at (298.2, 313.2 and 328.2) K, *Fluid Phase Equilib.* **282**(2), 117–120 (2009).

70   Arce, A., Earle, M.J., Rodríguez, H., and Seddon, K.R., Separation of aromatic hydrocarbons from alkanes using the ionic liquid 1-ethyl-3-methylimidazolium bis{(trifluoromethyl) sulfonyl}amide, *Green Chem.* **9**(1), 70–74 (2007).

71   Arce, A., Earle, M.J., Rodríguez, H., Seddon, K.R., and Soto, A., 1-Ethyl-3-methylimidazolium bis{(trifluoromethyl)sulfonyl}amide as solvent for the separation of aromatic and aliphatic hydrocarbons by liquid extraction—extension to C_7- and C_8-fractions, *Green Chem.* **10**(12), 1294–1300 (2008).

72   Arce, A., Earle, M.J., Rodríguez, H., Seddon, K.R., and Soto, A., Isomer effect in the separation of octane and xylenes using the ionic liquid 1-ethyl-3-methylimidazolium bis{(trifluoromethyl)sulfonyl}amide, *Fluid Phase Equilib.* **294**(1–2), 180–186 (2010).

73   Selvan, M.S., McKinley, M.D., Dubois, R.H., and Atwood, J.L., Liquid-liquid equilibria for toluene + heptane + 1-ethyl-3-methylimidazolium triiodide and toluene + heptane + 1-butyl-3-methylimidazolium triiodide, *J. Chem. Eng. Data* **45**(5), 841–845 (2000).

74   Meindersma, G.W., Galán Sánchez, L.M., Hansmeier, A.R., and de Haan, A.B., Invited review. application of task-specific Ionic liquids for intensified separations, *Monatsh. Chem.* **138**(11), 1125–1136 (2007).

75   Zhou, Q., and Wang, L.S., Activity coefficients at infinite dilution of alkanes, alkenes, and alkyl benzenes in 1-butyl-3-methylimidazolium tetrafluoroborate using gas-liquid chromatography, *J. Chem. Eng. Data* **51**(5), 1698–1701 (2006).

76   García, J., Fernández, A., Torrecilla, J.S., Oliet, M., and Rodríguez, F., Ternary liquid-liquid equilibria measurement for hexane and benzene with the ionic liquid 1-butyl-3-methylimidazolium methylsulfate at T = (298.2, 313.2, and 328.2) K, *J. Chem. Eng. Data* **55**(1), 258–261 (2009).

77   Hansmeier, A.R., Minoves Ruiz, M., Meindersma, G.W., and de Haan, A.B., Liquid-liquid equilibria for the three ternary systems (3-methyl-*N*-butylpyridinium dicyanamide + toluene + *n*-heptane), (1-butyl-3-methylimidazolium dicyanamide + toluene + *n*-heptane) and (1-butyl-3-methylimidazolium thiocyanate + toluene + *n*-heptane) at T = (313.15 and 348.15) K and p = 0.1 MPa, *J. Chem. Eng. Data* **55**(2), 708–713 (2010).

78   Zhang, J., Huang, C., Chen, B., Ren, P., and Lei, Z., Extraction of aromatic hydrocarbons from aromatic/aliphatic mixtures using chloroaluminate room-temperature ionic liquids as extractants, *Energy Fuels* **21**(3), 1724–1730 (2007).

79   Arce, A., Earle, M.J., Rodríguez, H., Seddon, K.R., and Soto, A., Bis{(trifluoromethyl) sulfonyl}amide ionic liquids as solvents for the extraction of aromatic hydrocarbons from their mixtures with alkanes: effect of the nature of the cation, *Green Chem.* **11**(3), 365–372 (2009).

80   Kato, R., and Gmehling, J., Activity coefficients at infinite dilution of various solutes in the ionic liquids $[MMIM]^+[CH_3SO_4]^-$, $[MMIM]^+[CH_3OC_2H_4SO_4]^-$, $[MMIM]^+[(CH_3)_2PO_4]^-$, $[C_5H_5NC_2H_5]^+[(CF_3SO_2)_2N]^-$ and $[C_5H_5NH]^+$ $[C_2H_5OC_2H_4OSO_3]^-$, *Fluid Phase Equilib.* **226**, 37–44 (2004).

81   García, J., García, S., Torrecilla, J.S., Oliet, M., and Rodríguez, F., Liquid–liquid equilibria for the ternary systems {heptane + toluene + *N*-butylpyridinium tetrafluoroborate or *N*-hexylpyridinium tetrafluoroborate} at T = 313.2 K, *J. Chem. Eng. Data* **55**(8), 2862–2865 (2010).

82   García, J., García, S., Torrecilla, J.S., Oliet, M., and Rodríguez, F., Separation of toluene and heptane by liquid-liquid extraction using z-methyl-N-butylpyridinium tetrafluoroborate isomers (z = 2, 3, or 4) at T = 313.2 K, *J. Chem. Thermodyn.* **42**(8), 1004–1008 (2010).

83   Meindersma, G.W., Podt, A., and de Haan, A.B., Ternary liquid-liquid equilibria for mixtures of an aromatic + an aliphatic hydrocarbon + 4-methyl-*N*-butylpyridinium tetrafluoroborate, *J. Chem. Eng. Data* **51**(5), 1814–1819 (2006).

84   Marciniak, A., and Wlazło, M., Activity coefficients at infinite dilution measurements for organic solutes and water in the ionic liquid 1-(3-hydroxypropyl) pyridinium trifluorotris(perfluoroethyl)phosphate,*J. Phys. Chem. B* **114**(20),6990–6994 (2010).

85   Domańska, U., Redhi, G.G., and Marciniak, A., Activity coefficients at infinite dilution measurements for organic solutes and water in the ionic liquid 1-butyl-1-methylpyrrolidinium trifluoromethanesulfonate using GLC, *Fluid Phase Equilib.* **278**(1–2), 97–102 (2009).

86   Domańska, U., Pobudkowska, A., and Krolikowski, M., Separation of aromatic hydrocarbons from alkanes using ammonium ionic liquid C_2NTf_2 at T = 298.15 K, *Fluid Phase Equilib.* **259**(2), 173–179 (2007).

87   Domańska, U., and Marciniak, A., Activity coefficients at infinite dilution measurements for organic solutes and water in the ionic liquid triethylsulphonium bis(trifluoromethylsulfonyl)imide, *J. Chem. Thermodyn.* **41**(6), 754–758 (2009).

88   Meindersma, G.W., and de Haan, A.B., Conceptual process design for aromatic/aliphatic separation with ionic liquids, *Chem. Eng. Res. Des.* **86**(7), 745–752 (2008).

89   Meindersma, G.W., and de Haan, A.B., Separation of aromatic and aliphatic hydrocarbons with ionic liquids: a conceptual process design, in: *Ionic liquids: from knowledge to application*, Vol. 1030, eds. N.V. Plechkova, R.D. Rogers, and K.R. Seddon, American Chemical Society, Washington, DC (2009), pp. 255–272.

90   Zimmermann, H., and Walzl, R., Ethylene, in: *Ullmann's encyclopedia of industrial chemistry*, Vol. 13, 7th ed., ed. B. Elvers, Wiley-VCH, Weinheim (2007), pp. 465–530.

91   Wasserscheid, P., and Welton, T., Outlook, in: *Ionic liquids in synthesis, second edition*, Vol. 2, 2nd ed., eds. P. Wasserscheid, and T. Welton, Wiley-VCH, Weinheim (2008), pp. 689–704.

92   Maase, M., Ionic liquids on a large scale, . . . how they can help to improve chemical processes, in *Ionic Liquids—A Road-Map to Commercialisation*, London, UK (2004).

93   Maase, M., Cosi fan tutte (They all can do it) an improved way of doing it, in *Proceedings 1st International Congress on Ionic Liquids (COIL)*, Salzburg, A (2005), p. 37.

94   Ito, E., and van Veen, J.A.R., On novel processes for removing sulphur from refinery streams, *Catal. Today* **116**(4), 446–460 (2006).

95   Bösmann, A., Datsevich, L., Jess, A., Lauter, A., Schmitz, C., and Wasserscheid, P., Deep desulfurization of diesel fuel by extraction with ionic liquids, *Chem. Commun.* **23**, 2494–2495 (2001).

96   Schmidt, R., [bmim]AlCl₄ ionic liquid for deep desulfurization of real fuels, *Energy Fuels* **22**(3), 1774–1778 (2008).

97   Huang, C., Chen, B., Zhang, J., Liu, Z., and Li, Y., Desulfurization of gasoline by extraction with new ionic liquids, *Energy Fuels* **18**(6), 1862–1864 (2004).

98   Xie, L.L., Favre-Reguillon, A., Wang, X.X., Fu, X., Pellet-Rostaing, E., Toussaint, G., Geantet, C., Vrinat, M., and Lemaire, M., Selective extraction of neutral nitrogen compounds found in diesel feed by 1-butyl-3-methylimidazolium chloride, *Green Chem.* **10**(5), 524–531 (2008).

99   Ko, N.H., Lee, J.S., Huh, E.S., Lee, H., Jung, K.D., Kim, H.S., and Cheong, M., Extractive desulfurization using Fe-containing ionic liquids, *Energy Fuels* **22**(3), 1687–1690 (2008).

100 Zhang, S.G., and Zhang, Z.C., Novel properties of ionic liquids in selective sulfur removal from fuels at room temperature, *Green Chem.* **4**(4), 376–379 (2002).

101 Zhang, S., Zhang, Q., and Zhang, Z.C., Extractive desulfurization and denitrogenation of fuels using ionic liquids, *Ind. Eng. Chem. Res.* **43**(2), 614–622 (2004).

102 Alonso, L., Arce, A., Francisco, M., Rodríguez, O., and Soto, A., Gasoline desulfurization using extraction with [C$_8$mim][BF$_4$] ionic liquid, *Am. Inst. Chem. Eng.* **53**(12), 3108–3115 (2007).

103 Liu, D., Gui, J., Song, L., Zhang, X., and Sun, Z., Deep desulfurization of diesel fuel by extraction with task-specific ionic liquids, *Pet. Sci. Technol.* **26**(9), 973–982 (2008).

104 Nie, Y., Li, C., Sun, A., Meng, H., and Wang, Z., Extractive desulfurization of gasoline using imidazolium-based phosphoric ionic liquids, *Energy Fuels* **20**(5), 2083–2087 (2006).

105 Nie, Y., Li, C.-X., and Wang, Z.-H., Extractive desulfurization of fuel oil using alkylimidazole and its mixture with dialkylphosphate ionic liquids, *Ind. Eng. Chem. Res.* **46**(15), 5108–5112 (2007).

106 Jiang, X., Nie, Y., Li, C., and Wang, Z., Imidazolium-based alkylphosphate ionic liquids—a potential solvent for extractive desulfurization of fuel, *Fuel* **87**(1), 79–84 (2008).

107 Nie, Y., Li, C.X., Meng, H., and Wang, Z.H., N,N-dialkylimidazolium dialkylphosphate ionic liquids: their extractive performance for thiophene series compounds from fuel oils versus the length of alkyl group, *Fuel Process. Technol.* **89**(10), 978–983 (2008).

108 Wang, J.-L., Zhao, D.-S., Zhou, E.-P., and Dong, Z., Desulfurization of gasoline by extraction with N-alkyl-pyridinium-based ionic liquids, *J. Fuel Chem. Technol.* **35**(3), 293–296 (2007).

109 Holbrey, J.D., Lopez-Martin, I., Rothenberg, G., Seddon, K.R., Silvero, G., and Zheng, X., Desulfurization of oils using ionic liquids: selection of cationic and anionic components to enhance extraction efficiency, *Green Chem.* **10**(1), 87–92 (2008).

110 Esser, J., Wasserscheid, P., and Jess, A., Deep desulfurization of oil refinery streams by extraction with ionic liquids, *Green Chem.* **6**(7), 316–322 (2004).

111 Olivier-Bourbigou, H., Uzio, D., Diehl, F., and Magna, L., Process for removal of sulfur and nitrogen compounds from hydrocarbon fractions, Fr. Patent FR 2840916 A1 (2003).

112 Lo, W.-H., Yang, H.-Y., and Wei, G.-T., One-pot desulfurization of light oils by chemical oxidation and solvent extraction with room temperature ionic liquids, *Green Chem.* **5**(5), 639–642 (2003).

113   Zhao, D., Wang, J., and Zhou, E., Oxidative desulfurization of diesel fuel using a Bronsted acid room temperature ionic liquid in the presence of H_2O_2, *Green Chem.* **9**(11), 1219–1222 (2007).

114   Zhao, D., Sun, Z., Li, F., Liu, R., and Shan, H., Oxidative desulfurization of thiophene catalyzed by $(C_4H_9)_4NBr + 2C_6H_{11}NO$ coordinated ionic liquid, *Energy Fuels* **22**(5), 3065–3069 (2008).

115   Zhao, D., Wang, Y., Duan, E., and Zhang, J., Oxidation desulfurization of fuel using pyridinium-based ionic liquids as phase-transfer catalysts, *Fuel Process. Technol.* **91**(12), 1803–1806 (2010).

116   Gui, J., Liu, D., Sun, Z., Liu, D., Min, D., Song, B., and Peng, X., Deep oxidative desulfurization with task-specific ionic liquids: an experimental and computational study, *J. Mol. Catal., A-Chem.* **331**(1–2), 64–70 (2010).

117   Huang, W., Zhu, W., Li, H., Shi, H., Zhu, G., Liu, H., and Chen, G., Heteropolyanion-based ionic liquid for deep desulfurization of fuels in ionic liquids, *Ind. Eng. Chem. Res.* **49**(19), 8998–9003 (2010).

118   Zhu, W., Li, H., Jiang, X., Yan, Y., Lu, J., He, L., and Xia, J., Commercially available molybdic compound-catalyzed ultra-deep desulfurization of fuels in ionic liquids, *Green Chem.* **10**(6), 641–646 (2008).

119   Gargano, G.J., and Ruether, T., Process for removing sulphur from liquid hydrocarbons, WO2007106943 Patent 2007-AU350, WO2007106943 (2007).

120   Johnson, R.D., The processing of biomacromolecules: a challenge for the eighties, *Fluid Phase Equilib.* **29**, 109–123 (1986).

121   Vernau, J., and Kula, M.R., Extraction of proteins from biological raw material using aqueous polyethylene glycol-citrate phase systems, *Biotechnol. Appl. Biochem.* **12**(4), 397–404 (1990).

122   Creagh, A.L., Hasenack, B.B.E., Van der Padt, A., Sudhoelter, E.J.R., and Van't Riet, K., Separation of amino-acid enantiomers using micellar-enhanced ultrafiltration, *Biotechnol. Bioeng.* **44**(6), 690–698 (1994).

123   Riedl, W., and Raiser, T., Membrane-supported extraction of biomolecules with aqueous two-phase systems, *Desalination* **224**(1–3), 160–167 (2008).

124   Hatti-Kaul, R., Aqueous two-phase systems: a general overview, *Methods Biotechnol.* **11**, Aqueous Two-Phase Systems, 1–10 (2000).

125   Andrews, B.A., Schmidt, A.S., and Asenjo, J.A., Correlation for the partition behavior of proteins in aqueous two-phase systems: effect of surface hydrophobicity and charge, *Biotechnol. Bioeng.* **90**(3), 380–390 (2005).

126   Costa, M.J.L., Cunha, M.T., Cabral, J.M.S., and Aires-Barros, M.R., Scale-up of recombinant cutinase recovery by whole broth extraction with PEG-phosphate aqueous two-phase, *Bioseparation* **9**(4), 231–238 (2000).

127   Martínez-Aragón, M., Burghoff, S., Goetheer, E.L.V., and de Haan, A.B., Guidelines for solvent selection for carrier mediated extraction of proteins, *Sep. Purif. Technol.* **65**(1), 65–72 (2009).

128   Pei, Y., Wanga, J., Wua, K., Xuana, X., and Lu, X., Ionic liquid-based aqueous two-phase extraction of selected proteins, *Sep. Purif. Technol.* **64**(3), 288–295 (2009).

129   Pei, Y., Li, Z., Liu, L., Wang, J., and Wang, H., Selective separation of protein and saccharides by ionic liquids aqueous two-phase systems, *Sci. China, Ser. B Chem.* **53**(7), 1554–1560 (2010).

130   Jiang, Y., Xia, H., Guo, C., Mahmood, I., and Liu, H., Phenomena and mechanism for separation and recovery of penicillin in ionic liquids aqueous solution, *Ind. Eng. Chem. Res.* **46**(19), 6303–6312 (2007).

131   Du, Z., Yu, Y.-L., and Wang, J.-H., Extraction of proteins from biological fluids by use of an ionic liquid/aqueous two-phase system, *Chem. Eur. J.* **13**(7), 2130–2137 (2007).

132   Dreyer, S., and Kragl, U., Ionic liquids for aqueous two-phase extraction and stabilization of enzymes, *Biotechnol. Bioeng.* **99**(6), 1416–1424 (2008).

133   Tzeng, Y.-P., Shen, C.-W., and Yu, T., Liquid-liquid extraction of lysozyme using a dye-modified ionic liquid, *J. Chromatogr. A* **1193**(1–2), 1–6 (2008).

134   Lei, Z., Li, C., and Chen, B., Extractive distillation: a review, *Sep. Purif. Rev.* **32**(2), 121–213 (2003).

135   Wu, W.Z., Zhang, J.M., Han, B.X., Chen, J.W., Liu, Z.M., Jiang, T., He, J., and Li, W.J., Solubility of room-temperature ionic liquid in supercritical CO_2 with and without organic compounds, *Chem. Commun.* **12**, 1412–1413 (2003).

136   Furter, W.F., and Cook, R.A., Salt effect in distillation: a literature review, *Int. J. Heat Mass Trans.* **10**(1), 23–36 (1967).

137   Pinto, R.T.P., Wolf-Maciel, M.R., and Lintomen, L., Saline extractive distillation process for ethanol purification, *Comput. Chem. Eng.* **24**(2–7), 1689–1694 (2000).

138   Liao, B., Lei, Z., Xu, Z., Zhou, R., and Duan, Z., New process for separating propylene and propane by extractive distillation with aqueous acetonitrile, *Chem. Eng. J.* **84**(3), 581–586 (2001).

139   Lei, Z., Wang, H., Zhou, R., and Duan, Z., Influence of salt added to solvent on extractive distillation, *Chem. Eng. J.* **87**(2), 149–156 (2002).

140   Lei, Z., Zhou, R., and Duan, Z., Application of scaled particle theory in extractive distillation with salt, *Fluid Phase Equilib.* **200**(1), 187–201 (2002).

141   Ligero, E.L., and Ravagnani, T.M.K., Dehydration of ethanol with salt extractive distillation—a comparative analysis between processes with salt recovery, *Chem. Eng. Process.* **42**(7), 543–552 (2003).

142   Furter, W.F., Extractive distillation by salt effect, *Chem. Eng. Commun.* **116**, 35–40 (1992).

143   Beste, Y.A., Schoenmakers, H., Arlt, W., Seiler, M., and Jork, C., Recycling of ionic liquids with extractive distillation, Germany Patent WO2005016484 (2005).

144   Arlt, W., Seiler, M., Jork, C., and Schneider, T., Ionic liquids as selective additives for the separation of close-boiling or azeotropic mixtures, DE Patent WO02074718 A2; also published as: WO02074718 (A3); EP1372807 (A3); EP1372807 (A2) (2002).

145   Gmehling, J., and Krummen, M., Use of ionic liquids as entraining agents and selective solvents for separation of aromatic hydrocarbons in aromatic petroleum streams, Application: DE Patent 2001-10154052, 10154052 (2003).

146   Jork, C., Seiler, M., Beste, Y.-A., and Arlt, W., Influence of ionic liquids on the phase behavior of aqueous azeotropic systems, *J. Chem. Eng. Data* **49**(4), 852–857 (2004).

147   Seiler, M., Jork, C., Kavarnou, A., Arlt, W., and Hirsch, R., Separation of azeotropic mixtures using hyperbranched polymers or ionic liquids, *Am. Inst. Chem. Eng.* **50**(10), 2439–2454 (2004).

148 Seiler, M., Jork, C., and Arlt, W., Phasenverhalten von hochselektiven nichtflüchtigen Flüssigkeiten mit designbarem Eigenschaftsprofil und neue Anwendungen in der thermischen Verfahrenstechnik, *Chem. Ing. Tech.* **76**(6), 735–744 (2004).

149 Beste, Y.A., and Schoenmakers, H., Distillative method for separating narrow boiling or azeotropic mixtures using ionic liquids, WO2005016483 Patent 2004-EP7869, 2005016483 (2005).

150 Beste, Y., Eggersmann, M., and Schoenmakers, H., Extraktivdestillation mit ionischen Flüssigkeiten, *Chem. Ing. Tech.* **77**(11), 1800–1808 (2005).

151 Beste, Y.A., Eggersmann, M., and Schoenmakers, H., Ionic liquids: breaking azeotrops efficiently by extractive distillation, in *Sustainable (Bio)Chemical Process Technology—Incorporating the 6th Intenational Conference on Process Intensification* (2005), pp. 5–8.

152 Lei, Z., Arlt, W., and Wasserscheid, P., Separation of 1-hexene and *n*-hexane with ionic liquids, *Fluid Phase Equilib.* **241**(1–2), 290–299 (2006).

153 Zhu, J., Chen, J., Li, C., and Fei, W., Study on the separation of 1-hexene and trans-3-hexene using ionic liquids, *Fluid Phase Equilib.* **247**(1–2), 102–106 (2006).

154 Lei, Z., Arlt, W., and Wasserscheid, P., Selection of entrainers in the 1-hexene/n-hexane system with a limited solubility, *Fluid Phase Equilib.* **260**(1), 29–35 (2007).

155 Liebert, V., Nebig, S., and Gmehling, J., Experimental and predicted phase equilibria and excess properties for systems with ionic liquids, *Fluid Phase Equilib.* **268**(1–2), 14–20 (2008).

156 Jork, C., Kristen, C., Pieraccini, D., Stark, A., Chiappe, C., Beste, Y.A., and Arlt, W., Tailor-made ionic liquids, *J. Chem. Thermodyn.* **37**(6), 537–558 (2005).

157 Möllmann, C., and Gmehling, J., Measurement of activity coefficients at infinite dilution using gas-liquid chromatography. 5. Results for N-methylacetamide, N,N-dimethylacetamide, N,N-dibutylformamide, and sulfolane as stationary phases, *J. Chem. Eng. Data* **42**(1), 35–40 (1997).

158 Weidlich, U., Roehm, H.J., and Gmehling, J., Measurement of $\gamma\infty$ using GLC. 2. Results for the stationary phases N-formylmorpholine and N-methylpyrrolidone, *J. Chem. Eng. Data* **32**(4), 450–453 (1987).

159 Santacesaria, E., Berlendis, D., and Carrà, S., Measurement of activity coefficients at infinite dilution by stripping and retention time methods, *Fluid Phase Equilib.* **3**(2–3), 167–176 (1979).

160 Maase, M., Industrial applications of ionic liquids, in: *Ionic liquids in synthesis. Second completely revised and enlarged edition*, Vol. 2, eds. P. Wasserscheid, and T. Welton, Wiley-VCH, Weinheim (2008), pp. 663–687.

161 Chakraborty, M., and Bart, H.-J., Highly selective and efficient transport of toluene in bulk ionic liquid membranes containing Ag$^+$ as carrier, *Fuel Process. Technol.* **88**(1), 43–49 (2007).

162 Branco, L.C., Crespo, J.G., and Afonso, C.A.M., Ionic liquids as an efficient bulk membrane for the selective transport of organic compounds, *J. Phys. Org. Chem.* **21**(7–8), 718–723 (2008).

163 Matsumoto, M., Mikami, M., and Kondo, K., Separation of organic nitrogen compounds by supported liquid membranes based on ionic liquids, *J. Jpn. Pet. Inst.* **49**(5), 256–261 (2006).

164   Branco, L.C., Crespo, J.G., and Afonso, C.A.M., Highly selective transport of organic compounds by using supported liquid membranes based on ionic liquids, *Angew. Chem. Int. Ed.* **41**(15), 2771–2773 (2002).

165   Branco, L.C., Crespo, J.G., and Afonso, C.A.M., Studies on the selective transport of organic compounds by using ionic liquids as novel supported liquid membranes, *Chem. Eur. J.* **8**(17), 3865–3871 (2002).

166   Matsumoto, M., Inomoto, Y., and Kondo, K., Selective separation of aromatic hydrocarbons through supported liquid membranes based on ionic liquids, *J. Membr. Sci.* **246**(1), 77–81 (2005).

167   De Jong, F., and De With, J., Process for the separation of olefins from paraffins using a supported ionic liquid membrane, WO2005061422 (A1) Patent 2004-EP53605 (2005).

168   Schaefer, T., Branco, L.C., Fortunato, R., Izak, P., Rodrigues, C.M., Afonso, C.A.M., and Crespo, J.G., Opportunities for membrane separation processes using ionic liquids, *ACS Symp. Ser.* **902**, Ionic Liquids IIIB: fundamentals, progress, challenges, and opportunities, 97–110 (2005).

169   Kröckel, J., and Kragl, U., Nanofiltration for the separation of nonvolatile products from solutions containing ionic liquids, *Chem. Eng. Technol.* **26**(11), 1166–1168 (2003).

170   Han, S., Wong, H.T., and Livingston, A.G., Application of organic solvent nanofiltration to separation of ionic liquids and products from ionic liquid mediated reactions, *Chem. Eng. Res. Des.* **83**(A3), 309–316 (2005).

171   Wasserscheid, P., Kragl, U., and Kröckel, J., Method for separating substances from solutions containing ionic liquids by means of a membrane, WO 2003039719 A2 Patent WO 03/039719 A2, 15 May 2003 (2003).

172   Schäfer, T., and Goulao Crespo, J.P.S., Removal and recovery of solutes present in ionic liquids by pervaporation, WO2003013685 A1 Patent (2003).

173   David, W., Letcher, T.M., Ramjugernath, D., and Raal, D.J., Activity coefficients of hydrocarbon solutes at infinite dilution in the ionic liquid, 1-methyl-3-octyl-imidazolium chloride from gas-liquid chromatography, *J. Chem. Thermodyn.* **35**(8), 1335–1341 (2003).

174   Won, J., Kim, D.B., Kang, Y.S., Choi, D.K., Kim, H.S., Kim, C.K., and Kim, C.K., An ab initio study of ionic liquid silver complexes as carriers in facilitated olefin transport membranes, *J. Membr. Sci.* **260**(1–2), 37–44 (2005).

175   Kang, S.W., Char, K., Kim, J.H., and Kang, Y.S., Ionic liquid as a solvent and the long-term separation performance in a polymer/silver salt complex membrane, *Macromol. Res.* **15**(2), 167–172 (2007).

176   Meisen, A., and Shuai, X., Research and development issues in CO_2 capture, *Energy Convers. Mgmt.* **38**(Suppl.), Proceedings of the Third International Conference on Carbon Dioxide Removal, 1996, S37–S42 (1997).

177   Rao, A.B., and Rubin, E.S., A technical, economic, and environmental assessment of amine-based CO_2 capture technology for power plant greenhouse gas control, *Environ. Sci. Technol.* **36**(20), 4467–4475 (2002).

178   Anthony, J.L., Maginn, E.J., and Brennecke, J.F., Solubilities and thermodynamic properties of gases in the ionic liquid 1-*n*-butyl-3-methylimidazolium hexafluorophosphate, *J. Phys. Chem. B* **106**(29), 7315–7320 (2002).

179 Pérez-Salado Kamps, Á., Tuma, D., Xia, J., and Maurer, G., Solubility of CO_2 in the ionic liquid [bmim][PF_6], *J. Chem. Eng. Data* **48**(3), 746–749 (2003).

180 Aki, S.N.V.K., Mellein, B.R., Saurer, E.M., and Brennecke, J.F., High-pressure phase behavior of carbon dioxide with imidazolium-based ionic liquids, *J. Phys. Chem. B* **108**(52), 20355–20365 (2004).

181 Cadena, C., Anthony, J.L., Shah, J.K., Morrow, T.I., Brennecke, J.F., and Maginn, E.J., Why is CO_2 so soluble in imidazolium-based ionic liquids? *J. Am. Chem. Soc.* **126**(16), 5300–5308 (2004).

182 Camper, D., Scovazzo, P., Koval, C., and Noble, R., Gas solubilities in room-temperature ionic liquids, *Ind. Eng. Chem. Res.* **43**(12), 3049–3054 (2004).

183 Baltus, R.E., Culbertson, B.H., Dai, S., Luo, H., and DePaoli, D.W., Low-pressure solubility of carbon dioxide in room-temperature ionic liquids measured with a quartz crystal microbalance, *J. Phys. Chem. B* **108**(2), 721–727 (2004).

184 Scovazzo, P., Camper, D., Kieft, J., Poshusta, J., Koval, C., and Noble, R., Regular solution theory and CO_2 gas solubility in room-temperature ionic liquids, *Ind. Eng. Chem. Res.* **43**(21), 6855–6860 (2004).

185 Anthony, J.L., Anderson, J.L., Maginn, E.J., and Brennecke, J.F., Anion effects on gas solubility in ionic liquids, *J. Phys. Chem. B* **109**(13), 6366–6374 (2005).

186 Shariati, A., and Peters, C.J., High-pressure phase equilibria of systems with ionic liquids, *J. Supercrit. Fluids* **34**(2), 171–176 (2005).

187 Shiflett, M.B., and Yokozeki, A., Solubilities and diffusivities of carbon dioxide in ionic liquids: [bmim][PF_6] and [bmim][BF_4], *Ind. Eng. Chem. Res.* **44**(12), 4453–4464 (2005).

188 Anderson, J.L., Dixon, J.K., and Brennecke, J.F., Solubility of CO_2, CH_4, C_2H_6, C_2H_4, O_2, and N_2 in 1-hexyl-3-methylpyridinium bis(trifluoromethylsulfonyl) imide: comparison to other ionic liquids, *Acc. Chem. Res.* **40**(11), 1208–1216 (2007).

189 Galán Sánchez, L.M., Meindersma, G.W., and de Haan, A.B., Solvent properties of functionalized ionic liquids for CO_2 absorption, *Chem. Eng. Res. Des.* **85**(1), 31–39 (2007).

190 Muldoon, M.J., Aki, S.N.V.K., Anderson, J.L., Dixon, J.K., and Brennecke, J.F., Improving carbon dioxide solubility in ionic liquids, *J. Phys. Chem. B* **111**(30), 9001–9009 (2007).

191 Galán Sánchez, L.M., Meindersma, G.W., and de Haan, A.B., Ionic liquid solvents for gas sweetening operations, in *Greenhouse Gases*, Kingston, Canada (2007).

192 Bates, E.D., Mayton, R.D., Ntai, I., and Davis, J.H., Jr, CO_2 capture by a task-specific ionic liquid, *J. Am. Chem. Soc.* **124**(6), 926–927 (2002).

193 Davis, J.H., Jr, Task-specific ionic liquids for separations of petrochemical relevance: reactive capture of CO_2 using amine-incorporating ions, *ACS Symp. Ser.* **902**, Ionic liquids IIIB: fundamentals, progress, challenges, and opportunities, 49–56 (2005).

194 Camper, D., Bara, J.E., Gin, D.L., and Noble, R.D., Room-temperature ionic liquid-amine solutions: tunable solvents for efficient and reversible capture of CO_2, *Ind. Eng. Chem. Res.* **47**(21), 8496–8498 (2008).

195 Bara, J.E., Gabriel, C.J., Lessmann, S., Carlisle, T.K., Finotello, A., Gin, D.L., and Noble, R.D., Enhanced CO_2 separation selectivity in oligo(ethylene glycol)

functionalized room-temperature ionic liquids, *Ind. Eng. Chem. Res.* **46**(16), 5380–5386 (2007).

196 Bara, J.E., Gabriel, C.J., Hatakeyama, E.S., Carlisle, T.K., Lessmann, S., Noble, R.D., and Gin, D.L., Improving CO_2 selectivity in polymerized room-temperature ionic liquid gas separation membranes through incorporation of polar substituents, *J. Membr. Sci.* **321**(1), 3–7 (2008).

197 Baltus, R.E., Counce, R.M., Culbertson, B.H., Luo, H.M., DePaoli, D.W., Dai, S., and Duckworth, D.C., Examination of the potential of ionic liquids for gas separations, *Sep. Sci. Technol.* **40**(1–3), 525–541 (2005).

198 Hanioka, S., Maruyama, T., Sotani, T., Teramoto, M., Matsuyama, H., Nakashima, K., Hanaki, M., Kubota, F., and Goto, M., CO_2 separation facilitated by task-specific ionic liquids using a supported liquid membrane, *J. Membr. Sci.* **314**(1–2), 1–4 (2008).

199 Scovazzo, P., Kieft, J., Finan, D.A., Koval, C., DuBois, D., and Noble, R., Gas separations using non-hexafluorophosphate $[PF_6]^-$ anion supported ionic liquid membranes, *J. Membr. Sci.* **238**(1–2), 57–63 (2004).

200 Kang, S.W., Char, K., Kim, J.H., Kim, C.K., and Kang, Y.S., Control of ionic interactions in silver salt-polymer complexes with ionic liquids: implications for facilitated olefin transport, *Chem. Mater.* **18**(7), 1789–1794 (2006).

201 Munson, C.L., Boudreau, L.C., Driver, M.S., and Schinski, W., Separation of olefins from paraffins using ionic liquid solutions, US6623659 Patent (2003).

202 Camper, D., Becker, C., Koval, C., and Noble, R., Low pressure hydrocarbon solubility in room temperature ionic liquids containing imidazolium rings interpreted using regular solution theory, *Ind. Eng. Chem. Res.* **44**(6), 1928–1933 (2005).

203 Morgan, D., Ferguson, L., and Scovazzo, P., Diffusivities of gases in room-temperature ionic liquids: data and correlations obtained using a lag-time technique, *Ind. Eng. Chem. Res.* **44**(13), 4815–4823 (2005).

204 Gutmann, M., Mueller, W., and Zeppenfeld, R., Verfahren zur Olefinabtrannung aus Spaltgasen von Olefinanlagen mittels ionischer Flüssigkeiten, DE10333546 (A1) Patent 10333546 (2005).

205 Camper, D., Becker, C., Koval, C., and Noble, R., Diffusion and solubility measurements in room temperature ionic liquids, *Ind. Eng. Chem. Res.* **45**(1), 445–450 (2006).

206 Ferguson, L., and Scovazzo, P., Solubility, diffusivity, and permeability of gases in phosphonium-based room temperature ionic liquids: data and correlations, *Ind. Eng. Chem. Res.* **46**(4), 1369–1374 (2007).

207 Huang, J.-F., Luo, H., Liang, C., Jiang, D.-E., and Dai, S., Advanced liquid membranes based on novel ionic liquids for selective separation of olefin/paraffin via olefin-facilitated transport, *Ind. Eng. Chem. Res.* **47**(3), 881–888 (2008).

208 Kang, Y.S., Jung, B., Kim, J.H., Won, J., Char, K.H., and Kang, S.W., Facilitated transport membranes for an alkene hydrocarbon separation, EP1552875 (A1) Patent 2004-30656, EP1552875 (2005).

209 Kang, S.W., Hong, J., Char, K., Kim, J.H., Kim, J., and Kang, Y.S., Correlation between anions of ionic liquids and reduction of silver ions in facilitated olefin transport membranes, *Desalination* **233**(1–3), 327–332 (2008).

210 Galán Sánchez, L.M., Meindersma, G.W., and de Haan, A.B., Potential of silver-based room-temperature ionic liquids for ethylene/ethane separation, *Ind. Eng. Chem. Res.* **48**(23), 10650–10656 (2009).

211 Ortiz, A., Gorri, D., Irabien, Á., and Ortiz, I., Separation of propylene/propane mixtures using Ag^+-RTIL solutions. Evaluation and comparison of the performance of gas-liquid contactors, *J. Membr. Sci.* **360**(1–2), 130–141 (2010).

212 Ortiz, A., Ruiz, A., Gorri, D., and Ortiz, I., Room temperature ionic liquid with silver salt as efficient reaction media for propylene/propane separation: absorption equilibrium, *Sep. Purif. Technol.* **63**(2), 311–318 (2008).

213 Zhu, J.-Q., Chen, J., Li, C.-Y., and Fei, W.-Y., Centrifugal extraction for separation of ethylbenzene and octane using 1-butyl-3-methylimidazolium hexafluorophosphate ionic liquid as extractant, *Sep. Purif. Technol.* **56**(2), 237–240 (2007).

214 Onink, S.A.F., Meindersma, G.W., and de Haan, A.B., Ionic liquids in extraction operations: comparison of Rotating Disc Contactor performance between [4-mebupy]BF$_4$ and sulfolane for aromatics extraction, in *ISEC 2008*, Tucson, AZ, USA (2008), pp. 1337–1342.

215 Godfrey, J.C., and Slater, M.J., *Liquid-liquid extraction equipment*, John Wiley & Sons, Inc., New York (1994).

216 Lo, T.C., Baird, M.H.I., and Hanson, C., *Handbook of solvent extraction*, John Wiley & Sons, New York, NY (1983).

217 Müller, E., Berger, R., Blass, E., Sluyts, D., and Pfennig, A., Liquid–liquid extraction, in: *Ullmann's encyclopedia of industrial chemistry*, Vol. 21, Wiley-VCH, Weinheim (2008), pp. 250–307.

218 Kamath, M.S., Rao, K.L., Jayabalou, R., Karanth, P.K., and Rau, M.G.S., Holdup studies in a rotary disc contactor, *Indian J. Technol.* **14**(1), 1–5 (1976).

219 Chiappe, C., and Pieraccini, D., Ionic liquids: solvent properties and organic reactivity, *J. Phys. Org. Chem.* **18**(4), 275–297 (2005).

220 Klamt, A., and Eckert, F., COSMO-RS: a novel and efficient method for the a priori prediction of thermophysical data of liquids, *Fluid Phase Equilib.* **172**(1), 43–72 (2000).

221 Klamt, A., *COSMO-RS, from quantum chemistry to fluid phase thermodynamics and drug design*, Elsevier, Amsterdam (2005).

222 Klamt, A., and Eckert, F., Prediction of vapor liquid equilibria using COSMOtherm, *Fluid Phase Equilib.* **217**(1), 53–57 (2004).

223 Putnam, R., Taylor, R., Klamt, A., Eckert, F., and Schiller, M., Prediction of infinite dilution activity coefficients using COSMO-RS, *Ind. Eng. Chem. Res.* **42**(15), 3635–3641 (2003).

224 Banerjee, T., and Khanna, A., Infinite dilution activity coefficients for trihexyltetradecyl phosphonium ionic liquids: measurements and COSMO-RS prediction, *J. Chem. Eng. Data* **51**(6), 2170–2177 (2006).

225 Diedenhofen, M., and Klamt, A., COSMO-RS as a tool for property prediction of IL mixtures—A review, *Fluid Phase Equilib.* **294**(1–2), 31–38 (2010).

226 Domańska, U., Pobudkowska, A., and Eckert, F., (Liquid + liquid) phase equilibria of 1-alkyl-3-methylimidazolium methylsulfate with alcohols, or ethers, or ketones, *J. Chem. Thermodyn.* **38**(6), 685 (2006).

227   Domańska, U., Pobudkowska, A., and Eckert, F., Liquid-liquid equilibria in the binary systems (1,3-dimethylimidazolium, or 1-butyl-3-methylimidazolium methylsulfate + hydrocarbons), *Green Chem.* **8**(3), 268–276 (2006).

228   Freire, M.G., Santos, L.M.N.B.F., Marrucho, I.M., and Coutinho, J.A.P., Evaluation of COSMO-RS for the prediction of LLE and VLE of alcohols + ionic liquids, *Fluid Phase Equilib.* **255**(2), 167–178 (2007).

229   Freire, M.G., Ventura, S.P.M., Santos, L.M.N.B.F., Marrucho, I.M., and Coutinho, J.A.P., Evaluation of COSMO-RS for the prediction of LLE and VLE of water and ionic liquids binary systems, *Fluid Phase Equilib.* **268**(1–2), 74–84 (2008).

230   Banerjee, T., Sahoo, R.K., Rath, S.S., Kumar, R., and Khanna, A., Multicomponent liquid-liquid equilibria prediction for aromatic extraction systems using COSMO-RS, *Ind. Eng. Chem. Res.* **46**(4), 1292–1304 (2007).

231   Banerjee, T., Singh, M.K., and Khanna, A., Prediction of binary VLE for imidazolium based ionic liquid systems using COSMO-RS, *Ind. Eng. Chem. Res.* **45**(9), 3207–3219 (2006).

232   Banerjee, T., Verma, K.K., and Khanna, A., Liquid-liquid equilibrium for ionic liquid systems using COSMO-RS: effect of cation and anion dissociation, *Am. Inst. Chem. Eng.* **54**(7), 1874–1885 (2008).

233   Hansmeier, A.R., Broersen, A.C., Meindersma, G.W., and de Haan, A.B., COSMO-RS supported design of task specific ionic liquids for aromatic/aliphatic separations, in *22nd European Symposium on Applied Thermodynamics, ESAT 2006*, Elsinore, Denmark (2006), pp. 38–41.

234   Anantharaj, R., and Banerjee, T., COSMO-RS based predictions for the desulphurization of diesel oil using ionic liquids: effect of cation and anion combination, *Fuel Process. Technol.* **92**(1), 39–52 (2010).

235   Anantharaj, R., and Banerjee, T., COSMO-RS-based screening of ionic liquids as green solvents in denitrification studies, *Ind. Eng. Chem. Res.* **49**(18), 8705–8725 (2010).

236   Varma, N.R., Ramalingam, A., and Banerjee, T., Experiments; correlations and COSMO-RS predictions for the extraction of benzothiophene from n-hexane using imidazolium-based ionic liquids, *Chem. Eng. J.* **166**(1), 30–39 (2010).

237   Diedenhofen, M., Klamt, A., Marsh, K., and Schaefer, A., Prediction of the vapor pressure and vaporization enthalpy of 1-*n*-alkyl-3-methylimidazolium-bis-(trifluoromethanesulfonyl) amide ionic liquids, *Phys. Chem. Chem. Phys.* **9**(33), 4653–4656 (2007).

238   Palomar, J., Ferro, V.R., Torrecilla, J.S., and Rodriguez, F., Density and molar volume predictions using COSMO-RS for ionic liquids. an approach to solvent design, *Ind. Eng. Chem. Res.* **46**(18), 6041–6048 (2007).

239   Freire, M.G., Neves, C.M.S.S., Carvalho, P.J., Gardas, R.L., Fernandes, A.M., Marrucho, I.M., Santos, L.M.N.B.F., and Coutinho, J.A.P., Mutual solubilities of water and hydrophobic ionic liquids, *J. Phys. Chem. B* **111**(45), 13082–13089 (2007).

240   Freire, M.G., Carvalho, P.J., Gardas, R.L., Santos, L.M.N.B.F., Marrucho, I.M., and Coutinho, J.A.P., Solubility of water in tetradecyltrihexylphosphonium-based ionic liquids, *J. Chem. Eng. Data* **53**(10), 2378–2382 (2008).

241   Bosse, D., and Bart, H.J., Binary vapor-liquid equilibrium predictions with COSMOSPACE, *Ind. Eng. Chem. Res.* **44**(23), 8873–8882 (2005).

242 Kato, R., and Gmehling, J., Systems with ionic liquids: measurement of VLE and $\gamma\infty$ data and prediction of their thermodynamic behavior using original UNIFAC, mod. UNIFAC(Do) and COSMO-RS(Ol), *J. Chem. Thermodyn.* **37**(6), 603–619 (2005).

243 Sahandzhieva, K., Tuma, D., Breyer, S., Kamps, A.P.S., and Maurer, G., Liquid-liquid equilibrium in mixtures of the ionic liquid 1-*n*-butyl-3-methylimidazolium hexafluorophosphate and an alkanol, *J. Chem. Eng. Data* **51**(5), 1516–1525 (2006).

244 Nebig, S., Bolts, R., and Gmehling, J., Measurement of vapor-liquid equilibria (VLE) and excess enthalpies (HE) of binary systems with 1-alkyl-3-methylimidazolium bis(trifluoromethylsulfonyl)imide and prediction of these properties and [gamma][infinity] using modified UNIFAC (Dortmund), *Fluid Phase Equilib.* **258**(2), 168–178 (2007).

245 Zhang, X., Liu, Z., and Wang, W., Screening of ionic liquids to capture CO_2 by COSMO-RS and experiments, *Am. Inst. Chem. Eng.* **54**(10), 2717–2728 (2008).

246 Zhang, S.J., Yuan, X.L., Chen, Y.H., and Zhang X.P., Solubilities of CO_2 in 1-butyl-3-methylimidazolium hexafluorophosphate and 1,1,3,3-tetramethylguanidium lactate at elevated pressures, *J. Chem. Eng. Data* **50,** 1582–1585 (2005).

247 Jacquemin, J., Husson, P., Majer V., and Gomes, M.F.C., Low-pressure solubilities and thermodynamics of solvation of eight gases in 1-butyl-3-methylimidazoliumhexafluorophosphate, *Fluid Phase Equilib.* **240**, 87–95 (2006).

2DOM Si$_x$Ge$_{1-x}$

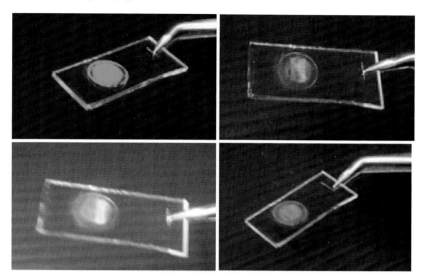

Figure 1.14 (b) Photographs of 2DOM Si$_x$Ge$_{1-x}$ (pore size ~370 nm) on ITO–glass substrate showing a colour change with changing the angle of the incident visible white light (GeCl$_4$ + SiCl$_4$) (0.1 + 0.1 M) in [C$_4$mim][NTf$_2$]. See text for full caption.

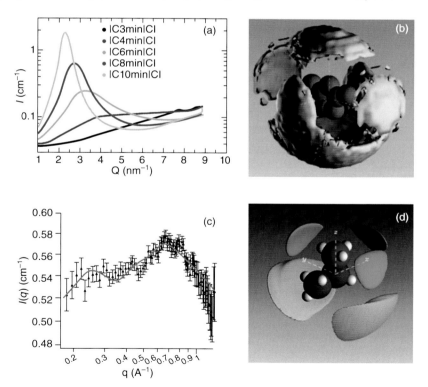

Figure 3.1 (A) X-ray diffraction spectra for a series of [C$_n$mim]Cl ionic liquids (used with permission from Ref. [14]). See text for full caption.

Ionic Liquids UnCOILed: Critical Expert Overviews, First Edition. Edited by Natalia V. Plechkova and Kenneth R. Seddon.

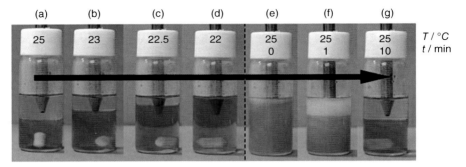

Figure 7.6 Temperature dependence of the phase behaviour of a $[P_{4\ 4\ 4\ 4}][Tf\text{-}Leu]/$ water mixture [27].

Figure 8.1 (a) First successful thermal vaporisation and recondensation of a pure aprotic ionic liquid, $[C_{10}mim][NTf_2]$, *in vacuo*. Insets show the ESI-MS spectra of the recondensed liquid and residue, which denotes the absence of decomposition. (b) Snapshot showing a distillation of $[C_4mim][NTf_2]$. Note the colour-free distillate. Reprinted with permission from Reference [23]. Copyright 2007, The American Chemical Society.

6 Theoretical Approaches to Ionic Liquids: From Past History to Future Directions

EKATERINA I. IZGORODINA

School of Chemistry, Monash University, Clayton, Victoria, Australia

ABSTRACT

In this chapter a critical overview of the theoretical approaches for predicting thermodynamic and transport properties of ionic liquids is given. These approaches can be divided into five distinct groups: (1) volume-based approaches, (2) quantum structure–property relationship (QSPR) approaches, (3) molecular dynamics (MD) simulations, (4) approaches from first principles based on *ab initio* theory and density functional theory, and (5) *ab initio* MD simulations. For each approach progress, limitations and directions are described in detail, with the emphasis on their predictive power for thermodynamic (melting point and enthalpy of vapourisation) and transport (conductivity and viscosity) properties. An outlook on why large-scale calculations of ionic liquids from first principles are becoming necessary for reliable predictions of their physical properties is discussed. Strategies that can be adopted to make these calculations feasible are outlined.

6.1 INTRODUCTION

The liquid character of ionic liquids at ambient temperatures stems from the properties of particular cations and anions and their long-range electrostatic and short-range dispersion interactions. Over the last decade, ionic liquids have been widely studied as replacements for traditional volatile and flammable organic solvents in a variety of applications from organic synthesis and catalysis to separation technologies. Due to their superior electrochemical and thermal stability over existing electrolytes, ionic liquids have attracted much attention for use in devices such as lithium-metal batteries, high-temperature fuel cells, dye-sensitised solar cells, electrochemical actuators, and numeric displays. These electrochemical devices possess huge potential to address the problems of sustainable means of energy generation and storage. The electrolyte materials used in these devices are required to have the following properties: (1) high conductivity, (2) low volatility (negligible vapour pressure) for safety reasons, (3) low viscosity, and (4) high electrochemical stability. Low melting point ionic liquids are of particular interest as the electrochemical devices can sustain high performance over a wider range of temperatures, including below zero. Ionic liquids represent an excellent replacement for traditional electrolytes as their thermodynamic and transport properties can be tailored by varying cations and anions to satisfy the mentioned criteria [1]. According to the latest estimation based on the existing pool of cations and anions, about 10^{19} ionic liquids can be potentially synthesised in the lab. Out of this enormous number of ionic liquids, only a fraction will be suitable for application in electrochemical devices. One can only guess how many hundreds of years the task of synthesising these ionic liquids might take, with some of the ionic liquids being left in a fume hood for good due to their unsuitable properties. At the moment, we cannot predict physical and chemical properties of ionic liquids by considering their chemical structure alone, as long-range Coulomb and short-range dispersion interactions between ions appear to affect their properties in a rather complex way. Therefore, theoretical approaches for accurate prediction of thermodynamic and transport properties of varied cation–anion combinations, *before synthesis*, are needed in order to design novel ionic liquids with tuned properties.

Theoretical approaches used for predictions of thermodynamic and transport properties of ionic liquids can be divided into four distinct groups (shown in Fig. 6.1): (1) volume-based approaches, (2) quantum structure–property relationship (QSPR) approaches, (3) molecular dynamics (MD) simulations, and (4) approaches *from first principles* based on *ab initio* theory and density functional theory (DFT). A fifth group, *ab initio* MD, has recently emerged that represents a symbiosis of *ab initio*- or DFT-based methods with the classical Newtonian mechanics for describing dynamics, thus offering an exciting opportunity to look into the dynamic behaviour and bulk properties of ionic liquids at various temperatures *from first principles*. This book chapter consists of six sections, and the first five sections describe progress, limitations, and directions of the five groups of theoretical approaches. In Section 6.7, an outlook on large-scale calculations of ionic liquids from first principles is given.

6.2 VOLUME-BASED APPROACH

The volume-based approach focuses on establishing correlations between thermodynamic and transport properties of materials and either their molecular

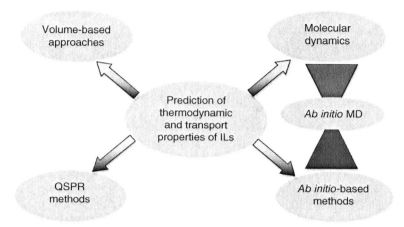

Figure 6.1 Theoretical approaches for prediction of thermodynamic and transport properties of ionic liquids (ILs).

volume, V_m, or density. Here we describe a well-known and widely used approach by Jenkins and Glasser for predicting lattice energies and standard entropies of ionic liquids in more detail.

6.2.1 Thermodynamic Properties: Lattice Energy and Standard Absolute Entropy

Jenkins et al. [2] empirically established relationship between the thermodynamic quantities such as lattice energy, U_L, and entropy of inorganic salt crystals and their molecular volume, V_m, defined as the sum of volumes of individual cations and anions. For an ionic salt, M_pX_q, the lattice energy at a given temperature can be written as:

$$U_L(T) = U_{POT} + \left[p\left(\frac{n_M}{2} - 2\right) + q\left(\frac{n_X}{2} - 2\right) \right] RT \qquad (6.1)$$

where U_{POT} is the lattice potential energy that depends solely on the molecular volume. The last term in Equation (6.1) represents the temperature correction to the lattice potential energy. The n_X and n_M parameters depend on the nature of the salt: (1) $n_M = n_X = 3$ for univalent salts, MX, (2) $n_M = n_X = 5$ for linear polyatomic ions, and (3) $n_M = n_X = 6$ for non-linear polyatomic ions. The Jenkins approach is based on Bartlett's relationship that established the $V^{1/3}$ dependence of the lattice energy in the MX salts, which is both dimensionally and conceptually equivalent to the sum of ionic radii of the cation and anion. The proportionality factor between the lattice energy and $V^{1/3}$ was shown to lie within 4% of the electrostatic interaction conversion factor for NaCl. Jenkins generalised Bartlett's relationship for other types of salts, including divalent salts. For univalent salts of the MX type, the following relationship was found:

$$U_{POT} = 2\left(\frac{\alpha}{\sqrt[3]{V_m}} + \beta\right) \tag{6.2}$$

where $\alpha = 117.3$ kJ mol^{-1} and $\beta = 51.9$ kJ mol^{-1}. The simplicity of this approach comes from the fact that the overall long-range electrostatic interactions (attractive and repulsive) in inorganic salts converge to a number known as the Madelung constant [3, 4]. For example, for halides of alkali metals such as NaCl [5], this number was calculated to be 1.748. The molecular volume can be easily obtained from X-ray crystal structures determined by X-ray, and therefore the lattice energy of inorganic salts can be estimated without performing elaborate calculations.

In a similar fashion, Glasser and Jenkins [6] established a relationship between the standard absolute entropy, S_{298}^0, of inorganic salts at ambient conditions and their molecular volume, V_m:

$$S_{298}^0 (\text{J K}^{-1}\text{mol}^{-1}) = 1360(\text{J K}^{-1}\text{mol}^{-1}\text{nm}^3) \cdot V_m(\text{nm}^3) + 15(\text{J K}^{-1}\text{mol}^{-1}) \tag{6.3}$$

Later, Glasser [7] published new correlation constants of Equation (6.3) for ionic liquids that were calculated as an average of the constants established for ionic solids [6] and organic liquids [8]:

$$S_{298}^0 (\text{J K}^{-1}\text{mol}^{-1}) = 1246.5(\text{J K}^{-1}\text{mol}^{-1}\text{nm}^3) \cdot V_m(\text{nm}^3) + 29.5(\text{J K}^{-1}\text{mol}^{-1}) \tag{6.4}$$

It has to be noted that the Jenkins approach accounts only for Coulomb interactions between ions, as it was established for inorganic salts whose ions are held together by strong electrostatic interactions. For organic ionic systems, in which dispersion interactions such as the π–π stacking interactions (e.g., between imidazolium rings), hydrogen bonding, and van der Waals interactions (e.g., between alkyl chains on cations) dominate the lattice energy, the approach is expected to show significant deviations [9]. Small ions, such as chloride, might also have a greater chance of establishing strong hydrogen bonds with cations and, therefore, the formation of ion pairs and aggregates might be expected. For example, dimer formation was observed between two imidazolium cations and two chloride anions in the crystal structure of 1-ethyl-2,3-dimethylimidazolium chloride, [C$_2$dmim]Cl [10].

In the literature, both Equations (6.3) and (6.4) are used to calculate the standard absolute entropy of ionic solids. Many researchers utilised the Jenkins and Glasser approach to predict melting points and heats of formation of ionic liquids, and examples of these applications are given in more detail in Section 6.2.2.

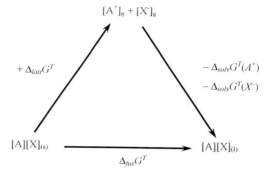

Figure 6.2 Born–Fajans–Haber cycle for calculating $\Delta_{fus}G^T$.

6.2.2 Thermodynamic Properties: Melting Points

Krossing et al. [11] exploited the application of the Jenkins approach to ionic liquids and proposed a thermodynamics-based approach that represents a combination of the volume-based approach by Jenkins and a thermodynamic cycle that connects three phases of ionic liquids: solid, liquid, and gas. Relatively low melting points of ionic liquids were explained using a simple thermodynamic Born–Fajans–Haber cycle (Fig. 6.2) that relates the Gibbs free energy of fusion, $\Delta_{fus}G$, the free lattice energy, $\Delta_{latt}G$, and the free energy of solvation, $\Delta_{solv}G$. This relationship at a given temperature, T, is given in Equation (6.5):

$$\Delta_{fus}G^T = \Delta_{latt}G^T - \Delta_{latt}G^{298} + \Delta_{corr}G^T \qquad (6.5)$$

where the lattice energy at a given temperature, T, is defined as:

$$\Delta_{latt}G^T = \Delta_{latt}H^T - T\Delta_{latt}S. \qquad (6.6)$$

A temperature correction factor, $\Delta_{corr}G^T$, is determined as $\Delta_{latt}G^T - \Delta_{latt}G^{298}$, whereas $\Delta_{solv}G^{298}$ is calculated as the sum of the free energies of solvation for individual ions using the **CO**nductorlike**S**creening**MO**del (COSMO) solvation model, that is, $\Delta_{solv}G^{298} = \Sigma \Delta_{solv}G^{298}(ions)$. The change in the entropy on going from the solid ionic liquid to individual ions in gas phase was calculated by Equation (6.7):

$$\Delta_{latt}S = S_{solid} - \Sigma S_{gas}(ions) \qquad (6.7)$$

where S_{solid} was approximated using the Jenkins and Glasser approach by Equation (6.3), where S_{gas} of constituent ions is taken from gas-phase quantum-chemical calculations and includes translational, rotational, and vibrational contributions as implemented in TURBOMOLE.† The solvation energies of

† For more information, see http://www.turbomole.com/.

individual ions in Equation (6.5) were calculated using the COSMO solvation model that is based on a quantum-chemical calculation of charge surface densities of individual ions and experimental dielectric constants [12]. In this model, the solvent represents a continuum of a finite dielectric constant that is placed outside the molecular cavity of a solute. The charge densities on the solute molecule are used to determine the overall solvation energy originating only from the electrostatic interaction between the solute and the solvent.

In their work, Krossing et al. studied 14 ionic liquids that melt below room temperature. According to their calculations, $\Delta_{fus}G^T$ turned out to be negative for these ionic liquids, indicating that there was a thermodynamic explanation for them being liquid at room temperature. At the melting point, $\Delta_{fus}G$ becomes zero and, therefore, the corresponding melting point can be easily calculated from Equation (6.5). The success of this thermodynamic approach was based on the fact that the molecular volume could either be extracted from the crystallographic data or computed using quantum-chemical calculations.

Considering that the melting points of ionic liquids included in the study fell in a rather narrow range between −19 and 23 °C, the average error in the melting point prediction of 8 °C was calculated to be about 20% within the range. One of the significant conclusions of this study is that a change in $\Delta_{fus}G$ by 10 kJ mol^{-1} leads to a significant change in the melting point in the range of 10–15 °C. Therefore, $\Delta_{fus}G$ energies have to be accurately calculated for the accurate prediction of corresponding melting points. Any significant non-electrostatic interactions in ionic liquids will lead to underestimation of the lattice energy, U_L, and therefore, result in a lower melting point.

Markusson et al. [13] applied the thermodynamic-based approach of Krossing for prediction of melting points of protic ionic liquids that are formed by a proton transfer reaction from a Brønsted acid to a Brønsted base. They studied a variety of acids whose pK_a^{aq} values ranged from −16 (CF$_3$SO$_3$H) to 3.75 (HCOOH), coupled with ammonia and its derivatives, $C_nH_{2n+1}NH_2$ (with $n = 2, 3$, or 4). The melting points calculated as described above—see Equation (6.5)—did not yield strong correlations with experimentally measured melting points, the data being fairly scattered. A conductor-like polarisable continuum model, CPCM (analogous to the COSMO model discussed above) [14], with a predefined solvent with the *static* dielectric constant of 10.36, was employed to calculate the free energies of solvation of individual ions. It was noted that when the dielectric constant changed from 10 to 20, a systematic shift of about 20 kJ mol^{-1} in $\Delta_{solv}G$ (defined as the sum of the free energies of the two counterions) was observed. A change in the final $\Delta_{fus}G^T$ (see Eq. 6.5) of 20 kJ mol^{-1} could potentially result in over- or underestimation of the melting point by 20–30 K, thus highlighting the importance of the dielectric constant itself in the calculations of free energies of solvation. One of the possible explanations for the failure of the Krossing approach might lie in the fact that the lattice energy of the protic ionic liquid could be strongly dominated by hydrogen bonding between ions, which was not included in the lattice energy calculations. For the ethylammonium protic ionic liquids, the authors found other interesting linear correlations between the melting point and (1) the molecular mass density, determined as the ratio of the molecular mass of ions divided by the calculated volume of the ion pair with a standard deviation of 47 K, and (2) the total free energy of solvation, determined as the sum of the free energies of solvation of individual ions with a standard deviation of 39 K. In the latter case, the change in the free energy of solvation by about 70 kJ mol^{-1} resulted in a substantial change in the melting point by as much as 245 K, which is equivalent to as little as 3 kJ mol^{-1} of change in the lattice energy per 10 K. The found correlation between melting point of protic ionic liquids and the total free energy of solvation of the constituent ions is certainly worth exploring for other ionic liquids (protic as well as aprotic).

Rogers et al. calculated the lattice energies using the Jenkins approach for 1,3-dialkylimidazolium ionic liquids coupled with the [PF$_6$]$^-$ anion [15]. The lattice energies of these salts changed only slightly with either increasing alkyl chain from methyl to tetradecyl on the cation or methylation of the C2 position on the imidazolium ring, whereas the melting points changed in a much wider range from 10 °C for [C$_4$mim][PF$_6$] to 201 °C for [C$_2$dmim][PF$_6$]. When entropic factors were included, that is, the $-T\Delta S$ term calculated according to Equation (6.3), the overall free Gibbs energies changed in the range of 77 kJ mol^{-1}, twice as much compared with the corresponding lattice energy change (about 39 kJ mol^{-1}). The authors concluded that since relatively small changes in the free energy were indicative of much larger changes in melting point, the accuracy of these calculations prevents a direct correlation between the lattice energy and the corresponding melting point.

6.2.3 Thermodynamic Properties: Free Energies of Reactions

Gutowski et al. [16] established a methodology for predicting the formation and stability of 1,3-dialkylimidazolium salts that are liquid at room temperature using the quaternisation reaction between the solid imidazole base and liquid substituted hydrocarbons, RX. The reaction enthalpy was calculated based on a Born–Haber cycle shown Figure 6.3 using Equation (6.8):

$$\Delta H_{rxn}(salt) = \Delta H_1 + \Delta H_2 + \Delta H_3 \tag{6.8}$$

where ΔH_3 corresponds to the lattice energy in the solid state, which was calculated using the volume-based approach by Jenkins (from Eqs. 6.1 and 6.2). ΔH_1 is the enthalpy of the heterolytic cleavage of the R–X bond, that is, the proton (for R = H) or methyl (for R = Me) acidity of RX, whereas ΔH_2 corresponds to the affinity of imidazole for R$^+$, that is, the proton or methyl cation affinity. The last two reaction enthalpies, ΔH_2 and ΔH_3, were calculated by extrapolating the MP2 energies to the complete basis set (CBS) using the

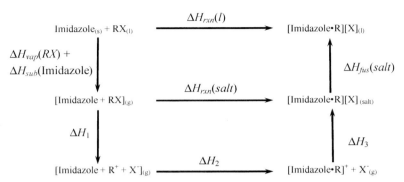

Figure 6.3   Born–Haber cycle for calculating heats of formation of imidazolium-based ionic liquids.

B3LYP† optimised geometries. Molecular volumes were calculated using the B3LYP functional and a double-zeta quality basis set (aug-cc-pVDZ for cations, and DZVP2 for anions), and were in good agreement with experimental crystallographic values. In order to estimate the Gibbs free energy of the reaction in Equation (6.8), the entropic contributions, $-T\Delta S$, for gas-phase reactants, imidazole, and RX, were calculated using the B3LYP frequency calculations. The entropy of solid imidazole (i.e., organic solid) and liquid RX (i.e., organic liquid) were calculated using the volume-based Jenkins approach established for organic solids and organic liquids, respectively [6, 7]; for ionic liquids, the standard entropy in the liquid state was calculated using Equation (6.4). The entropic contribution calculated this way was negligible for most of the imidazolium salts studied, indicating that the enthalpy of the reaction, ΔH_{rxn}, was the driving force behind the formation of these ionic liquids. The authors showed that nitro- and cyano-substituted imidazolium ionic liquids, coupled with the halide and triflate anions, had negative free Gibbs energy of formation. The methylation of the imidazolium ring in various positions was also predicted to result in stable ionic liquids incorporating all the anions studied, which was consistent with experimental results. Considering only enthalpies of the reaction, the cut-off criterion for prediction of the formation of stable salts was suggested to be below -54 kJ mol^{-1}. Some of the stable ionic liquids predicted using the described methodology were successfully synthesised in the lab [17, 18]. Therefore, the established methodology was shown to be reliable for the prediction of formation and stability of imidazolium-based ionic liquids using the quaternisation reaction between the solid imidazole base and substituted hydrocarbons.

---

† The B3LYP functional belongs to a family of hybrid functional used in density functional theory. Hybrid in this context means that a functional contains a fraction of the exact Hartree-Fock (HF) exchange energy. In the case of B3LYP, 20% of the HF exchange energy is included in its formulation.

6.2.4 Thermodynamic Properties: Heats of Formation

Gao et al. [19] proposed a method for estimation of the condensed phase heats of formation of ionic liquids with the view of designing energetic salts based on the following expression:

$$\Delta H_f^{298}(ionic\ salt) = \Delta H_f^{298}(cation) + \Delta H_f^{298}(anion) - U_L(T) \qquad (6.9)$$

The enthalpy of formation defined in Equation (6.9) indicates that the higher the enthalpy, the more energetic the formed salt. The total lattice energies at 298 K were calculated using the Jenkins approach (see Eqs. 6.1 and 6.2), whereas the heats of formation of individual ions were computed at the G2 [20] level of theory† based on the isodesmic (bond-type conserving) reactions. For organic salts, including ammonium-based salts whose heats of formation were measured experimentally, the absolute errors in the calculated heats of formation, ΔH_f^0, ranged from −159 to +87 kJ mol^{-1}. In most cases, the heats of formation were significantly underestimated. The authors established that the uncertainty of the lattice energies ranged from 2 to 150 kJ mol^{-1}, thus being the major source of error in the heats of formation in Equation (6.9).

Gutowski et al. [21] proposed a new parameterisation of Equation (6.2) using the experimental lattice energies of ammonium, hydrazinium, methyl-ammonium (mono-, di-, and tri-), and ethylammonium (di- and tri-) salts coupled with inorganic anions that resulted in the α and β values of 83.2 and 157.3 kJ mol^{-1}, respectively. The revised parameters now produced a smaller average error in the lattice energies with a standard deviation of only 15.9 ± 10.9 kJ mol^{-1} as compared to 66.9 ± 26.4 kJ mol^{-1} using the original parameters (see Eq. 6.2). The corrected lattice energies, U_L^T, varied only slightly within 15 kJ mol^{-1} for homologous series of salts with a common anion (e.g., perchlorate, nitrate, picrate, or nitro-substituted azolates). A new set of the α and β parameters was used to estimate heats of formation of energetic salts using Equation (6.9). Large differences were found between the predicted and experimental heats of formation for the majority of the studied salts; in some cases the average deviation was well above 400 kJ mol^{-1}. Using the experimental heats of formation of organic salts and rearranging Equation (6.9), one could easily calculate their lattice energies. The lattice energies appeared to significantly vary even within homologous series of salts with a common anion and two salts turned out to even have negative lattice energies. Therefore, it was suggested some of the experimental heats of formation might be questionable due to the following reasons: (1) difficulties in making accurate measurements, and (2) the solid-state materials might be molecular rather than ionic in nature. Recalculated free Gibbs energies used for the prediction of

---

† The G2 method (Gaussian-2) is a composite method that aims to approximate a high correlated level of *ab initio* theory such as QCISD(T) by performing a series of quantum-chemical calculations at lower levels of theory using large basis sets.

formation and stability of a series of nitro-, cyano-, and methyl-substituted imidazolium salts using the Born–Haber cycle (see Fig. 6.3) made the final ΔG_{rxn} values negative (originally positive) in the cases of the 1,3-dimethyl-4-nitroimidazolium iodide and methylsulfate formation, which is now in agreement with experiment, as these two salts were successfully synthesised in the laboratory [17]. The inability to synthesise dinitro-substituted methylsulfate salts is again consistent with the current prediction that the Gibbs free energies are very close to zero [21].

Recently Krossing et al. [22] introduced another way of predicting melting points of ionic liquids based on a simple thermodynamic approach by Yalkowsky et al., who proposed to approximate the enthalpy and entropy of fusion separately in order to estimate the melting point [23]. Using the Gibbs–Hemholtz equation ($\Delta G = \Delta H - T\Delta S$) and the fact that at the melting point $\Delta G = 0$, we arrive at the following Equation (6.10) for melting point, T_m:

$$T_m = \frac{\Delta H_{fus}}{\Delta S_{fus}} \tag{6.10}$$

Yalkowsky et al. [23] proposed to estimate the entropy of fusion of neutral organic molecules by using the simple relationship shown in Equation (6.11):

$$\Delta S_{fus} = a \log \sigma + b\tau + c \tag{6.11}$$

where σ is a rotational symmetry number that reflects the probability of a molecule being properly oriented for incorporation into the crystal lattice, and τ is the conformational degree of freedom that depends on the number of bonds that are free to rotate. The Yalkowsky approach was applied for prediction of melting points for ionic liquids with one exception: the principles of the volume-based approach were used to approximate ΔH_{fus}, according to Equation (6.12):

$$\Delta H_{fus} = cr_m^3 \tag{6.12}$$

The only difference in the proposed approach is that the molecular volume is now accounted for through r_m^3 rather than the molecular volume, where r_m is the sum of cationic and anionic radii calculated quantum-chemically at the BP86/TZVP† level of theory, in combination with the COSMO model. The final equation for the melting point is therefore as follows:

$$T_m = \frac{c \cdot r_m^3}{a \ln \sigma + b\tau + 1} \tag{6.13}$$

Based on Equation (6.13), a few correlations were established with respect to either similar cations (ammonium based, imidazolium based, etc.) or similar

† BP86 is a GGA functional in density functional theory. GGA stands for generalised gradient approximations that include the gradient of the electron density in a DFT functional. TZVP is an Ahlrichs-type triple-ζ quality basis set.

anions (bistriflamide based, aluminate based, etc.). For example, for the aluminate, $[Al(OR)_4]^-$, and borate, $[B(OR)_4]^-$, ionic liquids spanning 320 K of the melting point range, the established correlation predicted melting points within 21.2 K on average. To combine ionic liquids with various types of cations and anions in one correlation, certain modifications to Equation (6.13) were needed because that specific dispersion interactions (such as hydrogen bonding) could also affect the melting point. It was suggested to use the **CO**nductorlike**S**creening**MO**del for Real Solvents (COSMO-RS) calculations to estimate these directional interactions as a sum of the single-ion enthalpies arising from van der Waals interactions, H_{vdW}^0, and a small correction for the ring structure, H_{ring}, in a 1:1 mixture of the cation and the anion, thus arriving at the following expression for melting point:

$$T_m = \frac{c \cdot r_m^3 + d \cdot H_{vdW}^0 + e \cdot H_{ring}}{a \ln \sigma + b \tau + 1} \tag{6.14}$$

The resulting equation incorporating five optimised parameters produced a correlation with a correlation coefficient of about 0.9, and an average error in the melting point of 24.5 K for the 67 ionic liquids, whose melting points span a range of 337 K, thus providing a necessary link between molecular solvents and inorganic salts to predict melting points of ionic liquids as intermediates between the two classes. In this study the most energetically stable conformations of cations and anions were considered. Since the actual conformations of ions in the liquid state are mostly unknown, more studies on the influence of other energetically stable conformations taken, relative weights of the corresponding Boltzmann distribution on melting point are needed to further exploit the proposed approach.

6.2.5 Thermodynamic Properties: Enthalpies of Vapourisation and Heat Capacities

Although various attempts to correlate enthalpies of vapourisation and molecular volume, V_m, have been recently presented [24–27], these schemes indicate that there is no straightforward correlation between the enthalpy of vapourisation and molecular volume.

Krossing et al. [28] showed that a linear relationship existed between the heat capacity, C_p, of ionic liquids and molecular volume, V_m. Thirty-four ionic liquids were studied at two temperatures: 298 K and 323 K. These ionic liquids were divided into two sets: (1) set 1 contained only ionic liquids with low water content (<500 ppm), and (2) set 2 contained all 34 ionic liquids studied. The set 1 was used to determine parameters i and j in the linear correlation given in Equation (6.15):

$$C_p = i V_m + j \tag{6.15}$$

The correlation was proposed based on the volume-based approach by Jenkins and the fact that the entropy of the solid crystal correlates linearly with molecular volume. The average error in set 2 was 1.2%, with the data at 323 K having smaller errors than those of 298 K, presumably due to lower water content at higher temperatures. There were two outliers with errors of about 20% for salts, 1-hexyl-3-methylpyridinium bromide and ECOENG 500, that could potentially form mesophases due to the presence of long alkyl chain on the cations, thus limiting the applications of this approach to ionic systems that exhibit strong non-ionic-type interactions. As a result, the limitations of the volume-based approach by Jenkins appear to apply here.

6.2.6   Transport Properties: Conductivity and Viscosity

The volume-based approach was taken further by Slattery et al. [9, 29], who established correlations with transport properties of ionic liquids, such as conductivity and viscosity. Linear relationships between V_m and transport properties, such as conductivity and viscosity, were observed for a variety of ionic liquids including imidazolium-, pyrrolidinium-, and ammonium-based ionic liquids as shown below:

$$\eta = a \cdot \exp(b V_m) \tag{6.16}$$

$$\sigma = c \cdot \exp(-d V_m) \tag{6.17}$$

The a, b, c, and d coefficients in Equations (6.16) and (6.17) were found to be anion-dependent and as a result, for each series of ionic liquids coupled with the same anion, a new set of these parameters was generated. The sets of parameters were established for the following series of anions: $[BF_4]^-$, $[PF_6]^-$, $[NTf_2]^-$, and $[N(CN)_2]^-$. According to the established correlations the average errors were 26 cP and 0.2 mS cm^{-1} for viscosity and conductivity, respectively. Two outliers, $[SMePh_2][NTf_2]$ and $[N_{1114}][NTf_2]$, were found in the proposed correlations, and these findings could be explained by either improper measurements of physical properties or the presence of impurities that are known to strongly affect transport properties of ionic liquids [30]. The sets of ionic liquids used to establish the correlations above were close-to-ideal ionic liquids that lie close to the ideal line on the Walden plot (see Fig. 6.4), indicating that the mobility of ions is not limited by the ion pair or higher-order cluster formation in the liquid state.

The empirical relationship found between the molecular volume and viscosity in Equation (6.16) was applied for a series of $[C_n mim][NTf_2]$ and $[C_n dmim][NTf_2]$ [31]. In the latter series, the hydrogen atom in the C2 position in the imidazolium ring was replaced by the methyl group. Although the additional methyl group increases V_m only slightly, this small structural change of the ring has a more drastic effect on viscosity, making the $[C_n dmim]^+$ ionic liquids more viscous than the original $[C_n mim]^+$ ionic liquids. Using the a and b parameter established for the $[NTf_2]^-$ ionic liquids previously [29], the

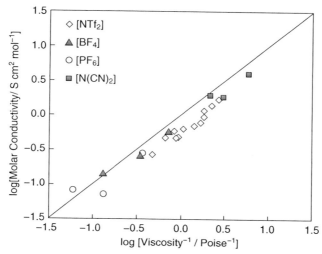

Figure 6.4   Walden plot for ionic liquids studied in Reference [11]. Ionic liquids were grouped with respect to anions.

Slattery approach tends to underestimate viscosities of the methylated imidazolium ionic liquids methylated in the C2 position by as much as 38 cP, thus not fully accounting for the methylation effect. Obviously, the methylation of the C2 position alters the ionic arrangement in the liquid state, and these subtle changes cannot be easily reproduced by the volume-based correlations.

To summarise, the volume-based approach by Jenkins and Glasser allows us to estimate the lattice energies and standard entropies of organic salts at various temperatures based on the common property that organic salts share with inorganic salts, that their long-range Coulomb interactions converge to a number known as the Madelung constant [32]. Krossing proposed to combine the Jenkins approach with a Born–Fajans–Haber cycle that thermodynamically connects three phases of ionic liquids: solid, liquid, and gas to calculate free Gibbs energies of fusion of ionic liquids and therefore predict their melting points. The thermodynamic approach by Krossing was successful in determining melting points within 8 °C on average for ionic liquids that are liquid at room temperature. It was established that free Gibbs energies of fusion needed to be calculated accurately as a slight deviation of 10 kJ mol^{-1} leads to a significant deviation in the melting point of 10–20 °C. Therefore, special care should be taken in calculations of lattice energies as well as free energies of solvation of individual ions, for which accurate dielectric constants of corresponding ionic liquids are not readily available. Shortcomings of the thermodynamic-based approach by Krossing can be foreseen for two groups of ionic liquids: (1) ionic liquids with strong dispersion interactions between ions and/or ion association/aggregation, and (2) protic ionic liquids forming extended hydrogen-bonded networks of interactions. The volume-based approach has been successfully applied to correlate transport properties of ionic liquids that exhibit close-to-ideal behaviour on the Walden plot. Clearly, more studies are needed to produce general correlations that are valid for ionic liquids with a varied degree of ion association in the liquid state.

6.3   QUANTITATIVE STRUCTURE–PROPERTY RELATIONSHIP METHODS

The quantitative structure–property relationship (QSPR) model introduces an alternative to predicting thermodynamic and transport properties of ionic liquids. The QSPR approach is based on the assumption that physical properties of a molecular compound should relate to its chemical structure, and as a result these properties can be correlated to so-called molecular descriptors that can be divided into the following categories: topological, geometrical, electronic, quantum chemical, and thermodynamic. Topological and geometrical descriptors reflect the geometry and the connectivity (i.e., overall topology) of a molecule. Electronic descriptors account for charge distribution and surface charge densities of the molecule, whereas quantum-chemical descriptors arise from quantum-chemical calculations of orbital energies (e.g., HOMO and LUMO). Thermodynamic descriptors can be also extracted from quantum-chemical calculations of vibrational frequencies in the form of entropic contributions. The process of identifying important descriptors is based on the family of so-called heuristic methods (HMs) and normally proceeds via two steps. At first the descriptors are included in the calculation for a training set of compounds containing experimentally available data for the target property. Then multi-linear regression (MLR) methods are applied to select descriptors that estimate the target property with the least error. Once the descriptors are chosen, the established linear correlation can be used to predict the target property for a test set of compounds, for which the target property is unknown. The biggest limitation of the multi-linear QSPR approach lies in the assumption that there is a direct linear correlation between the chosen molecular descriptors and the target property, thus introducing significant constraints at establishing reliable correlations.

6.3.1   Multi-Linear QSPR Approach: Prediction of Melting Points

Early studies of Katritzky et al. [33, 34] and Eike et al. [35] identified that *electronic* descriptors needed to be included in the QSPR approach to better reproduce melting points of bromide ionic liquids. Good correlations for the prediction of melting points of pyridinium bromides [34] were achieved by inclusion of electronic and quantum-chemical descriptors such as the average nucleophilic reactivity index reflecting electrostatic intermolecular interactions, and the total entropy per atom related to conformational and rotational degrees of freedom [34]. To obtain better correlations for melting points of the imidazolium and benzimidazolium bromides, quantum-chemical descriptors needed to also reflect the difference in size and electrostatic interactions in the cation [33]. In the case of the (hydroxyalkyl)trialkylammonium bromides, inclusion of electronic descriptors such as positive and negative charged partial surface areas significantly improved the correlation with the melting point [35]. Although the established correlation did not agree quantitatively with the experimental values (with the deviation of up to 70 K in some cases), qualitatively the correlation was able to identify ionic liquids with the highest and lowest melting points. In these studies semi-empirical methods such as AM1 [36] and PM3 [37, 38] were used to calculate electronic and quantum-chemical descriptors.

Good correlations between experimental melting points of 1,2,4-triazolium-based bromide and nitrate ionic liquids and those obtained with the QSPR approach were found by using electronic, electrostatic, and thermodynamic descriptors obtained from quantum-chemical calculations at the HF/6-31G** level of theory* [39]. Three-descriptor models were established for both bromide and nitrate ionic liquids, with the resulting correlation coefficient (r^2) exceeding 0.9. The descriptors included in the correlation for the bromide salts included the nucleophilic reactivity index for the amine nitrogen, the weighted surface charge area of atoms acting as hydrogen-bond acceptors, and the reciprocal lowest unoccupied orbital (LUMO) energy that served to quantify the cation–anion interactions. In the case of nitrate ionic liquids, the descriptors were slightly different and included measures of the hydrogen-bond donating ability of the cation and the minimum nucleophilic reactivity index for carbon atoms. Surprisingly, inclusion of descriptors derived from quantum-chemical calculations at a higher level of theory (such as MP2/6-31G**[†] only made the correlations poorer [40]. In the study of López-Martin et al. [41], good correlations of melting points for a test set containing 1-alkyl-3-methylimidazolium ionic liquids with various anions (including $[NTf_2]^-$, $[C_nF_{2n+1}BF_3]^-$, $[CH_3SO_4]^-$, $[BF_4]^-$, $[NO_3]^-$, and others) were obtained using topological, geometrical, and quantum-chemical descriptors. Two descriptors, the geometric descriptor relating to the anion symmetry and the electronic descriptor reflecting the charge distribution, were necessary to obtain the correlation coefficient of over 0.9. In the study of López-Martin et al., the structural variation of the anions was encoded only by one descriptor that related to the degree of sphericity of the anion and its charge. Using the established correlation, melting points of 62 imidazolium-based ionic liquids coupled with $[BF_4]^-$, $[NO_3]^-$, $[AlCl_4]^-$, $[PF_6]^-$, $[NO_2]^-$, $[SbF_6]^-$, $[ClO_4]^-$, $[AsF_6]^-$, and halides were

---

* HF stands for the Hartree-Fock method, the lowest quantum-chemical method among wavefunction-based methods, whereas 6-31G** represents a split-valence basis set with additional polarisation functions, denoted as *, on all atoms, including hydrogen.

[†] MP2 stands for a second-order Møller-Plesset perturbation theory and represents the lowest of the wavefunction-based methods that account for electron correlation effects.

obtained, with good agreement with experimental data, although in a couple of cases the error in the predicted melting points was about 50 K.

The anions in ionic liquids were proven to be crucial in determining descriptors needed to provide a good correlation for melting point. In the study of Sun et al. [42], the 1,3-dialkylimidazolium tetrafluoroborate and hexafluorophosphate ionic liquids were compared with each other. Although both $[BF_4]^-$ and $[PF_6]^-$ anions can be considered relatively spherical, the descriptors necessary to predict their melting points are of a different nature. In the case of the $[BF_4]^-$ ionic liquids, three descriptors representing the H-donors surface area, conformational changes in the alkyl chain of the imidazolium cation, and the Onsager-Kirkwook solvation energy were sufficient to provide the correlation coefficient of over 0.9. For the $[PF_6]^-$-based ionic liquids, three additional descriptors were needed to be included in the correlation to produce a correlation coefficient of over 0.9. These descriptors reflected the resonance energy of C–H bonds, the distribution of charge on the hydrogen atom in the C–H bonds, as well as the relative negative charged surface area responsible for polar interactions between ions.

To improve the flexibility of the QSPR approach based on the multi-linear regression (MLR) presented above, methods incorporating non-linear algorithms have been recently considered. These methods can be divided into two groups: (1) methods based on non-linear regression analysis such as a projection pursuit regression (PPR) [43], and (2) non-linear machine-learning methods such as decision trees [44], or those based on artificial neural networks such as the multilayer perceptron neural network [45], the back-propagation neural network (BPNN) [46, 47], the recursive neural network [48], the counter-propagation neural network [49], the support vector network (SVN) [46], and the associative neural network (ANN) [46]. Flexibility of these approaches comes from the fact that a non-linear-type correlation is assumed between molecular descriptors and the target property. In order to establish this correlation, a number of molecular descriptors from the conventional QSPR approach can be easily fed to either a neural network [45–49] or an ensemble of decision trees [44].

6.3.2   Non-Linear Machine-Leaning Methods: Prediction of Melting Points

The PPR approach is based on projections of high-dimensional data to low-dimensional data by numerically maximising a certain objective function called the *projective index* by means of a non-parametric fitting [50]. Non-parametric in this context means that parameters used in the approach have to be estimated from the available experimental data. The PPR approach can be considered a refined concept of the linear regression approach as it allows for (1) non-linearity by using general smooth functions of descriptors, and, more important, (2) interactions between descriptors through the smooth functions [51].

In the study of Ren et al. [43], the PPR approach was used to correlate melting points from a diverse set of 288 bromides of pyridinium-, imidazolium-, benzimidazolium-, and 4-amino-1,2,4-triazolium-based cations by using eight descriptors. The chosen descriptors contained information of the ring structure, bonds with low-barrier rotations, branching, symmetry, and intramolecular electronic effects. The proposed model produced a correlation coefficient (r^2) of 0.810 and an average error of 17.75%. The conventional heuristic method, using the same number of descriptors, yielded worse correlations, with a correlation coefficient and an average error of 0.712 and 24.33%, respectively. Obviously, the PPR approach shows a better predictive ability of melting points by assuming non-linear relationships between molecular descriptors and the melting point. Due to the diversity and complexity of the studied cations, no straightforward correlation was found between the descriptors and melting points, indicating that melting points are influenced by a complex (non-linear) relationship between various descriptors. Therefore, the MLR methods are less likely to capture this complex behaviour.

Artificial neural networks, on the other hand, represent an emulation of a biological neural system. Essentially, a neural network is a parallel system consisting of multiple layers of interconnected nodes called artificial neurons that solve a problem collectively rather than individually. The neurons communicate with each other through mathematical algorithms combined into a network function that can be as complex as needed. The main advantage of the parallel and collective nature of the neural network lies in its ability to successfully perform a task, or solve a problem, using non-linear correlations, otherwise not affordable by traditional linear-based methods. In the event of one neuron failing, the network will continue working due to the parallelism between the other neurons. Before the artificial neural network can be used for any prediction, it needs to be trained with respect to the target property. In the beginning, each neuron has multiple inputs corresponding to randomly weighted molecular descriptors. The neuron uses a selected mathematical (training) function to compute the output. During the training or learning process, the network readjusts the weights of the inputs to obtain the desired output (i.e., the target property). The success of the learning process relies on the type of the training function chosen [45]. The extensive learning process limits efficiency of the neural networks if a large number of molecular descriptors are used. Larger neural networks also require higher processing times. Depending on the algorithms and training functions used by the neurons, a number of artificial neural networks can be generated such as multilayer perceptron networks, back-propagation networks, recursive networks, and counter-propagation networks. Torrecilla et al. [45] showed that a multilayer perceptron neural network was successful in modelling melting points of 97 imidazolium-based ionic liquids with varied anions. The network predicted melting points with a regression coefficient of 0.99 and a mean prediction error of 1.30%. Fourteen molecular descriptors were used to imitate the complexity of different structures of ionic liquids. The performance of the network was found to strongly depend on the anion size, with the prediction error decreasing with increasing anion size. It has to be noted that the chosen molecular descriptors reflect molecular structures of imidazolium-based cations and, therefore, the optimised network can only be applied for prediction of melting points of imidazolium-based ionic liquids. Further optimisations are required to ascertain a wider application of the network to non-imidazolium-based ionic liquids.

The machine-learning methods appear to show a lot of potential in improving the accuracy of predicted melting points by outperforming linear-based methods [46, 47]. In the study by Varnek et al. [46], a comprehensive comparison between the traditional MLR analysis method and various non-linear machine-learning methods, including SVN, ANN, and BPNN, was performed for correlation of melting points of 717 bromides coupled with pyridinium, imidazolium, benzimidazolium, and ammonium cations. The predictive ability of non-linear SVN, ANN, and BPNN techniques was slightly better, regardless of the molecular descriptors used. The standard deviation error was determined to be still relatively high in the range of 37.5–46.4 K. The moderate accuracy was related to the quality of some experimental data, and difficulty to take into account specific interactions occurring in the solid state of ionic liquids. In the study of Yan et al. [47], the melting points of a smaller set containing 50 imidazolium bromides and chlorides were correlated using the MLR and BPNN methods. The non-linear BPNN method outperformed the MLR method, reducing the mean absolute error to the range of 5–9 K. Although the prediction of the melting points of ionic liquids with neural networks is more promising, compared with the conventional multi-linear models, more studies are needed to validate the applicability of this approach for a wider range of ionic liquids.

Another machine-learning method, the decision tree approach, generates a tree model, with each branch following a logical rule with respect to a chosen descriptor. The approach is very robust, as it can consider a large number of descriptors at once and disregard the most irrelevant descriptors from the start. It was shown that an ensemble of decision trees (random forest), generated in a sequential manner, significantly improved the prediction accuracy [52]. Carrera et al. [44] utilised the decision tree approach for prediction of melting points of 126 ionic liquids. The first regression tree was constructed using an entire pool of 1085 available molecular descriptors. The chosen descriptors were then deleted from the pool, which was subsequently used to generate a second tree. The procedure was repeated until n regression trees were generated. The advantage of using an ensemble of trees lies in their ability to easily screen a vast number of molecular descriptors quite quickly, and identify those that are most likely to influence the target property. The decision tree model generated this way resulted in the r^2 value of 0.95 for melting points of 126 ionic liquids, which was relatively high compared with the accuracy of multi-linear methods [34, 35]. An ensemble of four trees was found to produce the lowest errors, with no improvement being achieved for larger ensembles. A cross-validation test against the data set used by Katritzky et al. [34] in the MLR QSPR studies yielded slightly better correlations. Overall, the quality of the regression tree model was similar to that of MLR methods, thus offering a faster alternative to conventional QSPR studies utilising linear algorithms, as it allows for efficient screening of a much larger number of descriptors.

A totally different angle from the QSPR approach was recently presented by Bini et al. [48], who proposed an RNN that did not require any molecular descriptor and was based solely on the molecular structure of compounds. In this approach, each chemical structure is represented by trees, in which each atom is considered as a node and chemical bonds represent edges that start from these nodes. The RNN approach exploits a recursive process that computes a numerical code based on the chemical structure, and then the code is mapped out to the target property. Using a training set of experimental data, the RNN algorithm generates a learning process that aims at finding a direct relationship between the structure and the target property. The proposed approach was cross-validated using the data sets from studies by Katritzky et al. [34] and Carrera et al. [44], and produced the standard deviation of about 30 K, which is only slightly worse than these studies that included proper molecular descriptors. It has to be noted that this result is rather impressive, as no molecular descriptors other than the actual chemical structure were used to predict melting points of ionic liquids.

6.3.3   Prediction of Transport Properties

Only one study has been published so far that used the QSPR approach to predict transport properties of ionic liquids, such as conductivity and viscosity [53]. A polynomial expansion method using the genetic algorithm was applied to both properties, by incorporating a number of descriptors related to ionic volume, shape, and dipole moment. The main conclusion of the study was that viscosity followed conductivity and, therefore, lower viscosity resulted in higher conductivity. This finding is basically the justification of the Walden rule, which states that the product of conductivity and viscosity remains constant with increasing temperature [54]. Some ionic liquids studied fell out of the found trends, especially ionic liquids that are composed of small aromatic cations and anions of zero dipole moment. Since it is the only study on the prediction of transport properties, further studies of ionic liquid systems that do not follow the Walden rule are needed to make general conclusions on the ability of the QSPR approach to predict transport properties of ionic liquids.

To summarise, the latest trend in the QSPR approach successfully utilises non-linear machine-learning algorithms such as neural networks and PPR that show more predictive ability for the melting points of a diverse range of ionic liquids by assuming non-linear relationships between molecular descriptors and the target property. The correlations for melting points of ionic liquids were generally better compared with linear-based methods, emphasising a complex relationship between molecular descriptors and melting points. Although more flexible, machine-learning methods are more expensive due to the learning process involved, and depending on the algorithms implemented, they are also subject to high processing times. As a result, one of the main limitations is the total number of descriptors that can be used in the process. In contrast, a machine-learning method based on the decision tree approach represents a much faster alternative to MLR methods, as it can easily screen a large number of molecular descriptors and disregard the most irrelevant ones at the start, producing similar quality to conventional linear-based methods in prediction performance. More systematic and comparative studies are necessary to ascertain the predictive ability of non-linear machine-learning methods for melting points of ionic liquids, and especially their transport properties.

6.4   MOLECULAR DYNAMICS SIMULATIONS

A few comprehensive reviews have been written addressing problems associated with slow dynamics, simulations of ionic liquids at interfaces and with solutes, as well as development of new force fields and coarse-grained models for ionic liquids [55–64]. Of particular interest is the prediction of thermodynamic and transport properties [57], and future opportunities for MD simulations [61, 63].

Although there are many properties that can be extracted from MD simulations, such as the liquid structure of ionic liquids at various temperatures, and the ability of ionic liquids to solvate polar and non-polar solutes, there are certain thermodynamic and transport properties that still represent a challenge for MD simulations. These properties include melting point, enthalpy of vapourisation, and transport properties such as self-diffusion coefficients and viscosity.

Melting point is one of the most challenging properties to predict using MD simulations. The melting occurs when the free energy of the liquid becomes equal to that of the solid. A new free-energy-based method by Maginn emphasised the importance of a force field or an *ab initio* method to provide an appreciable level of accuracy for melting points. For example, Maginn et al. showed that a change in free energy of 6 kJ mol^{-1} translated into a 20 K difference in melting point for [C$_4$mim]Cl [65]. Therefore, it is not surprising that the error in melting point calculations of 30 K is considered to be in a very good agreement with experiment. A smaller change in free energy, of 4 kJ mol^{-1}, appeared to result in an almost 50 K difference in the melting points of alkali nitrates [66]. In the latter case, it was estimated that in order to predict melting point within 10 K of the experiment, the free energy must be calculated with an accuracy of 1 kJ mol^{-1}. It is quite disturbing that, in order to predict melting points within 10 K of accuracy, the free energy of the ionic system has to be treated within the spectroscopic accuracy of 1 kJ mol^{-1}. This outcome introduces a serious demand on the quality of force fields used in MD simulations.

The slow dynamics of ionic liquids has been thoroughly studied using MD simulations [67–69]. This phenomenon is due to large errors in calculated values of viscosity, especially when the sub-diffusive régimes from short-run MD simulations are used [70]. Usually, an overestimation in viscosity by an order of magnitude is observed [61]. As a result, the ionic systems have to be equilibrated for a very long time on the scale of 10 ns to reduce errors in diffusion coefficients and viscosities. Although longer simulations times are necessary, these are not easily achievable due to computational restraints.

6.4.1   What Are the Current Trends in MD Simulations?

1. More accurate, possibly generalised, force fields for various ionic liquid ions are needed to give MD simulations more predictive power for tailoring thermodynamic and transport properties of ionic liquids. Although a few of force fields have been developed for various ionic liquid ions by Canongia Lopes and Pádua [71, 72], current classic MD simulations can be considered mainly "post-prediction" simulations [61], as many of these MD simulations are conducted to reproduce already available experimental data on thermodynamic and transport properties of well-characterised ionic liquids. Therefore, there is an urgent need to move towards predictive MD simulations of novel ionic liquids prior to their synthesis. This, in turn, requires robust validation procedures of the developed force fields. Most of the current force fields are validated against the experimental values of liquid density, which can be insensitive to inter-ionic interactions and, hence, obscure the validation process [62]. Melting points were shown to be very sensitive to the force field [65, 66] and thus could be used as benchmarks for developing more accurate force fields. Recently, it was also shown that a force field could be significantly improved by validating it against enthalpies of vapourisation [73].

2. Longer simulations runs need to be carried out from 10 to 100 ns to ensure proper equilibration of ionic systems and, therefore, reliable accuracy for diffusion coefficients and viscosity. One of the current solutions is to use coarse-grained models that divide a molecule into beads composed of a few atoms rather than considering individual atoms separately as in the classical all-atom MD approach, thus decreasing the number of rotational and vibrational degrees of freedom and offering a possibility of running longer simulations [61, 64]. These models are yet to be thoroughly studied in terms of their accuracy with regard to thermodynamic and transport properties.

3. Currently, the majority of the force fields are parameterised against gas-phase quantum-chemical calculations. Force field matching and/or

parameterisation approaches against quantum-chemical condensed phase calculations are needed to develop more accurate force fields.

4. More studies are needed to determine whether classical force fields including partial (or reduced) charges on ions can compete in accuracy with fully polarisable force fields. The former force fields certainly deserve particular attention, as MD simulations become computationally less demanding t those incorporating polarisable force fields.

5. *Ab initio* (or, more strictly speaking, DFT) MD simulations are becoming very popular as they better reproduce the microstructure of ionic liquids, and therefore can be used to parameterise existing force fields, as well as study ionic systems that may be reactive under certain conditions (e.g., $[N(CN)_2]^-$ ionic liquids that exhibit catalytic properties [74]), or be subject to a not-yet-well-understood proton conductivity (e.g., the proton hopping mechanism in protic ionic liquids [75]).

Although these trends have been discussed in great detail elsewhere [58, 62, 64], we will focus on the last three trends that deserve particular attention due to the increased number of publications in recent years.

6.4.1.1 *Polarisable Force Field versus Force Field with Reduced Charges*

Ab initio MD simulations of the $[C_1mim]Cl$ ionic liquid by Lynden-Bell et al. [76] indicated a significant fluctuation in electron distribution of ions resulting in non-negligible fluctuations of their dipole moments. Even for anions with a static dipole moment of zero, such as chloride, a fluctuation of up to 0.5 D could be observed. There are a growing number of publications introducing flexible force fields that allow for fluctuations in the electronic polarisability of ions, hence the term *polarisable force fields* [70, 77]. One of the earliest approaches was to introduce dipole–dipole interactions in a classical force field that were parameterised using electronic polarisabilities calculated at a correlated level of *ab initio* theory such as MP2. For the ionic liquid, $[C_2mim][NO_3]$, it was shown that inclusion of electronic polarisability produced better agreement with experiment for viscosity, as polarisation effects resulted in ions being more mobile and, hence, having higher diffusion coefficients [78]. A more robust approach is to include a many-body polarisable force field, which was achieved by Borodin and Smith [79] and tested for $[C_3mim][NTf_2]$ at two temperatures, 393 K and 303 K. The potential energy of this polarisable force field contained energies of multipole–multipole (charge–charge, charge–dipole, etc.) Coulomb interactions, and the potential energy due to polarisation. In Borodin and Smith's study, a good agreement between calculated and experimental data of diffusion coefficients, conductivity, and viscosity was observed. It has to be noted that extra care in equilibrating the ionic system was taken to ensure more reliable data for viscosity due to the slow dynamics of ions. For example, MD simulations of 13.2 ns and 16.2 ns were performed at 393 K and 303 K, respectively, to achieve the desired accuracy.

A recent study by Borodin proposed a generalised polarisable force field that can be potentially applied to a wide class of ionic liquids by allowing transferability of the repulsion–dispersion parameters for certain functional groups, such as the cyano group, $-CN$, and the $-CF_3$ group [70]. For example, many well-characterised ionic liquids contain anions with these electron-withdrawing groups, such as $[N(CN)_2]^-$, $[B(CN)_4]^-$, $[BF_3CF_3]^-$, $[NTf_2]^-$, and $[SO_3CF_3]^-$, due to the fact that the corresponding ionic liquids are more likely to have low melting points and be relatively fluid. Therefore, repulsion–dispersion parameters were shown to be transferrable from one ionic liquid to another. Thirty ionic liquids were extensively studied in Borodin's paper, including imidazolium-, pyrrolidinium-, pyridinium-, morpholinium-, and piperidinium-based ionic liquids coupled with various anions such as $[NTf_2]^-$, $[BF_4]^-$, $[PF_6]^-$, $[CH_3BF_3]^-$, $[CF_3BF_3]^-$, $[SO_3CF_3]^-$, and $[NO_3]^-$. The proposed polarisable force field produced diffusion coefficients for 30 ionic liquids under study within 20–40% of the experimental data. Due to the slow dynamics of the ions, the MD simulations needed to be sufficiently equilibrated. When the diffusion coefficients were extracted from the sub-diffusive régime occurring between 100 and 150 ps, these were overestimated by at least 50% compared with those extracted from the correctly equilibrated simulations. The conductivity and viscosity data were also in good agreement with experiment, the absolute deviation being 23% and 33%, respectively. The calculated heats of vapourisation of these ionic liquids were in excellent agreement with available experimental data. Although relatively expensive, the approach by Borodin utilising a transferable polarisable force field certainly opens more possibilities for prediction of thermodynamic and transport properties of ionic liquids that are yet to be synthesised.

In the study of Borodin and Smith, it was noted that non-polarisable force fields could potentially provide adequate descriptions of the thermodynamic properties (such as enthalpies of vapourisation) and transport properties (such as self-diffusion coefficient) [79] if polarisation in the description of the force field could be represented by effective partial charge on the ions. Morrow and Maginn [80] showed that ion-pair calculations of $[C_4mim][PF_6]$ indicated a charge transfer between the cation and anion, yielding the total charge on the ions of $0.904e$. In the study of Bhargava and Balasubramanian [81], the force field parameters were obtained by tuning charges on the cation and anion so that the refined classical force field reproduced the pair correlation functions from previous *ab initio* MD simulations. The resulting uniformly scaled charge on the cation and anion was determined to be $0.8e$ for $[C_4mim][PF_6]$. The reduced overall charge led to reduced electrostatic interactions between ions and, thus, increased diffusion coefficients that reproduced experimental data within 20% for a number of temperatures from 300 K to 500 K. The diffusion coefficients obtained using the classical force field by Lopes [72, 82] with the total charge of $1e$ on both cation, $[C_4mim]^+$, and anion, $[PF_6]^-$, had larger deviations from experimental data and for some temperatures deviations of up to one order of magnitude were observed. Youngs and Hardacre [83] conducted a similar study for the $[C_1mim]Cl$ ionic liquid by scaling the total charge from 0.5e to 1.0e, thus reducing the electro-static interactions by as much as a factor of 2. Not surprisingly, diffusion coefficients increased by three orders of magnitude going from the unity charges on ions unity to 0.5e, with the cationic diffusion increasing at a faster rate compared with that of the anion. It was established that the total ionic charge of 0.6–0.7e provided excellent structural agreement with the previous *ab initio* MD simulations [84], and the ratio of the cation and anionic diffusion coefficients was also in excellent agreement with the experimental value. In the recent study by Lynden-Bell and Youngs [85], a comprehensive comparison of a number of force field models, including those with explicit polarisation effects as well as reduced charges on ions, was performed for the [C$_1$mim]Cl ionic liquid. It was shown that in order to accurately predict dynamic properties of this ionic liquid (such as diffusion coefficients and cohesive energy), a charge reduction on ions or explicit polarisation effects are necessary to be included in the force field formulation.

6.4.1.2   *Refinement of Force Fields*   It was noticed that the quality of predicted enthalpies of vapourisation often correlated with excellent prediction of self-diffusion coefficients, conductivity, and viscosity [70]. Rebelo et al. [86] calculated the components of enthalpies of vapourisation at room temperature for a series of [C$_n$mim][NTf$_2$], with n changing from 2 to 8. They concluded that, in this homologous series, the Coulombic contribution to the enthalpy is almost constant (varying slightly from 62 to 75 kJ mol^{-1}), whereas the van der Waals component steadily increases with increasing chain length on the cation from 91 kJ mol^{-1} for ethyl to 120 kJ mol^{-1} for octyl. The classical force field used by Rebelo and coworkers overestimated the experimental heats of vapourisation due to overestimation of the dispersion component [87]. The significance of these results lies in the fact that dispersion interactions are non-negligible in ionic liquids, and appear to dominate over Coulomb interactions in the enthalpy of vapourisation with increasing alkyl chain on the cation. Consequently, a force field that is capable of capturing this effect performs better in predicting thermodynamic and transport properties. Köddermann et al. [73] refined the existing the all-atom force field by Canongia Lopes and coworkers [72] by optimising the Lennard-Jones parameters that account for dispersion interactions to better reproduce the diffusion coefficient of the anion in the [C$_2$mim][NTf$_2$] ionic liquid. As a result of this refinement, the calculated enthalpies of vapourisation, self-diffusion coefficients, and viscosity in the [C$_n$mim][NTf$_2$] series (with n = 2, 4, 6 or 8) were in excellent agreement with experimental data. This offers new avenue for refining force fields against experimental enthalpies of vapourisation with a view to improving accuracy in calculations of transport properties, such as self-diffusion coefficient and viscosity.

A more robust but computationally more demanding method of refining a force field is to use a force matching approach, in which trajectories generated by *ab initio* MD simulations are used to refine a classical force field for a specific ionic liquid [77]. In the early study of Youngs et al., the force field matching approach was used in MD simulations of [C$_1$mim]Cl [77]. While the charges on the cation and anion were set to $+1e$ and $-1e$, respectively, individual atomic charges were allowed to vary, thus allowing for the cationic polarisation. The generated force field better reproduced the *ab initio* MD radial distribution functions, favouring strong hydrogen bonding between the C^2–H bond and the chloride anion. Charges of $\pm1e$ on the cation and anion obviously favour very strong inter-ionic interactions, thus leading to the over-structuring of the liquid state. When the charges on the cation and anion were allowed to vary throughout the minimisation procedure in the same ionic liquid (to keep neutrality the charge on the anion was taken opposite to that on the cation), the total charge on the cation was optimised to be about 0.8, making the approach of reducing charges on ions quite realistic [83]. Inclusion of both cationic and anionic polarisation yielded the self-diffusion of the system almost identical to that of the original force field that included the scaled charges on both ions of 0.8, thus indicating that the ion dynamics was strongly influenced by the total charge on the ions (especially on the chloride), rather than by the charge distribution on the cation. A comparison was made between the two generated force fields for the same ionic liquid, [C_1mim]Cl: one with the unity charge on ions and one with reduced charges of $0.8e$ on ions. Although both models gave good agreement with the experimental crystal structure, the force field with reduced charge favoured faster ion dynamics by at least two orders of magnitude, but resulted in a significantly underestimated liquid cohesive energy per ion pair in both crystal and liquid states [85]. Clearly, the ion dynamics in ionic liquids are strongly influenced by the overall charges of the cation and anion and, therefore, more studies on the polarisation of more complex anions are necessary to identify parameters that lead to more accurate prediction of self-diffusion coefficients and, hence, viscosity.

To summarise, accurate and (of equal importance) generalised force fields for various ionic liquid ions are needed to make MD simulations more predictive and, therefore, suitable for tailoring thermodynamic and transport properties of novel ionic liquids. In order to develop more accurate force fields, robust validation procedures are required to ensure their predictive power. For example, properties such as melting point and enthalpy of vapourisation have been shown to be more sensitive to the quality of the force fields than the liquid density that is often used in the validation process. Force fields with reduced charges on ions, and force fields that explicitly allow for ionic polarisation, even for ions with zero-dipole moments, performed better for dynamic properties such as self-diffusion coefficients and viscosity. This finding indicates that ionic polarisation needs to be treated accurately for accurate predictions of dynamic properties of ionic liquids.

6.5 APPROACHES *FROM FIRST PRINCIPLES* BASED ON *ab initio* AND DENSITY FUNCTIONAL THEORY

Quantum-chemical methods based on standard molecular orbital theory offer a few advantages over the volume-based and QSPR approaches, as these methods rely only on fundamental constants and, therefore, provide a thorough insight into structural, molecular, and energetic properties of ionic liquids *from first principles*. It has to be emphasised that most of the quantum-chemical studies performed to date have aimed at predictions of physical properties of ionic liquids based on calculations of either individual ions or single ion pairs. There have been only a few reports in the literature that adopted a cluster approach incorporating a number of ion pairs in the calculations. In the following sections, an overview of the progress achieved in understanding key factors influencing thermodynamic and transport properties of ionic liquids *from first principles* is presented.

6.5.1   Thermodynamic Properties: Enthalpies of Vapourisation and Vapour Pressure

Emel'yanenko et al. [88–90] proposed to combine high-level *ab initio* calculations with combustion calorimetry to predict enthalpies of vapourisation. The molar enthalpy of vapourisation of an ionic liquid can be evaluated using the following thermodynamic relationship:

$$\Delta_l^g H_m(298\ K) = \Delta_f H_m^0(g) - \Delta_f H_m^0(l). \tag{6.18}$$

where $\Delta_f H_m^0(l)$ is the molar enthalpy of the ionic liquid formation in the liquid state obtained by high-precision combustion calorimetry, and $\Delta_f H_m^0(g)$ is the gas-phase enthalpy of the ionic liquid formation that can be easily calculated at a suitable *ab initio* level of theory, assuming that the ionic liquid exists as a neutral ion pair in gas phase. The heats of formation were calculated based on the atomisation reactions and "bond separation" reactions. The authors established that the G3MP2 composite method [91] was already sufficient to accurately predict the heats of formation of ionic liquids [90]. This approach was applied to evaluation of enthalpies of vapourisation of imidazolium-based ionic liquids such as [C_2mim][NO_3], [C_4mim][NO_3] and 1-methylimidazolium nitrate. The results were in good agreement with available experimental data. In the case of 1-methylimidazolium nitrate, a more reliable enthalpy of vapourisation was calculated assuming that the ion pair underwent a proton transfer, thus dissociating into two neutral species, the acid and the base, in gas phase.

A totally different approach was introduced by Diedenhofen et al. [92], who employed the COSMO-RS method to calculate the enthalpies of vapourisation and vapour pressure. The COSMO-RS approach [12] is different from the COSMO solvation model mentioned earlier as a quantum-chemical calculation of the charge distribution on the molecular surface of the solute and solvent is performed in a conductor material with an *infinitely large dielectric constant*, which fully screens the charges on their surfaces, creating a molecule-shaped surface of the counter-charges (i.e., the COSMO surface). The interactions between the solute and the solvent are expressed as a sum of pair-wise surface contact energies by bringing the generated COSMO surfaces in contact. These contact energies arise from electrostatic and hydrogen-bonding interactions, with the hydrogen bond energy being described as an interaction between highly polar and opposite charged surface segments. It has to be noted that, in contrast, the COMSO solvation model accounts only for electrostatic interactions. Principles of statistical thermodynamics are applied to the COSMO surfaces to calculate chemical potentials, free energies of solvation, enthalpies of vapourisation, vapour pressure, and other thermodynamic quantities. In the study of Diedenhofen et al. [92], the vapour pressure and the enthalpies of vapourisation were calculated assuming the presence of ion pairs in both phases: vapour and liquid. The predicted theoretical enthalpies of vapourisation were found to be 164.6 kJ mol^{-1} for [C$_2$mim][EtSO$_4$] and 159.3 kJ mol^{-1} for [C$_4$mim][N(CN)$_2$], in very good agreement with experimental data of 164 ± 4 kJ mol^{-1} and 157.2 ± 1.1 kJ mol^{-1}, respectively. The vapour pressure results had uncertainty of 10%, assuming that ionic liquids existed as ion pairs. If individual ions were considered in the COSMO-RS approach, the vapour pressures were negligible (in the order of 10^{-7} to 10^{-4} Pa) even at elevated temperatures, which are about three orders of magnitude lower than the experimental ones. Thus, the COSMO-RS approach combining interacting charge surfaces from quantum-chemical calculations with principles of statistical thermodynamics represents an exciting and easy way of calculating thermodynamic quantities of ionic liquids such as enthalpy of vapourisation.

6.5.2 Melting Point and Transport Properties, Such as Conductivity and Viscosity

The early calculations studied simple building blocks in ionic liquids: single ion pairs. They carry information about the strength of the interaction between the cation and anion and thus have potential in identifying key factors influencing thermodynamic and transport properties of the bulk ionic liquids. Ion pairs of the imidazolium-, phosphonium-, and pyrrolidinium-based ionic liquids have been extensively studied at various levels of *ab initio* theory and DFT [94–108].

Hunt et al. calculated the ionic strength of inter-ionic interactions in 1-butyl-3-methylimidazolium ionic liquids coupled with the Cl$^-$, [BF$_4$]$^-$, and [NTf$_2$]$^-$ anions [93]. In this homologous series, the dissociation energy (which is equivalent to the energy released by separating the ion pair into its constituent ions) steadily decreases, going from Cl$^-$ to [BF$_4$]$^-$ to [NTf$_2$]$^-$, following the experimental trend of decreasing melting point and viscosity. A quantity called "connectivity index" was proposed to define a number of possible "internal" ion-pair cation–anion contacts and "external" contacts that each anion makes with surrounding cations, thus forming an external network. The connectivity index is based on identifying energetically favourable positions of the anion around the imidazolium ring and the symmetry (or the lack of thereof) of the anion. For example, only two bulky [NTf$_2$]$^-$ anions can physically interact with the cation at once, with the nitrogen and three oxygen atoms on the anion involved in the cation–anion ion-pair contacts, leaving only one oxygen atom per each anion to form an external link. This reasoning makes the connectivity index for the $[NTf_2]^-$-based ionic liquid as low as $2 \times 1 = 2$. Information on the connectivity index and the ionic strength of the ion-pair interactions allows one to partially rationalise why $[C_4mim]Cl$, with a connectivity index of 20 and the largest dissociation energy, has high viscosity and high melting point, whereas $[C_4mim][NTf_2]$, with a connectivity of 2 and the weakest interaction, forms a low-melting and fluid ionic liquid.

The strength of ion-pair interactions alone cannot explain the substantial difference in experimental viscosity between the 1-butyl-3-methylimidazolium ($[C_4mim]^+$) and 1-butyl-2,3-dimethylimidazolium ($[C_4dmim]^+$) ionic liquids [94]. The substitution of the hydrogen atom at the C2 position in the imidazolium ring by the methyl group reduces the dissociation energy of the ion pair only slightly, whereas the corresponding viscosity and melting point increase significantly. For example, the $[C_4dmim]Cl$ ion pair was found to be 10 kJ mol^{-1} less stable than the $[C_4mim]Cl$ ion pair. A similar difference of about 16 kJ mol^{-1} was observed for binding energies of the $[C_2mim][BF_4]$ and $[C_2dmim][BF_4]$ ion pairs [95]. In addition, the replacement of the C2 hydrogen by the methyl group blocks in-plane hydrogen-bonding interactions in the front of the imidazolium ring, thus reducing the number of possible interactions of the anion around the ring. It was hypothesised that a loss in entropy in the $[C_4dmim]^+$ ionic liquids might play a rôle. A significant reduction in entropy was attributed to the decreased connectivity index and a restricted rotation of the butyl chain due to the steric hindrance enforced by the methyl group in the C2 position. One of the limitations of this approach is that the ion-pair information was used to explain melting points and viscosities of ionic liquids governed by long-range electrostatic interactions and thus the obtained results can only be used in combination with experimentally available data.

The C^2–H strong interactions were found to be of a different nature from traditional hydrogen bonding driven mainly by electrostatic interactions [96, 97], with the orientation of the anion relative to the C^2–H bond changing the interaction energy only slightly [96]. Recently, Zahn et al. calculated the energy path that the chloride anion followed in order to move from the above of the imidazolium ring all the way across to below the ring [98]. The two energy paths were generated at the MP2/aug-cc-pVTZ† level of theory for imidazolium-based ionic liquids: non-methylated, $[C_1mim]Cl$, and methylated at the C2 position, $[C_1dmim]Cl$. A barrier of more than 40 kJ mol^{-1} was found on the potential energy surface of the methylated ionic liquid, whereas in the non-methylated ionic liquid two barriers separated by less than 10 kJ mol^{-1} were located. These findings supported the energy landscape paradigm first proposed by Goldstein

---

† aug-cc-pVTZ stands for a correlated consistent polarised triple-ζ quality basis set augmented by a set of diffuse functions. Diffuse functions are usually included for a better description of charged species.

[99], according to which shallow energy minima and low transition barriers (within the thermal energy) between minima on the potential energy surface allow the molecule to sample a large part of the conformational space at a little cost in energy, thus resulting in a low melting point. As a result, the presence of extra hydrogen-bonding between the C^2–H bond and the anion leads to a larger conformational space that spans the energy level within the thermal energy and thus, lower melting points. The presence of these interactions in imidazolium-based ionic liquids is also attributed to the disruption of the Coulomb network existing between the ions and, as a result, reducing their melting point and fluidity [100, 101]. On the other hand, blocking the C2 position for the in-plane hydrogen-bonding formation results in a smaller conformational space of similar energy and, therefore, reduction in hydrogen bonding is responsible for higher melting points of methylated ionic liquids.

In our group, the concept of proton affinity was proposed in an attempt to correlate melting points of ionic liquids and charge delocalisation on ionic liquid anions [102]. In this work, proton affinity was defined as the amount of energy (i.e., the reaction enthalpy, $\Delta_r H$) required for a probe proton to transfer to an anion and thus represented the degree of charge delocalisation on the anion:

$$A^- + [H_3O]^+ = A\text{-}H + H_2O; \Delta_r H \tag{6.19}$$

As a result, the lower proton affinity values correspond to a higher degree of charge delocalisation, thus leading to lower melting points. The proton affinities of routinely used ionic liquid anions were calculated at a high correlated level of theory, CCSD(T)† /aug-cc-pVTZ, which is considered the benchmark method in *ab initio* theory [103]. It was shown that proton affinities indeed correlated well with melting points (see Fig. 6.5), with the exception of the tosylate anions; these may potentially favour π–π stacking interactions, thus increasing the overall lattice energy and, hence, melting point. The proton affinity approach can be used as a starting point for screening novel ionic liquid anions that could potentially result in a low melting point ionic liquid. It has to be noted that this approach does not provide any guidance on the effect of different cations on melting point, as well as specific interactions such as van der Waals and hydrogen bonding that may occur between ions in the liquid state.

Interaction energies of ion pairs of [$C_n$mim]X salts ($n = 2$, 3 or 4; X = Cl, Br, or I) calculated at the MP2/6-31+G* level of theory were correlated with corresponding melting points [104]. Close-to-linear correlations were found between the melting points of the chloride and iodide ionic liquids and their corresponding interaction energies, with the bromide falling out of the trend.

---

† CCSD(T) belongs to a group of coupled cluster (CC) theory methods and incorporates singly and doubly excited electronic configurations to account for electron correlation, with the triple excitations being treated non-iteratively.

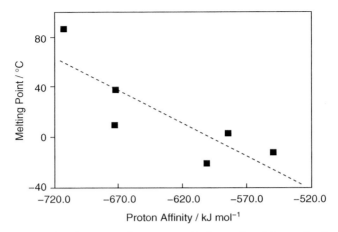

Figure 6.5   Correlation between the proton affinity and melting point for $[C_2mim]Y$ ($Y^- = [NTf_2]^-$, $[N(CN)_2]^-$, Cl^-, $[NO_3]^-$, $[CF_3SO_3]^-$, or $[(SO_2CH_3)N(SO_2CF_3)]^-$.

These results indicated an anion dependence of interaction energy on melting point, although the trends had different slopes. It was noticed that the difference between successive melting points with increasing alkyl chain length on the cation and keeping the anion constant was much greater than the difference between corresponding interaction energies, with the largest difference being only 5 kJ mol^{-1}, within the systematic error of the calculations. Considering a wider range of anions, $[AlCl_4]^-$, $[N(CN)_2]^-$, $[BF_4]^-$, $[SCN]^-$, $[CF_3CO_2]^-$, $[CF_3SO_3]^-$, $[NTf_2]^-$, and $[PF_6]^-$, for the $[C_2mim]^+$ ionic liquids, no obvious correlation between ion-pair binding energies (IPBEs) and melting points were found [95, 105]. Although the electrostatic interaction is the main interaction in ion pairs of ionic liquids, the potential energy surfaces around the equilibrium geometry are quite shallow, due to pronounced contributions from dispersion forces such as van der Waals and induction interactions [95, 105]. Recently, in our group, we showed that the dispersion component to the IPBE (approximated as the difference between the MP2 and HF binding energies) was also significant for pyrrolidinium-based ionic liquids and varied, depending on the anion, from 7% (20 kJ mol^{-1}) for $[BF_4]^-$ to 14% (45 kJ mol^{-1}) for the $[NTf_2]^-$ anion [106]. It was observed that the more electron-rich was the anion, the larger the dispersion component, thus further supporting the importance of dispersion (i.e., non-electrostatic) interactions in ionic liquids between cations and anions.

Early studies did not find any correlation between IPBEs (in some papers this quantity is referred to as interaction energy) and molar conductivity of ionic liquids. For imidazolium-based ionic liquids, Tsuzuki et al. [95] found a correlation between interaction energies in ion pairs and the ionic dissociation/association dynamics, defined by the Watanabe approach as a ratio between molar conductivity obtained from impedance measurements (λ_{imp}) and molar conductivity obtained from ion diffusivity measurements (λ_{NMR}) [107]. The $\lambda_{imp}/\lambda_{NMR}$ ratio was shown to provide insight into "ionicity" of ionic liquids that indicates how dissociated, that is, ionic, a certain ionic liquid is. Clearly, the values closer to 1 indicate that ionic liquids do not participate in association processes and behave almost as independent ions. On the other hand, smaller values of this ratio correspond to more association dynamics in ionic liquids and, not surprisingly, to stronger IPBEs in imidazolium-based ionic liquids in the electric field. In the homologous series of the $[C_4mim]^+$ ionic liquids, the $\lambda_{imp}/\lambda_{NMR}$ values decreased in the order $[PF_6]^- > [BF_4]^- > [NTf_2]^- > [CF_3SO_3]^- > [CF_3CO_2]^-$ and followed the analogous trend in IPBEs. In our group, we also showed a similar correlation for phosphonium-based ionic liquids, using the deviation from the ideal line on the Walden plot as a measure of ionicity [108, 109]. The deviation from the ideal line on the Walden plot (ΔW) strongly correlated with IPBEs for phosphonium-based ionic liquids coupled with various traditional anions, such as $[NTf_2]^-$, $[N(CN)_2]^-$, and Cl^- and pharmaceutical anions such as acesulfamate, saccharinate, and cyclamate, with the stronger IPBEs also corresponding to larger deviations, ΔW.

Recently, Kobrak et al. [110] proposed a charge lever momentum (CLM) approach that is based on a sophisticated formulation combining the centre of charge with the ionic moments of inertia and the asymmetry of the ion. This approach accounts for a combined effect of cations and anions on electrostatic interactions in ionic liquids and, thus, aims at predicting their transport properties such as viscosity. The methodology was developed for asymmetric ions, whose centre of charge does not coincide with the centre of mass (i.e., ions of low or no symmetry) resulting in the rotational/librational motion of the ions, altering their electrostatic interactions with neighbouring ions. A three-dimensional quantity called "charge lever momentum" is defined such that it accounts for the degree to which electrostatic interaction induce torque about a given axis, and the rate of the dynamic response in the ion through the moments of inertia. The argument behind the proposed quantity is as follows. The larger the CLM values, the faster the dynamic response of the ion. The fast response to the field of neighbouring ions will allow the ion to sample the configurational space of the liquid more rapidly and, hence, will result in lower viscosities. The B3LYP/6-31G(d,p) level of theory was used to perform geometry optimisations of ions and the **CH**arges from **E**lectrostatic **P**otentials using a **G**rid based (CHELPG) algorithm was used to fit atomic charges to reproduce the electrostatic potential calculated with B3LYP. The authors suggest that a function combining all three CLMs must correlate with viscosity but the form of this functional is not clear. As a result, it was proposed that the ions with the largest CLM that is perpendicular to the direction of the applied field would have the most rapid response to the field. Since a second function combining CLMs of cations and anions is also unknown, the correlation between the maximum CLM and viscosity was explored for a series of ionic liquids, in which one ion (be it cation or anion) was held constant and the other ion belonged to a structurally homologous series (i.e., imidazolium-based cations). Taking a series of 1-ethoxymethyl-1-methylpyrrolidinium,1-ethoxymethyl-1-methylpyrrolidinium, and 1-ethoxymethyl-1-methylpiperidinium couples with tetrafluoroborate and (perfluoroalkyl)trifluoroborates ($[(C_nF_{2n+1})BF_3]^-$), a correlation between the CLM values of the anions and viscosity was observed. As was expected, viscosity decreased with increasing CLM. A similar trend was observed for imidazolium-based ionic liquids coupled with the $[NTf_2]^-$ anion. The shortcoming of the proposed approach came for imidazolium ionic liquids with branched alkyl chains, as the approach failed to account for a significant increase in viscosity.

Recently, in our group, we found that the ratio of the IPBE and the dispersion component to IPBE correlated well with melting point, whereas the dispersion component to IPBE correlated better with transport properties, such as conductivity and viscosity [111]. As shown previously, melting point, T_m, is defined as the ratio between the enthalpy of fusion and the entropy of fusion (see Eq. 6.10), and it was hypothesised that the enthalpy of fusion should correlate with the total IPBE, whereas the entropy of fusion should correlate with the dispersion component with the total IPBE, thus leading to Equation (6.20):

$$T_m = f\left(\frac{IPBE_{total}}{IPBE_{disp}}\right) \tag{6.20}$$

The rationale behind Equation (6.20) was as follows: ΔH_{fus} in Equation (6.10) is related to the change in electrostatic energy (per mol), E_{es}, on going from the crystalline lattice to the liquid state and, therefore, is related to the lattice energy through the Madelung constant and IPBE:

$$U_L = E_{es} + E_{SR} \tag{6.21}$$

$$E_{es} = M \cdot IPBE \tag{6.22}$$

where M is the Madelung constant and E_{SR} is the energy of short-range interactions that, as a first approximation, was assumed to contribute only a small percentage to the lattice energy. On melting, the E_{es} energy will change as a result of changes in IPBE, as average distances between ions expand slightly. Since volume changes on melting of many crystalline salts are relatively similar, this relative change in IPBE we assume to be of the same order of magnitude for most salts:

$$\Delta H_f \propto k \cdot IPBE \tag{6.23}$$

where k is the change on melting due to the expansion of the lattice. Recently, we showed that the long-range electrostatic interactions indeed converge for a number of ionic liquid forming organic salts [32], further supporting the proposed relation between ΔH_{fus} and the IPBE in Equation (6.23).

On the other hand, the increase in entropy that occurs upon melting arises from translational and rotational motions that are allowed by a decrease in dispersion interactions; a large gain in entropy at melting should produce a lower melting point. As ions move apart on melting, the decrease in short-range dispersion interactions is much greater than long-range electrostatic interactions and, therefore, the weakening of these short-range interactions is expected to be primarily responsible for the increase in motions of the ions and thus correlate with the entropy of fusion.

The IPBEs were calculated at the MP2/6-311+G(3df,2p)† level of theory and were also corrected for the basis set superposition error arising from use of finite basis sets in molecular complexes with non-covalent interactions such as ion pairs of ionic liquids. Dispersion interactions originate entirely from electron correlation effects and, therefore, the dispersion component to IPBE was estimated as the difference between the MP2 and HF total electronic energies.

Linear correlations were indeed observed between T_m and the ratio in Equation (6.20) for a series of $[C_n\text{mim}]^+$ and $[C_n\text{mpyr}]^+$ ionic liquids, particularly when considering trends for individual anions, with the ratio decreasing in the sequence $[C_n\text{mpyr}][NTf_2]$, $[C_n\text{mpyr}][N(CN)_2]$, $[C_n\text{mim}]Br$, $[C_n\text{mim}][NTf_2]$, $[C_n\text{mpyr}][CH_3C_6H_4SO_3]$, $[C_n\text{mpyr}][CH_3SO_3]$, and $[C_n\text{mpyr}][BF_4]$, where n was selected such that the melting point still decreased with increasing alkyl chain in these ionic liquids. As a result, the application of the proposed approach is limited to ionic liquids that are mainly governed by inter-ionic interactions between cations and anions. Once the melting point increases, other interactions such as van der Waals interactions between longer alkyl chain cations start to dominate the ionic system and, therefore, the melting point increases with increasing alkyl chain [15, 112]. The approach is not expected to hold for these systems (as e.g., in the case of the $[C_n\text{mim}][BF_4]$ ionic liquids with alkyl chains longer than octyl), as the melting point increases drastically for n above 8 [113]. For systems dominated by other specific interactions, such as $\pi-\pi$ stacking interactions between imidazolium cations [114] or hydrogen bonding [100, 115], the approach would not account for these extra contributions.

In the case of transport properties, the highly directional nature of dispersion forces and their short-range nature present the greatest barrier to relative motions of the ions with respect to each other. Short-range dispersion interactions rely on instantaneous dipole moments created in nearby electron clouds and, therefore, the motion of the ions may diminish these interactions quite substantially. Thus, the dispersion component of the IPBE is more likely to

---

† 6-311+G(3df,2p) is a Pople-type triple-ζ quality basis set with additional diffuse and polarisation functions on all atoms. It was shown that MP2 calculations needed to be performed with at least a triple-quality basis set to produce reliable results for reaction energies [99].

correlate with the transport properties such as conductivity and viscosity. Linear correlations were indeed observed between the dispersion component to the IPBE and conductivity and viscosity for ionic liquids that exhibit a close-to-ideal behaviour on the Walden plot. This correlation is not surprising if one considers that transport properties such as conductivity and viscosity are related to the ion mobility. For ions to be mobile, the short-range interactions are being broken since the long-range electrostatic interactions change only slightly upon ion movements [95]. These findings do not contradict correlations found in the previous studies [95, 108, 109], indicating that the deviation from the ideal line correlated with the IPBE for more associated ionic liquids when the deviation from the ideal line was larger than half an order of magnitude [108, 109]. Obviously, transport properties in highly associated ionic liquids are governed by strong IPBEs and not just the dispersion component to IPBE.

So far, most of the quantum-chemical calculations have been performed on single ion pairs of ionic liquids with the view of identifying factors affecting their thermodynamic and transport properties. Counterintuitively, dispersion interactions between same-charged species, other than van der Waals interactions between long alkyl chains on cations, can also be quite stable as observed in crystal structures of various ionic liquids [116, 117]. In the study by Li et al. [114], a π–π stacking interaction between two imidazolium rings in a system consisting of two imidazolium chloride ion pairs were found to be energetically favourable, contributing about 28 kJ mol^{-1} to the overall interaction energy. Thus, inclusion of multiple ion pairs in the calculation is important to account for *all possible* interactions, electrostatic and dispersion, in ionic liquids. It has to be noted that extra care should be taken in considering isolated ion pairs for studying physical properties of protic ionic liquids that are obtained by a proton transfer from a Brønsted acid (AH) to a Brønsted base (B):

$$AH + B = [HB]^+ + A^- \qquad (6.24)$$

It was shown that gas-phase quantum-chemical calculations favoured the stability of complexes between neutral species, AH and B, over ion pairs, [HB]$^+$ and A$^-$, in triazolium-, tetrazolium-, and pentazolium-based energetic protic ionic liquids [118–120]. Recently, a study by Stoimenovski et al. [121] showed that combining primary and tertiary amines of similar pK_a^{aq} values with a common acid (ethanoic acid) resulted in ionic liquids of various degrees of ionisation. Protic ionic liquids with primary amines exhibited close-to-ideal behaviour on the Walden plot, whereas tertiary amines formed fluid mixtures with a very low degree of ionisation. It was suggested that, due to additional hydrogen bonding sites, primary ammonium ions provided a good solvating environment for ions obtained by shifting the thermodynamic equilibrium towards the formation of ionised species in Equation (6.24). Thus, understanding and (equally importantly) quantifying the extent of the proton transfer and its influence on transport properties of protic ionic liquids can only be achieved from large-scale *ab initio*-based calculations that accurately describe the condensed phase in these ionic systems.

Recently, Ludwig proposed a cluster approach that incorporated frequency calculations of [C_1mim][SCN] clusters of various size on their optimised geometries at the HF/3-21G level of theory [122]. The standard formulae of statistical thermodynamics were applied to subsequently calculate thermodynamic properties of the ionic liquid, such as enthalpies of vapourisation and boiling points. Taking into account larger clusters containing more than six ion pairs, the enthalpy of vapourisation was predicted within 5 kJ mol^{-1} of the experimental value. The proposed cluster approach was also used to study the formation of positively charged large aggregates in protic ionic liquids, as detected in electrospray ionisation mass spectra [123]. For the [$C_2H_5NH_3$][NO_3] ionic liquid, it was shown that the {C_8A_7}$^+$ aggregates were the most energetically stable clusters and exhibited extended hydrogen bonding [124]. Although the Ludwig approach represents an interesting way of predicting thermodynamic properties of ionic liquids, two immediate limitations can be foreseen. First, the HF level of theory used in these studies only accounts for electrostatic interactions, ignoring any interaction that originates directly from electron correlation, such as van der Waals interactions and π–π stacking interactions, or even a part of the IPBE. Second, frequency calculations at any higher correlated level of theory, such as MP2, will be computationally infeasible as these do not scale efficiently with the number of CPUs used, and require substantial computer resources. For example, an MP2/6-311+G(3df,2p) frequency calculation for a cluster of four ion pairs of [NMe_4][BF_4] requires more than 400GB of hard disk!

Although quite successful in predicting thermodynamic and transport properties of ionic liquids, the correlated *ab initio*-based methods are falling behind because of their cost and, hence, infeasibility of large-scale calculations. To put this into perspective, the HF method and most of the DFT methods scale as N^3 (where N is the number of basis functions that reflects the number of electrons in the molecular system), whereas MP2, the lowest wavefunction-based method including electron correlation, scales as N^5. Any of the coupled cluster methods that offer chemical accuracy (i.e., within *ca.* 4 kJ mol^{-1}) for energy calculations scales from N^6, depending on the type of single, double, triple, and so on excitations included in the formulation, thus making these types of calculations, although accurate, hardly affordable for systems larger than a single ion pair. Specific interactions occurring in ionic liquids, such as hydrogen bonding, van der Waals interactions, and π–π stacking interactions (e.g., between two imidazolium rings [114]), originate from electron correlation effects and, therefore, need to be treated at a correlated level of theory, at least MP2. Although some DFT functionals such as M05-2X [125],† and

† M05-2X is a hybrid meta-GGA functional that was specifically designed to describe non-covalent interactions. 2X in the abbreviation stands for 54% of the exact HF exchange included in its formulation.

those functionals including the dispersion interaction correction [126]† explicitly in their formulation, show promising results with respect to IPBEs, none of these functionals produced reliable binding energies across a wide range of anions [106]. The contribution from dispersion interactions in IPBEs was shown to vary, depending on the anion, from 7% (equivalent to 20 kJ mol^{-1}) for $[BF_4]^-$ to 14% (equivalent to 45 kJ mol^{-1}) for the $[NTf_2]^-$ anion in ion pairs of pyrrolidinium-based ionic liquids. Ballone et al. showed that the dispersion forces changed much faster than the Coulomb forces with the increasing size of an ionic cluster [127]. In the case of the $[C_4mim][CF_3SO_3]$ ionic liquid, the binding energy of a single ion pair was found to consist of a large Coulomb component of 343 kJ mol^{-1} and a much smaller dispersion component of only 13.5 kJ mol^{-1}. The energetics of the system changed when two single ion pairs were brought together, with the dispersion interaction now contributing 41 kJ mol^{-1} to the overall energy. It was estimated that the addition of a single ion pair to a neutral cluster consisting of n ion pairs (with $n > 4$) resulted in an increase in the dispersion contribution that was equivalent to about 75% of the corresponding increase in the Coulomb energy. In a similar study by Kossman et al. [128], linear chain and ring aggregates of the $[C_1mim]Cl$ ionic liquid were studied at the BP86/TZVPP level of theory. Interaction energies of the linear-chained oligomers with respect to a single ion pair increased linearly with the number of ion pairs in the chain from −49.4 kJ mol^{-1} for a dimer to −576.2 kJ mol^{-1} for a nonamer. The corresponding relative interaction energies calculated per number of ion pairs in the chain grew from −24.7 kJ mol^{-1} to −64 kJ mol^{-1}, thus indicating strong cooperativity effects between ions, which obviously play a more important rôle than Coulomb interactions in influencing their thermodynamic and transport properties. Therefore, levels of theory used to study ionic liquids need to be able to capture the complexity of interactions occurring within them.

To summarise, from the presented overview on the approaches *from first principles*, the emerging trend is very clear: large-scale calculations of ionic liquids that describe the condensed phase more accurately are becoming more of a necessity in the future, due to the rapidly increasing need for developing predictive approaches *from first principles* for studying thermodynamic and transport properties of novel ionic liquids. Due to the complex nature of interactions in ionic liquids, correlated levels of *ab initio* theory have to be employed for reliable predictions, as a DFT functional that is capable of the accurate description of these interactions is yet to be determined.

---

† The empirical correction (second term in the equation below) accounting for dispersion interactions is explicitly included into the total electronic energy:

$$E_{DFT-D} = E_{DFT} + E_{disp} = E_{DFT} - S_6 \sum_i \sum_j \frac{C_6^{ij}}{R_{ij}^6} f_{dump}(R_{ij})$$

6.6   *Ab initio* MD SIMULATIONS

Ab initio MD simulations are becoming more popular due to their rigorous formulation that does not require additional parameterisation [75, 84, 129–132]. In 1985, Car and Parrinello [133] proposed to marry ground state electronic structure calculations using density functional theory and MD based on classical Newtonian mechanics, hence the term Car–Parrinello molecular dynamics (CPMD). In the proposed approach, the force field potential was replaced by actual electronic structure DFT-based calculations of a molecular system under study, thus avoiding parameterisation of the force field and making MD simulations more robust. Strictly speaking, DFT methods cannot be considered *ab initio*, as they are based on the electron density formulation rather than on the electronic wavefunction, as any true *ab initio* method is. Therefore, a clear distinction should be made between DFT-based MD simulations and fully *ab initio* MD simulations incorporating wavefunction-based methods such as HF, methods of perturbation theory (e.g., MP2), or coupled cluster theory (e.g., CCSD). So far, DFT functionals such as BLYP, B3P86, and PBE have been utilised in *ab initio* MD simulations of ionic liquids, and none of the wavefunction-based methods incorporated in the MD formulation have been reported to date.

Although quite expensive, these *ab initio* (strictly speaking DFT) MD simulations better reproduce the microstructure in ionic liquids. For example, in 1,3-dialkylimidazolium ionic liquids, a strong preference for the unique and directional hydrogen bond between the C^2–H bond on the imidazolium cation and the chloride anion was observed [84, 129, 134], whereas classical MD simulations suggested that this hydrogen bond was only weakly present [134]. As the processes related to bond cleavage and bond formation can be easily studied by *ab initio* methods, *ab initio* MD simulations have become invaluable for studying protic ionic liquids, in which the proton transfer and proton transport processes are not well understood [75, 132]. For example, in the work of Del Pópolo et al. [75], a single molecule of HCl in the $[C_1mim]Cl$ ionic liquid was found to form a stable anion, $[ClHCl]^-$. By forcing one of the H–Cl bonds in $[ClHCl]^-$ to break, a new $[ClHCl]^-$ ion is formed by borrowing one of the neighbouring chloride anions, thus allowing for an easy hopping mechanism of the protons in this particular ionic liquid. Recently, Zahn et al. [132] applied a dispersion corrected atom-centre potential [135, 136] that explicitly includes dispersion corrections in the conventional CPMD formulation. A significant finding lied in the fact that no reverse proton transfer between the cation and anion in the $[CH_3NH_3][NO_3]$ ionic liquid was observed, keeping the ionic nature intact. The liquid structure of this ionic liquid appeared to have long-lived ion pairs with a rather directional hydrogen bond between the N–H fragment of the $[RNH_3]^+$ group and one of the oxygen atoms in the nitrate anion, imitating the conformation of the initial proton transfer between methylamine and nitric acid. These ion pairs stayed "connected" for over

14.5 ps, thus indicating that the liquid state could be best described as ions rattling in cages formed by long-lived ion pairs. A major limitation of the current *ab initio* MD simulations are short running times accounting for the sub-diffusive region, which may not allow for accurate calculations of self-diffusion coefficients and consequently, viscosity. The question still remains unanswered regarding the accuracy of the chosen DFT functionals in *ab initio* MD simulations, as it was shown that the dispersion interactions are non-negligible in ion pairs of ionic liquids [106, 137]. The DFT functionals such as BLYP-D, PBE-D, and B3P86-D that explicitly include the dispersion correction in their formulation show improved performance in energetics of ionic liquids, with electron-rich anions such as $[NTf_2]^-$, $[N(CN)_2]^-$, mesylate, and tosylate, whereas for spherical anions such as Cl^-, $[BF_4]^-$, and $[PF_6]^-$, the performance of these functionals was surprisingly relatively poor [106]. As a result, care must be taken in applying these functionals with the explicit dispersion component for ionic liquids containing Cl^-, $[BF_4]^-$ and $[PF_6]^-$ anions. More systematic studies on the performance of these functionals for a series of ionic liquid anions are needed before some of these functionals can be accepted as "universal" functionals for studying ionic liquids.

To summarise, *ab initio* (strictly speaking DFT) MD simulations represent a robust way of tailoring physical properties of ionic liquids, as these simulations avoid problems with the validation process, the charge transfer between ions, as well as ionic polarisation, thus resulting in a better description of the liquid structure. The main limitation of *ab initio* MD lies in the fact that longer running times cannot be afforded at the moment, thus hindering the predictive ability of dynamic properties of ionic liquids.

6.7   OUTLOOK: TOWARDS LARGE-SCALE CALCULATIONS OF IONIC LIQUIDS *FROM FIRST PRINCIPLES*

The theoretical approaches described in this Chapter have been relatively successful in studying physical properties of ionic liquids, although various limitations are yet to be overcome. The current trend has been expressed by many researchers in the field that *before synthesis* predictive theoretical methods must be developed to enable us to tailor physical properties of novel ionic liquids, thus leading to easy screening of potentially viable ionic liquids for applications in electrochemical devices. Large-scale calculations *from first principles* utilising correlated levels of *ab initio* theory represent an excellent approach that accounts for all possible interactions (Coulomb and dispersion) in ionic liquids, and does not require additional parameterisation and/or prior knowledge of experimental data. It has to be noted that these calculations are versatile and can be applied to any type of ionic liquids. In order to study bulk properties of ionic liquids, dynamic simulations at various temperatures have to be performed and therefore large-scale *ab initio* calculations in combination with the classical Newtonian mechanics will pave the way towards fully *ab initio* (non-DFT) MD simulations, which can be unarguably considered the ultimate approach for accurate studies and predictions of physical properties of ionic liquids. In order to move towards accurate large-scale calculations and, consequently, fully *ab initio* MD, two major issues have to be addressed: first the level of *ab initio* theory and its scalability with molecular size, and second the efficient use of parallelism to afford longer running times for MD simulations.

One of the issues that need particular attention is whether we can afford to compromise the level of theory for accurate predictions of physical properties of ionic liquids. Both MD simulations and quantum-chemical calculations presented above have shown that dispersion interactions between ions cannot be neglected, and for accurate prediction of thermodynamic (especially, melting point) and transport (especially self-diffusion coefficients and viscosity) properties, the energetics of the solid and liquid states have to be treated with moderate accuracy, at least within chemical accuracy of *ca.* 4 kJ mol^{-1}. DFT functionals have been, in some way, invaluable in the field of ionic liquids due to their linear scalability with molecular size. Although some of these functionals (especially those with explicit terms for dispersion interactions) describe the energetics of ionic liquids quite well, more research into their versatility for a wide range of ionic liquids has to be carried out. Perhaps, we have exhausted our current strategy, and we might need to consider changing it. Instead of approximating highly accurate correlated levels of theory by lower, cost-effective but less accurate methods, current methods in *ab initio* theory could be modified such that they scale linearly with molecular size and, thus, can be easily parallelised to run large-scale calculations on massively parallel computers. In the last decade, massively parallel computers have become available to researchers throughout the world. Use of a thousand CPUs at once is becoming more of a reality than a long-term goal, thus making large-scale *ab initio* calculations of ionic liquids more affordable. Hutter et al. identified that the current Car–Parrinello MD formulation could be easily modified to run efficiently on massively parallel computers [138, 139], offering a unique opportunity to perform longer computer simulations and, thus, accurately predict *dynamic* properties of ionic liquids such as self-diffusion and viscosity *from first principles*.

A recent comprehensive review by Gordon et al. [140] highlighted approaches *from first principles* (i.e., fully *ab initio*) that satisfied the proposed strategy of linear scalability with molecular size: hybrid methods combining quantum mechanics with molecular mechanics (QM/MM) [141, 142], the systematic fragmentation method (SFM) [143–145], the effective fragment potential method (EFM) [146, 147], and the fragment MO method (FMO) [148–151]. Among these methods, the FMO method deserves special consideration, as it takes full advantage of massively parallel computers due to their linear scalability with molecular size (for more detail regarding other methods, see the review). The desired scalability of the FMO method is achieved by dividing the system into smaller fragments (so-called monomers) that can easily be tackled at any traditional correlated levels of theory such as perturbation theory (e.g., MP2 [152]) and coupled cluster theory (e.g., CCSD(T) [153]) and in combination with a polarised continuum solvation model [154]. The FMO method calculates the total energy of the molecule as a sum of the electronic energies of the "monomers," with corrections for two-body [155] and three-body [156, 157] interactions between the fragments being included for improved accuracy. At each stage of the calculation, the reminder of the system is represented by a Coulomb bath that is determined by distance cutoffs. In ionic liquids, the building blocks (fragments) can be considered to be individual ions or ion pairs of oppositely charged ions, as the FMO method is very flexible with respect to the selection of fragments. The FMO method dramatically improved the computer time requirements for ionic liquid systems consisting of tetramers and hexamers, without sacrificing accuracy [140]. For example, for the 1-amino-4-H-1,2,4-triazolium dinitramide ionic liquid, the FMO method in combination with the MP2/6-31+G(d) level of theory and three-body corrections produced absolute errors below 4 kJ mol^{-1} with respect to the conventional MP2 calculation. The FMO method was also shown to efficiently utilise multilevel parallelism through the generalised distributed data interface (GDDI), thus further decreasing running times [158]. As an example, on 128 nodes the running time of the FMO calculation using the GDDI was 93 times faster than that of the traditional distributed data interface (DDI) for a cluster consisting of 256 water molecules. Using the GDDI the FMO method delivered scalability in the range of 80–90% on 128 nodes, depending on the molecular system (be it a water cluster or a protein molecule).

A new generation of the computer architecture such as Blue Gene was designed to reach operating speeds in the petaFLOPS (PFLOPS) range up to 500 PFLOPS. A combination of this peta-scale computer architecture with the excellent scalability of the FMO method makes *ab initio* large-scale calculations feasible and ultimately, paves the way towards fully *ab initio* MD simulations of ionic liquids on 1000 CPUs!

REFERENCES

1 Davis, J.H., Task-specific ionic liquids, *Chem. Lett.* **33**, 1072–1077 (2004).

2 Jenkins, H.D.B., Roobottom, H.K., Passmore, J., and Glasser, L., Relationships among ionic lattice energies, molecular (formula unit) volumes, and thermochemical radii, *Inorg. Chem.* **38**, 3609–3620 (1999).

3 Evjen, H.M., On the stability of certain heteropolar crystals, *Phys. Rev.* **39**, 675–687 (1932).

4 Ewald, P.P., The Berechnung optischer und elektrostatischer Gitterpotentiale, *Ann. Phys.* **369**, 253–287 (1921).

5 Madelung, E., Das elektrische Feld in Systemen von regelmäßig angeordneten Punktladungen, *Phys. Z.* **19**, 524–533 (1918).

6   Jenkins, H.D.B., and Glasser, L., Standard absolute entropy, S-298°, values from volume or density. 1. Inorganic materials, *Inorg. Chem.* **42**, 8702–8708 (2003).

7   Glasser, L., Lattice and phase transition thermodynamics of ionic liquids, *Thermochim. Acta* **421**, 87–93 (2004).

8   Glasser, L., and Jenkins, H.D.B., Standard absolute entropies, S-298°, from volume or density—Part II. Organic liquids and solids, *Themochim. Acta* **414**, 125–130 (2004).

9   Krossing, I., and Slattery, J.M., Semi-empirical methods to predict the physical properties of ionic liquids: an overview of recent developments, *Z. Phys. Chem.* **220**, 1343–1359 (2006).

10   Abdul-Sada, A.K., Al-Juaid, S., Greenway, A.M., Hitchcock, P.B., Howells, M.J., Seddon, K.R., and Welton, T., Upon the structure of room-temperature halogenoaluminate ionic liquids, *Struct. Chem.* **1**, 391–394 (1989).

11   Krossing, I., Slattery, J.M., Daguenet, C., Dyson, P.J., Oleinikova, A., and Weingärtner, H., Why are ionic liquids liquid? A simple explanation based on lattice and solvation energies, *J. Am. Chem. Soc.* **128**, 13427–13434 (2006).

12   Klamt, A., Eckert, F., and Arlt, W., COSMO-RS: an alternative to simulation for calculating thermodynamic properties of liquid mixtures, *Annu. Rev. Chem. Biomol. Eng.* **1**, 101–122 (2009).

13   Markusson, H., Belieres, J.-P., Johansson, P., Angell, C.A., and Jacobsson, P., Prediction of macroscopic properties of protic ionic liquids by *ab initio* calculations, *J. Phys. Chem. A* **111**, 8717–8723 (2007).

14   Barone, V., Cossi, M., and Tomasi, J., Geometry optimization of molecular structures in solution by the polarizable continuum model, *J. Comput. Chem.* **19**, 404–417 (1998).

15   Reichert, W.M., Holbrey, J.D., Swatloski, R.P., Gutowski, K.E., Visser, A.E., Nieuwenhuyzen, M., Seddon, K.R., and Rogers, R.D., Solid-state analysis of low-melting 1,3-dialkylimidazolium hexafluorophosphate salts (ionic liquids) by combined x-ray crystallographic and computational analyses, *Cryst. Growth Des.* **7**, 1106–1114 (2007).

16   Gutowski, K.E., Holbrey, J.D., Rogers, R.D., and Dixon, D.A., Prediction of the formation and stabilities of energetic salts and ionic liquids based on ab initio electronic structure calculations, *J. Phys. Chem. B* **109**, 23196–23208 (2005).

17   Katritzky, A.R., Yang, H., Zhang, D., Kirichenko, K., Smiglak, M., Holbrey, J.D., Reichert, W.M., and Rogers, R.D., Strategies toward the design of energetic ionic liquids: nitro- and nitrile-substituted N,N´-dialkylimidazolium salts, *New J. Chem.* **30**, 349–358 (2006).

18   Xue, H., Gao, Y., Twamley, B., and Shreeve, J.M., New energetic salts based on nitrogen-containing heterocycles, *Chem. Mater.* **17**, 191–198 (2005).

19   Gao, H., Ye, C., Piekarski, C.M., and Shreeve, J.M., Computational characterization of energetic salts, *J. Phys. Chem. C* **111**, 10718–10731 (2007).

20   Curtiss, L.A., Raghavachari, K., Trucks, G.W., and Pople, J.A., Gaussian-2 theory for molecular-energies of first- and second-row compounds, *J. Chem. Phys.* **94**, 7221–7230 (1991).

21   Gutowski, K.E., Rogers, R.D., and Dixon, D.A., Accurate thermochemical properties for energetic materials applications. II. Heats of formation of imidazolium-,

1,2,4-triazolium-, and tetrazolium-based energetic salts from isodesmic and lattice energy calculations, *J. Phys. Chem. B* **111**, 4788–4800 (2007).

22   Preiss, U., Bulut, S., and Krossing, I., In silico prediction of the melting points of ionic liquids from thermodynamic considerations: a case study on 67 salts with a melting point range of 337 degrees C, *J. Phys. Chem. B* **114**, 11133–11140 (2010).

23   Jain, A., and Yalkowsky, S.H., Estimation of melting points of organic compounds-II, *J. Pharm. Sci.* **95**, 2562–2618 (2006).

24   Zaitsau, D.H., Kabo, G.J., Strechan, A.A., Paulechka, Y.U., Tschersich, A., Verevkin, S.P., and Heintz, A., Experimental vapor pressures of 1-alkyl-3-methylimidazolium bis(trifluoromethylsulfonyl) imides and a correlation scheme for estimation of vaporization enthalpies of ionic liquids, *J. Phys. Chem. A* **110**, 7303–7306 (2006).

25   Borodin, O., Relation between heat of vaporization, ion transport, molar volume, and cation–anion binding energy for ionic liquids, *J. Phys. Chem. B* **113**, 12353–12357 (2009).

26   Deyko, A., Lovelock, K.R.J., Corfield, J.-A., Taylor, A.W., Gooden, P.N., Villar-Garcia, I.L., Licence, P., Jones, R.G., Krasovskiy, V.G., Chernikova, E.A., and Kustov, L.M., Measuring and predicting $\Delta_{vap}H(298)$ values of ionic liquids, *Phys. Chem. Chem. Phys.* **11**, 8544–8555 (2009).

27   Armstrong, J.P., Hurst, C., Jones, R.G., Licence, P., Lovelock, K.R.J., Satterley, C.J., and Villar-Garcia, I.J., Vapourisation of ionic liquids, *Phys. Chem. Chem. Phys.* **9**, 982–990 (2007).

28   Preiss, U.P.R.M., Slattery, J.M., and Krossing, I., In silico prediction of molecular volumes, heat capacities, and temperature-dependent densities of ionic liquids, *Ind. Eng. Chem. Res.* **48**, 2290–2296 (2009).

29   Slattery, J.M., Daguenet, C., Dyson, P.J., Schubert, T.J.S., and Krossing, I., How to predict the physical properties of ionic liquids: a volume-based approach, *Angew. Chem. Int. Ed.* **46**, 5384–5388 (2007).

30   Huddleston, J.G., Visser, A.E., Reichert, W.M., Willauer, H.D., Broker, G.A., and Rogers, R.D., Characterization and comparison of hydrophilic and hydrophobic room temperature ionic liquids incorporating the imidazolium cation, *Green Chem.* **3**, 156–164 (2001).

31   Ludwig, R., and Paschek, D., Applying the inductive effect for synthesizing low-melting and low-viscosity imidazolium-based ionic liquids, *Chem. Phys. Chem.* **10**, 516–519 (2009).

32   Izgorodina, E.I., Bernard, U.L., Dean, P.M., Pringle, J.M., and MacFarlane, D.R., The Madelung constant of organic salts, *Cryst. Growth. Des.* **9**, 4834–4839 (2009).

33   Katritzky, A.R., Jain, R., Lomaka, A., Petrukhin, R., Karelson, M., Vesser, A.E., and Rogers, R.D., Correlation of the melting points of potential ionic liquids (imidazolium bromides and benzimidazolium bromides) using the CODESSA program, *J. Chem. Inf. Comput. Sci.* **42**, 225–231 (2002).

34   Katritzky, A.R., Lomaka, A., Petrukhin, R., Jain, R., Karelson, M., Vesser, A.E., and Rogers, R.D., QSPR correlation of the melting point for pyridinium bromides, potential ionic liquids, *J. Chem. Inf. Comput. Sci.* **42**, 71–74 (2002).

35   Eike, D., Brennecke, J.F., and Maginn, E.J., Predicting melting points of quaternary ammonium ionic liquids, *Green Chem.* **5**, 323–328 (2003).

36   Dewar, M.J.S., Zoebisch, E.G., Healy, E.F., and Stewart, J.J.P., The development and use of quantum-mechanical molecular-models. 76. AM1: a new general-purpose quantum-mechanical molecular-model, *J. Am. Chem. Soc.* **107**, 3902–3909 (1985).

37   Stewart, J.J.P., Optimization of parameters for semiempirical methods. 2. Applications, *J. Comput. Chem.* **10**, 221–264 (1989).

38   Stewart, J.J.P., Optimization of parameters for semiempirical methods. 1. Method, *J. Comput. Chem.* **10**, 209–220 (1989).

39   Trohalaki, S., Pachter, R., Drake, G.W., and Hawkins, T., Quantitative structure-property relationships for melting points and densities of ionic liquids, *Energ. Fuel.* **19**, 279–284 (2005).

40   Trohalaki, S., and Pachter, R., Prediction of melting points for ionic liquids, *QSAR Comb. Sci.* **24**, 485–490 (2005).

41   López-Martin, I., Burello, E., Davey, P.N., Seddon, K.R., and Rothenberg, G., Anion and cation effects on imidazolium salt melting points: a descriptor modelling study, *Chem. Phys. Chem.* **8**, 690–695 (2007).

42   Sun, N., He, X., Dong, K., Zhang, X., Liu, X., He, H., and Zhang, S., Prediction of the melting points for two kinds of room temperature ionic liquids, *Fluid Phase Equilibr.* **246**, 137–142 (2006).

43   Ren, Y., Qin, J., Liu, H., Yao, X., and Liu, M., QSPR study on the melting points of a diverse set of potential ionic liquids by projection pursuit regression, *QSAR Comb. Sci.* **28**, 1237–1244 (2009).

44   Carrera, G., and Aires-De-Sousa, J., Estimation of melting points of pyridinium bromide ionic liquids with decision trees and neural networks, *Green Chem.* **7**, 20–27 (2005).

45   Torrecilla, J.S., Rodriguez, F., Bravo, J.L., Rothenberg, G.R., Seddon, K.R., and Lopez-Martin, I., Optimising an artificial neural network for predicting the melting point of ionic liquids, *Phys. Chem. Chem. Phys.* **10**, 5826–5831 (2008).

46   Varnek, A., Kireeva, N., Tetko, I.V., Baskin, I.I., and Solov'ev, V.P., Exhaustive QSPR studies of a large diverse set of ionic liquids: how accurately can we predict melting points? *J. Chem. Inf. Modal.* **47**, 1111–1122 (2007).

47   Yan, C., Han, M., Wan, H., and Guan, G., QSAR correlation of the melting points for imidazolium bromides and imidazolium chlorides ionic liquids, *Fluid Phase Equilib.* **292**, 104–109 (2010).

48   Bini, R., Chiappe, C., Duce, C., Micheli, A., Solaro, R., Starita, A., and Tine, M.R., Ionic liquids: prediction of their melting points by a recursive neural network model, *Green Chem.* **10**, 306–309 (2008).

49   Carrera, G.V.S.M., Branco, L.C., Aires-De-Sousa, J., and Afonso, C.A.M., Exploration of quantitative structure-property relationships (QSPR) for the design of new guanidinium ionic liquids, *Tetrahedron* **64**, 2216–2224 (2008).

50   Huber, P.J., Projection pursuit, *Ann. Stat.* **13**, 435–475 (1985).

51   Friedman, J.H., and Stuetzle, W., Projection pursuit regression, *J. Am. Statis. Assoc.* **76**, 817–823 (1981).

52   Svetnik, V., Liaw, A., Tong, C., Culberson, J.C., Sheridan, R.P., and Feuston, B.P., Random forest: a classification and regression tool for compound classification and QSAR modeling, *J. Chem. Inf. Comput. Sci.* **43**, 1947–1958 (2003).

53   Tochigi, K., and Yamamoto, H., Estimation of ionic conductivity and viscosity of ionic liquids using a QSPR model, *J Phys. Chem. C* **111**, 15989–15994 (2007).

54   Walden, P., Über organische Lösungs- und Ionisierungsmttel. III Teil: innere Reibung, und deren Zusammenhang mit dem Leitvermögen, *Z. Phys. Chem.* **55**, 207–249 (1906).

55   Del Pópolo, M.G., Kohanoff, J., Lynden-Bell, R.M., and Pinilla, C., Clusters, liquids, and crystals of dialkyimidazolium salts. A combined perspective from ab initio and classical computer simulations, *Acc. Chem. Res.* **40**, 1156–1164 (2007).

56   Hu, Z., and Margulis, C.J., Room-temperature ionic liquids: slow dynamics, viscosity, and the red edge effect, *Acc. Chem. Res.* **40**, 1097–1105 (2007).

57   Lynden-Bell, R.M., Del Pópolo, M.G., Youngs, T.G.A., Kohanoff, J., Hanke, C.G., Harper, J.B., and Pinilla, C.C., Simulations of ionic liquids, solutions, and surfaces, *Acc. Chem. Res.* **40**, 1138–1145 (2007).

58   Maginn, E.J., Atomistic simulation of the thermodynamic and transport properties of ionic liquids, *Acc. Chem. Res.* **40**, 1200–1207 (2007).

59   Pádua, A.A.H., Costa Gomes, M.F., and Canongia Lopes, J.N.A., Molecular solutes in ionic liquids: a structural, perspective, *Acc. Chem. Res.* **40**, 1087–1096 (2007).

60   Rebelo, L.P.N., Lopes, J.N.C., Esperança, J.M.S.S., Guedes, H.J.R., Łachwa, J., Najdanovic-Visak, V., and Visak, Z.P., Accounting for the unique, doubly dual nature of ionic liquids from a molecular thermodynamic, and modeling standpoint, *Acc. Chem. Res.* **40**, 1114–1121 (2007).

61   Wang, Y., Jiang, W., Yan, T., and Voth, G.A., Understanding ionic liquids through atomistic and coarse-grained molecular dynamics simulations, *Acc. Chem. Res.* **40**, 1193–1199 (2007).

62   Maginn, E., Molecular simulation of ionic liquids: current status and future opportunities, *J. Phys. Condens. Matter* **21**, Article ID 373101 (2009).

63   Hunt, P., The simulation of imidazolium based ionic liquids, *Mol. Simul.* **32**, 1–10 (2006).

64   Bhargava, B.L., Balasubramanian, S., and Klein, M.L., Modelling room temperature ionic liquids, *Chem. Commun.*, 3339–3351 (2008).

65   Jayaraman, S., and Maginn, E.J., Computing the melting point and thermodynamic stability of the orthorhombic and monoclinic crystalline polymorphs of the ionic liquid 1-n-butyl-3-methylimidazolium chloride, *J. Chem. Phys.* **127**, Article ID 214504 (2007).

66   Jayaraman, S., Thompson, A.P., von Lilienfield, A.A., and Maginn, E.J., Molecular simulation of the thermal and transport properties of three alkali nitrate salts, *Ind. Eng. Chem. Res.* **49**, 559–571 (2010).

67   Del Pópolo, M.G., and Voth, G.A., On the structure and dynamics of ionic liquids, *J. Phys. Chem. B* **108**, 1744–1752 (2004).

68   Margulis, C.J., Stern, H.A., and Berne, B.J., Computer simulation of a "green chemistry" room-temperature ionic solvent, *J. Phys. Chem. B* **106**, 12017–12021 (2002).

69   Hu, Z., and Margulis, C.J., Heterogeneity in a room-temperature ionic liquid: persistent local environments and the red-edge effect, *P. Natl. Acad. Sci.* **103**, 831–836 (2006).

70 Borodin, O., Polarizable force field development and molecular dynamics simulations of ionic liquids, *J. Phys. Chem. B* **113**, 11463–11478 (2009).

71 Canongia Lopes, J.N., and Pádua, A.A.H., Molecular force field for ionic liquids composed of the triflate or bistriflylimide anions, *J. Phys. Chem. B* **108**, 16893–16898 (2004).

72 Canongia Lopes, J.N., Deschamps, J., and Pádua, A.A.H., Modeling ionic liquids using a systematic all-atom force field, *J. Phys. Chem. B* **108**, 2038–2047 (2004).

73 Köddermann, T., Paschek, D., and Ludwig, R., Ionic liquids: dissecting the enthalpies of vaporization, *Phys. Chem. Chem. Phys.* **9**, 549–555 (2008).

74 MacFarlane, D.R., Pringle, J.M., Johansson, K.M., Forsyth, S.A., and Forsyth, M., Lewis base ionic liquids, *Chem. Commun.*, 1905–1917 (2006).

75 Del Pópolo, M.G., Kohanoff, J., and Lynden-Bell, R.M., Solvation structure and transport of acidic protons in ionic liquids: a first-principles simulation study, *J. Phys. Chem. B* **110**, 8798–8803 (2006).

76 Prado, C.E.R., Del Pópolo, M.G., Youngs, T.G.A., Kohanoff, J., and Lynden-Bell, R.M., Molecular electrostatic properties of ions in an ionic liquid, *Mol. Phys.* **104**, 2477–2483 (2006).

77 Youngs, T.G.A., Del Pópolo, M.G., and Kohanoff, J., Development of complex classical force fields through force matching to ab initio data: application to a room-temperature ionic liquid, *J. Phys. Chem. B* **110**, 5697–5707 (2006).

78 Yan, T., Burnham, C.J., Del Pópolo, M.G., and Voth, G.A., Molecular dynamics simulation of ionic liquids: the effect of electronic polarizability, *J. Phys. Chem. B* **108**, 11877–11881 (2004).

79 Borodin, O., and Smith, G.D., Structure and dynamics of N-methyl-N-propylpyrrolidinium bis(trifluoromethanesulfonyl)imide ionic liquid from molecular dynamics simulations, *J. Phys. Chem. B* **110**, 11481–11490 (2006).

80 Morrow, T.J., and Maginn, E.J., Molecular dynamics study of the ionic liquid 1-n-butyl-3-methylimidazolium hexafluorophosphate, *J. Phys. Chem. B* **106**, 12807–12813 (2002).

81 Bhargava, B.L., and Balasubramanian, S., Refined potential model for atomistic simulations of ionic liquid [bmim][PF6], *J. Chem. Phys.* **127**, Article ID 114510 (2007).

82 Canongia Lopes, J.N., Deschamps, J., and Padua, A.A.H., Modeling ionic liquids using a systematic all-atom force field, *J. Phys. Chem. B* **108**(6), 2038–2047, 11250 (2004).

83 Youngs, T.G.A., and Hardacre, C., Application of static charge transfer within an ionic-liquid force field and its effect on structure and dynamics, *Chem. Phys. Chem.* **9**, 1548–1558 (2008).

84 Del Pópolo, M.G., Lynden-Bell, R.N., and Kohanoff, J., Ab initio molecular dynamics simulation of a room temperature ionic liquid, *J. Phys. Chem. B* **109**, 5895–5902 (2005).

85 Lynden-Bell, R.M., and Youngs, T.G.A., Simulations of imidazolium ionic liquids: when does the cation charge distribution matter? *J. Phys. Condens. Matter* **21**, Article ID 424120 (2009).

86 Santos, L.M.N.B.F., Canongia-Lopes, J.N., Coutinho, J.A.P., Esperança, J.M.S.S., Gomes, L.R., Marrucho, I.M., and Rebelo, L.P.N., Ionic liquids: first direct determination of their cohesive energy, *J. Am. Chem. Soc.* **129**, 284–285 (2007).

87   Kelkar, M.S., and Maginn, E.J., Calculating the enthalpy of vaporization for ionic liquid clusters, *J. Phys. Chem. B* **111**, 9424–9427 (2007).

88   Emel'yanenko, V.N., Verevkin, S.P., Heintz, A., Corfield, J.-A., Deyko, A., Lovelock, K.R.J., Licence, P., and Jones, R.G., Pyrrolidinium-based ionic liquids. 1-butyl-1-methyl pyrrolidinium dicyanoamide: thermochemical measurement, mass spectrometry, and ab initio calculations, *J. Phys. Chem. B* **112**, 11734–11742 (2008).

89   Emel'yanenko, V.N., Verevkin, S.P., Heintz, A., and Schick, C., Ionic liquids. Combination of combustion calorimetry with high-level quantum chemical calculations for deriving vaporization enthalpies, *J. Phys. Chem. B* **112**, 8095–8098 (2008).

90   Emel'yanenko, V.N., Verevkin, S.P., Heintz, A., Voss, K., and Schulz, A., Imidazolium-based ionic liquids. 1-methyl imidazolium nitrate: thermochemical measurements and Ab initio calculations, *J. Phys. Chem. B* **113**, 9871–9876 (2009).

91   Curtiss, L.A., Redfern, P.C., Raghavachari, K., Rassolov, V., and Pople, J.A., Gaussian-3 theory using reduced Moller-Plesset order, *J. Chem. Phys.* **110**, 4703–4709 (1999).

92   Diedenhofen, M., Klamt, A., Marsh, K., and Schaefer, A., Prediction of the vapor pressure and vaporization enthalpy of 1-n-alkyl-3-methylimidazolium-bis-(trifluoromethanesulfonyl) amide ionic liquids, *Phys. Chem. Chem. Phys.* **9**, 4653–4656 (2007).

93   Hunt, P.A., and Gould, I.R., Structural characterization of the 1-butyl-3-methylimidazolium chloride ion pair using a initio methods, *J. Phys. Chem. A* **110**, 2269–2282 (2006).

94   Hunt, P.A., Why does a reduction in hydrogen bonding lead to an increase in viscosity for the 1-butyl-2,3-dimethyl-imidazolium-based ionic liquids? *J. Phys. Chem. B* **111**, 4844–4853 (2007).

95   Tsuzuki, S., Tokuda, H., Hayamizu, K., and Watanabe, M., Magnitude and directionality of interaction in ion pairs of ionic liquids: relationship with ionic conductivity, *J. Phys. Chem. B* **109**, 16474–16481 (2005).

96   Tsuzuki, S., Tokuda, H., and Mikami, M., Theoretical analysis of the hydrogen bond of imidazolium C-2-H with anions, *Phys. Chem. Chem. Phys.* **9**, 4780–4784 (2007).

97   Hunt, P.A., Kirchner, B., and Welton, T., Characterising the electronic structure of ionic liquids: an examination of the 1-butyl-3-methylimidazolium chloride ion pair, *Chem. Eur. J.* **12**, 6762–6775 (2006).

98   Zahn, S., Bruns, G., Thar, J., and Kirchner, B., What keeps ionic liquids in flow? *Phys. Chem. Chem. Phys.* **10**, 6921–6924 (2008).

99   Goldstein, M., Viscous liquids and the glass transition: a potential energy barrier picture, *J. Chem. Phys.* **51**, 3728–3739 (1969).

100 Fumino, K., Wulf, A., and Ludwig, R., Strong, localized, and directional hydrogen bonds fluidize ionic liquids, *Angew. Chem. Int. Ed.* **47**, 8731–8734 (2008).

101 Fumino, K., Wulf, A., and Ludwig, R., The potential role of hydrogen bonding in aprotic and protic ionic liquids, *Phys. Chem. Chem. Phys.* **11**, 8790–8794 (2009).

102 Izgorodina, E.I., Forsyth, M., and MacFarlane, D.R., Towards a better understanding of "delocalized charge" in ionic liquid anions, *Aust. J. Chem.* **60**, 15–20 (2007).

103 Helgaker, T., Jørgensen, P., and Olsen, J., *Molecular electronic-structure theory*, John Wiley and Sons Ltd., New York (2000).

104 Turner, E.A., Pye, C.C., and Singer, R.D., Use of ab initio calculations toward the rational design of room temperature ionic liquids, *J. Phys. Chem. A* **107**, 2277–2288 (2003).

105 Zahn, S., Uhlig, F., Thar, J., Spickerman, C., and Kirchner, B., Intermolecular forces in an ionic liquid ([Mmim][Cl]) versus those in a typical salt (NaCl), *Angew. Chem. Int. Ed.* **47**, 3639–3641 (2008).

106 Izgorodina, E.I., Bernard, U.L., and MacFarlane, D.R., Ion-pair binding energies of ionic liquids: can DFT compete with Ab initio-based methods? *J. Phys. Chem. A* **113**, 7064–7072 (2009).

107 Tokuda, H., Ishii, K., Susan, M.A.B.H., and Watanabe, M., Physicochemical properties and structures of room temperature ionic liquids. 1. Variation of anionic species, *J. Phys. Chem. B* **108**, 16593–16600 (2004).

108 Fraser, K.J., Izgorodina, E.I., Forsyth, M., Scott, J.L., and MacFarlane, D.R., Liquids intermediate between "molecular" and "ionic" liquids: Liquid ion pairs? *Chem. Commun.*, 3817–3819 (2007).

109 MacFarlane, D.R., Forsyth, M., Izgorodina, E.I., Abbott, A.P., Annat, G., and Fraser, K., On the concept of ionicity in ionic liquids, *Phys. Chem. Chem. Phys.* **11**, 4962–4967 (2009).

110 Li, H., Ibrahim, M., Agberemi, I., and Kobrak, M.N., The relationship between ionic structure and viscosity in room-temperature ionic liquids, *J. Chem. Phys.* **129**, Article ID 124507 (2008).

111 Bernard, U.L., Izgorodina, E.I., and MacFarlane, D.R., New insights into the relationship between ion-pair binding energy and thermodynamic and transport properties of ionic liquids, *J. Phys. Chem. C* **114**, 20472–20478 (2010).

112 Holbrey, J.D., and Seddon, K.R., The phase behaviour of 1-alkyl-3-methylimidazolium tetrafluoroborates; ionic liquids and ionic liquid crystals, *J. Chem. Soc., Dalton Trans.*, 2133–2139 (1999).

113 Urahata, S.M., and Ribeiro, M.C.C., Structure of ionic liquids of 1-alkyl-3-methylimidazolium cations: a systematic computer simulation study, *J. Chem. Phys.* **120**, 1855–1863 (2004).

114 Li, H., Boatz, J.A., and Gordon, M.S., Cation–cation $\pi-\pi$ sacking in small ionic clusters of 1,2,4-triazolium, *J. Am. Chem. Soc.* **130**, 392–393 (2008).

115 Thar, J., Brehm, M., Seitsonen, A.P., and Kirchner, B., Unexpected hydrogen bond dynamics in imidazolium-based ionic liquids, *J. Phys. Chem. B* **113**, 15129–15132 (2009).

116 Belanger, S., and Beauchamp, A.L., 2,2´-Bi(1H-imidazolium) dichloride, *Acta Cryst.* **C52**, 2588–2590 (1996).

117 Dean, P.M., Pringle, J.M., and MacFarlane, D.R., CCDC, Refcode: 713021, 50, (2008).

118 Schmidt, M.W., Gordon, M.S., and Boatz, J.A., Triazolium-based energetic ionic liquids, *J. Phys. Chem. A* **109**, 7285–7295 (2005).

119 Zorn, D.D., Boatz, J.A., and Gordon, M.S., Electronic structure studies of tetrazolium-based ionic liquids, *J. Phys. Chem. B* **110**, 11110–11119 (2006).

120   Pimienta, I.S.O., Elzey, S., Boatz, J.A., and Gordon, M.S., Pentazole-based energetic ionic liquids: a computational study, *J. Phys. Chem. A* **111**, 691–703 (2007).

121   Stoimenovski, J., Izgorodina, E.I., and MacFarlane, D.R., Ionicity and proton transfer in protic ionic liquids, *Phys. Chem. Chem. Phys.* **12**, 10341–10347 (2010).

122   Ludwig, R., Thermodynamic properties of ionic liquids—a cluster approach, *Phys. Chem. Chem. Phys.* **10**, 4333–4339 (2008).

123   Kennedy, D.F., and Drummond, C.J., Large aggregated ions found in some protic ionic liquids, *J. Phys. Chem. B* **113**, 5690–5693 (2009).

124   Ludwig, R., and Simple, A., Geometrical explanation for the occurrence of specific large aggregated ions in some protic ionic liquids, *J. Phys. Chem. B* **113**, 15419–15422 (2009).

125   Zhao, Y., Schultz, N.E., and Truhlar, D.G., Design of density functionals by combining the method of constraint satisfaction with parametrization for thermochemistry, thermochemical kinetics, and noncovalent interactions, *J. Chem. Theory Comput.* **2**, 364–382 (2006).

126   Grimme, S., Accurate description of van der Waals complexes by density functional theory including empirical corrections, *J. Comput. Chem.* **25**, 1463–1473 (2004).

127   Ballone, P., Pinilla, C., Kohanoff, J., and Del Pópolo, M.G., Neutral and charged 1-butyl-3-methylimidazolium triflate clusters: equilibrium concentration in the vapor phase and thermal properties of nanometric droplets, *J. Phys. Chem. B* **111**, 4938–4950 (2007).

128   Kossman, S., Thar, J., Kirchner, B., Hunt, P.A., and Welton, T., Cooperativity in ionic liquids, *J. Chem. Phys.* **124**, Article ID 174506 (2006).

129   Bühl, M., Chaumont, A., Schurhammer, R., and Wipff, G., Ab initio molecular dynamics of liquid 1,3-dimethylimidazolium chloride, *J. Phys. Chem. B* **109**, 18591–18599 (2005).

130   Ghatee, M.H., and Ansari, Y., Ab initio molecular dynamics simulation of ionic liquids, *J. Chem. Phys.* **126**, Article ID 154502 (2007).

131   Kirchner, B., and Seitsonen, A.P., Ionic liquids from Car-Parrinello simulations. 2. Structural diffusion leading to large anions in chloraluminate ionic liquids, *Inorg. Chem.* **46**, 2751–2754 (2007).

132   Zahn, S., Thar, J., and Kirchner, B., Structure and dynamics of the protic ionic liquid monomethylammonium nitrate ($[CH_3NH_3][NO_3]$) from *ab initio* molecular dynamics simulations, *J. Chem. Phys.* **132**, 124506 (2010).

133   Car, R., and Parrinello, M., Unified approach for molecular-dynamics and density-functional theory, *Phys. Rev. Lett.* **55**, 2471–2474 (1985).

134   Bhargava, B.L., and Balasubramanian, S., Intermolecular structure and dynamics in an ionic liquid: a Car-Parrinello molecular dynamics simulation study of 1,3-dimethylimidazolium chloride, *Chem. Phys. Lett.* **417**, 486–491 (2006).

135   von Lilienfeld, O.A., Tavernelli, I., Rothlisberger, U., and Sebastiani, D., Optimization of effective atom centered potentials for London dispersion forces in density functional theory, *Phys. Rev. Lett.* **93**, Article ID 153004 (2004).

136   Zimmerli, U., Parrinello, M., and Koumoutsakos, P., Dispersion corrections to density functionals for water aromatic interactions, *J. Chem. Phys.* **120**, 2693–2699 (2004).

137   Zahn, S., and Kirchner, B., Validation of dispersion-corrected density functional theory approaches for ionic liquid systems, *J. Phys. Chem. A* **112**, 8430–8435 (2008).

138   Hutter, J., and Curioni, A., Car-Parrinello molecular dynamics on massively parallel computers, *Chem. Phys. Chem.* **6**, 1788–1793 (2005).

139   Hutter, J., and Curioni, A., Dual-level parallelism for ab initio molecular dynamics: reaching teraflop performance with the CPMD code, *Parallel Comput.* **31**, 1–17 (2005).

140   Gordon, M.S., Mullin, J.M., Pruitt, S.R., Roskop, L.B., Slipchenko, L.V., and Boatz, J.A., Accurate methods for large molecular systems, *J. Phys. Chem. B* **113**, 9646–9663 (2009).

141   Gao, J., and Truhlar, D.G., Quantum mechanical methods for enzyme kinetics, *Annu. Rev. Phys. Chem.* **53**, 467–505 (2002).

142   Friesner, R.A., and Guallar, V., Ab initio quantum chemical and mixed quantum mechanics/molecular mechanics (QM/MM) methods for studying enzymatic catalysis, *Annu. Rev. Phys. Chem.* **56**, 389–427 (2005).

143   Deev, V., and Collins, M.A., Approximate ab initio energies by systematic molecular fragmentation, *J. Chem. Phys.* **122**, Article ID 154102 (2005).

144   Collins, M.A., and Deev, V.A., Accuracy and efficiency of electronic energies from systematic molecular fragmentation, *J. Chem. Phys.* **125**, 104104 (2006).

145   Collins, M.A., Molecular potential energy surfaces constructed from interpolation of systematic fragment surfaces, *J. Chem. Phys.* **127**, Article ID 024104 (2007).

146   Gordon, M.S., Freitag, M.A., Bandyopadhyay, P., Jensen, J.H., Kairys, V., and Stevens, W.J., The effective fragment potential method: a QM-based MM approach to modeling environmental effects in chemistry, *J. Phys. Chem. A* **105**, 293–307 (2001).

147   Gordon, M.S., Slipchenko, L., Li, H., and Jensen, J.H, Chapter 10 The effective fragment potential: a general method for predicting intermolecular interactions, *Annu. Rep. Comput. Chem.* **3**, 177–193 (2003).

148   Kitaura, K., Ikeo, E., Asada, T., Nakano, T., and Uebayasi, M., Fragment molecular orbital method: an approximate computational method for large molecules, *Chem. Phys. Lett.* **313**, 701–706 (1999).

149   Kitaura, K., Sugiki, S.-I., Nakano, T., Komeiji, Y., and Uebayasi, M., Fragment molecular orbital method: analytical energy gradients, *Chem. Phys. Lett.* **336**, 163–170 (2001).

150   Nakano, T., Kaminuma, T., Sato, T., Akiyama, Y., Uebayasi, M., and Kitaura, K., Fragment molecular orbital method: application to polypeptides, *Chem. Phys. Lett.* **318**, 614–618 (2000).

151   Komeiji, Y., Nakano, T., Fukuzawa, K., Ueno, Y., Inadomi, Y., Nemoto, T., Uebayasi, M., Fedorov, D.G., and Kitaura, K., Fragment molecular orbital method: application to molecular dynamics simulation, "ab initio FMO-MD", *Chem. Phys. Lett.* **372**, 342–347 (2003).

152   Fedorov, D.G., and Kitaura, K., Second order Moller-Plesset perturbation theory based upon the fragment molecular orbital method, *J. Chem. Phys.* **121**, 2483–2490 (2004).

153 Fedorov, D.G., and Kitaura, K., Coupled-cluster theory based upon the fragment molecular-orbital method, *J. Chem. Phys.* **123**, 134103–134111 (2005).

154 Fedorov, D.G., Kitaura, K., Li, H., Jensen, J.H., and Gordon, M.S., The polarizable continuum model (PCM) interfaced with the fragment molecular orbital method (FMO), *J. Comp. Chem.* **27**, 976–985 (2006).

155 Fedorov, D.G., and Kitaura, K., Pair interaction energy decomposition analysis, *J. Comput. Chem.* **28**, 222–237 (2007).

156 Fedorov, D.G., and Kitaura, K., The importance of three-body terms in the fragment molecular orbital method, *J. Chem. Phys.* **120**, 6832–6840 (2004).

157 Fedorov, D.G., Ishimura, K., Ishida, T., Kitaura, K., Pulay, P., and Nagase, S., Accuracy of the three-body fragment molecular orbital method applied to Moller-Plesset perturbation theory, *J. Comput. Chem.* **28**, 1476–1484 (2007).

158 Fedorov, D.G., Olson, R.M., Kitaura, K., Gordon, M.S., and Koseki, S., A new hierarchical parallelization scheme: generalized distributed data interface (GDDI), and an application to the fragment molecular orbital method (FMO), *J. Comput. Chem.* **25**, 872–880 (2004).

7 Ionic Liquids Derived from Natural Sources

JUNKO KAGIMOTO and HIROYUKI OHNO

Department of Biotechnology, Tokyo University of Agriculture and Technology, Koganei, Tokyo, Japan

ABSTRACT

Many ions exist in nature, and some of them have been used to prepare novel ionic liquids. The preparation, physicochemical properties, and applications of, for example, ionic liquids that are derived from naturally occurring amino acids, are discussed and explained here. We also mention a few interesting findings, such as temperature-sensitive phase separation of these ionic liquids after mixing with water. Some of these ionic liquids are miscible with water at low temperatures and yet exhibit clear phase separation induced by a small increase in temperature. This temperature-sensitive phase behaviour has been used, for example, to extract proteins. In addition, some unique applications, such as biodegradable ionic liquids, are also discussed as a future target.

7.1 INTRODUCTION

There are a great variety of ions that can be transformed into ionic liquids. Of course, a few requirements should be satisfied to prepare salts with low melting points. Despite these constraints, there are still many candidate ions that can be found in nature. In this chapter, we summarise the latest data on ionic liquids derived from natural sources.

Some ionic liquids are expected to be reusable "green solvents" due to their low vapour pressure and excellent chemical stability [1]. However, component ions of ionic liquids are not necessarily environmentally friendly ions. The electron-withdrawing effect of halogen atoms, such as fluorine, leads to

Ionic Liquids UnCOILed: Critical Expert Overviews, First Edition. Edited by Natalia V. Plechkova and Kenneth R. Seddon.

delocalisation of the negative charge. Therefore, anions containing fluorine atoms, such as hexafluorophosphate, $[PF_6]^-$ [2], or bis{(trifluoromethyl)sulfonyl}amide, $[NTf_2]^-$ [3], are frequently used as component ions of ionic liquids. These ionic liquids with fluorinated anions exhibit excellent liquid properties, for example, low melting point and low viscosity. On the other hand, there is fear that these ionic liquids can cause environmental pollution due to ambient but long-term exposure, so the development of halogen-free ionic liquids is one of the important tasks of ionic liquid science. Some halogen-free ionic liquids have already been reported, containing anions such as lactate [4, 5], saccharinate, acesulfame [6], methanoate [7], and ascorbate [8], derived from naturally occurring substances. Additionally, ionic liquids containing lactate were reported as chiral catalysts [4], and ionic liquids containing ascorbate were used as media for the *in situ* preparation of gold nanoparticles [8]. These ionic liquids have attracted much attention not only as halogen-free ionic liquids but also as designer ionic liquids. However, halogen-free ionic liquids have yet to be developed further.

To better understand ionic liquids from nature, we focus on amino acids as component ions of halogen-free ionic liquids. Amino acids contain both a carboxylic acid residue and an amino group in a single molecule. In a neutral aqueous solution, they exist as zwitterions, containing protonated amino groups and dissociated carboxylic acids. Under acidic conditions, they act as protonated cations, and under basic conditions they act as deprotonated anions. Thus, amino acids can be used as either anions [9] or cations [10, 11]. Previously, we first reported a series of ionic liquids containing amino acids as anions; these ionic liquids were called "amino acid ionic liquids" [9]. Amino acid ionic liquids are attracting attention not only as halogen-free ionic liquids but also as ionic liquids having many advantages. For example, amino acids are cheap, and amino acid ionic liquids are easily synthesised. These amino acids are archetypal examples of ions derived from natural sources.

Amino acid ionic liquids may also be biodegradable ionic liquids depending on the partner cation structure. It should be noted here that all molecules and/or ions derived from natural sources are simply considered non-toxic, but this is not true. Without citing many toxic molecules and ions from nature, there is no guarantee of the safety or non-toxicity of ionic liquids derived from natural sources. Since the side chain of amino acids is rich in diversity, the effect of anion structure on ionic liquid properties may be easily studied with these amino acids. The chemical modifications of either the amino group or the carboxylate group, or both, also contribute to the functional design of amino acid ionic liquids. Wide development of amino acid ionic liquids is expected because of the great potential for property design. Since we first reported amino acid ionic liquids in 2005 [9], many papers about amino acid ionic liquids have been published. In this chapter, we will review the amino acid ionic liquids from physicochemical properties to application. Many amino acid ionic liquids have been prepared using amino acids as anions, but it is also possible to prepare ionic liquids with amino acids as cations. Ionic

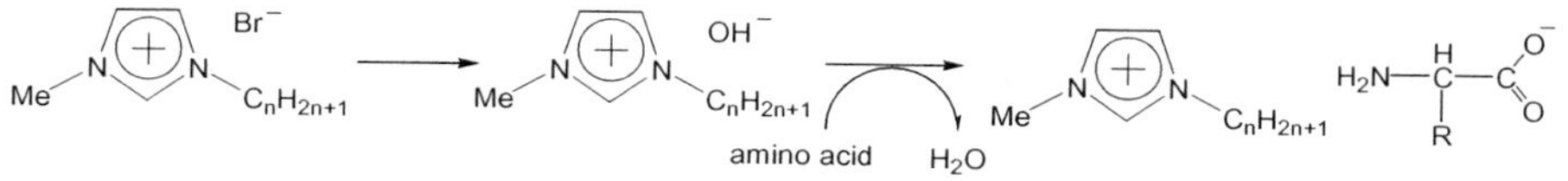

Figure 7.1 Typical preparation of amino acid ionic liquids, $[C_n\text{mim}][H_2NCHRCO_2]$, by neutralisation.

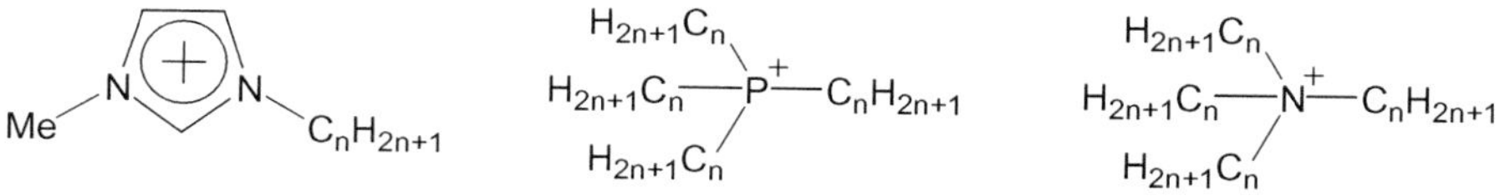

Figure 7.2 Typical cations: 1-alkyl-3-methylimidazolium ($[C_n\text{mim}]^+$; left), tetraalkyl-phosphonium ($[P_{nnnn}]^+$; centre), and tetraalkylammonium ($[N_{nnnn}]^+$; right), frequently used to prepare amino acid ionic liquids.

liquids composed of other ions derived from natural sources will also be mentioned.

7.2 SYNTHESIS OF AMINO ACID IONIC LIQUIDS

Many routes for ionic liquid synthesis have been reported, and most of these techniques may be applicable for the synthesis of amino acid ionic liquids. However, it is important to carefully select the solvent for reactions. We synthesised amino acid ionic liquids, [cation][deprotonated amino acid], by neutralising cation hydroxide ([cation][OH]) with amino acids. In this section, amino acids were used as a source of anions for ionic liquid synthesis. Figure 7.1 shows the synthetic procedure for imidazolium-derived amino acid ionic liquids as an example of the synthesis of amino acid ionic liquids. Previously, many amino acid ionic liquids, containing frequently used cations (Fig. 7.2), such as 1-alkyl-3-methylimidazolium [9], tetraalkylammonium [12], and tetraalkylphosphonium [13], have been synthesised, and their properties have been reported. Details of the synthetic procedures have been described in previously published papers [9, 12, 13].

7.3 PHYSICOCHEMICAL PROPERTIES

The properties of amino acid ionic liquids are summarised in Table 7.1. Viscosity (η), ionic conductivity (σ_i), and density (ρ) of reported amino acid ionic liquids are summarised, as well as their thermal properties. Data for amino acid ionic liquids, in which amino acids are used as cations, are not summarised in this table.

TABLE 7.1 Some Physicochemical Properties of Amino Acid Ionic Liquids

Ionic liquid	$T_m/°C$	$T_g/°C$	$T_d/°C$	η/cP (25°C)	σ_i/S cm^{-1} (25°C)	ρ/g cm^{-3} (25°C)	Reference
[C$_2$mim][Ala]	ND	−57	212		6.4×10^{-4}	1.12	[9, 14]
[C$_2$mim][Met]	ND	−57	209		2.4×10^{-4}		[9]
[C$_2$mim][Gly]	ND	−65	207		5.7×10^{-4}		[9]
[C$_2$mim][Leu]	ND	−51	214		8.1×10^{-5}		[9]
[C$_2$mim][Ile]	ND	−52	220		6.9×10^{-5}		[9]
[C$_2$mim][Ser]	ND	−49	218		6.5×10^{-4}		[9]
[C$_2$mim][Val]	ND	−52	218		8.8×10^{-5}		[9]
[C$_2$mim][Lys]	ND	−47	206		7.8×10^{-5}		[9]
[C$_2$mim][Pro]	ND	−48	226		1.6×10^{-4}		[9]
[C$_2$mim][Thr]	ND	−40	213		1.0×10^{-4}		[9]
[C$_2$mim][Phe]	ND	−36	222		6.0×10^{-5}		[9]
[C$_2$mim][Arg]	ND	−18	212		9.0×10^{-7}		[9]
[C$_2$mim][Trp]	ND	−31	237		9.1×10^{-9}		[9]
[C$_2$mim][Gln]	ND	6	250		1.7×10^{-7}		[9]
[C$_2$mim][Glu]	ND	−12	250		5.0×10^{-7}		[9]
[C$_2$mim][Cys]	ND	−19	173		3.5×10^{-5}		[9]
[C$_2$mim][Asp]	ND	5	227		1.7×10^{-9}		[9]
[C$_2$mim][Tyr]	ND	−23	230		4.0×10^{-8}		[9]
[C$_2$mim][Asn]	ND	−16	205		1.1×10^{-6}		[9]
[C$_2$mim][His]	ND	−24	220		1.0×10^{-7}		[9]
[C$_3$mim][Ala]						1.1	[14]
[C$_4$mim][Ala]						1.08	[14]
[C$_5$mim][Ala]						1.06	[14]
[C$_6$mim][Ala]						1.042	[14]
[P$_{4444}$][Ala]	ND	−70	286	340			[13]
		−75	202	227	4.18×10^{-4}	0.95	[15]
[P$_{4444}$][β-Ala]		−77	203	245	4.04×10^{-4}	0.959	[15]

Ionic liquid					κ	ratio	Ref
[P4444][Met]	ND	−64	217	370			[13]
[P4444][Gly]	13.6	−64	293	420			[13]
[P4444][Leu]	30	−75	200	233	4.85×10^{-4}	0.963	[15]
[P4444][Ile]	ND	−63	292	390			[13]
[P4444][Ser]	ND	−61	294	610			[13]
[P4444][Val]	26	−60	243	900	1.68×10^{-4}	0.991	[15]
[P4444][Lys]	ND	−61	220	734			[13]
[P4444][Pro]	25.4	−59	286	420	1.0×10^{-4}	0.973	[15]
[P4444][Thr]	ND	−59	277	780			[13]
[P4444][Phe]	8.1	−65	225	745			[13]
[P4444][Arg]	30.7	−58	314	850			[13]
[P4444][Trp]	ND	−56	223	970			[13]
[P4444][Gln]	ND	−53	288	930			[13]
[P4444][Glu]	101.7	−36	286	ND			[13]
[P4444][Cys]	ND	−26	316	ND			[13]
[P4444][Asp]	ND	−25	311	ND			[13]
[P4444][Tyr]	ND	−23	319	3030			[13]
[P4444][Asn]	83	−21	190	7440 (60°C)		[13]	[13]
[P4444][His]	85.9	−8	246	ND			[13]
[P4441][Ala]	ND	−46	294	6930			[13]
[P4448][Ala]	ND	−74	224	377			[13]
[P44412][Ala]	ND	−67	299	619			[13]
[P4444][Nle]	ND	−70	188	330			
[P4444][H2N(CH2)5CO2]	21.3	−67	282	630			
[P4444][Suc]	ND	−47	296	6300		0.93	
[P5555][Leu]	ND	−66	250	1400			[16]

(Continued)

TABLE 7.1 *(Continued)*

Ionic liquid	$T_m/°C$	$T_g/°C$	$T_d/°C$	η/cP (25°C)	$\sigma_i/S\ cm^{-1}$ (25°C)	$\rho/g\ cm^{-3}$ (25°C)	Reference
[P$_{6666}$][Leu]	ND	−80	291	550		0.908	[16]
[P$_{8886}$][Leu]	ND	−84	268	1630		0.9	[16]
[P$_{8888}$][Leu]	ND	−81	266	800		0.9	[16]
[P$_{8888}$][Ala]	ND	−78	288	1620		0.903	[16]
[P$_{8888}$][Gly]	ND	−80	286	1830		0.908	[16]
[P$_{8888}$][Ile]	ND	−82	264	1950		0.901	[16]
[P$_{8888}$][Ser]	ND	−79	252	1160		0.92	[16]
[P$_{8888}$][Val]	ND	−81	285	1480		0.902	[16]
[P$_{8888}$][Lys]	ND	−83	224	2770		0.913	[16]
[P$_{8888}$][Pro]	ND	−82	308	670		0.915	[16]
[P$_{8888}$][Phe]	ND	ND	299	610		0.927	[16]
[P$_{8888}$][Glu]	86	3	313	S			[16]
[P$_{8888}$][Asp]	ND	ND	239	2540		0.945	[16]
[P$_{8889}$][Leu]	ND	−84	268	1630		0.894	[16]
[P$_{88810}$][Leu]	ND	−85	272	1400		0.897	[16]
[P$_{88812}$][Leu]	ND	−87	304	1230		0.894	[16]
[P$_{88818}$][Leu]	−38	ND	310	900		0.886	[16]
[P$_{10\,10\,10\,10}$][Leu]	ND	−78	310	6470		0.897	[16]
[P$_{12\,12\,12\,12}$][Leu]	ND	ND	326	S			[16]
[P$_{14\,14\,14\,14}$][Leu]	14	−8.5	315	S			[16]
[P$_{444\,3a}$][Ala]		−67	283	758	1.1×10^{-4}	0.986	[17]
[P$_{444\,3a}$][Met]		−65	275	767	8.0×10^{-5}	1.018	[17]
[P$_{444\,3a}$][Gly]		−70	281	714	1.4×10^{-4}	0.997	[17]
[P$_{444\,3a}$][Leu]		−61	283	1194	5.0×10^{-5}	0.966	[17]
[P$_{444\,3a}$][Ile]		−58	284	1408	3.1×10^{-5}	0.974	[17]
[P$_{444\,3a}$][Ser]		−53	254	1342	5.4×10^{-5}	1.026	[17]
[P$_{444\,3a}$][Val]		−63	293	888	5.2×10^{-5}	0.975	[17]
[P$_{444\,3a}$][Lys]		−61	292	1432	3.8×10^{-5}	0.999	[17]

Compound							Ref.
[P₄₄₄₃ₐ][Pro]		−55	280	1772	3.1×10^{-5}	1.005	[17]
[P₄₄₄₃ₐ][Thr]		−56	245	1791	4.3×10^{-5}	1.013	[17]
[P₄₄₄₃ₐ][Phe]		−51	282	1985	1.9×10^{-5}	1.022	[17]
[P₄₄₄₃ₐ][Arg]		−56	228	360 (55 °C)	1.5×10^{-5}	0.994	[17]
[P₄₄₄₃ₐ][Trp]		−30	277		1.9×10^{-7}	1.06	[17]
[P₄₄₄₃ₐ][Gln]		−34	297	1653 (55 °C)	1.1×10^{-6}	1.057	[17]
[P₄₄₄₃ₐ][Glu]		−52	262	1418 (55 °C)	7.2×10^{-6}	1.016	[17]
[P₄₄₄₃ₐ][Cys]		−36	202	1544 (65 °C)	1.1×10^{-6}	1.05	[17]
[P₄₄₄₃ₐ][Asp]		−54	254	1701 (55 °C)			[17]
[P₄₄₄₃ₐ][Tyr]		−41	231	1291 (75 °C)	8.9×10^{-7}	1.043	[17]
[P₄₄₄₃ₐ][Asn]		−42	230	1701 (55 °C)	1.8×10^{-6}	1.046	[17]
[P₄₄₄₃ₐ][His]		−52	300	1050 (55 °C)	5.6×10^{-6}	0.999	[17]
[N₁₁₁₁][Ala]	42	ND	219	s			[12]
[N₁₁₁₁][Gly]	ND	−68	181	304		1.089	[12]
[N₁₁₁₁][Val]	40	ND	223	s			[12]
[N₁₁₁₁][βAla]	ND	−82	193	668		1.088	[12]
[N₂₂₂₂][Ala]	ND	−80	185	81		0.998	[12]
[N₂₂₂₂][βAla]	ND	−85	184	132		1.013	[12]
[N₄₄₄₄][Ala]	76	ND	162				[13]
[N₄₄₄₄][Gly]	16	−71	179	214		0.941	[12]
[N₄₄₄₄][Val]	25	−69	185	660		0.947	[12]
[N₄₄₄₄][βAla]	ND	−73	180	465		0.9545	[12]
[N₂₂₂₆][Ala]	ND	−40	150				[13]

ND, not detected; s, solid or amorphous glass at 25 °C; [P₄₄₄₃ₐ]: (3-aminopropyl)tributylphosphonium; HAla: Alanine; HArg: Arginine; HHis: Histidine; HLys: Lysine; HAsp: Asparatic acid; HGlu: Glutamic acid; HSer: Serine; HThr: Threonine; HAsn: Asparagine; HGln: Glutamine; HCys: Cysteine; HGly: Glycine; HPro: Proline; HIle: Isoleucine; HLeu: Leucine; HMet: HMethionine; HPhe: Phenylalanine; HTrp: Tryptophan; HTyr: Tyrosine; HVal: Valine; HNle: Norleucine; $H_2N(CH_2)_5CO_2H$: 6-aminohexanoic acid; HSuc: Succinic acid.

The ionic liquid containing the 1-ethyl-3-methylimidazolium cation ($[C_2mim]^+$) and the $[NTf_2]^-$ anion shows a low glass transition temperature and low viscosity. Based on this background, $[C_2mim]^+$ was fixed as a cation to synthesise and characterise amino acid ionic liquids containing 20 different natural amino acids, all of which have been prepared as isotropic liquids at room temperature. Physicochemical properties depend on the amino acid structure, and their glass transition temperature (T_g) was found to be between $-65\,^{\circ}C$ and $6\,^{\circ}C$. They are relatively viscous, and $[C_2mim][Gly]$ shows the lowest viscosity of 490 cP at $25\,^{\circ}C$.

Considering the relation between the side chain structure of amino acids and the T_g of the corresponding amino acid ionic liquids, high viscosity may be attributed to hydrogen bonding between anion and cation and/or anion and anion. Woo and coworkers showed the existence of hydrogen bonds between ions in $[C_2mim]$[deprotonated amino acid] by computational study [18]. On the other hand, most amino acid ionic liquids containing phosphonium ($[P_{nnnn}]^+$) or ammonium ($[N_{nnnn}]^+$) cations are liquid, with lower T_g and lower viscosity than those of their $[C_2mim]^+$ analogues. In $[C_2mim]^+$ salts, there are hydrogen bonds not only between anion and anion but also between anion and cation. In contrast, both phosphonium and ammonium salts can show hydrogen bonds only between anions, with no hydrogen bonding between anion and cation. This was easily understood from their ion structure, and was actually confirmed by the X-ray analysis of $[P_{4444}][Glu]$ [19]. Additionally, $[P_{4443a}]$[deprotonated amino acid] have a propylamino group on the phosphonium cation, which certainly will contribute to anion–cation hydrogen bonding, and accordingly they show higher viscosity than their tetrabutylphosphonium analogues [17]. The viscosities and glass transition temperatures of amino acid ionic liquids containing amino acid side chain, such as found in Lys and Asp, are also higher than those of amino acid ionic liquids containing simple alkyl chains such as Ala and Leu. Therefore, the viscosity and T_g of ionic liquids are considerably influenced by the functional groups on the component ions. This result is important for the design of functional ionic liquids because introduction of functions onto the ionic liquids usually induces an increase in viscosity and an elevation of T_g.

Decomposition temperature (T_d) is another important property of ionic liquids. The T_d of a series of amino acid ionic liquids increased in the following order, for a given anion:

$$[N_{nnnn}]^+ < [C_nmim]^+ < [P_{nnnn}]^+.$$

Indeed, the decomposition temperature of the tetraalkylphosphonium amino acid ionic liquids was similar to that of the corresponding pure amino acid. This strongly suggested that amino acid anions started to decompose before the cation. On the other hand, the thermal decomposition of both ammonium cation and imidazolium cations occurred at temperatures below that of the amino acid anions. It is known that the thermal decomposition of ionic liquids containing anions having nucleophilicity is initiated by dealkylation of the cations [20]. The activation energy the dealkylation of the cations, estimated by quantum-chemical calculations, increased in the following order [21]:

$$[N_{nnnn}]^+ < [C_n mim]^+ < [P_{nnnn}]^+.$$

These data support the above-mentioned thermal decomposition mechanism. Additionally, the chiral stability of $[P_{4\,4\,4\,4}]$[deprotonated amino acid] against heating has been analysed. The chirality of $[P_{4\,4\,4\,4}]$[L-Val] was retained at almost 100% after heating at 100 °C for 10 hours [22]. When the amino group of the amino acid was acetylated, the corresponding ionic liquid showed better thermal stability, and no racemisation was found after heating it at 150 °C for 10 hours [23].

There are some reports on the density of amino acid ionic liquids: the density of the tetraalkylphosphonium ionic liquids is lower than that of their 1-alkyl-3-methylimidazolium analogues; this trend is also seen in ionic liquids having other anions [24, 25]. In the case of tetraalkylphosphonium amino acid ionic liquids, the density mainly depends on the number of carbon atoms, regardless of the symmetry of the cations. For these ionic liquids, the density decreased with an increase in alkyl chain length. Since the density of molecular liquids, such as hydrocarbons, is governed by van der Waals forces and molecular packing, their density also increased with increasing the carbon number. Against this, the density of ionic liquids is mainly governed by electrostatic interaction between anion and cation, whose tendency can also be seen in the case of amino acid ionic liquids. Among many amino acid ionic liquids, phosphonium salts are potential candidates in the preparation of less dense ionic liquids: some of them are less dense than water. Although 1-ethyl-3-methylimidazolium amino acid ionic liquids are hydrophilic, some tetraalkylphosphonium amino acid ionic liquids, with n larger than 6, are hydrophobic ionic liquids. These ionic liquids showed phase separation when they were mixed with water [16]. Since these hydrophobic amino acid ionic liquids are halogen-free, they are applicable as eco-friendly ionic liquids in many fields. As already mentioned, the density of these hydrophobic amino acid ionic liquids with tetraalkylphosphonium cations was lower than 1.00 g cm^{-3}. Hence, in contrast to general hydrophobic ionic liquids, these tetraalkylphosphonium amino acid ionic liquids form the upper phase after mixing with water. These hydrophobic and low density amino acid ionic liquids could be useful as new functional solvents for various applications such as separation–extraction solvents for materials in an aqueous phase, as well as for covering liquid membranes of aqueous solution.

7.4   AMINO ACID IONIC LIQUIDS CONTAINING DERIVATISED AMINO ACIDS

7.4.1   Hydrophobic Amino Acid Ionic Liquids Containing Derivatised Amino Acids

Amino acids contain both amino groups and carboxylic groups on the α-carbon, and functionalisation of amino acid ionic liquids is available by modification of these residues. As mentioned previously, addition of hydrophobicity to amino acid ionic liquids is easy, just by introducing long alkyl chains onto the amino acids. However, introduction of long alkyl chains reduces the ion density. On the other hand, introduction of fluorine atom(s) onto the anion frequently contributes to increasing hydrophobicity. We have also reported the formation of hydrophobic amino acid ionic liquids using derivatised amino acids in which the (trifluoromethyl)sulfonyl group was introduced onto the amino group [26]. Clear phase separation after mixing with water was proof of the hydrophobic properties [26].

Figure 7.3 shows the synthetic route to hydrophobic derivatised amino acid salts. After protecting the carboxylic group of each amino acid (alanine (HAla), leucine (HLeu), valine (HVal)) as a methyl ester, (trifluoromethyl) sulfonyl anhydride was added to obtain hydrophobic derivatised amino acids ([HI-AA]s). Some physicochemical properties of [cation][I-AA] are summarised in Table 7.2. Some [C$_4$mim][I-AA] salts have been obtained as liquids at room temperature, showing only T_g. Similarly, most [P$_{4\,4\,4\,4}$][I-AA] salts, except for [P$_{4\,4\,4\,4}$][I-Val], are liquid at room temperature. The T_g and T_m of these [cation][I-AA] salts are higher than those of [C$_4$mim][NTf$_2$] (T_m: $-4\,^{\circ}$C). [P$_{4444}$][I-AA] showed better properties than [P$_{4\,4\,4\,4}$][NTf$_2$] (T_m: 86 $^{\circ}$C). Since [cation][I-AA] salts also retain chirality, as shown in Table 7.2 (as [α]: specific rotatory power), they are expected as stable chiral ionic liquids.

The phase behaviour of the mixture of [cation][I-AA] and water has been studied. The [C$_4$mim][I-AA]s were all miscible with water. However, all [P$_{4\,4\,4\,4}$][I-AA] were immiscible with water as shown in Figure 7.4. Although there is hydrogen bonding between the imidazolium cation and water molecules, the phosphonium cation has no possibility of forming hydrogen bonds. Additionally, phosphonium cations are more hydrophobic than imidazolium

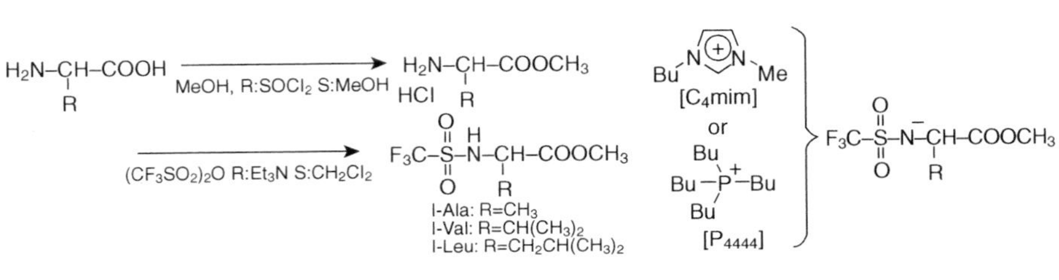

Figure 7.3   Preparation of I-AA and their salts.

TABLE 7.2   Physicochemical Properties of Representative [Cation][I-AA]

	$T_m/°C$	$T_g/°C$	$T_d/°C$	$[\alpha]_D^{22}\,a/°$
[C$_4$mim][I-Ala]	ND	−62.8	260	−24.6
[C$_4$mim][I-Val]	ND	−42.2	253	−24.5
[C$_4$mim][I-Leu]	ND	−37.5	237	−22.1
[P$_{4\,4\,4\,4}$][I-Ala]	ND	−54.0	253	−8.4
[P$_{4\,4\,4\,4}$][I-Val]	61.2	−69.2	277	−18.9
[P$_{4\,4\,4\,4}$][I-Leu]	13.8	ND	264	−16.3

$[\alpha]$, specific rotatory power; a, at a concentration of 1.00 g (100 cm^3 MeOH)$^{-1}$.

Figure 7.4   Mixtures of ionic liquids with water after vortex mixing for 2 minutes. (Left to right): [C$_4$mim][I-Ala], [C$_4$mim][I-Val], [C$_4$mim][I-Leu], [P$_{4\,4\,4\,4}$][I-Ala], [P$_{4\,4\,4\,4}$][I-Val], and [P$_{4\,4\,4\,4}$][I-Leu]. The upper phase of [P$_{4\,4\,4\,4}$][I-AA] systems is an ionic liquid-rich phase and the bottom phase is a water-rich phase [26]. Phosphonium salts generally have lower densities than imidazolium ones [16, 26].

cations. After the phase separation, the ionic liquid-rich phase was found to contain a small amount of water. The water content of [P$_{4\,4\,4\,4}$][I-Ala] and [P$_{4\,4\,4\,4}$][I-Leu] was 5.1 and 2.9 wt%, respectively.

7.4.2   Amino Acid Ionic Liquids Containing Derivatised Amino Acid Showing Lower Critical Separation Temperature Behaviour

[P$_{4\,4\,4\,4}$][I-AA]s are immiscible with water due to introduction of the (trifluoro-methyl)sulfonyl group onto the amino group of amino acid. However, even after phase separation, the ionic liquid rich phase was confirmed to contain a small amount of water (2–5 wt%). The separated phase of [P$_{4\,4\,4\,4}$][I-AA]/water was stable regardless of temperature. If the phase behaviour of ionic liquid/water mixtures were controllable by external stimuli, such as temperature change, many interesting applications could be proposed, such as one-pot

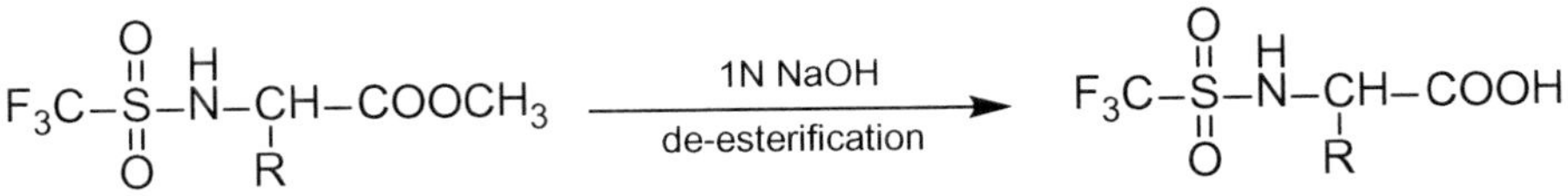

$$F_3C-\overset{\overset{O}{\|}}{\underset{\overset{\|}{O}}{S}}-\overset{H}{N}-\underset{R}{CH}-COOCH_3 \quad \xrightarrow[\text{de-esterification}]{\text{1N NaOH}} \quad F_3C-\overset{\overset{O}{\|}}{\underset{\overset{\|}{O}}{S}}-\overset{H}{N}-\underset{R}{CH}-COOH$$

HI-AA   HTf-AA

Figure 7.5 Preparation of N-(trifluoromethyl)sulfonylated amino acids.

TABLE 7.3 Physicochemical Properties of [P$_{444\,n}$][Tf-AA]s

	$T_m/°C$	$T_d/°C$	$[\alpha]_D^{22}\,a/°$
[P$_{4444}$][Tf-Val]	51	274	4.5
[P$_{4444}$][Tf-Leu]	64	257	9.8
[P$_{4444}$][Tf-Ile]	51	267	3.6
[P$_{4444}$][Tf-Phe]	64	240	1.5
[P$_{4448}$][Tf-Leu]	−50 (T_g)	259	7.0

$[\alpha]$, specific rotatory power; a, at a concentration of 1.00 g (100 cm^3 MeOH)$^{-1}$.

reaction/extraction systems. To reduce the hydrophobicity of the [I-AA] anion, the methyl ester of the carboxylic acid group of the I-AA was deprotected, to obtain a less hydrophobic derivatised amino acid ([HTf-AA]), as shown in Figure 7.5. The [P$_{444n}$][Tf-AA] salts were obtained by neutralisation of HTf-AA with [P$_{4444}$][OH] or [P$_{4448}$][OH] [27]. [P$_{4444}$][Tf-AA]s were found to be solids, with melting points in the range of 50–60 °C, but [P$_{4448}$][Tf-AA]s were liquid at room temperature, showing only glass transition temperatures. Table 7.3 summarises the physicochemical properties of [P$_{444\,n}$][Tf-AA].

As expected, mixtures of [P$_{4444}$][Tf-AA] and water showed temperature-sensitive phase transitions. Furthermore, they showed an unexpectedly unique phase change, associated with "lower critical separation temperature" (LCST) behaviour. [P$_{4444}$][Tf-AA]s formed homogeneous mixtures with water at low temperatures, which phase-separated at higher temperatures. Figure 7.6 shows the temperature dependence of the phase behaviour of a [P$_{4444}$][Tf-Leu]/water mixture. To demonstrate this, the ionic liquid phase was coloured with Nile Red. Addition of Nile Red was confirmed not to affect the phase behaviour. The mixture phase-separated at 25 °C or higher (Fig. 7.6a). Upon recooling to 22 °C, the mixture formed again a completely homogeneous phase (Fig. 7.6d). It became turbid, as seen in Figure 7.6e by heating again to 25 °C, and it produced a clear phase separation by holding it at 25 °C, as seen in Figure 7.6g. Thus, this phase transition behaviour is thermally reversible and induced by a very small temperature change, as little as 3 °C. This highly temperature-sensitive phase change could be useful in many processes.

There are two types of temperature-driven phase separation behaviour, namely, LCST and UCST (upper critical solution temperature). The latter,

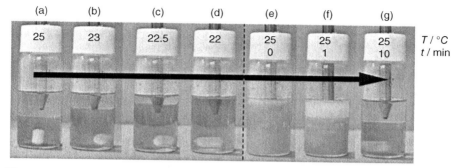

Figure 7.6 Temperature dependence of the phase behaviour of a $[P_{4\,4\,4\,4}][Tf\text{-}Leu]$/ water mixture [27]. See colour insert.

UCST, is generally seen in a mixture where it becomes miscible at a higher temperature, while it is immiscible at a lower temperature. There are many UCST-type behaviours seen in mixtures of low molecular weight compounds. In the case of ionic liquid/water mixtures, all reported examples have been classified into UCST-type phase behaviour. For example, [cholinium][NTf$_2$]/ water shows UCST behaviour [28]. On the other hand, only a few LCST-type phase behaviours were found in mixtures of low molecular weight compounds, such as triethylamine/water mixtures. In the case of ionic liquids, Rebelo and coworkers have reported that ionic liquid/chloroalkane mixtures showed both UCST and LCST transitions [29]. However, there was no report on the LCST behaviour of ionic liquid/water mixture. From the viewpoint of wide application, especially for bioscience, an ionic liquid/water mixed system is more favourable than an ionic liquid/organic solvent mixture. This $[P_{4\,4\,4\,n}][Tf\text{-}AA]$/ water mixture is the first reported as a unique mixed system to show LCST behaviour.

Additionally, the phase separation temperature (T_c) of a $[P_{4\,4\,4\,n}][Tf\text{-}AA]$/ water mixture was controlled by added water content, cation structure, and anion structure. The T_c shifted to lower temperatures by increasing water content in the mixture. This is because the larger amount of the added water needs a lower temperature to fully solubilise it in the ionic liquid phase in the case of an LCST system. Elongation of the alkyl chain on the amino acid or phosphonium cation also lowered the T_c. This is also understandable due to the lower solubility of water in more hydrophobic ionic liquids. These results suggest that both the hydrophobicity of $[P_{4\,4\,4\,n}][Tf\text{-}AA]$ and the amount of water present are important factors in directly controlling the T_c. Although the water content of the ionic liquid phase for the [C$_2$mim][NTf$_2$]/water system is about 4 wt%, that of the $[P_{4\,4\,4\,n}][Tf\text{-}Leu]$/water system (in the mixing ratio of 50:50 wt% at 50 °C) is 26 wt%. It should be noted that there are two different expressions of water content. To avoid confusion, it should clearly be mentioned here. One is "added amount of water" to the ionic liquid. This is

TABLE 7.4 Some Examples of the Application of Amino Acid Ionic Liquids

Application	AA ionic liquids	Reference
CO_2 capture	$[P_{4\,4\,4\,4}][AA]$ $[P_{4\,4\,4\,3a}][AA]$ $[N_{1\,1\,1\,1}][AA]$ $[P_{3\,3\,3\,3}][Pro]$	[15] [17] [12] [32]
N-Benzyloxycarbonylation of amines 		
Michael addition 	$[C_2mim][Pro]$ (solvent: organic solvent)	[33]
N-arylation 	Poly-$[D_{mvim}][Pro] + Cu^I$ (as solid catalyst)	[34]
Synthesis of NCA 	$[C_2mim][AA]$	[31]
Chiral selectors	$[C_nmim][Pro] + Cu^{\alpha}$ $[C_4mim][AA]$	[35] [35]
Additives for lubricant	$[P_{4\,4\,4\,4}][AA]$ (AA = Asp, Glu)	[36]
Gene delivery vector	Poly-$[C_4vim][Pro]$	[37]

$[D_{mvim}]$, 1,2-dimethyl-3-(4-vinylbenzyl)imidazolium; $[C_4vim]$, 3-butyl-1-vinylimidazolium.

regarded as the initial mixing ratio of ionic liquid to water, and is controlled by the amounts of liquids mixed. In contrast, the other is "water content in the ionic liquid phase." This is the amount of water dissolved specifically in the ionic liquid phase—the solubility of water in the ionic liquid.

Here, it is important to analyse the water content in the ionic liquid phase, especially its temperature dependence. By cooling the $[P_{444\,n}][Tf\text{-}Leu]$/water mixture, the water content in the ionic liquid phase increased, and the mixture became completely homogeneous at 25 °C or below. As already mentioned, the water content of the ionic liquid phase of the mixed $[P_{444\,n}][Tf\text{-}AA]$/water system was controlled by a small temperature change. These are expected to find application as novel solvent systems, not only for separation or extraction processes, but also in the bioscience field.

7.5   APPLICATION OF AMINO ACID IONIC LIQUIDS

Since amino acid ionic liquids were reported first in 2005, many papers have been published not only about basic research, but also about applications of amino acid ionic liquids. Table 7.4 summarises the major applications of amino acid ionic liquids reported in the literature.

7.5.1   CO$_2$ Capture

It is well known that aqueous amine solutions are used to chemically trap carbon dioxide. Davis and coworkers reported that amine-functionalised ionic liquids reacted reversibly with CO_2 [30]. Based on this report, Zhang et al. have focussed on the amino group of amino acid ionic liquids to trap CO_2. $[P_{4444}][AA]$ was supported on porous silica gel, and the CO_2 capture ability has been studied [15]. The CO_2 capture ability of amino acid ionic liquids supported on porous silica gel was in the same range as amine-functionalised ionic liquids with one CO_2 molecule per two ionic liquid ion pairs. Additionally, amino acid salts of (3-aminopropyl)tributylphosphonium, $[P_{4443a}][AA]$, have been synthesised. It is reported that the chemical absorption of CO_2 by $[P_{4443a}][AA]$ goes up to an equimolar ratio of CO_2 and amino acid ionic liquids [17]. These amino acid ionic liquids can be cycled repeatedly for CO_2 uptake–release, and no change of absorption capacity was found after four cycles of absorption/desorption [17].

7.5.2   Synthesis of *N*-Carboxyamino Acid Anhydrides

An *N*-carboxyamino acid anhydride (NCA) is an inevitable monomer for the chemical synthesis of a polypeptide. An NCA is generally prepared by the reaction of an amino acid with phosgene. However, phosgene is highly toxic, and HCl is evolved as by-product: safer synthetic processes are required. Endo and Sudo have developed a novel process for NCA synthesis, focussing on diphenyl carbonate (DPC) [31]. However, since the amino group of the amino acid was dehydrogenated by the carboxyl group in the organic solvent, no reaction proceeded due to loss of nucleophilicity. They focussed on the presence of the free amino group on $[C_2mim][AA]$, and succeeded in preparing NCA by the reaction of $[C_2mim][AA]$ and DPC at room temperature and in a short period. Thus, amino acid ionic liquids can be used to synthesise NCA without phosgene, in high yield and safely.

7.5.3   Catalytic Reactions Using Amino Acid Ionic Liquids

Proline is often used as a catalyst in organic reactions. Suryakiran et al. reported that the *N*-benzyloxycarbonylation of amines in tetrapropylammonium L-prolinate ($[N_{3\,3\,3\,3}][Pro]$) proceeds under mild conditions with excellent yields (90–96%) [32]. An amino acid may also be used as a reaction medium or a catalyst for asymmetric synthesis. Qian et al. reported Michael addition of cyclohexanone to chalcones, in the presence of $[C_2mim][Pro]$ as an asymmetric catalyst [33]. In the presence of $[C_2mim][Pro]$ in methanol, Michael addition has proceeded with an excellent yield (98%) and high enantiomeric excess (% ee) value (up to 86%) [33]. Chen et al. synthesised a Cu^I-saturated catalyst using polymerised 1,2-dimethyl-3-(4-vinylbenzyl)imidazolium prolinate (Poly-$[D_{mvim}][Pro]$) [34]. Using this Cu^I-saturated catalyst, *N*-arylation of nitrogen-containing heterocycles proceeded in high yield: this process is the most efficient approach to the *N*-arylation of imidazoles with haloarenes so far reported.

7.5.4   Application for Chiral Selectors

In preparation of medicine and biologically active agents, it is important to get chiral molecules in high % ee. High performance liquid chromatography (HPLC) or capillary electrophoresis (CE) are popular ways of preparing chiral molecules in high % ee. Chiral ionic liquids have been synthesised and used as chiral selectors in CE. Liu et al. reported the application of amino acid ionic liquids in chiral separation, where they functioned as chiral selectors. By using $[C_nmim][Pro]$ as a source of a chiral ligand to coordinate to Cu^α, higher enantioselectivity in HPLC and CE has been achieved than with conventional amino acid ligands [38]. Amino acid ionic liquids are expected, therefore, to act not only as catalysts but also as a source of chiral ligands.

7.6   CONCLUSION

Amino acid ionic liquids, developed in our laboratory, are expected to attract much attention as a platform for developing halogen-free and designer ionic liquids. Since amino acid ionic liquids are halogen-free, they are expected not only to be eco-friendly but also to have low toxicity and to be biodegradable, depending on the cation structure. As mentioned in the introduction, although there is not enough proof that ionic liquids derived from natural sources have low toxicity, they still have a high probability of being safe compared with ordinary chemicals. Although there are some studies on the toxicity of amino acid ionic liquids, detailed studies should be performed for the generalisation of ion structure and toxicity. Major toxicity tests have generally been carried out with microorganisms, or fish, in aqueous environments. However, adding ionic liquids to water may result in them performing as simple salts, and osmotic pressure may be a principal factor affecting their toxic properties.

Additionally, amino acid ionic liquids have the potential to be applied in a wide range of scientific and industrial areas because they have many advantages such as stable chiral centres, diversity of amino acid side chains, low cost, facile synthesis, and a vast possibility of chemical modification of amino and carboxylic groups. After the discovery of amino acid ionic liquids in 2005 [9], many papers on their basic physicochemical properties have been published. In spite of the many studies on the synthesis of novel amino acid ionic liquids, detailed physicochemical properties, and ion–ion interactions analysed by computer simulation, there are still many unresolved problems. For example, the viscosity of amino acid ionic liquids is generally higher than that of [C₂mim][NTf₂], thermal stability is not high enough (the upper limit of T_d is approximately 300 °C), there are no detailed data on toxicity, biodegradability, and so on.

The biodegradability of ionic liquids is a crucial area of study, if they are to be widely used in many fields. Ionic liquids derived from natural sources are promising candidates for a future sustainable world. In the future, the study on amino acid ionic liquids could greatly expand the application areas, and some fascinating properties would also be created.

ACKNOWLEDGEMENTS

The authors acknowledge Dr. Kenta Fukumoto for his great contribution to the science of amino acid ionic liquids. Our study of amino acid ionic liquids was mainly supported by a Grant-in-Aid for Scientific Research from the Japan Society for the Promotion of Science (No. 21225007) and a Grant-in-Aid for Scientific Research on Priority Areas from the Ministry of Education, Culture, Sports, Science, and Technology of Japan (No. 17073005).

REFERENCES

1 Wasserscheid, P., and Welton, T., *Ionic liquids in synthesis*, 2nd ed., Wiley-VCH, Weinheim (2007).

2 Fuller, J., Carlin, R., Delong, H.C., and Haworth, D., Structure of 1-ethyl-3-methylimidazolium hexafluorophosphate: model for room temperature molten salts, *J. Chem. Soc. Chem. Commun.*, 299–300 (1994).

3 Bonhôte, P., Dias, A.P., Papageorgiou, N., Kalyanasundaram, K., and Grätzel, M., Hydrophobic, highly conductive ambient-temperature molten salts, *Inorg. Chem.* **35**, 1168–1178 (1996).

4 Earle, M.J., McCormac, P.B., and Seddon, K.R., Diels-Alder reactions in ionic liquids: a safe recyclable green alternative to lithium perchlorate-diethyl ether mixtures, *Green Chem.* **1**, 23–25 (1999).

5 Pernak, J., Goc, I., and Mirska, I., Antimicrobial activities of protic ionic liquids with lactate anion, *Green Chem.* **6**, 323–329 (2004).

6 Carter, E.B., Culver, S.L., Fox, P.A., Goode, R.D., Ntai, I., Tickell, M.D., Traylor, R.K., Hoffman, N.W., and Davis, J.H., Sweet success: ionic liquids derived from non-nutritive sweeteners, *Chem. Commun.*, 630–631 (2004).

7 Fukaya, Y., Sugimoto, A., and Ohno, H., Superior solubility of polysaccharides in low viscosity, polar, and halogen-free 1,3-dialkylimidazolium formates, *Biomacromolecules* **7**, 3295–3297 (2006).

8 Dinda, E., Si, S., Kotal, A., and Mandal, T.K., Novel ascorbic acid-based ionic liquids for the *in situ* synthesis of quasi-spherical and anisotropic gold nanostructures in aqueous medium, *Chem. Eur. J.* **14**, 5528–5537 (2008).

9 Fukumoto, K., Yoshizawa, M., and Ohno, H., Room temperature ionic liquids from 20 natural amino acids, *J. Am. Chem. Soc.* **127**, 2398–2399 (2005).

10 Tao, G.H., He, L., Sun, N., and Kou, Y., New generation ionic liquids: cations derived from amino acids, *Chem. Commun.*, 3562–3564 (2005).

11 Plaquevent, J.C., Levillain, J., Guillen, F., Malhiac, C., and Gaumont, A.C., Ionic liquids: new targets and media for α-amino acid and peptide chemistry, *Chem. Rev.* **108**, 5035–5060 (2008).

12 Jiang, Y., Wang, G., Zhou, Z., Wu, Y., Geng, J., and Zhang, Z.B., Tetraalkylammonium amino acids as functionalized ionic liquids of low viscosity, *Chem. Commun.*, 505–507 (2008).

13 Kagimoto, J., Fukumoto, K., and Ohno, H., Effect of tetrabutylphosphonium cation on the physico-chemical properties of amino acid ionic liquids, *Chem. Commun.*, 2254–2256 (2006).

14 Fang, D.W., Guan, W., Tong, J., Wang, Z.W., and Yang, J.Z., Study on physicochemical properties of ionic liquids based on alanine [$C_n$mim][Ala] (n = 2, 3, 4, 5, 6), *J. Phys. Chem. B* **112**, 7499–7505 (2008).

15 Zhang, J.M., Zhang, S.J., Dong, K., Zhang, Y.Q., Shen, Y.Q., and Lv, X.M., Supported absorption of CO_2 by tetrabutylphosphonium amino acid ionic liquids, *Chem. Eur. J.* **12**, 4021–4026 (2006).

16 Kagimoto, J., Taguchi, S., Fukumoto, K., and Ohno, H., Hydrophobic and low-density amino acid ionic liquids, *J. Mol. Liq.* **153**, 133–138 (2010).

17 Zhang, Y., Zhang, S., Lu, X., Zhou, Q., Fan, W., and Zhang, X., Dual amino-functionalised phosphonium ionic liquids for CO_2 capture, *Chem. Eur. J.* **15**, 3003–3011 (2009).

18 Sirjoosingh, A., Alavi, S., and Woo, T.K., Molecular dynamics simulations of equilibrium and transport properties of amino acid-based room temperature ionic liquids, *J. Phys. Chem. B* **113**, 8103–8113 (2009).

19   Kagimoto, J., Noguchi, K., Murata, K., Fukumoto, K., Nakamura, N., and Ohno, H., Polar and low viscosity ionic liquid mixtures from amino acids, *Chem. Lett.* **37**, 1026–1027 (2008).

20   Chan, B.K.M., Chang, N.H., and Grimmett, M.R., The synthesis and thermolysis of imidazole quaternary salts, *Aust. J. Chem.* **30**, 2005–2013 (1977).

21   Kroon, M.C., Buijs, W., Peters, C.J., and Witkamp, G.J., Quantum chemical aided prediction of the thermal decomposition mechanisms and temperatures of ionic liquids, *Thermochim. Acta* **465**, 40–47 (2007).

22   Fukumoto, K., Kohno, Y., and Ohno, H., Chiral stability of phosphonium-type amino acid ionic liquids, *Chem. Lett.* **35**, 1252–1253 (2006).

23   Ohno, H., and Fukumoto, K., Amino acid ionic liquids, *Acc. Chem. Res.* **40**, 1122–1129 (2007).

24   Del Sesto, R.E., Corley, C., Robertson, A., and Wilkes, J.S., Tetraalkylphosphonium-based ionic liquids, *J. Organomet. Chem.* **690**, 2536–2542 (2005).

25   Huddleston, J.G., Visser, A.E., Reichert, W.M., Willauer, H.D., Broker, G.A., and Rogers, R.D., Characterization and comparison of hydrophilic and hydrophobic room temperature ionic liquids incorporating the imidazolium cation, *Green Chem.* **3**, 156–164 (2001).

26   Fukumoto, K., and Ohno, H., Design and synthesis of hydrophobic and chiral anions from amino acids as precursor for functional ionic liquids, *Chem. Commun.*, 3081–3083 (2006).

27   Fukumoto, K., and Ohno, H., LCST-type phase changes of a mixture of water and ionic liquids derived from amino acids, *Angew. Chem. Int. Edit.* **46**, 1852–1855 (2007).

28   Nockemann, P., Binnemans, K., Thijs, B., Parac-Vogt, T.N., Merz, K., Mudring, A., Menon, P., Rajesh, R., Cordoyiannis, G., Thoen, J., Leys, J., and Glorieux, C., Temperature-driven mixing-demixing behavior of binary mixtures of the ionic liquid choline bis(trifluoromethylsulfonyl)imide and water, *J. Phys. Chem. B* **113**, 1429–1437 (2009).

29   Łachwa, J., Szydlowski, J., Najdanovic-Visak, V., Rebelo, L.P.N., Seddon, K.R., Ponte, M.N., Esperança, J.M.S.S., and Guedes, H.J.R., Evidence for lower critical solution behavior in ionic liquid solutions, *J. Am. Chem. Soc.* **127**, 6542–6543 (2005).

30   Bates, E.D., Mayton, R.D., Ntai, I., and Davis, J.H., Jr., CO_2 capture by a task-specific ionic liquid, *J. Am. Chem. Soc.* **124**, 926–927 (2002).

31   Endo, T., and Sudo, A., Development of monomers contributing to useful applications to functional materials, *Kobunshi* **58**, 411–416 (2009).

32   Suryakiran, N., Mahesh, K.C., Ramesh, D., Selvam, J.J.P., and Venkateswarlu, Y., N-Benzyloxycarbonylation of amines in the ionic liquid [TPA][L-Pro] as an efficient reaction medium, *Tetrahedron Lett.* **49**, 2607–2610 (2008).

33   Qian, Y.B., Xiao, S.Y., Liu, L., and Wang, Y.M., A mild and efficient procedure for asymmetric Michael additions of cyclohexanone to chalcones catalyzed by an amino acid ionic liquid, *Tetrahedron-Asymmetr* **19**, 1515–1518 (2008).

34   Chen, W., Zhang, Y.Y., Zhu, L.B., Lan, J.B., Xie, R.G., and You, J.S., A concept of supported amino acid ionic liquids and their application in metal scavenging and heterogeneous catalysis, *J. Am. Chem. Soc.* **129**, 13879–13886 (2007).

35   Tang, F., Wu, K.K., Nie, Z., Ding, L., Liu, Q., Yuan, J.B., Guo, M.L., and Yao, S.Z., Quantification of amino acid liquids using liquid chromatography mass spectrometry, *J. Chromatogr. A* **1208**, 175–181 (2008).

36   Minami, I., Watanabe, N., Nanao, H., Mori, S., Fukumoto, K., and Ohno, H., Improvement in the tribological properties of imidazolium-derived ionic liquids by additive technology, *Chem. Lett.* **37**, 300–301 (2008).

37   Zhang, Y.Y., Chen, X., Lan, J.B., You, J.S., and Chen, L.J., Synthesis and biological applications of imidazolium-based polymerized ionic liquid as a gene delivery vector, *Chem. Biol. Drug Des.* **74**, 282–288 (2009).

38   Liu, Q., Wu, K., Tang, F., Yao, L., Yang, F., Nie, Z., and Yao, S., Amino acid ionic liquids as chiral ligands in ligand-exchange chiral separations, *Chemistry* **15**, 9889–9896 (2009).

8 Ionic Liquids Studied at Ultra-High Vacuum

KEVIN R.J. LOVELOCK and PETER LICENCE

School of Chemistry, The University of Nottingham, Nottingham, UK

ABSTRACT

Ionic liquids offer huge opportunities in the ultra-high vacuum (UHV) environment, both from an academic stance of studying their physical and chemical properties and from the industrial standpoint of materials and process development. The low vapour pressures associated with many ionic liquids mean that they can be studied using many of the UHV techniques developed by the surface science community. It has been shown that, at room temperature, ionic liquids evaporate slowly enough such that the condensed phase can be easily studied at UHV without the modifications to experimental kit necessary for molecular liquids. However, at elevated temperatures, evaporation occurs at sufficient rates that the nature of the vapour phase and the thermodynamics of vaporisation can also be studied. The establishment of an ultra-clean, collision-free environment that uniquely combines a liquid-based substrate with a wide range of traditional UHV tools offers significant opportunities in the controlled preparation of advanced materials. In this chapter, we review the rôle that UHV techniques have played in ionic liquid surface science, and highlight the most exciting recent breakthroughs.

8.1 INTRODUCTION

In everyday terms, the word *vacuum* is generally accepted to mean a volume of space that is devoid of matter, such that the measured pressure is less than that of the atmosphere, that is, pressure tends towards zero as the quality of

Ionic Liquids UnCOILed: Critical Expert Overviews, First Edition. Edited by Natalia V. Plechkova and Kenneth R. Seddon.
© 2013 John Wiley & Sons, Inc. Published 2013 by John Wiley & Sons, Inc.

TABLE 8.1   Conversion Factors for Commonly Used Units of Pressure

mbar	Pa	Torr[a]	Bar	atm	% vac
1,013	101,300	760.000	1.01	1.000	0.00
1,000	100,000	750.000	1.00	0.987	1.30
100	10,000	75.000	0.10	0.099	90.10
10	1,000	7.500	0.01	0.010	99.00
1	100	0.750	0.001	0.001	99.90
0.1	10	0.075	1×10^{-4}	1×10^{-4}	99.99
1×10^{-n}	$1 \times 10^{-(n+2)}$	$7.5 \times 10^{-(n+2)}$	$1 \times 10^{-(n+3)}$	$1 \times 10^{-(n+3)}$	
0	0	0	0	0	100.00

[a] 1 Torr = 1 mm Hg.

TABLE 8.2   Characteristics of Systems with Different Pressures

p/Pa	p/mbar	n/m^{-3}	λ^a	Z_{wall}/cm^2 s^{-1} [b]	t_{ML}[c]	Vacuum Quality
10^5	10^3	2.4×10^{25}	68 nm	2.9×10^{23}	3.5 ns	Atmospheric
10^2	1	2.4×10^{22}	68 μm	2.9×10^{20}	3.5 μs	Low
10^{-1}	10^{-3}	2.4×10^{19}	68 mm	2.9×10^{17}	3.5 ms	Medium
10^{-4}	10^{-6}	2.4×10^{16}	68 m	2.9×10^{14}	3.5 s	High
10^{-7}	10^{-9}	2.4×10^{13}	68 km	2.9×10^{11}	3.5 ks	Ultra-high

[a] Estimated for N_2 gas, using a diameter of 3.7 Å.
[b] All estimates at 298 K.
[c] ML = monolayer; a monolayer is assumed to consist of 10^{15} atoms cm^{-2}.

the vacuum improves. The quality of a vacuum is a direct indicator of how closely it approaches so-called perfect vacuum (or free space); however the concept of perfect vacuum, where the measured pressure within a system would be exactly equal to zero, is purely a philosophical scenario and is never encountered in an experimental context. The SI unit used in the measurement of pressure is the Pascal (Pa), which is also the SI unit for its complement; however vacuum is most commonly measured in either mbar or torr. The conversion factors for a range of units commonly used in the description and measurement of vacuum systems are given in Table 8.1.

In simple terms, vacuum systems may be described as a function of how much matter is still present within the system of interest. The residual gas pressure, within a given system, is the primary indicator of vacuum, and is consequently used to differentiate régimes of vacuum quality (see Table 8.2). Ultra-high vacuum (UHV) is defined as the vacuum régime where the residual pressure within the system is in the range of 10^{-8} to 10^{-12} mbar. UHV is achieved generally using either diffusion or turbomolecular pumps, which are capable of routinely achieving a base pressure (the pressure of the system when no experiments are being carried out) of $<10^{-9}$ mbar.

The primary rôle of a vacuum system is to generate a controlled environment in which chemical processes can be investigated in isolation. As such, the aim is to reduce the number density of particles (n) in any given volume, that is, the vacuum chamber. n is clearly related to both the pressure (p) and the temperature (T) of the system under consideration by the kinetic theory of gases. The relationship is commonly written as:

$$p = nkT \tag{8.1}$$

where k is the Boltzmann constant and T is the absolute (or thermodynamic) temperature.

UHV systems offer clear advantages to the experimental scientist, many of which are dependent upon the magnitude of n. The first clear advantage is the relatively large mean free path, λ. For example, for N_2 at 298 K at 10^{-9} mbar, $\lambda = 68$ km, whereas at 10^{-3} mbar, $\lambda = 68$ mm. In practical terms, at 10^{-9} mbar, any molecule that is in the vapour phase will hit the wall of the UHV chamber long before a collision occurs; hence, UHV is often referred to as a collision-free environment in the vapour phase. Therefore, it is safe to neglect intermolecular collisions in the vapour phase at UHV pressures. This occurrence is directly relevant to vapour phase studies of ionic liquids and will be discussed further in Section 8.2. The large magnitude of λ allows the use of techniques that are not possible at higher pressures, that is, those that require probing by, or detection of, particles (electrons, ions, or molecules) rather than photons, which can be detected at any pressure. The second major advantage for UHV science of the low n is the low Z_{wall}, the impingement rate of a gas or vapour on a surface. At 10^{-9} mbar $Z_{wall} \approx 2.9 \times 10^{11}$ cm^{-2} s^{-1}, whereas at 10^{-6} mbar, $Z_{wall} \approx 2.9 \times 10^{14}$ cm^{-2} s^{-1}. Assuming that a monolayer (ML) is 10^{15} atoms cm^{-2} at 10^{-9} mbar it would take almost an hour for a monolayer to form (assuming a sticking probability of unity), whereas at 10^{-6} mbar it would take ≈ 4 seconds. Therefore, for surface science studies which require a clean surface, $p < 10^{-9}$ mbar is required; such estimates are directly relevant to the results discussed in Section 8.3.

The relatively low Z_{wall} also means that vaporisation is more likely to occur at reduced system pressure, as less work is required to overcome collisions of gases or vapours with the surface. At UHV, there are so few collisions that water, for example, sublimes from ice to vapour at ≈ 160 K without forming liquid [1]. Most liquids evaporate at room temperature at UHV in <1 second; therefore, typical molecular liquids cannot be studied in their liquid state at UHV using standard equipment. Throughout the history of UHV science, liquids have generally required specialised, expensive equipment for study [2–5].

Ionic liquids have long been known to have very low vapour pressures. Over the years ionic liquids have been placed in high-vacuum (HV) environments; indeed the salt [N_{2222}][CF_3SO_3], with a melting point of 160 °C, was analysed by Wagner et al. using X-ray photoelectron spectroscopy (XPS) in

1980 [6], and Osteryoung and coworkers used HV to remove hydrogen chloride from chloroaluminate ionic liquids in 1991 [7]. Although the earliest studies appeared some 30 years ago, it should be noted that these applications of vacuum-based technologies are somewhat isolated. The cost and inherent difficulties of maintaining UHV-based systems, which are in general terms experiment specific, mean that many researchers outside of the field of classical surface science simply do not have access to UHV equipment. Similarly, those active in surface science have tended to be "reluctant" to place liquids into ultra-clean UHV chambers, so access to existing equipment has not been simple. Accordingly, the number of groups active in the area of liquid surface science is small. It was not until 2005 that the HV compatibility of ionic liquids was fully acknowledged, leading to the more widespread application of techniques including XPS, ultraviolet photoelectron spectroscopy (UPS), secondary ion mass spectrometry (SIMS), temperature programmed desorption (TPD), scanning electron microscopy (SEM), transmission electron microscopy (TEM), and synchrotron-based experiments [8–12]. There is an ever-expanding range of UHV techniques and experiments that may be used in the investigation of ionic liquid systems; sadly, we do not have the time or space in this chapter to cover each and every one of them in detail. This chapter highlights a number of advances in this evolving field; we have focussed on two key areas of study, the vapour phase and the liquid vacuum interface. Furthermore, in the context of this chapter, we have restricted our discussion to the study of aprotic ionic liquids, where the equilibrium reaction between starting materials and the ionic liquid product lies firmly to the right, that is, minimising thermal decomposition of the ionic liquid yielding volatile molecular starting materials, under reduced pressures or increased temperatures. This point is particularly important when considering the nature of the vapour phase.

8.2 THE VAPOUR PHASE

To fully appreciate current developments and investigations on the vapour phase of ionic liquids, it is necessary to first revisit the history of heating ionic liquids at atmospheric pressure. It had been repeatedly stated for many years that ionic liquids have no vapour pressure, or at least no detectable vapour pressure [13–15]. This received wisdom was based on a large number of studies that showed that heating ionic liquids at atmospheric pressure leads to thermal decomposition rather than vaporisation [16–20]. The products observed in the decomposition of 1,3-dialkylimidazolium ionic liquids are a function of the anion that is present in the ionic liquid before thermolysis [21]. The thermal decomposition temperature, T_d, is typically in the range 400–700 K and is dependent on both the composition of the ionic liquid studied and the method used to determine T_d. As a general statement, data obtained from thermogravimetric analysis (TGA) should be considered with care; readers should

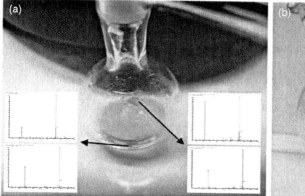

Figure 8.1   (a) First successful thermal vaporisation and recondensation of a pure aprotic ionic liquid, [C$_{10}$mim][NTf$_2$], *in vacuo*. Insets show the ESI-MS spectra of the recondensed liquid and residue, which denotes the absence of decomposition. (b) Snapshot showing a distillation of [C$_4$mim][NTf$_2$]. Note the colour-free distillate. Reprinted with permission from Reference [23]. Copyright 2007, The American Chemical Society. See colour insert.

pay particular attention to the experimental conditions used in measuring TGA data.

The pioneering breakthrough in the study of ionic liquid vapours using vacuum techniques came in 2006, with the demonstration of reduced pressure distillation for a variety of ionic liquids (see Fig. 8.1) [22, 23]. This work was carried out by two common laboratory techniques: (1) using a bulb-to-bulb distillation apparatus, where the distillation pressure p_{dist} of the system was in the range $0.05 < p_{dist} < 8$ mbar (1 mm $> \lambda > 9$ μm), and (2) using a short path sublimation apparatus with pressures ≤ 0.001 mbar ($\lambda > 7$ cm). Under these conditions, and certainly in the case of the bulb-to-bulb distillation, the apparatus is not "collision-free," and the ionic liquid vapour will have undergone a significant number of intermolecular collisions before condensation. For identification of the nature of the vapour phase, significantly lower pressure, that is, UHV, is required.

The vapour phases of a wide range of ionic liquids have since been investigated *in situ* using two groups of techniques, both of which employ UHV conditions: mass spectrometry (MS) [24, 25] and photoelectron spectroscopy (PES) [26]. In all studies carried out to date, the vapour phase was produced by direct heating of the bulk ionic liquid and the resulting vapour was the primary subject of study, rather than any condensate that may be present elsewhere within the vacuum chamber.

Experiments based on both MS and PES suggest that the vapour phases of those ionic liquids studied are composed of neutral ion pairs (NIPs). NIPs have been demonstrated for ionic liquids containing cation components of common ionic liquids, that is, 1,3-dialkylimidazolium, 1,1-dialkylpyrrolidinium, 1-alkylpyridinium, tetraalkylphosphonium, and alkylisouronium [27, 28]. As expected, the vapour phase of dicationic ionic liquids of the type $C^{2+}(A^-)_2$, that is, containing two anions and a single cation bearing two discrete charge centres, for example, two imidazolium functionalities joined by a non-conjugated linker, have been shown to consist of neutral ion triplets (NITs) [29].

MS has given unequivocal evidence regarding the nature of ionic liquids in the vapour phase. In the absence of any ionising source, that is, electron ionisation (EI), or photoionisation, etc., no ions are observed [24, 25]. These data suggest that the vapour phase consists only of neutrals, or clusters of ions that bear no nett charge: if ions were vaporised individually, then they would be detected even without ionisation.

When ionisation is employed (i.e., EI or photoionisation), the intact parent cation, C^+, of the ionic liquid (CA), is clearly observed in positive-mode MS. Figure 8.2a shows the mass spectrum of the vapour of $[C_8mim][NTf_2]$ at 503 K, with the base peak at m/z 195, corresponding to the parent cation $[C_8mim]^+$ [25]. It should be noted that the NIP is not directly observed, the parent cation being the dominant component in the positive ion mode MS (see Figure 8.3). Similarly, in the negative ion mode, Leal et al. have observed intact parent anions (see Figure 8.2b) where the peak at m/z 280 corresponds to intact $[NTf_2]^-$ in $[C_8mim][NTf_2]$ at 503 K [25].

The experimental data highlighted so far have only evidenced the formation of NIPs, and do not consider the existence of higher m/z clusters of the form $[C_mA_{m-1}]^+$ (where $m \geq 2$). Clusters of this nature have been observed for ionic liquids using electrospray ionisation MS [30], matrix-assisted laser desorption (MALDI) [31, 32], and secondary ion mass spectrometry (SIMS) [33]; however, there is no evidence for the existence of higher m/z clusters in the thermalised vapour phase. Cumulatively, these observations strongly support that vaporisation of many ionic liquids gives rise to isolated, intact NIPs or NITs. However, this must only be considered as a "simple rule of thumb," as it completely ignores any opportunities for thermal decomposition prior to evaporation. Indeed, when studies using MS at UHV have been performed, both evaporation and thermal decomposition products were observed when 1-butyl-1-methylpyrrolidinium dicyanamide, $[C_4mpyr][N(CN)_2]$ was heated at UHV [34]. This result suggests that, in the case of ionic liquids where one of the components may be thermally labile or perhaps more nucleophilic in nature [35], decomposition can occur at UHV if the thermodynamics of decomposition are sufficiently favourable. Zaitsau et al. heated 1-butyl-3-methylimidazolium hexafluorophosphate, $[C_4mim][PF_6]$, at a base pressure of $<10^{-3}$ Pa ($\lambda > 6.8$ m); they analysed the condensate *ex situ* by infrared (IR) spectroscopy and found that it did not match that of the ionic liquid starting material (the condensate was not positively identified) [36]. However, Armstrong et al. observed NIPs for $[C_8mim][PF_6]$ using MS when heated in their UHV chamber (base pressure $\approx 5 \times 10^{-8}$ Pa, $\lambda \approx 140$ km) [24]. These

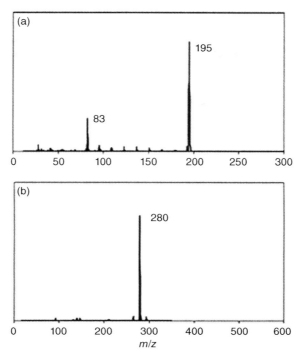

Figure 8.2 Fourier transform ion cyclotron resonance (FTICR) mass spectra of [C$_8$mim][NTf$_2$] evaporating at 503 K. (a) The +ve ion spectrum: the peaks at m/z = 83 and 195 correspond to [Hmim]$^+$ and [C$_8$mim]$^+$, respectively; the peaks in between are due to the fragmentation of the C$_8$ alkyl side chain. (b) The –ve ion spectrum: the peak at m/z = 280 corresponds to [NTf$_2$]$^-$. Reprinted with permission from Reference [25]. Copyright 2007, The American Chemical Society.

$$[C^+A^-] + e^- \rightarrow \{C^+A^-\}^{\bullet+} + 2e^- \qquad (i)$$

$$\{C^+A^-\}^{\bullet+} \rightarrow \{C^+A^{\bullet}\} \qquad (ii)$$

$$\{C^+A^{\bullet}\} \rightarrow C^+ + A^{\bullet} \qquad (iii)$$

Figure 8.3 Proposed mechanism for the collapse of the vapour phase NIP, [C$^+$A$^-$], upon ionisation (i.e., 70 eV incident electron energy) yielding the transient excited radical cation, Reaction (i), which collapses via charge combination, Reaction (ii) and dissociation, Reaction (iii), to yield the parent cation (C$^+$) and the non-charged radical A$^{\bullet}$, which is not detected. Reproduced with permission from Reference [27]. Copyright 2010, The Royal Society of Chemistry.

observations clearly indicate that the system pressure can affect the products of heating, and indeed the nature of the vapour phase produced, even when both systems use reduced pressure. If one considers this statement in terms of λ, it would suggest that for both systems no intermolecular gas phase collisions can occur; however, for the higher pressure system, the significantly larger number of collisions with the ionic liquid surface (i.e., larger Z_{wall}) lead to vaporisation becoming less favourable than thermal decomposition. These observations confirm that *in situ* analysis techniques are the most appropriate in determining the products of heating and therefore the nature of the vapour phase; MS adds the significant advantage that vaporisation and thermal decomposition can be distinguished.

The vaporisation temperature, T_{vap}, for most ionic liquids in UHV studies is of the range 420 K $< T_{vap} <$ 620 K [27, 37, 38]. Data obtained from higher temperatures must be treated with a degree of caution, as it has been shown experimentally that thermal decomposition of ionic liquids is more common at much higher temperatures [17]. Recently, it was demonstrated that ionic liquid vapour can be detected at temperatures $\leq$ 373 K. [C_2mim][NTf_2] vapour was detected as NIPs when $T_{vap} \approx$ 340 K. These experiments suggest that some ionic liquid vapours should, in principle, be detectable at room temperature at UHV if suitably sensitive detection systems are employed [38].

Having identified the nature of the vapour phase, it is possible to measure thermodynamic quantities of the vapour phase. Such studies have been carried out both at UHV and at higher vacuum pressures all the way up to atmospheric pressure [24, 39–44]. The two quantities that have generally been measured are the enthalpy of vaporisation at 298 K, $\Delta_{vap}H_{298}$, and the vapour pressure. $\Delta_{vap}H_{298}$ is of particular interest for the validation of molecular dynamics simulations [45, 46], and for determination of the intermolecular energy of ionic liquids. The vapour pressure is crucial for estimation of the normal boiling temperature and critical temperature, which are vital in the determination and modelling of ionic liquid corresponding states [47].

Excellent agreement for $\Delta_{vap}H_{298}$ has been achieved across a number of techniques, both UHV and non-UHV, for a small range of ionic liquids [$C_n$mim][NTf_2] (where n = 2–8), as shown in Figure 8.4 [24, 39–44]. To date, although $\Delta_{vap}H_{298}$ has been determined for an ever-increasing range of liquids, very few values have been measured using more than one technique. Accordingly, $\Delta_{vap}H_{298}$ needs to be measured for a wide range of ionic liquids using complementary techniques so greater confidence can be developed in the measurements. It must be noted that the majority of the values for $\Delta_{vap}H_{298}$ measured at UHV have been determined using MS-based techniques [24, 27, 29, 34, 38]. At the time of writing, only three values for $\Delta_{vap}H_{298}$ were determined using PES-based methods [26].

The vapour pressure, p_{vp}, has been successfully measured at UHV for only two ionic liquids [38]. In fact, p_{vp} values for only six ionic liquids have been determined in total, indicating the difficulty of such measurements (Table 8.3) [39, 42, 44].

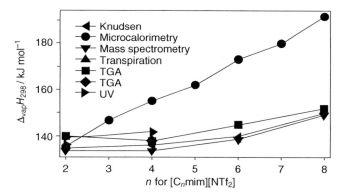

Figure 8.4 Comparison of all literature values of $\Delta_{vap}H_{298}$ for ionic liquids of the general form, [C$_n$mim][NTf$_2$], where n = 2–8 [24, 39–44].

For measurements of both $\Delta_{vap}H_{298}$ and p_{vp}, it is important to bear in mind the benefits of measuring at UHV compared with non-UHV methods: first, the probability of thermal decomposition is greatly reduced; second, the purity of the ionic liquid is likely to be greater; and third, *in situ* vapour phase identification is possible. It is clear that far more vapour pressure measurements are required, primarily to confirm the accuracy of the measurements made to date, and then to extend the range of ionic liquids studied [37].

Although not strictly a probe of the vapour phase, it is possible to distil ionic liquids at UHV and condense any thermally produced vapour phase, thereby facilitating the production of ultra-high-purity liquids for further applications. Using a custom-built distillation apparatus connected to a UHV pumping system, Taylor et al. successfully distilled [C$_2$mim][NTf$_2$] and [C$_8$mim][NTf$_2$], and also mixtures of the two ionic liquids [48]. The distillates recovered were colourless (see Fig. 8.5), and trace impurities of water and 1-methylimidazole had been removed. It is important to note that for 1-methylimidazole, removal occurred at $T \approx 450$ K and $p \approx 6 \times 10^{-6}$ mbar, showing that 1-methylimidazole is strongly bound within the ionic liquid and is unlikely to be removed by heating on a rotary pumped Schlenk line or other conventional degassing apparatus. Preparation of extremely high purity ionic liquids is particularly difficult, especially removal of water. Heating at UHV will remove most volatile impurities such as water, but separation of less volatile components, including Li[NTf$_2$] [49], is more challenging. It is envisaged that distillation at UHV, using a diffusion or turbomolecular pump, will be a viable method for producing relatively small quantities of extremely high-purity ionic liquids.

The distillation of ionic liquids is an interesting and potentially important area of activity; however, it must be stressed that direct distillation of ionic

TABLE 8.3   Measured Values of p_{vp} for Ionic Liquids

Ionic Liquid	$\Delta_{vap}H_{298}$/kJ mol^{-1}	$p_{vp(T)}$/Pa	T_{vap}/k	$p_{vp(298)}$/Pa^a	Technique; Ref.
[C$_2$mim] [EtOSO$_3$]	153 (3)	1.1×10^{-2}	490	6×10^{-12}	UHV-MS [38]
[C$_2$mim] [NTf$_2$]	134 (2)	5.9×10^{-3}	425	4×10^{-9}	UHV-MS [38]
[C$_2$mim] [NTf$_2$]	136.7 (3.4)	0.396	516.2	1×10^{-10}	Transpiration [39]
[C$_2$mim] [NTf$_2$]	135.3 (1.3)	0.0319	464.97	1×10^{-9}	Knudsen [44]
[C$_4$mim] [NTf$_2$]	138^b	20	600	6×10^{-9}	TGA [42]
[C$_4$mim] [NTf$_2$]	136.2 (1.7)	0.0521	477.68	4×10^{-10}	Knudsen [44]
[C$_4$mim] [N(CN)$_2$]	157.2 (1.1)	0.614	463.2	3×10^{-10}	Transpiration [39]
[C$_6$mim] [NTf$_2$]	139.8 (0.8)	0.0141	455.46	5×10^{-10}	Knudsen [44]
[C$_8$mim] [NTf$_2$]	150.0 (0.8)	0.0322	474.44	8×10^{-11}	Knudsen [44]

^{a} Extrapolated to $T = 298$ K from measurements made at higher T.
^{b} $\Delta_l^g C_p = -100$ J K^{-1} mol^{-1} used to obtain $\Delta_{vap}H_{298}$.

liquids is not always possible. Direct chemical reaction and thermal decomposition within the anionic and cationic components of a sample may also compete. Recently, the attempted distillation of tetrafluoroborate ionic liquids was reported. Distillates were recovered in high yield and analytical purity; however, the structure of the distillates was not consistent with that of the original starting material. The products of this reactive distillation were confirmed as "condensation" products, borane-substituted imidazol-2-ylidenes (Fig. 8.6), with the elimination of HF, which was removed from the system by the UHV pumps. Although [BF$_4$]$^-$ has been shown to undergo hydrolysis in the presence of water, it should be noted that this reaction system was completely degassed ($T = 500$–550 K, $p \approx 1$–5×10^{-5} mbar) prior to distillation. Clearly, this reaction is driven by the formation of the strong H–F bond (bond energy ≈ 570 kJ mol^{-1}), assisted by the rapid removal of HF [50].

Clearly, this reaction highlights a rather unusual opportunity that is afforded by the rapid pumping systems that are required to maintain UHV conditions within experimental apparatus. UHV systems are routinely used in the preparation of solid materials, semi-conductors, catalytic surfaces, and the like. However, the use of UHV systems to mediate or control reaction chemistries in the liquid state is entirely uninvestigated: to the best of our knowledge, this example is the first reported.

Figure 8.5 (A) [C$_2$mim][NTf$_2$] collecting in the receiving arm of a UHV distillation apparatus (T = 588 K, p = 5 × 10^{-5} mbar), (B) the ionic liquid as prepared, (C) the distillate, and (D) the residue. Reprinted with permission from Reference [48]. Copyright 2010, The Royal Society of Chemistry.

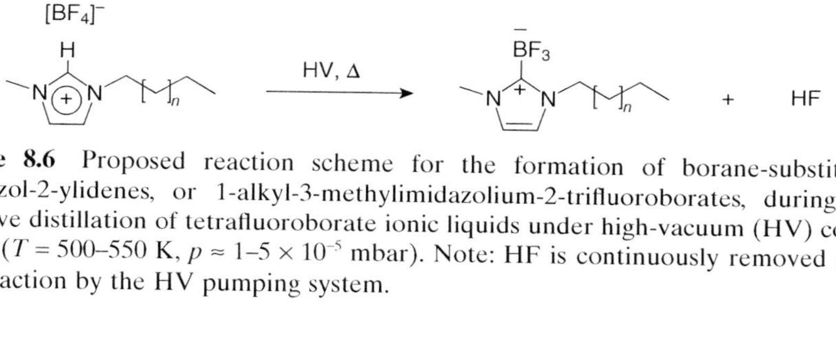

Figure 8.6 Proposed reaction scheme for the formation of borane-substituted imidazol-2-ylidenes, or 1-alkyl-3-methylimidazolium-2-trifluoroborates, during the reactive distillation of tetrafluoroborate ionic liquids under high-vacuum (HV) conditions, (T = 500–550 K, p ≈ 1–5 × 10^{-5} mbar). Note: HF is continuously removed from the reaction by the HV pumping system.

8.3 THE CONDENSED PHASE

UHV studies of ionic liquids in the condensed phase are more diverse than those in the vapour phase, because the range of techniques used to date is much wider. This section will concentrate on areas where either the properties of ionic liquids, or ionic liquid-based systems, have been probed using UHV techniques, particularly where ionic liquids are vital to the experimental method used. Consequently, the characterisation of materials made in ionic liquids at atmospheric pressure will not be considered in this chapter.

As explained previously, salts have been investigated at UHV at room temperature for many years. However, it was only recognised in 2005–2006 that ionic liquids could be studied with the full range of UHV techniques [51–53]. Before moving our discussion onwards, we should perhaps consider the general principles by which UHV-based techniques are used to gather

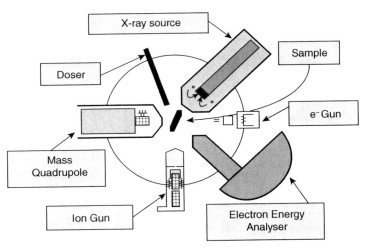

Figure 8.7 A schematic representation of a typical UHV apparatus used for the study of ionic liquid surface chemistry. The system consists of a stainless steel welded bell-jar, evacuated to pressures on the order of 10^{-9} mbar, and equipped with several surface-sensitive techniques such as XPS, Auger electron spectroscopy (AES), ion scattering spectroscopy (ISS), and temperature programmed desorption (TPD). This diagram is not meant to be comprehensive: additional probes are available, and (as a general comment) most systems are custom designed and built to a user's specifications. Diagram adapted with permission from Reference [55]. Copyright 2006, Elsevier.

information about a sample. As a general comment, most UHV techniques involve irradiating a sample with some form of an energetic probe beam (Fig. 8.7). The nature of the incoming probe varies between techniques; typically, it can be a photon of electromagnetic radiation of known energy, for example, X-rays, UV, and IR; a projectile of known properties and energies, for example, neutrons, or ion beams; or electrons of a known energy.

The information returned from experiments is the summation of all outgoing energies or fragments that are generated by the irradiation or collision of the incident probe; typically UHV techniques require the detection of electrons or ions. Such particles have relatively short mean free paths in the condensed phase and, hence, most UHV techniques are termed "surface sensitive." For example, an electron with associated energy of 1200 eV will have a mean free path, in typical organic compounds, of ≈3 nm [54].

Obviously, all incident probes are energetic in nature and penetrate much deeper into the sample than the measured information depth [54]. Consequently, chemical damage can be noted not only in the information depth but also much deeper into the bulk sample. Techniques that involve electrons have been shown to cause more surface damage than photons: irradiation with focussed ion beams also causes irreparable damage to the sample surface [6].

In the case of ionic liquids, the surface is mobile, which in principle, renders liquid samples self-healing as the products either diffuse out of the analysis region or evaporate into the vacuum. For ionic liquids, sample damage has not yet been thoroughly investigated, although it has been noted that metastable atoms, low-energy photons, and electrons with a kinetic energy of $\approx\Delta 10$ eV cause only minor changes to the liquid surface, in contrast to high-energy photons and electrons (>1 keV), which have been shown to cause significant changes [56].

As described in Section 8.1, the low magnitude of Z_{wall} at UHV pressures means that relatively few gas molecules will collide with a substrate's surface at UHV, essentially ensuring an ultra-clean surface is investigated throughout the experiment. Consequently, investigations of ionic liquids at UHV are typically carried out on very high-purity ionic liquids. UHV-based techniques can be employed in the measurement of physical properties of ionic liquids, including for example, $\Delta_{vap}H_{298}$ and p_{vp} (as described in Section 8.2) and surface-related properties including surface potentials, surface tensions [57], and related wetting phenomena [58]. However, to date the most common application of UHV has been the use of PES and related techniques to investigate the structure of both ionic liquids and solutes therein [12].

8.3.1 Photoelectron Spectroscopy and Related Techniques

XPS, UPS, and also the closely related technique, metastable impact electron spectroscopy (MIES), have all been used to investigate the electronic structure of ionic liquids. XPS is typically used to probe the electronic environment of core orbitals, whereas UPS and MIES are used to probe the valence band structure of samples, primarily to support and inform density of states (DOS) calculations.

One of the key advantages offered by XPS is the unambiguous identification of all elements (heavier than helium) within a sample based on the core level binding energies (BE), which are distinct for each atomic species (see Figure 8.8) [59]. More subtly, variations in the local electron density, at a particular atom, can contribute to shifts in the corresponding core levels in XP spectra; such shifts are manifested as binding energy shifts (see Fig. 8.9) [60].

Due to the element-specific nature of XPS, it is possible to confirm sample stoicheiometry and identify impurities that are present in the near-surface region of the sample. It must be noted that the nature of any impurities is limited to those of high molecular mass and low volatility. As described earlier, small molecule impurities, including organic precursors and water, evaporate at room temperature while samples are pumped to UHV. A common impurity, particularly in early XPS studies, was silicon [52, 61], this observation being interpreted as the segregation of silicon, in the form of silicone-based greases left over from synthesis, towards the sample surface (Fig. 8.10). It is expected that the presence of non-volatile surfactants at the surface of ionic liquids will have a significant impact on the surface properties and chemical processes

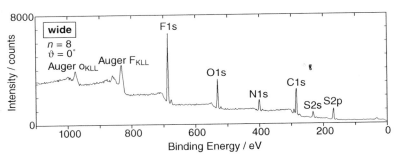

Figure 8.8   XP spectrum (survey scan) of [C$_8$mim][NTf$_2$] recorded at 0° PE emission angle with respect to the surface normal. Reprinted with permission from Reference [59]. Copyright 2009, The American Chemical Society.

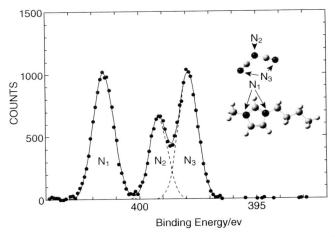

Figure 8.9   XP spectrum of N 1s for [C$_4$mim][N(CN)$_2$]. The spectrum consists of three peaks. The intensity of these peaks is in agreement with the stoicheiometric ratio of N in both the imidazolium cation and the dicyanamide anion. Reprinted with permission from Reference [60]. Copyright 2010, Elsevier.

occurring at that interface. However, no direct evidence has yet been found to support this supposition. Surface-concentrated impurities can be removed from the sample surface by Ar$^+$ bombardment, a routine procedure used by solid surface scientists, where energetic Ar$^+$ is directed at the surface. The collision of the incident Ar$^+$ with the sample surface is sufficient to dislodge surface contaminants, including silicon and adventitious carbon, thereby generating an ultra-clean surface.

At this time, there is just one example of surface segregation by design. Maier et al. published an angle resolved XPS (ARXPS) analysis of a platinum

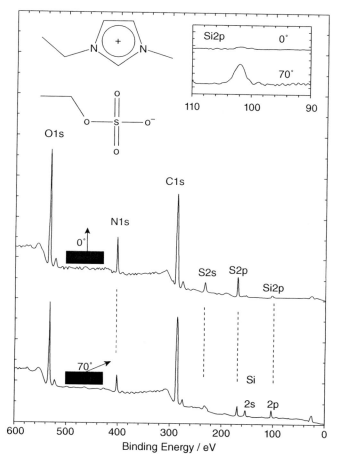

Figure 8.10   Wide scan XP spectrum of [C₂mim][EtOSO₃]; the inset shows evidence of surface segregated silicon-containing impurities; recorded using non-monochromatised Al Kα radiation. Reprinted with permission from Reference [61]. Copyright 2006, Oldenbourg Wissenschaftsverlag.

complex, [Pt(NH₃)₄]Cl₂, dissolved in [C₂mim][EtOSO₃] [62]. ARXPS data revealed that the metal complex was highly enriched at the surface and that chloride was not observed at the surface. These observations were attributed to the different polarisability (mainly size) of the two ions. However, in the case of a homogeneous solution (i.e., without residual undissolved salt), the XPS intensity of the metal complex rapidly decreased over exposure time, indicating beam damage.

In a typical XPS experiment, irradiation of the sample by X-rays causes the emission of photoelectrons from the sample, leaving the sample positively charged as a consequence of the loss of electrons. In the case of earthed conducting samples, for example, metals, their conductivity is sufficiently large to ensure that any positive charge is effectively neutralised within the time scale of the XPS experiment and no change in binding energy (BE) is observed. Under these conditions, the spectrometer's calibration itself should be sufficient to acquire reliable and stable binding energies. However, when semi- or non-conducting samples are irradiated with X-rays, the positive charge generated at the surface cannot be neutralised quickly enough and a positive potential is created between the sample surface and the instrument stage/sample holder. This process is called surface charging and is manifested in XP spectra as a shift in the BE of the photoelectron peaks to higher BEs, often by several eV. Although conducting, ionic liquids are intermediate between metallic conductors and insulators, that is, they conduct, but poorly when compared with metallic systems. This phenomenon may be explained by significant differences in model charge conduction/migration mechanisms in the two very different systems. Charging becomes more pronounced as the viscosity of the substrate increases, or as the analysis temperature approaches the glass transition temperature (T_g) of the ionic liquid being studied (Fig. 8.11) [33].

Over the past five years, independent researchers have noted the apparent variability of BEs recorded for a range of common ionic liquids, highlighting

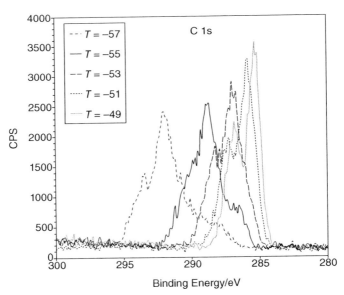

Figure 8.11 High-resolution XPS spectra of [C₂mim][EtOSO₃] showing the measured BE of the C 1s photoemission as a function of temperature (°C). At lower temperatures, when the sample becomes frozen, surface charging occurs, and the peak shifts to a higher binding energy. Reprinted with permission from Reference [33]. Copyright 2006, The American Chemical Society.

the requirement for a system of charge correction, or energy scale calibration [33, 59, 63–66]. The simplest method of charge correction is internal charge referencing using a common chemical group that does not change electronic environment across a range of different samples. It has been shown that for $[C_n mim]^+$ ionic liquids, where $n \geq 8$, the C atoms in the second, longer, alkyl group can be used as an internal charge reference, allowing direct comparisons of BE across structurally similar ionic liquids and solutions [66]. However, more appropriate charge reference methods are required for $[C_n mim]^+$ ionic liquids, where $n < 8$, and also for ionic liquids containing functionalised side chains.

Based on this developing knowledge of sample charging, it has been shown that for ionic liquids of the general formula, $[C_8 mim][A]$, after charge referencing, small differences in BE can be observed and used to probe the interaction between the anion and cation [63, 66]. As a general comment, reduced BEs are indicative of an increase in electron density at the atom being studied, and vice versa. These data indicate that the nature of the anion has an impact on the charge density and electronic structure of the imidazolium ring. For ionic liquids containing small, highly charged anions, such as halides and nitrate, the carbon and nitrogen atoms in the imidazolium ring have lower BEs relative to ionic liquids containing larger anions such as $[NTf_2]^-$ and $[PF_3(C_2F_5)_3]^-$ (or FAP).

These observations are supported by both nuclear magnetic resonance (NMR) experiments and theoretical calculations based on the same ionic liquid systems. This work appears to contain the first directly measured experimental data that give insight into the tunable nature of the Coulombic interactions that exist between the two components of an ionic liquid. Unsurprisingly, this interaction is dependent on the structure and charge density of the ions of concern (see Fig. 8.12). This study reveals the power and potential of XPS as a tool for further investigation of the physical chemistry of ionic liquids.

XPS can also be used to probe the electronic environment of solutes in ionic liquids. However, a major drawback of this technique is that the solute must have a low enough vapour pressure to be stable at room temperature at UHV. To date, the primary focus of research has been on the variable oxidation states of dissolved metal complexes and catalysts [52, 62, 67]. In the case of solute-based studies, the determination of accurate BE data for comparison purposes is vital; hence, precise charge referencing is crucial if chemistry is to be separated from surface-related charging. Building on simple monitoring studies, Licence and coworkers have developed *in situ* electrochemical techniques that allow the oxidation state of a redox active solute to be controlled or switched in real time during electrochemical XPS (EC-XPS) experiments. Initial studies focussed on oxidation state changes of $[FeCl_4]^-$ in $[C_2 mim][EtOSO_3]$ [68]; a peak-fitting model was used to differentiate between Fe(II) and Fe(III) during extended Coulometric experiments that allowed XPS data to be collected in real time. EC-XPS was later applied to the study of copper dissolution and plating in 1–(methyl ethanoate)-4-methylpyridinium bis{(trifluoromethyl)sulfonyl}amide [69] (see Fig. 8.13). The

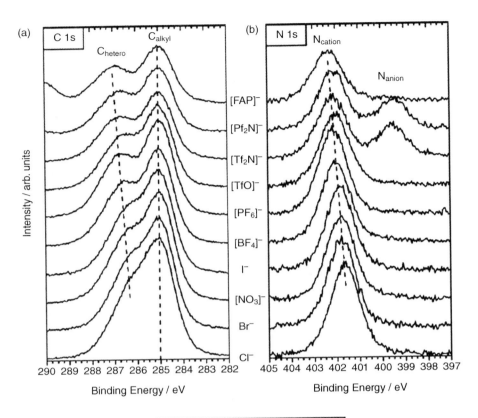

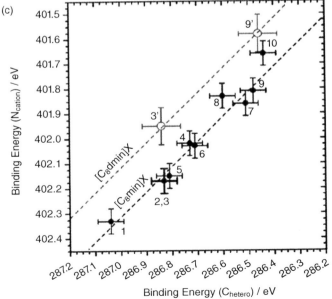

Figure 8.12   Detailed XP spectra of (a) the C 1s, and (b) the N 1s regions for a series of [C_8mim]$^+$ ionic liquids (the dashed lines are guide to the eyes for the BE changes of the imidazolium ring signals). (c) Correlation of BE positions for C_{hetero} and N_{cation} for 10 [C_8mim]$^+$ ionic liquids (●) and 2 [C_8dmim]$^+$ ionic liquids (○). The key for (c) is: 1 [$PF_3(C_2F_5)_3$]$^-$ or [FAP]; 2 [NPf$_2$]$^-$; 3,3′ [NTf$_2$]$^-$; 4 [OTf]$^-$; 5 [PF$_6$]$^-$; 6 [BF$_4$]$^-$; 7 I$^-$; 8 [NO$_3$]$^-$; 9,9′ Br$^-$; 10 Cl$^-$. Reprinted with permission from Reference [63]. Copyright 2010, Wiley VCH.

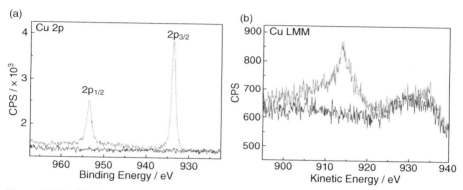

Figure 8.13   High-resolution XPS scans of dissolved Cu in the ionic liquid 1–(methyl ethanoate)-4-methylpyridinium bis{(trifluoromethyl)sulfonyl}amide. (a) Cu 2p and (b) Cu LMM Auger regions taken before (dark) and after (light) 120 minutes *in situ* electrolysis at +1.8 V (vs. stub). The working electrode (Cu source) is a 1-mm Cu wire and a Mo stub is acting as the counter/reference electrode. Reprinted with permission from Reference [69]. Copyright 2010, The Royal Society of Chemistry.

in situ XPS technique was used to decouple and characterise two distinct diffusion mechanisms, a slow bulk diffusion process, and a surface diffusion process which was much more rapid. There is enormous, and at present unexploited, potential for reaction monitoring in ionic liquids using UHV-based techniques.

As mentioned earlier, many UHV-based techniques are sensitive to chemistry and composition in the near-surface region. Consequently, the composition of the ionic liquid/vacuum interface can be investigated using a wide range of UHV techniques including: ARXPS [59, 64, 65, 70], MIES (plus DOS calculations) [71], high-resolution electron energy loss spectroscopy (HREELS), [56] and O(^{3}P) atoms as a chemical probe [72–74]. As highlighted earlier, a major advantage of using UHV for surface studies, aside from access to a wide variety of techniques, is the increased confidence in the cleanness of the surface due to lower values of Z_{wall}. A thorough understanding and knowledge of surface composition is vital for many areas of materials-based science, particularly catalysis and electrochemistry/sensors [8, 9]. Using sum frequency generation (SFG) spectroscopy, Baldelli and coworkers showed that water has a significant impact on the surface structure of [C$_4$mim][NTf$_2$] [75], an observation that reinforces the fact that high purity ionic liquids are vital for surface studies, and the UHV environment provides the best possible environment for maintaining high-purity ionic liquids. The interfaces of ionic liquids are discussed in more detail in Chapters 2 and 3 of this book. UHV data, predominantly from ARXPS, suggest that alkyl chain substituents are located at the vacuum interface, with a layer of anions and cationic head groups just below the "surface." These findings are in general agreement with complementary non-UHV techniques including SFG spectroscopy and X-ray reflectivity.

The preparation of ionic liquid surfaces by evaporation has many advantages over non-UHV based techniques, especially in the study of more volatile samples [55]. Researchers from the University of Erlangen-Nuremberg have demonstrated that ionic liquid surfaces of controlled thickness can be prepared via direct vaporisation onto a suitable sample holder whilst at UHV (see Fig. 8.14). Ionic liquid deposition has been studied using ARXPS [76], infrared reflection absorption spectroscopy (IRAS, also known as reflection absorption infrared spectroscopy, RAIRS) [77, 78] and time-of-flight secondary ion mass spectrometry (TOF-SIMS) [79–81]. To date, the aims of these studies have been twofold: primarily, to study ionic liquids at the solid/liquid interface, which is difficult using many surface science techniques due to the small mean free path of electrons/ions in the liquid (see Chapter 2 for more details on the solid/ionic liquid interface), and secondarily, for preparing model systems of more complex catalyst systems. At present, as with many areas discussed thus far, this area is in its infancy, and much potential exists for investigation and discovery.

8.3.2   Microscopy and Imaging

Electron microscopy offers the opportunity to investigate materials and composites on dimensions approaching that of small molecules. Both SEM and transmission electron microscopy (TEM) have been used for the characterisation of materials prepared at atmospheric pressure using ionic liquids. However, as this is not the primary focus of this chapter, the interested reader is directed to the excellent review on the use of ionic liquids and nanoparticles by Dupont and coworkers [82].

One of the simplest and most elegant UHV applications of ionic liquids to date is the use of low viscosity ionic liquids, as coatings to improve conductivity of insulating samples for SEM analysis (see Fig. 8.15). Typically, insulating samples require precoating with a conducting material; classically, this is achieved by the evaporation of gold onto the sample prior to observation. Although efficient, the deposition of gold can lead to loss of fine morphological/topographical detail; thermally labile materials can become damaged during the vaporisation process; and samples are permanently modified, as the gold

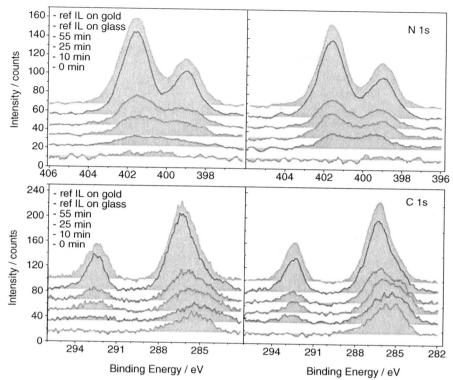

Figure 8.14   N 1s and C 1s XP spectra of [C₂mim][NTf₂] deposited on a glass substrate, at emission angles of 0° (left) and 70° (right). Spectra are shown for deposition times of 0, 10, 25, and 55 minutes and for a thick layer ("ref IL on glass"). Also shown are the corresponding spectra of the ionic liquid in the reservoir ("ref IL on gold"). In the case of the ionic liquid films on the glass substrate, charging was observed. To compensate for this, and to allow a comparison of the different layers, the spectra were shifted uniformly to lower binding energies (with maximum deviations of ±0.5 eV) by the following values (0°/70°): 0 minute: 9.2/6.9 eV; 10 minutes: 8.7/7.0 eV; 25 minutes: 9.2/7.1 eV; 55 minutes: 8.5/6.6 eV; thick: 0.5/0.4 eV. Note that the N 1s and C 1s spectra have been smoothed. Modified with permission from Reference [76]. Copyright 2008, Wiley-VCH.

may not be removed post observation. Ionic liquids can be applied at room temperature, wet the entire surface evenly, and can be washed off with solvent to allow sample recovery [83, 84].

The development, by Kuwabata and coworkers, of *in situ* electrochemical SEM (ECSEM) has added the ability to monitor electrochemical generation of new materials as reactions proceed. To date, studies have focussed on the growth of polypyrrole films [85] and metal nanoparticles/dendrites from ionic liquid-based electrolytes under electrochemical control [86] and electron

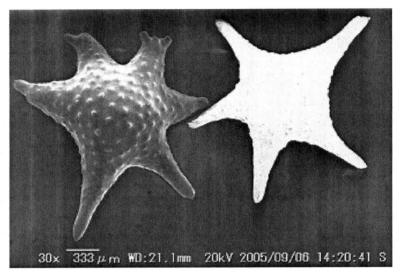

Figure 8.15 SEM image of two grains of star sand. The left grain was dipped in [C₂mim][NTf₂] and the right was subjected to no pre-treatment prior to observation. Reprinted with permission from Reference [84]. Copyright 2006, The Chemical Society of Japan.

beam irradiation [87, 88] (see Fig. 8.16). Nanoparticle synthesis in ionic liquids can be crudely differentiated into two discrete groups: those where metal is deposited onto the surface of the ionic liquid, or those where the source of metal is dissolved in the ionic liquid, either as an ionic liquid component or as a solute. Scherson and coworkers used an *in situ* UHV system for the electrodeposition of aluminium on tungsten and gold electrodes from chloroaluminate ionic liquids, which, although not air- and water-stable, can be successfully employed for deposition under an inert atmosphere of UHV [89]. The working electrodes were characterised *in situ,* both before and after deposition, using Auger electron spectroscopy. The use of UHV ensured that the working electrode was atomically clean prior to deposition; it was concluded that deposition at UHV gave distinct results from those obtained in an inert gas environment (glove box) when the working electrode was cleaned using chemical methods prior to deposition.

Real-time *in situ* monitoring of experiments using UHV-based techniques is clearly very useful in terms of studying reaction outcomes and mechanism. However, one must always ensure that the probe method used does not directly affect the outcome of the experiment being monitored. This is particularly relevant for techniques where the incident probe is high in energy, for example, XPS, where prolonged exposure to the X-ray beam has been shown to cause damage to the ionic liquid itself [64, 90]. Imaging experiments can

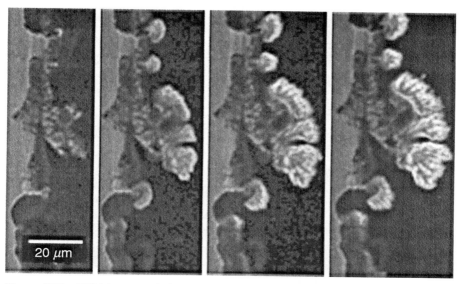

Figure 8.16 SEM images of silver dendrite growth polarised at -1.14 V versus Ag/Ag(I) for (left to right) 15, 60, 120, and 180 minutes. Reprinted with permission from Reference [86]. Copyright 2009, Elsevier.

also be conducted using software mapping modes that are common on many surface-related analysis techniques. Imaging may be used to investigate the distribution of solutes, or particularly distinct changes in chemistry across a surface; however, at the time of writing, this opportunity had yet to be realised. TOF-SIMS imaging has been carried out on the surface of frozen ionic liquid samples [33]. TOF-SIMS mapping gave no indication of changes of chemistry across the frozen surface of $[C_2mim][EtOSO_3]$; however, when a focussed Ga^+ ion beam (FIB), used as a primary-ion source, is scanned across the frozen sample, images are written on the sample surface (see Fig. 8.17) [91].

The darker areas, characteristic of a reduced flux of exiting secondary electrons, correspond to the original path of the incident ion beam. The reduction in brightness is proposed to be due to the development of charge at the surface. Close examination of MS data from across a light/dark junction shows that there is negligible damage to the surface, minimal topographical change, and undetectable incorporation of Ga^+ within the sample surface. The patterns are easily erased by melting, or by bathing the surface with a low-energy electron flux. In terms of characterisation of the ionic liquids themselves, the formation of higher molecular mass cluster ions was observed when glassy samples were irradiated by the Ga^+ beam, that is, clusters of the type $[C(CA)]^+$ are observed in the positive mode, and clusters of the type $[A(CA)]^-$ are observed in the negative mode.

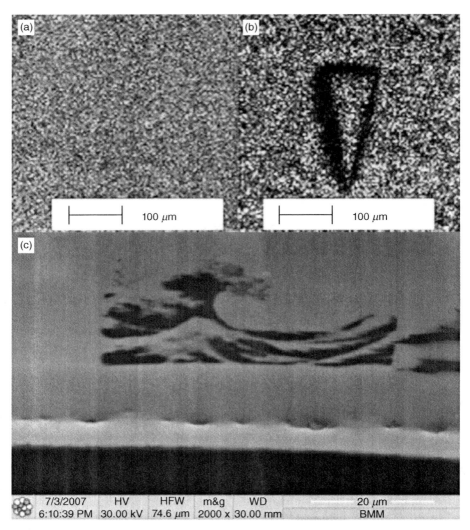

Figure 8.17 (A) Total negative ion SIMS map of [C$_2$mim][EtOSO$_3$] presented as a liquid ($T > 188$ K). The sample appears as a homogeneous surface with no contrasting areas observed. (B) Total negative ion SIMS map of [C$_2$mim][EtOSO$_3$] presented as a glass ($T < 188$ K). The sample shows an arrowhead-shaped contrast area against the lighter background. (C) Higher resolution charge-based image written using a high-resolution focussed ion beam (FIB): the image was prepared on a glassy surface of [C$_4$mim]Cl at room temperature ($T = 297$ K), the image was erased by flooding the surface with a flux of low energy electrons confirming a charge-based phenomena. This image is based upon a digital copy of "The Great Wave off Kanagawa" by Hokusai, the total width of image C is approximately 50 μm. (A) and (B) are reprinted with permission from Reference [91]. Copyright 2007, Wiley-VCH.

8.4   CONCLUSIONS AND OPPORTUNITIES FOR THE FUTURE

The past five years have been an exciting time in the field of UHV-based surface science; the award of the Nobel Prize in 2007 to Prof. Gerhard Ertl has once again focussed interest and stimulated activity in this field of study. It has also become clear over recent years that an intimate understanding of the interfacial areas surrounding ionic liquids will be crucial to the successful scale-up and application of ionic liquid-based processes in the areas of catalysis and advanced materials. If one considers any chemical processes, or materials-based devices, that employ ionic liquids, it becomes clear that phase boundaries, interfaces, and near-interfacial areas dominate both function and performance. The study of the ionic liquid/vacuum interface and the solid/ionic liquid interface is the subject of a rapidly expanding literature base. UHV-based techniques can provide unique data regarding chemical state identification, near-surface composition, and indeed the orientational structure of the ionic components at the surface.

As of September 2010, the development of ionic liquid surface science has been strong. UHV techniques have enabled both the characterisation of ionic liquids in the vapour phase and facilitated the preparation of structurally precise, ultra-clean surfaces on which the most fundamental of interactions and reactions can be studied. However, the range of UHV techniques that have been used to study both the vapour and the condensed phase is modest. There is an almost endless range of laboratory experiments that could be used to give insight into topics as diverse as thermophysical properties to inter-ionic interactions and solute solubility. High-intensity energy sources, that is, synchrotrons and linear accelerators, offer another dimension of experiments at both atmospheric pressure [92] and indeed under UHV. Until now, the application of synchrotron-based experiments is very limited: future opportunities are just mind-boggling. The future offers boundless opportunities for those brave enough to trust new physical data on volatility that may otherwise preclude investigation in a particular UHV instrument or facility.

Until now, many of those active in the field of liquid surface science have been fortunate enough to know a traditional solid-surface scientist that is either incredibly trusting of a "solution chemist," or genuinely intrigued about studying a new class of liquids that does not evaporate at room temperature whilst exposed to UHV. Either way, studies carried out to date have typically been performed on instruments and vacuum chambers that were designed for use with solid samples. In many cases, the orientation of the sample was simply decided by the geometry of the stainless steel bell jar and its appended flanges. As a consequence, experiment geometries may not be ideal for the study of liquids that may flow off the sample holder if inclined such that a particular source or detector may be given line of sight. The reconfiguration of experiments, particularly avoiding upright sample orientations, is required. The construction of a bespoke UHV apparatus for ionic liquid study, although costly, will ensure that all experimental opportunities may be realised in an efficient way. Soon, the complete toolbox of UHV-based surface science methods will be available to allow the investigation of ionic liquid surfaces with *atomic-level accuracy*.

REFERENCES

1 Henderson, M.A., The interaction of water with solid surfaces: fundamental aspects revisited, *Surf. Sci. Rep.* **46**, 5–308 (2002).

2 Fellner-Feldegg, H., Siegbahn, H., Asplund, L., Kelfve, P., and Siegbahn, K., ESCA applied to liquids. 4. Wire system for ESCA measurements on liquids, *J. Electron. Spectrosc.* **7**, 421–428 (1975).

3 Siegbahn, H., Svensson, S., and Lundholm, M., A new method for ESCA studies of liquid-phase studies, *J. Electron. Spectrosc.* **24**, 205–213 (1981).

4 Wilson, K.R., Rude, B.S., Catalano, T., Schaller, R.D., Tobin, J.G., Co, D.T., and Saykally, R.J., X-ray spectroscopy of liquid water microjets, *J. Phys. Chem. B* **105**, 3346–3349 (2001).

5 Lundholm, M., Siegbahn, H., Holmberg, S., and Arbman, M., Core electron-spectroscopy of water solutions, *J. Electron. Spectrosc.* **40**, 163–180 (1986).

6 Wagner, C.D., Zatko, D.A., and Raymond, R.H., Use of the oxygen KLL Auger lines in identification of surface chemical-states by electron-spectroscopy for chemical analysis, *Anal. Chem.* **52**, 1445–1451 (1980).

7 Noel, M.A.M., Trulove, P.C., and Osteryoung, R.A., Removal of protons from ambient-temperature chloroaluminate ionic liquids, *Anal. Chem.* **63**, 2892–2896 (1991).

8 Aliaga, C., Santos, C.S., and Baldelli, S., Surface chemistry of room-temperature ionic liquids, *Phys. Chem. Chem. Phys.* **9**, 3683–3700 (2007).

9 Santos, C.S., and Baldelli, S., Gas-liquid interface of room-temperature ionic liquids, *Chem. Soc. Rev.* **39**, 2136–2145 (2010).

10 Steinrück, H.-P., Surface science goes liquid! *Surf. Sci.* **604**, 481–484 (2010).

11 Torimoto, T., Tsuda, T., Okazaki, K., and Kuwabata, S., New frontiers in materials science opened by ionic liquids, *Adv. Mater.* **22**, 1196–1221 (2010).

12 Lovelock, K.R.J., Villar-Garcia, I.J., Maier, F., Steinrück, H.-P., and Licence, P., Photoelectron spectroscopy of ionic liquid-based interfaces, *Chem. Rev.* **110**, 5158–5190 (2010).

13 Earle, M.J., and Seddon, K.R., Ionic liquids. Green solvents for the future, *Pure Appl. Chem.* **72**, 1391–1398 (2000).

14 Rogers, R.D., and Seddon, K.R., Ionic liquids—solvents of the future? *Science* **302**, 792–793 (2003).

15 Seddon, K.R., Ionic liquids for clean technology, *J. Chem. Technol. Biot.* **68**, 351–357 (1997).

16 Awad, W.H., Gilman, J.W., Nyden, M., Harris, R.H., Sutto, T.E., Callahan, J., Trulove, P.C., DeLong, H.C., and Fox, D.M., Thermal degradation studies of alkyl-imidazolium salts and their application in nanocomposites, *Thermochim. Acta* **409**, 3–11 (2004).

17 Baranyai, K.J., Deacon, G.B., MacFarlane, D.R., Pringle, J.M., and Scott, J.L., Thermal degradation of ionic liquids at elevated temperatures, *Aust. J. Chem.* **57**, 145–147 (2004).

18 Chan, B.K.M., Chang, N.H., and Grimmett, M.R., Synthesis and thermolysis of imidazole quaternary salts, *Aust. J. Chem.* **30**, 2005–2013 (1977).

19 Huddleston, J.G., Visser, A.E., Reichert, W.M., Willauer, H.D., Broker, G.A., and Rogers, R.D., Characterization and comparison of hydrophilic and hydrophobic room temperature ionic liquids incorporating the imidazolium cation, *Green Chem.* **3**, 156–164 (2001).

20 Veith, H.J., Mass spectrometry of ammonium and iminium salts, *Mass Spectrom. Rev.* **2**, 419–446 (1983).

21 Kroon, M.C., Buijs, W., Peters, C.J., and Witkamp, G.J., Quantum chemical aided prediction of the thermal decomposition mechanisms and temperatures of ionic liquids, *Thermochim. Acta* **465**, 40–47 (2007).

22 Earle, M.J., Esperança, J.M.S.S., Gilea, M.A., Lopes, J.N.C., Rebelo, L.P.N., Magee, J.W., Seddon, K.R., and Widegren, J.A., The distillation and volatility of ionic liquids, *Nature* **439**, 831–834 (2006).

23 Rebelo, L.P.N., Lopes, J.N.C., Esperança, J., Guedes, H.J.R., Lachwa, J., Najdanovic-Visak, V., and Visak, Z.P., Accounting for the unique, doubly dual nature of ionic liquids from a molecular thermodynamic, and modeling standpoint, *Acc. Chem. Res.* **40**, 1114–1121 (2007).

24 Armstrong, J.P., Hurst, C., Jones, R.G., Licence, P., Lovelock, K.R.J., Satterley, C.J., and Villar-Garcia, I.J., Vapourisation of ionic liquids, *Phys. Chem. Chem. Phys.* **9**, 982–990 (2007).

25 Leal, J.P., Esperança, J.M.S.S., Da Piedade, M.E.M., Canongia Lopes, J.N., Rebelo, L.P.N., and Seddon, K.R., The nature of ionic liquids in the gas phase, *J. Phys. Chem. A* **111**, 6176–6182 (2007).

26 Strasser, D., Goulay, F., Kelkar, M.S., Maginn, E.J., and Leone, S.R., Photoelectron spectrum of isolated ion-pairs in ionic liquid vapor, *J. Phys. Chem. A* **111**, 3191–3195 (2007).

27 Deyko, A., Lovelock, K.R.J., Corfield, J.A., Taylor, A.W., Gooden, P.N., Villar-Garcia, I.J., Licence, P., Jones, R.G., Krasovskiy, V.G., Chernikova, E.A., and Kustov, L.M., Measuring and predicting $\Delta_{vap}H_{298}$ values of ionic liquids, *Phys. Chem. Chem. Phys.* **11**, 8544–8555 (2009).

28 Deyko, A., Adsorption and desorption studies of ionic liquids. PhD thesis, University of Nottingham (2010).

29 Lovelock, K.R.J., Deyko, A., Corfield, J.A., Gooden, P.N., Licence, P., and Jones, R.G., Vaporisation of a dicationic ionic liquid, *ChemPhysChem* **10**, 337–340 (2009).

30 Dyson, P.J., Khalaila, I., Luettgen, S., McIndoe, J.S., and Zhao, D.B., Direct probe electrospray (and nanospray) ionization mass spectrometry of neat ionic liquids, *Chem. Commun.*, 2204–2205 (2004).

31 Armstrong, D.W., Zhang, L.K., He, L.F., and Gross, M.L., Ionic liquids as matrixes for matrix-assisted laser desorption/ionization mass spectrometry, *Anal. Chem.* **73**, 3679–3686 (2001).

32 Liu, J.F., Jonsson, J.A., and Jiang, G.B., Application of ionic liquids in analytical chemistry, *TRAC-Trend, Anal. Chem.* **24**, 20–27 (2005).

33 Smith, E.F., Rutten, F.J.M., Villar-Garcia, I.J., Briggs, D., and Licence, P., Ionic liquids in vacuo: analysis of liquid surfaces using ultra-high-vacuum techniques, *Langmuir* **22**, 9386–9392 (2006).

34 Emel'yanenko, V.N., Verevkin, S.P., Heintz, A., Corfield, J.A., Deyko, A., Lovelock, K.R.J., Licence, P., and Jones, R.G., Pyrrolidinium-based ionic liquids. 1-butyl-1-methyl pyrrolidinium dicyanoamide: thermochemical measurement, mass spectrometry, and ab initio calculations, *J. Phys. Chem. B* **112**, 11734–11742 (2008).

35 Lovelock, K.R.J., Ionic liquids: into the vapour phase. PhD thesis, University of Nottingham (2008).

36 Zaitsau, D.H., Paulechka, Y.U., and Kabo, G.J., The kinetics of thermal decomposition of 1-butyl-3-methylimidazolium hexafluorophosphate, *J. Phys. Chem. A* **110**, 11602–11604 (2006).

37 Esperança, J.M.S.S., Canongia Lopes, J.N., Tariq, M., Santos, L.M.N.B.F., Magee, J.W., and Rebelo, L.P.N., Volatility of aprotic ionic liquids—a review, *J. Chem. Eng. Data* **55**, 3–12 (2010).

38 Lovelock, K.R.J., Deyko, A., Licence, P., and Jones, R.G., Vaporisation of an ionic liquid near room temperature, *Phys. Chem. Chem. Phys.* **12**, 8893–8901 (2010).

39 Emel'yanenko, V.N., Verevkin, S.P., and Heintz, A., The gaseous enthalpy of formation of the ionic liquid 1-butyl-3methylimidazolium dicyanamide from combustion calorimetry, vapor pressure measurements, and ab initio calculations. *J. Am. Chem. Soc.* **129**(13), 3930–3937 (2007).

40 Luo, H.M., Baker, G.A., and Dai, S., Isothermogravimetric determination of the enthalpies of vaporization of 1-alkyl-3-methylimidazolium ionic liquids, *J. Phys. Chem. B* **112**, 10077–10081 (2008).

41 Santos, L.M.N.B.F., Canongia Lopes, J.N., Coutinho, J.A.P., Esperança, J.M.S.S., Gomes, L.R., Marrucho, I.M., and Rebelo, L.P.N., Ionic liquids: first direct determination of their cohesive energy, *J. Am. Chem. Soc.* **129**, 284–285 (2007).

42 Seeberger, A., Andresen, A.K., and Jess, A., Prediction of long-term stability of ionic liquids at elevated temperatures by means of non-isothermal thermogravimetrical analysis, *Phys. Chem. Chem. Phys.* **11**, 9375–9381 (2009).

43 Wang, C.M., Luo, H.M., Li, H.R., and Dai, S., Direct UV-spectroscopic measurement of selected ionic-liquid vapors, *Phys. Chem. Chem. Phys.* **12**, 7246–7250 (2010).

44 Zaitsau, D.H., Kabo, G.J., Strechan, A.A., Paulechka, Y.U., Tschersich, A., Verevkin, S.P., and Heintz, A., Experimental vapor pressures of 1-alkyl-3-methylimidazolium bis(trifluoromethylsulfonyl) imides and a correlation scheme for estimation of vaporization enthalpies of ionic liquids, *J. Phys. Chem. A* **110**, 7303–7306 (2006).

45 Köddermann, T., Paschek, D., and Ludwig, R., Molecular dynamic simulations of ionic liquids: a reliable description of structure, thermodynamics and dynamics, *ChemPhysChem.* **8**, 2464–2470 (2007).

46 Köddermann, T., Paschek, D., and Ludwig, R., Ionic liquids: dissecting the enthalpies of vaporization, *ChemPhysChem.* **9**, 549–555 (2008).

47 Weiss, V.C., Guggenheim's rule and the enthalpy of vaporization of simple and polar fluids, molten salts, and room temperature ionic liquids, *J. Phys. Chem. B* **114**, 9183–9194 (2010).

48 Taylor, A.W., Lovelock, K.R.J., Deyko, A., Licence, P., and Jones, R.G., High vacuum distillation of ionic liquids and separation of ionic liquid mixtures, *Phys. Chem. Chem. Phys.* **12**, 1772–1783 (2010).

49 Leal, J.P., Da Piedade, M.E.M., Canongia Lopes, J.N., Tomaszowska, A.A., Esperança, J.M.S.S., Rebelo, L.P.N., and Seddon, K.R., Bridging the gap between ionic liquids and molten salts: group 1 metal salts of the bistriflamide anion in the gas phase, *J. Phys. Chem. B* **113**, 3491–3498 (2009).

50 Taylor, A.W., Lovelock, K.R.J., Jones, R.G., and Licence, P., Borane-substituted imidazol-2-ylidenes: syntheses in vacuo, *Dalton Trans.* **40**, 1463–1470 (2011).

51 Caporali, S., Bardi, U., and Lavacchi, A., X-ray photoelectron spectroscopy and low energy ion scattering studies on 1-butyl-3-methyl-imidazolium bis(trifluoromethane) sulfonimide, *J. Electron. Spectrosc.* **151**, 4–8 (2006).

52 Smith, E.F., Villar Garcia, I.J., Briggs, D., and Licence, P., Ionic liquids in vacuo; solution-phase X-ray photoelectron spectroscopy, *Chem. Commun.*, 5633–5635 (2005).

53 Yoshimura, D., Yokoyama, T., Nishi, T., Ishii, H., Ozawa, R., Hamaguchi, H., and Seki, K., Electronic structure of ionic liquids at the surface studied by UV photoemission, *J. Electron. Spectrosc.* **144**, 319–322 (2005).

54 Roberts, R.F., Allara, D.L., Pryde, C.A., Buchanan, D.N.E., and Hobbins, N.D., Mean free path for inelastic scattering of 1.2 kev electrons in thin poly (methylmethacrylate) films, *Surf. Interf. Anal.* **2**, 5–10 (1980).

55 Ma, Z., and Zaera, F., Organic chemistry on solid surfaces, *Surf. Sci. Rep.* **61**, 229–281 (2006).

56 Krischok, S., Eremtchenko, M., Himmerlich, M., Lorenz, P., Uhlig, J., Neumann, A., Öttking, R., Beenken, W.J.D., Höfft, O., Bahr, S., Kempter, V., and Schaefer, J.A., Temperature-dependent electronic and vibrational structure of the 1-ethyl-3-methylimidazolium bis(trifluoromethylsulfonyl)amide room-temperature ionic liquid surface: a study with XPS, UPS, MIES, and HREELS, *J. Phys. Chem. B* **111**, 4801–4806 (2007).

57 Martinez, I.S., and Baldelli, S., High vacuum cells for classical surface techniques, *Rev. Sci. Instrum.* **81**, Article ID 044101 (2010).

58 Halka, V., and Freyland, W..Z., Electrowetting of ionic liquids under high vacuum conditions, *Z. Phys. Chem.* **222**, 117–127 (2008).

59 Lovelock, K.R.J., Kolbeck, C., Cremer, T., Paape, N., Schulz, P.S., Wasserscheid, P., Maier, F., and Steinrück, H.P., Influence of different substituents on the surface composition of ionic liquids studied using ARXPS, *J. Phys. Chem. B* **113**, 2854–2864 (2009).

60 Hashimoto, H., Ohno, A., Nakajima, K., Suzuki, M., Tsuji, H., and Kimura, K., Surface characterization of imidazolium ionic liquids by high-resolution Rutherford backscattering spectroscopy and X-ray photoelectron spectroscopy, *Surf. Sci.* **604**, 464–469 (2010).

61 Gottfried, J.M., Maier, F., Rossa, J., Gerhard, D., Schulz, P.S., Wasserscheid, P., and Steinrück, H.P., Surface studies on the ionic liquid 1-ethyl-3-methylimidazolium ethylsulfate using X-ray photoelectron spectroscopy (XPS), *Z. Phys. Chem.* **220**, 1439–1453 (2006).

62 Maier, F., Gottfried, J.M., Rossa, J., Gerhard, D., Schulz, P.S., Schwieger, W., Wasserscheid, P., and Steinrück, H.P., Surface enrichment and depletion effects of ions

dissolved in an ionic liquid: an X-ray photoelectron spectroscopy study, *Angew. Chem. Int. Edit.* **45**, 7778–7780 (2006).

63 Cremer, T., Kolbeck, C., Lovelock, K.R.J., Paape, N., Wolfel, R., Schulz, P.S., Wasserscheid, P., Weber, H., Thar, J., Kirchner, B., Maier, F., and Steinrück, H.P., Towards a molecular understanding of cation-anion interactions—probing the electronic structure of imidazolium ionic liquids by NMR spectroscopy, X-ray photoelectron spectroscopy and theoretical calculations, *Chem. Eur. J.* **16**, 9018–9033 (2010).

64 Kolbeck, C., Cremer, T., Lovelock, K.R.J., Paape, N., Schulz, P.S., Wasserscheid, P., Maier, F., and Steinrück, H.P., Influence of different anions on the surface composition of ionic liquids studied using ARXPS, *J. Phys. Chem.* B **113**, 8682–8688 (2009).

65 Maier, F., Cremer, T., Kolbeck, C., Lovelock, K.R.J., Paape, N., Schulz, P.S., Wasserscheid, P., and Steinruck, H.P., Insights into the surface composition and enrichment effects of ionic liquids and ionic liquid mixtures, *Phys. Chem. Chem. Phys.* **12**, 1905–1915 (2010).

66 Villar-Garcia, I.J., Smith, E.F., Taylor, A.W., Qiu, F.L., Lovelock, K.R.J., Jones, R.G., and Licence, P., Charging of ionic liquid surfaces under X-ray irradiation: the measurement of absolute binding energies by XPS, *Phys. Chem. Chem. Phys.* **13**, 2797–2808 (2011).

67 Chiappe, C., Malvaldi, M., Melai, B., Fantini, S., Bardi, U., and Caporali, S., An unusual common ion effect promotes dissolution of metal salts in room-temperature ionic liquids: a strategy to obtain ionic liquids having organic-inorganic mixed cations, *Green Chem.* **12**, 77–80 (2010).

68 Taylor, A.W., Qiu, F.L., Villar-Garcia, I.J., and Licence, P., Spectroelectrochemistry at ultrahigh vacuum: in situ monitoring of electrochemically generated species by X-ray photoelectron spectroscopy, *Chem. Commun.*, 5817–5819 (2009).

69 Qiu, F.L., Taylor, A.W., Men, S., Villar-Garcia, I.J., and Licence, P., An ultra high vacuum-spectroelectrochemical study of the dissolution of copper in the ionic liquid (N-methylacetate)-4-picolinium bis(trifluoromethylsulfonyl)imide, *Phys. Chem. Chem. Phys.* **12**, 1982–1990 (2010).

70 Lockett, V., Sedev, R., Bassell, C., and Ralston, J., Angle-resolved X-ray photoelectron spectroscopy of the surface of imidazolium ionic liquids, *Phys. Chem. Chem. Phys.* **10**, 1330–1335 (2008).

71 Iwahashi, T., Nishi, T., Yamane, H., Miyamae, T., Kanai, K., Seki, K., Kim, D., and Ouchi, Y., Surface structural study on ionic liquids using metastable atom electron spectroscopy, *J. Phys. Chem.* C **113**, 19237–19243 (2009).

72 Waring, C., Bagot, P.A.J., Slattery, J.M., Costen, M.L., and McKendrick, K.G., O(^{3}P) atoms as a probe of surface ordering in 1-alkyl-3-methylimidazolium-based ionic liquids, *J. Phys. Chem. Lett.* **1**, 429–433 (2010).

73 Waring, C., Bagot, P.A.J., Slattery, J.M., Costen, M.L., and McKendrick, K.G., O(^{3}P) atoms as a chemical probe of surface ordering in ionic liquids, *J. Phys. Chem.* A **114**, 4896–4904 (2010).

74 Wu, B.H., Zhang, J.M., Minton, T.K., McKendrick, K.G., Slattery, J.M., Yockel, S., and Schatz, G.C., Scattering dynamics of hyperthermal oxygen atoms on ionic liquid surfaces: [emim][NTf$_2$] and [C$_{12}$mim][NTf$_2$], *J. Phys. Chem.* C **114**, 4015–4027 (2010).

75 Rivera-Rubero, S., and Baldelli, S., Influence of water on the surface of hydrophilic and hydrophobic room-temperature ionic liquids, *J. Am. Chem. Soc.* **126**, 11788–11789 (2004).

76 Cremer, T., Killian, M., Gottfried, J.M., Paape, N., Wasserscheid, P., Maier, F., and Steinrück, H.P., Physical vapor deposition of [EMIM][Tf₂N]: a new approach to the modification of surface properties with ultrathin ionic liquid films, *ChemPhysChem.* **9**, 2185–2190 (2008).

77 Sobota, M., Nikiforidis, I., Hieringer, W., Paape, N., Happel, M., Steinrück, H.P., Gorling, A., Wasserscheid, P., Laurin, M., and Libuda, J., Toward ionic-liquid-based model catalysis: growth, orientation, conformation, and interaction mechanism of the [Tf₂N]⁻ anion in [BMIM][Tf₂N] thin films on a well-ordered alumina surface, *Langmuir* **26**, 7199–7207 (2010).

78 Sobota, M., Schmid, M., Happel, M., Amende, M., Maier, F., Steinruck, H.P., Paape, N., Wasserscheid, P., Laurin, M., Gottfried, J.M., and Libuda, J., Ionic liquid based model catalysis: interaction of [BMIM][Tf₂N] with Pd nanoparticles supported on an ordered alumina film, *Phys. Chem. Chem. Phys.* **12**, 10610–10621 (2010).

79 Souda, R., and Gunster, J., Temperature-programed time-of-flight secondary ion mass spectrometry study of 1-butyl-3-methylimidazolium trifluoromethanesulfonate during glass-liquid transition, crystallization, melting, and solvation, *J. Chem. Phys.* **129**, Article ID 094707 (2008).

80 Günster, J., Höfft, O., Krischok, S., and Souda, R., A time-of-flight secondary ion mass spectroscopy study of 1-ethyl-3-methylimidazolium bis (trifluoromethylsulfonyl)imide RT-ionic liquid, *Surf. Sci.* **602**, 3403–3407 (2008).

81 Souda, R., Glass-liquid transition, crystallization, and melting of a room temperature ionic liquid: thin films of 1-ethyl-3-methylimidazolium Bis [trifluoromethanesulfonyl]imide studied with TOF-SIMS, *J. Phys. Chem. B* **112**, 15349–15354 (2008).

82 Dupont, J., and Scholten, J.D., On the structural and surface properties of transition-metal nanoparticles in ionic liquids, *Chem. Soc. Rev.* **39**, 1780–1804 (2010).

83 Arimoto, S., Sugimura, M., Kageyama, H., Torimoto, T., and Kuwabata, S., Development of new techniques for scanning electron microscope observation using ionic liquid, *Electrochim. Acta* **53**, 6228–6234 (2008).

84 Kuwabata, S., Kongkanand, A., Oyamatsu, D., and Torimoto, T., Observation of ionic liquid by scanning electron microscope, *Chem. Lett.* **35**, 600–601 (2006).

85 Arimoto, S., Oyamatsu, D., Torimoto, T., and Kuwabata, S., Development of in situ electrochemical scanning electron microscopy with ionic liquids as electrolytes, *ChemPhysChem.* **9**, 763–767 (2008).

86 Arimoto, S., Kageyama, H., Torimoto, T., and Kuwabata, S., Development of in situ scanning electron microscope system for real time observation of metal deposition from ionic liquid, *Electrochem. Commun.* **10**, 1901–1904 (2008).

87 Imanishi, A., Tamura, M., and Kuwabata, S., Formation of Au nanoparticles in an ionic liquid by electron beam irradiation, *Chem. Commun.*, 1775–1777 (2009).

88 Tsuda, T., Seino, S., and Kuwabata, S., Gold nanoparticles prepared with a room-temperature ionic liquid-radiation irradiation method, *Chem. Commun.*, 6792–6794 (2009).

89 Johnston, M., Lee, J.J., Chottiner, G.S., Miller, B., Tsuda, T., Hussey, C.L., and Scherson, D.A., Electrochemistry in ultrahigh vacuum: underpotential deposition of Al on polycrystalline W and Au from room temperature AlCl₃/1-ethyl-3-methylimidazolium chloride melts, *J. Phys. Chem. B* **109**, 11296–11300 (2005).

90 Lovelock, K.R.J., Smith, E.F., Deyko, A., Villar-Garcia, I.J., Licence, P., and Jones, R.G., Water adsorption on a liquid surface, *Chem. Commun.*, 4866–4868 (2007).

91 Rutten, F.J.M., Tadesse, H., and Licence, P., Rewritable imaging on the surface of frozen ionic liquids, *Angew. Chem. Int. Edit.* **46**, 4163–4165 (2007).

92 Hardacre, C., Application of EXAFS to molten salts and ionic liquid technology, *Ann. Rev. Mater. Res.* **35**, 29–49 (2005).

9 Pioneering Biological Processes in the Presence of Ionic Liquids: The Potential of Filamentous Fungi

MARIJA PETKOVIC

Instituto de Tecnologia Química e Biológica, Universidade Nova de Lisboa, Oeiras, Portugal

CRISTINA SILVA PEREIRA

Instituto de Tecnologia Química e Biológica, Universidade Nova de Lisboa, Oeiras, Portugal
Instituto de Biologia Experimental e Tecnológica (IBET), Oeiras, Portugal

ABSTRACT

In the majority of current applications, ionic liquids are used to substitute for organic solvents and/or catalysts in a given process. They generally aim to increase efficiency and sustainability. Both academic and industrial data give unequivocal support to the utility of ionic liquids.

Our work is positioned at the interface between ionic liquid chemistry and fungal biology; initially we aimed to better understand ionic liquid toxicity and biodegradability. Filamentous fungi are found to have an extremely high capacity to tolerate ionic liquids. To date, they are the most resistant group of microorganisms identified. Filamentous fungi are also able to readily or partially degrade certain ionic liquids and are thus likely to play a central role in their biotic degradation in the environment.

Supplementing growth media with ionic liquids augments the diversity of low molecular-weight molecules synthesised by these fungi. These fungal metabolites are of paramount importance with widespread applications, ranging from the clinical to biotechnological usage. Any chemical causing the

Ionic Liquids UnCOILed: Critical Expert Overviews, First Edition. Edited by Natalia V. Plechkova and Kenneth R. Seddon.

same cellular stimulus would therefore be of interest; the dual nature of ionic liquids and their tuneability regarding both the cation and the anion, per se, offer a virtually unlimitedly source of potential stimuli. Transcriptomic, proteomic, and metabolomic approaches are being taken to understand comprehensively the biological response of fungal cells during exposure to ionic liquids.

Novel and/or more efficient fungal bio-refinery processes can be envisaged by combining the elevated catalytic potential of fungi—for production of anything from fine chemicals to pharmaceuticals—and the high biocompatibility and solvent quality of some ionic liquids.

9.1 INTRODUCTION

The number of publications and patents pertaining to ionic liquids has constantly increased in the last two decades. These solvents have rapidly evolved from academic curiosities to industrial reagents with a broad range of applications [1]. Several hundred ionic liquids are already commercially available and characterised [2]; one can reasonably estimate that millions of formulations are possible [3]. Ionic liquids are commonly defined as salts that are liquid below 100 °C [3]. They are often regarded as green solvents due to their excellent solvation capacity, negligible vapour pressure, and bulk non-flammability [4]. The outstanding solvation behaviour results from their Coulombic environment; that is, ionic liquids are composed of two (high and low) electrically charged nanodomains [5]. This is clearly advantageous compared with conventional molecular organic solvents. There is controversy regarding the greenness of ionic liquids [6]; many of these solvents do not fully address the principles of green chemistry [7], especially as regards toxicity and environmental persistence.

A precautionary approach towards environmental protection is critically important when investigating any chemical for potential large-scale usage. In Europe, chemicals produced in quantities exceeding 1 tonne per year are, since 2007, controlled by REACH (Registration, Evaluation, Authorisation and restriction of CHemical substances) [8]. It is impossible to predict ionic liquid concentrations that are likely to contaminate a specific environment upon accidental discharge or spillage. Attenuation will be controlled by numerous abiotic and biotic processes. The former depends both on the chemical stability and mobility of each particular ion (or ion pairs if formed) and the properties of that particular environmental setting [9, 10], including possible pre-pollution scenarios [11].

Ionic liquids comprise a very heterogeneous group of fluids that *a priori* cannot be considered benign. Extensive toxicological assessment is time-consuming and often expensive. A broad range of testing models, covering all five kingdoms of living organisms, has already been used to evaluate ecotoxicity of certain ionic liquid groups (for review, see Reference [6]). The majority of these studies considered, preferentially, the imidazolium family and lack systematisation. Whilst preliminary, this certainly provides a notable assessment of ionic liquid toxicity. It leads to the recommendation that ionic liquid synthesis is performed in conjunction with a timely evaluation of toxicity.

Ionic liquids carrying biocompatible ions are now being synthesised [12]. This includes chemical functions of non-nutritive sweeteners [13], amino acids [14], sugars [15], or carboxylic acids [16]. Increased ionic liquid biocompatibility may further stimulate their application in biological sciences. They have already been used as co-solvents in enzymatic [17] and whole-cell biocatalysis [18, 19], protein stabilisers [20–22], and potential active pharmaceutical ingredients [23, 24].

The data compiled here describe the impact of a variety of ionic liquids families on Ascomycota fungi growth behaviour, biodegradation ability, and cell biochemistry. The demonstration that sublethal concentrations of ionic liquids augment diversity of the biosynthesised low molecular weight fungal molecules was ground breaking. We provide a preliminary discussion as to how ionic liquids affect gene transcription, protein expression, and permeability of cellular membranes. This contributes, in our opinion, to a better understanding of structure-activity correlations in ionic liquids. It also presents an outlook that might encourage ionic liquid exploitation in novel bio-refinery processes.

9.2 TOLERANCE OF FILAMENTOUS FUNGI TO IONIC LIQUIDS

Our group first used filamentous fungi as eukaryotic model organisms to evaluate the ecotoxicity of several ionic liquids families. Filamentous fungi are ubiquitous in all environments [25]. They are critical soil colonisers, an environmental aspect likely to be affected by accidental ionic liquid contamination. The most commonly used ionic liquids are persistent in soil [26, 27]. The capacity of filamentous fungi to survive in contaminated environments, and play a major role in biotic decay of a broad range of organic and persistent pollutants [28], might be useful in defining the ecotoxicological risk of ionic liquids. This adds heightened ecological meaning to these analyses.

Figure 9.1 shows the structures of the ionic liquids initially tested on 14 fungal isolates (10 *Penicillium* isolates [29], *Trichoderma longibrachiatum*, *Mucor plumbens*, *Cladosporium herbarum*, and *Chrysonilia sitophila*). Ionic liquids comprised distinct cations, covering 1-methylimidazolium, pyridinium, 1-methylpyrrolidinium, 1-methylpiperidinium, and cholinium head groups, and anions, such as chloride, ethanoate, and DL-lactate (hereafter described just as lactate). Fungal growth in media supplemented with 50 mM of each compound to test was scored by measuring the absorbance of the medium at defined incubation time points. Increased absorbance was scored as growth, whilst spore formation and/or absorbance approaching a constant value indicated a stationary growth phase [29]. Figure 9.2 shows *Penicillium adametzii*

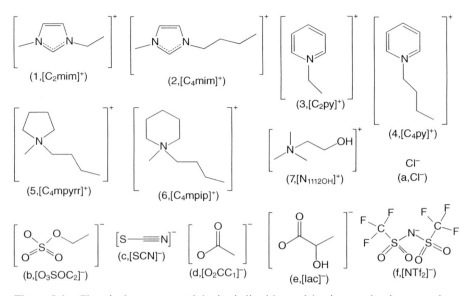

Figure 9.1 Chemical structures of the ionic liquids used (cations and anions, numbers, and letters, respectively). (1a) 1-ethyl-3-methylimidazolium chloride ([C_2mim]Cl); (1b) 1-ethyl-3-methylimidazolium ethyl sulfate ([C_2mim][O_3SOC_2]); (1c) 1-ethyl-3-methylimidazolium thiocyanate ([C_2mim][SCN]); (1d) 1-ethyl-3-methylimidazolium ethanoate ([C_2mim][O_2CC_1]); (1e) 1-ethyl-3-methylimidazolium DL-lactate ([C_2mim][lac]); (2a) 1-butyl-3-methylimidazolium chloride ([C_4mim]Cl); (3d) 1-ethylpyridinium ethanoate ([C_2py][O_2CC_1]); (3e) 1-ethylpyridinium DL-lactate ([C_2py][lac]); (4a) 1-butylpyridinium chloride ([C_4py]Cl); (5a) 1-butyl-1-methylpyrrolidinium chloride ([C_4mpyr]Cl); (5e) 1-butyl-1-methylpyrrolidinium DL-lactate ([C_4mpyr][lac]); (6d) 1-butyl-1-methylpiperidinium ethanoate ([C_4mpip][O_2CC_1]); (7a) cholinium chloride ([N_{1112OH}]Cl); (7d) cholinium ethanoate ([N_{1112OH}][O_2CC_1]); (7e) cholinium DL-lactate ([N_{1112OH}][lac]); (7f) cholinium bis{(trifluoromethyl)sulfonyl}amide ([N_{1112OH}][NTf$_2$]).

growth curves in media supplemented with 50 mM of either 1,3-dialkylimidazolium ionic liquids or 1-methylimidazole as an example.

This methodology measures the effect of a single compound concentration on fungal growth and rapidly evaluates their toxicity. As presented in Figure 9.3, toxic effects can be visualised in a bidimensional matrix of fungal growth inhibition, or lack thereof. Fungal growth was observed in the majority of cases: in only 16% of experiments was it completely inhibited. Testing higher concentrations, some filamentous fungi, for example, *Penicillium olsonii*, were observed to grow in media supplemented with 375 mM [C_2mim]Cl [30].

The higher resistance of filamentous fungi-relative to bacteria [31, 32]—was further demonstrated in an independent study by our group. Some fungal strains, amongst extreme soil biotypes sourced from locations of high salinity and high hydrocarbon load, were able to grow in solid media supplemented

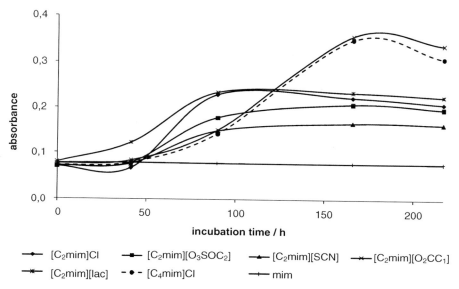

Figure 9.2   Growth curves of *Penicillium adametzii* in media containing imidazolium-based ionic liquids: 1-ethyl-3-methylimidazolium chloride ([C₂mim]Cl), 1-ethyl-3-methylimidazolium ethyl sulfate ([C₂mim][O₃SOC₂]), 1-ethyl-3-methylimidazolium thiocyanate ([C₂mim][SCN]), 1-ethyl-3-methylimidazolium ethanoate ([C₂mim][O₂CC₁]), 1-ethyl-3-methylimidazolium DL-lactate ([C₂mim][lac]), 1-butyl-3-methylimidazolium chloride ([C₄mim]Cl), and 1-methylimidazole (mim). Growth curves were obtained by measuring the absorbance (600 nm) of the growth media at different incubation times.

with 1 M 1,3-dialkylimidazolium ionic liquids [33]. The continuing exposure of fungal strains to petroleum hydrocarbons is likely to underlie their acquired resistance to the tested imidazolium salts.

Hierarchical cluster analysis of ionic liquid ecotoxicity data on *Penicillium* sp. strains identified four groups according to their distinct ionic liquid susceptibility: high (*P. variable, P. diversum*), intermediate (*P. corylophilum, P. restrictum*) and two of low susceptibility {(*P. brevicompactum, P. janczewskii, P. adametzii*) and (*P. olsonii, P. glandicola, P. glabrum*)}.

There is a high degree of correlation between clusters defined by ionic liquid susceptibility and their phylogeny (Table 9.1). Phylogeny clusters were defined based on calmodulin partial gene sequences [34], and included all *Penicillium* sp. tested here (except *P. diversum* and *P. olsonii*). One can rationalise future toxicological assessments, and to some extent, predict the behaviour of different fungal species when exposed to these chemicals.

On the basis of these data, it also became evident that the overall toxicity of the ionic liquids towards filamentous fungi is mainly defined by the head group in the cation. Furthermore, toxicity increases with longer substituted

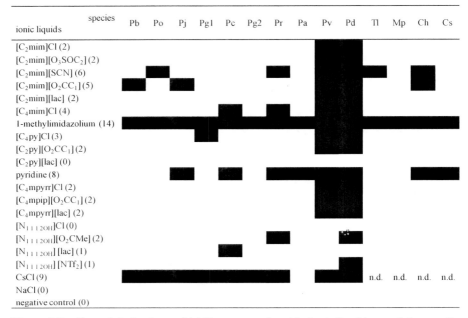

Figure 9.3 Growth behaviour of 14 filamentous fungi in ionic liquid containing media (50 mM) and control culture. Black fields show cases of growth inhibition. Numbers in brackets correspond to the toxicity ranking (0 to 14). Fungal isolate codes are as follows: Pb (*Penicillium brevicompactum*), Po (*P. olsonii*), Pj (*P. janczewskii*), Pg1 (*P. glandicola*), Pc (*P. corylophilum*), Pg2 (*P. glabrum*), Pr (*P. restrictum*), Pa (*P. adametzii*), Pv (*P. variabile*), Pd (*P. diversum*), Tl (*Trichoderma longibrachiatum*), Mp (*Mucor plumbens*), Ch (*Cladosporium herbarum*), and Cs (*Chrysonilia sitophila*).

TABLE 9.1 Comparison of Our Results with the Phylogenetic Study by Wang and Zhuang [34]

Susceptibility	Ionic Liquid Toxicity Study	Phylogenetic Study
High	(Pv)	(XI)
Intermediate	(Pc,Pr)	(VIII)
Low	(Pb,Pj,Pa), (Pg1,Pg2)	(I,II)

Species codes are as follows: Pb (*Penicillium brevicompactum*), Pj (*P. janczewskii*), Pg1 (*P. glandicola*), Pc (*P. corylophilum*), Pg2 (*P. glabrum*), Pr (*P. restrictum*), Pa (*P. adametzii*), and Pv (*P. variabile*).

alkyl chain length. These general trends have been reported in numerous studies using highly distinct testing models, ranging from microbial strains (e.g., *Vibrio fischeri*) [35] to mammalian cell lines [36]. The aromatic cations (containing imidazolium and pyridinium) are usually considered more toxic than alicyclic (piperidinium or pyrrolidinium) or quaternary ammonium cations [37].

Our data on filamentous fungi substantiate this idea. 1,3-Dialkylimidazolium and cholinium ionic liquids had the highest and the lowest toxicity, respectively. Toxicities of the remaining ionic liquids, containing cations with pyridinium, pyrrolidinium, or a piperidinium ring, were between these two extremes. The toxic effects of 1-methylimidazole and pyridine were determined to better evaluate the role of these aromatic head groups. Both free bases demonstrated stronger effects than the imidazolium and pyridinium ionic liquids tested, inhibiting fungal growth in 100% and 60% of the cases, respectively. Toxicity of 1-methylimidazole towards *Vibrio fischeri* was lower than that of [C$_4$mim]Cl [38]. Data discrepancies were probably due to the dissimilar model organisms used, the different cultivation media, and testing concentrations.

The data discussed here also support the theoretical model for toxicity prediction proposed by Couling et al. [35]. Despite numerous discrepancies seen in other studies [39, 40], this suggests that pyridinium salts are less toxic than their imidazolium counterparts. Computational methods are powerful tools for predicting ionic liquid toxicity, for example, by applying quantitative structure–activity relationship (QSAR) modelling to experimental data on their toxicity using either descriptors calculated at a low semi-empirical computational level [35], or descriptors calculated by group contribution methods [41], or by neural network models [42, 43]. More systematic sources of ecotoxicity data are still needed to properly implement such methods.

We recently initiated ecotoxicological assessment of a set of phosphonium ionic liquids that have until now been overlooked [39, 44–46]. These quaternary phosphonium cations provide a large collection of distinct formulations by independently varying each of the four chains in the cation. Phosphonium ionic liquids are generally considered chemically and thermally more stable than the corresponding quaternary ammonium salts [47], properties important for chemical processes and regeneration of the solvent.

We here present preliminary data on toxicity of tributyl(alkyl)phosphonium chlorides ([P$_{444n}$]Cl, where $n = 1$–8, 10, 12, or 14; Fig. 9.4a) to *Aspergillus nidulans* [48]. We decided to study *A. nidulans*, the genome of which has been fully sequenced; due to its high phylogenetic correlation with *Penicillium* sp., right up to the family level *Trichocomaceae* [49]. Effects of compounds were quantified as minimal inhibitory or fungicidal concentrations (MIC or MFC, respectively). The toxic effect of tributyl(alkyl)phosphonium chlorides was generally seen to increase with cation chain length (Fig. 9.4b). Exceptions observed to date are [P$_{444n}$]Cl, where $n = 3$ or 4, and [P$_{4442}$]Cl, showing lower and higher than expected toxicity, respectively. This is probably a consequence of the geometry of the tributyl(alkyl)phosphonium ion (steric effects), yet the higher toxicity of the latter cannot be explained easily. The trend in toxicity, however, matches that previously observed in numerous studies of toxicity of 1,3-dialkylimidazolium [32] or 1-alkylpyridinium [50] ionic liquids. It seems likely that the mode of toxicity of [P$_{444n}$]Cl is baseline toxicity or narcosis. This is supported by the evaluation of their ability to alter *in vivo* the permeability

(a)

(b)

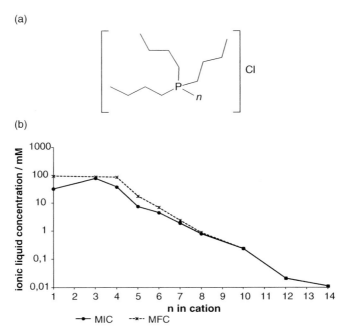

Figure 9.4   (a) General structure of tributyl(alkyl)phosphonium chlorides ([P_{444n}]Cl, where n = 1–8, 10, 12, or 14); (b) minimal inhibitory and fungicidal concentrations (MIC and MFC, respectively) (in mM) towards *Aspergillus nidulans*. MIC and MFC values are plotted on a logarithmic scale against the alkyl chain length of the substituent in the cation.

of the cellular membrane [48]. To date, most studies exploited *in vitro* measures, such as determining membrane/water partition coefficients [37], or artificial liposome leakage [51]. In our opinion, one of the most interesting approaches was the use of spectroscopic measurements to define the fraction of the ionic liquid biosorbed by the cells [52]. However, this method was unable to differentiate between bioaccumulated (i.e., up-taken by the cell) and bioadsorbed (i.e., onto the cell surface) fractions. In our opinion, *in vivo* observations are essential to a better understanding of the mode of ionic liquid interaction with biological membranes.

The anion effect has often been neglected in the context of ionic liquid toxicity, despite it generally having a decisive role in physicochemical properties. The anion is recognised as contributing in some unpredictable way to the overall toxicity of the ionic liquid [6]. This concept is most probably influenced by the range of chemically unrelated structures, the effects of which are difficult to compare. Only in important exceptions has this been comprehensively investigated [53]. To better define the hallmarks governing the anion effect, we decided to investigate the toxicity of several cholinium alkanoates [16].

Our aim was also to advance towards a more conscious design of ionic liquids. This justifies selection of cholinium ($[N_{1112OH}]^+$) as the cation. It is part of the vitamin B complex and generally regarded as both benign [54, 55] and readily biodegradable [56]. The low toxicity of several cholinium ionic liquids has already been proven, for example, saccharinate and acesulfamate (*Daphnia magna*) [57], dimethylphosphate (*Cladosporium sporogenes*) [58], and lactate (filamentous fungi) [29].

$[N_{1112OH}]^+$ was combined with a range of linear alkanoate anions ($[C_nH_{2n+1}CO_2]^-$, $n = 1$–5, 7, or 9; Fig. 9.5a), and two structural isomers (for $n = 3$ or 4) [16]. Their toxicity towards filamentous fungi was assessed on four fungal

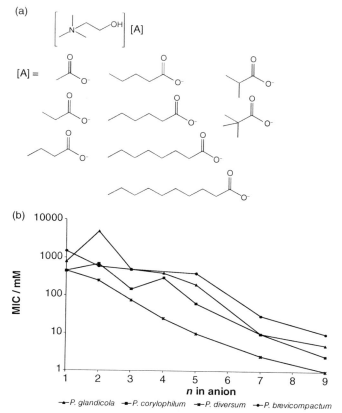

Figure 9.5 (a) General chemical structure of cholinium alkanoates and the structures of linear alkyl chain anions in $[N_{1112OH}][C_nH_{2n+1}CO_2]$, where $n = 1$–5, 7, or 9, and two structural isomers with $n = 3$ or 4; (b) comparison of minimal inhibitory concentrations (MIC) of $[N_{1112OH}][C_nH_{2n+1}CO_2]$, where $n = 1$–5, 7, or 9, in four fungal isolates, *Penicillium brevicompactum*, *P. glandicola*, *P. corylophilum*, and *P. diversum*. MIC values (in mM) plotted on a logarithmic scale.

TABLE 9.2   Comparison of Minimal Inhibitory and Fungicidal Concentrations (MIC and MFC, Respectively) (mM) of $[N_{1112OH}][C_nH_{2n+1}CO_2]$, $n = 3$ or 4 and Two Structural Isomers in Four Fungal Isolates

Fungal Isolates [N_{1112OH}][X]	Penicillium brevicompactum		Penicillium glandicola		Penicillium corylophilum		Penicillium diversum	
	MIC	MFC	MIC	MFC	MIC	MFC	MIC	MFC
Butanoate	500	1000	500	750	150	150	75	75
2-Methylpropanoate	500	1500	600	750	250	250	100	200
Pentanoate	n.d.	n.d.	400	600	300	300	25	50
2,2-Dimethylpropanoate	750	1500	400	700	750	1500	75	150

strains that cover the ionic liquid susceptibility groups defined previously. Fungal tolerance ranking defined here, with MICs varying from 2.5 to 1500 mM, was consistent with the susceptibility group classification (Fig. 9.5b). This clearly highlights the robustness and predictive value of the filamentous fungi testing system, irrespective of the chemical nature of the salt.

From the data, it was obvious that elongation of the anionic alkyl chain led to higher lipophilicity and toxicity, akin to the substituted alkyl chain effect in various cations [50, 59, 60].

2-Methylpropanoate and 2,2-dimethylpropanoate are branched isomers containing four and five carbon atoms, respectively. Relative to the linear isomers, they exhibited lower inhibitory and fungicidal effects. This was especially evident in the latter compound (Table 9.2). These observations reinforce the importance of both cation and anion in the conscious design of ionic liquids.

There is no doubt that one of the landmarks of our study of cholinium alkanoates was that it demonstrated that $[N_{1112OH}][C_nH_{2n+1}CO_2]$, $n = 5, 7$, or 9, are also remarkable solvents. Cholinium alkanoates can selectively dissolve suberin in cork [61]. This is a major observation; a benign solvent can efficiently disrupt the recalcitrant cork composite. This, *per se*, is groundbreaking research on fungal bio-refinery processes, especially when the aim is valorisation of agricultural and industrial wastes. Their environmental and industrial applications may be far ranging; for example, the substitution of the alkali cation by a cholinium cation increases the solubility of the respective fatty acid soap without lowering its biocompatibility [62].

9.3   IONIC LIQUID BIODEGRADATION BY FILAMENTOUS FUNGI

The biodegradability of an ionic liquid is a critical feature, especially when the aim is their conscious design. Biodegradability ultimately defines environmental persistence of a chemical and its long-term impact. Rapid and complete mineralisation indicates that the chemical is readily biodegradable. Weak or extensive molecular cleavage suggests primary or ultimate biodegradability potential, respectively. These definitions are generally labelled solely as biodegradability potential, tending to misinform the reader.

The design of biodegradable chemicals, whether ionic liquids or not, follows general criteria (for review, see Boethling et al. [56] and Coleman and Gathergood [26]). Several chemical functions enhance biodegradability, not necessarily resulting in products of lower toxicity. These include the presence of esters, amides, hydroxyl, aldehyde, carboxylic acid groups, or linear alkyl chains. The latter may significantly increase the toxicity of the chemical.

Looking at the specific case of 1,3-dialkylimidazolium ionic liquids, we have found no evidence to date that the imidazole core can be biodegraded. Formation of different hydroxyl and carboxyl functions in the substituted chain of the imidazolium cation has been described [63]. The biodegradability of 1,3-dialkylimidazolium ionic liquids is a critical parameter since currently this group is predominantly undergoing REACH registration. These include $[C_2mim]X$ (X = Cl$^-$, $[O_3SOC_2]^-$, $[C_1SO_3]^-$, $[O_2CC_1]^-$, and $[NTf_2]^-$) and $[C_4mim]$ Cl [64].

Toxicity and biodegradability of imidazolium and pyridinium ionic liquids are frequently compared, the latter normally showing higher biodegradability [65–67]. Seeking filamentous fungi that efficiently degrade these major groups of ionic liquids, we initially focussed on several *Penicillium* sp. The capacity of the fungi to degrade the $[C_2mim]^+$ (375 mM), $[C_4mim]^+$ (50 mM), and $[C_4m_\beta py]^+$ (50, 250, and 500 mM) cations was monitored, qualitatively by nuclear magnetic resonance (NMR) and quantitatively by liquid chromatography [30]. Different culture times and conditions have already been tested, yet no evidence that these cations can be degraded by *Penicillium* sp. has been observed. Currently, we are testing a set of bacterial and fungal strains, previously isolated from soils at locations of high salinity and high hydrocarbon load [33], on their ability to act as bioremediation agents of these ionic liquids.

Cholinium alkanoates were shown to display high biodegradability potential [16]. After two-week incubation in a static culture, alkanoate anions were only partially degraded by the filamentous fungi. Degradation was highly concentration dependent (see Table 9.3a). By altering the culture conditions (four weeks under agitation) in *P. corylophilum* (intermediate susceptibility group) these anions were found to be degraded more efficiently (Table 9.3b). The branched chain anions, 2,2-dimethylpropanoate and 2-methylpropanoate, both less toxic than their corresponding linear isomers, were more resistant to fungal attack, leading to no or only partial degradation, respectively (four weeks of incubation). Linear and branched chains are generally considered to be, respectively, more and less liable to biodegradation [56]. This is observed in the distinctive biodegradability potential of butanoate and 2-methylpropanoate (100 mM). However, NMR peaks attributed to the cholinium cation were conserved, suggesting only partial degradation, perhaps due to the high initial concentration in the medium (Table 9.3b). After two months, the depletion of

TABLE 9.3   Biodegradation of the Alkanoate Anions of $[N_{11120H}][C_nH_{2n+1}CO_2]$, where $n = 1$–5 or 7, when Growth Media of *Penicillium corylophilum* was Supplemented with Ionic Liquids: (a) in Static Culture, with Two Weeks Incubation, and (b) with Orbital Agitation and Four Weeks Incubation

(a)

$[C_nH_{2n+1}CO_2]^-$	Concentration/mM	Anion Biodegradation (UPLC Analyses)
Ethanoate	125	52%
	250	16%
Propanoate	62.5	63%
	125	44%
	250	16%

(b)

$[C_nH_{2n+1}CO_2]^-$	Concentration/mM	Anion Degradation	
		^{1}H NMR	UPLC Analyses
Ethanoate	375	n.d.	27%
Propanoate	500	n.d.	10%
Butanoate	100	c.d.	100%
2-Methylpropanoate	100	p.d.	72%
Pentanoate	100	c.d.	100%
2,2-Dimethylpropanoate	500	n.d.	0%
Hexanoate	50	c.d.	88%
Octanoate	10	c.d.	100%

c.d., complete degradation; p.d., partial degradation; n.d., not detected.

cholinium cation in cultures of soil microorganisms, at an initial concentration of 0.5 or 1 M, reached levels of only 20% and 5%, respectively [33]. The cholinium cation has been previously reported to be readily biodegraded, reaching almost complete biodegradation (93%) [56]. However, the concentration used in these standard bioassays is several orders of magnitude lower.

The capacity of filamentous fungi to degrade ionic liquids, both cation and anion, generally follows the expected trends. Despite the broad catalytic capacity of the tested filamentous fungal strains, we were unable to degrade both the imidazolium and the pyridinium rings under the conditions tested.

9.4   THE IMPACT OF IONIC LIQUIDS ON THE METABOLISM OF FILAMENTOUS FUNGI

One of the most remarkable aspects of fungal biology is the capacity to produce numerous, chemically diverse low molecular weight molecules. These secondary metabolites generally display sub-nanomolar potency and are highly specific. They are thought to be produced by filamentous fungi as a niche defence function. Fungi generally produce low yields of novel secondary metabolites when grown under normal laboratory conditions; this can be a major obstacle in screening efforts [69]. One way of stimulating fungi production is by promoting inter-species interaction [70]. A conclusive analysis of results from such systems is almost impossible due to its high degree of inherent complexity. A straightforward alternative strategy is to induce these metabolic pathways by means of secondary stimuli, such as supplementing growth media with exogenous agents.

To better understand the effect of sublethal concentrations of ionic liquids, we analysed fungal metabolomic footprints [29]. Spent media were filtered and low molecular weight metabolites extracted with ethyl ethanoate and analysed by electrospray ionisation–mass spectrometry ($\pm$ESI-MS; target mass between 150 and 1100). Spectra were initially analysed qualitatively only, aiming to record the footprints (presence or absence of peaks). Sodium chloride was also tested to identify if effects caused by the ionic liquids in the fungal metabolism were correlated with those of simple inorganic salts. Whilst analysing all spectra, we aimed to identify peaks that could be traced back to each specific exogenous agent. From the data, it is apparent that each ionic liquid induces a specific metabolomic footprint. Using methods of hierarchical cluster analysis (Fig. 9.6) we were able to demonstrate that ionic liquids induce alterations in the pattern of low molecular weight metabolites that cannot be correlated to that of sodium chloride. This dissimilar behaviour may be, in part, explained by ion pairing in solution. We recently observed formation of ions pairs in >0.06 M solutions of some cholinium alkanoates by Nuclear Overhauser Effect (NOE)[71]. This contradicts the prevailing opinion that ions behave as independent chemical entities in solution. Some ecotoxicological data suggest that ionic liquid toxicity cannot be systematically estimated by a summation of the independent effects of their ions [42], reinforcing the need to further understand their behaviour in solution.

The ionic liquid-induced metabolomic footprints might, in some cases, account for sub-products of their degradation. That was certainly not the case for [C$_2$mim]Cl. Filamentous fungi tested were unable to degrade this ionic liquid under the culture conditions used [30]. Whilst the most diverse metabolome was induced by the 1,3-dialkylimidazolium ionic liquids, it also becomes clear that the stimuli are not accounting for antifungal properties only. Both [C$_2$py][lac] and [N$_{1112OH}$]Cl have low antifungal activity but induced significant alterations in metabolomic footprints.

Low molecular weight fungal metabolites have tremendous potential. There is an enormous diversity amongst fungal species, 70,000 identified so far and counting; estimates point to a total in the range of 1.6 million [25]. Each fungus can produce a set of species-specific metabolites, some displaying interesting pharmacological properties [72, 73]. From 1993 to 2001, about 1500 fungal metabolites were reported to have antitumour or antibiotic activity [74]. We have made the preliminary observation that fungal metabolomic extracts

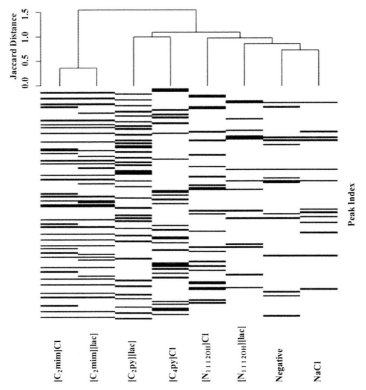

Figure 9.6 Hierarchical cluster analysis of the joint peak lists, uniting all individual peak matrices of positive and negative mode, detected by ESI–MS after growth of 10 *Penicillium* sp. in ionic liquid containing media and controls (m/z = 150 to 1100). Black fields indicate the presence of a peak at a given m/z value.

induced by some of the ionic liquids also display possible anticancer activity (unpublished data; collaboration with Thelial Technologies S.A., Portugal). To provide a more complete picture of the events caused by ionic liquids in the fungal metabolic network, we are now attempting to identify some of the inducible fungal metabolites.

Biosynthetic and regulatory pathways leading to the formation of low molecular weight fungal metabolites are regulated by both environmental and developmental signals. Genes required for a fungus to produce a given secondary metabolite are frequently clustered and co-regulated [75, 76]. In addition, most secondary metabolite biosynthetic gene clusters are known to be silent [77]. Using whole genome transcriptomic profiling, we have observed that ionic liquids provoke major alterations in transcript levels of genes, mediating both secondary metabolism and developmental processes in filamentous fungi [78]. Between 20% and 30% of genes with an assigned function were differentially expressed during fungal growth in media supplemented with ionic liquids, and ~20% of those genes belong to secondary metabolism categories. These include major changes in *laeA* transcript levels (global regulator of secondary metabolism). We have also observed major changes in the fungal proteome. Differentially expressed proteins might be exploited *in vitro* for the biosynthesis of new compounds.

The dual nature of ionic liquids, and their tuneability regarding both the cation and the anion, offer, *per se*, a virtually unlimited source of potential stimuli for boosting diversity of fungal low molecular weight molecules [29]. The creation of defined methodologies for stimulating new biosynthetic pathways in filamentous fungi is of extreme significance to the field of natural products. We have no doubt that these methods will lead to discovery of pharmacologically valuable novel natural compounds. Advanced functional level studies are still necessary to better understand how ionic liquids alter fungal metabolism.

9.5 CONCLUDING REMARKS

There are numerous demonstrations of high protein stability and high efficiency of enzymatic catalysis in "wet" ionic liquids [17, 79–82]. These solvents have proven to be remarkably useful in a wide range of fermentation processes. This includes the use of water-immiscible liquids to augment the yield of certain biphasic biotransformation processes. Partition to the ionic liquid phase prevents a toxic product [19], or substrate [44], accumulating in the aqueous growth media phase. The latter example is highlighting the potential of ionic liquids in bioremediation applications, which in the near future may include the complex scenario of *in situ* soil bioremediation. Application of surfactants to soil increases the bioavailability of persistent aromatic chemicals and consequently the efficiency of the bioremediation [83]. Some 1,3-dialkylimidazolium ionic liquids also enhance *in situ* photolytic degradation of chlorinated aromatic pollutants [84].

The studies discussed in this chapter should further highlight the unquestionable utility of ionic liquids in a broad biotechnological context. This is especially true when process efficiency and sustainability are taken into consideration.

Our data motivate the development of pioneering processes, especially considering the increased diversity of low molecular weight molecules produced by fungi grown in the presence of ionic liquids. Whilst still not comprehensively understood, this phenomenon might depend on properties exclusive to ionic liquids. Ionic liquid solvent qualities, together with their biocompatibility with filamentous fungi, even at high concentrations, may also be exploited for developing new types of biorefineries. This possibly constitutes an elegant way for producing diverse fine chemicals, and may define another useful aspect of ionic liquids regarding fungal biology.

The sum of ecotoxicological data on ionic liquids raises the question as to their impact on prokaryotic and eukaryotic microorganisms in both the short and long term—conscious ionic liquid design is a high priority. One can confidently predict an exponential increase in interest in biocompatible/degradable ionic liquids in the not too distant future. We anticipate their application in the development of novel biotechnological processes for industrial, pharmacological, and environmental uses.

ACKNOWLEDGEMENTS

M.P. is grateful to FC&T for the fellowship SFRH/BD/31451/2006. This work was partially supported by a grant from Iceland, Liechtenstein, and Norway through the EEA financial mechanism (Project PT015).

REFERENCES

1   Plechkova, N.V., and Seddon, K.R., Applications of ionic liquids in the chemical industry, *Chem. Soc. Rev.* **37**, 123–150 (2008).

2   Novionic Inc., delphIL database, http://www.delphil.net (2009).

3   Stark, A., and Seddon, K.R., *Kirk-Othmer encyclopaedia of chemical technology*, vol. 26, A. Seidel, John Wiley & Sons, Inc., Hoboken, NJ (2007), pp. 836–920.

4   Deetlefs, M., and Seddon, K.R., Ionic liquids: fact and fiction, *Chim. Oggi-Chem. Today* **24**(2), 16–23 (2006).

5   Rebelo, L.P.N., Lopes, J.N.C., Esperança, J.M.S.S., Guedes, H.J.R., Łachwa, J., Najdanovic-Visak, V., and Visak, Z.P., Accounting for the unique, doubly dual nature of ionic liquids from a molecular thermodynamic, and modeling standpoint, *Acc. Chem. Res.* **40**(11), 1114–1121 (2007).

6   Petkovic, M., Seddon, K.R., Rebelo, L.P.R., and Silva Pereira, C., Ionic liquids: a pathway to environmental acceptability, *Chem. Soc. Rev.* **40**, 1383–1403 (2011).

7   Anastas, P.T., and Kirchhoff, M.M., Origins, current status, and future challenges of green chemistry, *Acc. Chem. Res.* **35**(9), 686–694 (2002).

8   European Commission, REACH—registration, evaluation, authorisation and restriction of CHemicals, http://ec.europa.eu/enterprise/sectors/chemicals/reach/index_en.htm (2011).

9   Matzke, M., Thiele, K., Müller, A., and Filser, J., Sorption and desorption of imidazolium based ionic liquids in different soil types, *Chemosphere* **74**(4), 568–574 (2009).

10   Mrozik, W., Jungnickel, C., Ciborowski, T., Pitner, W.R., Kumirska, J., Kaczyński, Z., and Stepnowski, P., Predicting mobility of alkylimidazolium ionic liquids in soils, *J. Soils Sediments* **9**(3), 237–245 (2009).

11   Zhang, J., Liu, S.S., and Liu, H.L., Effect of ionic liquid on the toxicity of pesticide to *Vibrio-qinghaiensis sp.*-Q67, *J. Hazard. Mater.* **170**(2–3), 920–927 (2009).

12 Imperato, G., König, B., and Chiappe, C., Ionic green solvents from renewable resources, *Eur. J. Org. Chem.* **7**, 1049–1058 (2007).

13 Carter, E.B., Culver, S.L., Fox, P.A., Goode, R.D., Ntai, I., Tickell, M.D., Traylor, R.K., Hoffman, N.W., and Davis, J.H., Sweet success: ionic liquids derived from non-nutritive sweeteners, *Chem. Commun.*, 630–631 (2004).

14 Fukumoto, K., and Ohno, H., Design and synthesis of hydrophobic and chiral anions from amino acids as precursor for functional ionic liquids, *Chem. Commun.*, 3081–3083 (2006).

15 Poletti, L., Chiappe, C., Lay, L., Pieraccini, D., Polito, L., and Russo, G., Glucose-derived ionic liquids: exploring low-cost sources for novel chiral solvents, *Green Chem.* **9**, 337–341 (2007).

16 Petkovic, M., Ferguson, J.L., Gunaratne, H.Q.N., Ferreira, R., Leitão, M.C., Seddon, K.R., Rebelo, L.P.N., and Silva Pereira, C., Novel biocompatible cholinium-based ionic liquids—toxicity and biodegradability, *Green Chem.* **12**(4), 643–649 (2010).

17 van Rantwijk, F., and Sheldon, R.A., Biocatalysis in ionic liquids, *Chem. Rev.* **107**(6), 2757–2785 (2007).

18 Bräutigam, S., Bringer-Meyer, S., and Weuster-Botz, D., Asymmetric whole cell biotransformations in biphasic ionic liquid/water-systems by use of recombinant *Escherichia coli* with intracellular cofactor regeneration, *Tetrahedron: Asymmetry* **18**, 1883–1887 (2007).

19 Pfruender, H., Jones, R., and Weuster-Botz, D., Water immiscible ionic liquids as solvents for whole cell biocatalysis, *J. Biotechnol.* **124**(1), 182–190 (2006).

20 Fujita, K., MacFarlane, D.R., and Forsyth, M., Protein solubilising and stabilising ionic liquids, *Chem. Commun.*, 4804–4806 (2005).

21 Byrne, N., Wang, L.M., Belieres, J.P., and Angell, C.A., Reversible folding-unfolding, aggregation protection, and multi-year stabilization, in high concentration protein solutions, using ionic liquids, *Chem. Commun.*, 2714–2716 (2007).

22 Fujita, K., MacFarlane, D.R., Forsyth, M., Yoshizawa-Fujita, M., Murata, K., Nakamura, N., and Ohno, H., Solubility and stability of cytochrome c in hydrated ionic liquids: effect of oxo acid residues and kosmotropicity, *Biomacromolecules* **8**(7), 2080–2086 (2007).

23 Hough, W.L., and Rogers, R.D., Ionic liquids then and now: from solvents to materials to active pharmaceutical ingredients, *Bull. Chem. Soc. Jpn.* **80**(12), 2262–2269 (2007).

24 Bica, K., Rijksen, C., Nieuwenhuyzen, M., and Rogers, R.D., In search of pure liquid salt forms of aspirin: ionic liquid approaches with acetylsalicylic acid and salicylic acid, *Phys. Chem. Chem. Phys.* **12**, 2011–2017 (2010).

25 Carlile, M., Watkinson, S., and Gooday, G., *The fungi*, Elsevier Academic Press, Amsterdam (2001).

26 Coleman, D., and Gathergood, N., Biodegradation studies of ionic liquids, *Chem. Soc. Rev.* **39**(2), 600–637 (2010).

27 Studzińska, S., Kowalkowski, T., and Buszewski, B., Study of ionic liquid cations transport in soil, *J. Hazard. Mater.* **168**(2–3), 1542–1547 (2009).

28 Pinedo-Rivilla, C., Aleu, J., and Collado, I.G., Pollutants biodegradation by fungi, *Curr. Org. Chem.* **13**(12), 1194–1214 (2009).

29 Petkovic, M., Ferguson, J.L., Bohn, A., Trindade, J., Martins, I., Carvalho, M.B., Leitão, M.C., Rodrigues, C., Garcia, H., Ferreira, R., Seddon, K.R., Rebelo, L.P.N., and Silva Pereira, C., Exploring fungal activity in the presence of ionic liquids, *Green Chem.* **11**(6), 889–894 (2009).

30 Petkovic, M., and Silva Pereira, C., Unpublished data (2008).

31 Demberelnyamba, D., Kim, K.S., Choi, S.J., Park, S.Y., Lee, H., Kim, C.J., and Yoo, I.D., Synthesis and antimicrobial properties of imidazolium and pyrrolidinonium salts, *Bioorg. Med. Chem.* **12**(5), 853–857 (2004).

32 Pernak, J., Sobaszkiewicz, K., and Mirska, I., Anti-microbial activities of ionic liquids, *Green Chem.* **5**(1), 52–56 (2003).

33 Deive, F.J., Rodríguez, A., Varela, A., Rodrígues, C., Leitão, M.C., Houbraken, J.A.M.P., Pereiro, A.B., Longo, M.A., Ángeles Sanromán, M., Samson, R.A., Rebelo, L.P.N., and Silva Pereira, C., Impact of ionic liquids on extreme microbial biotypes from soil, *Green Chem.* **13**(3), 687–696 (2011).

34 Wang, L., and Zhuang, W.Y., Phylogenetic analyses of penicillia based on partial calmodulin gene sequences, *Biosystems* **88**(1–2), 113–126 (2007).

35 Couling, D.J., Bernot, R.J., Docherty, K.M., Dixon, J.K., and Maginn, E.J., Assessing the factors responsible for ionic liquid toxicity to aquatic organisms via quantitative structure-property relationship modeling, *Green Chem.* **8**(1), 82–90 (2006).

36 Ranke, J., Stolte, S., Stormann, R., Arning, J., and Jastorff, B., Design of sustainable chemical products—the example of ionic liquids, *Chem. Rev.* **107**(6), 2183–2206 (2007).

37 Stolte, S., Arning, J., Bottin-Weber, U., Müller, A., Pitner, W.R., Welz-Biermann, U., Jastorff, B., and Ranke, J., Effects of different head groups and functionalised side chains on the cytotoxicity of ionic liquids, *Green Chem.* **9**(7), 760–767 (2007).

38 Docherty, K.M., and Kulpa, C.F., Toxicity and antimicrobial activity of imidazolium and pyridinium ionic liquids, *Green Chem.* **7**(4), 185–189 (2005).

39 Wells, A.S., and Coombe, V.T., On the freshwater ecotoxicity and biodegradation properties of some common ionic liquids, *Org. Process Res. Dev.* **10**(4), 794–798 (2006).

40 Latała, A., Nędzi, M., and Stepnowski, P., Toxicity of imidazolium and pyridinium based ionic liquids towards algae. *Chlorella vulgaris, Oocystis submarina* (green algae) and *Cyclotella meneghiniana, Skeletonema marinoi* (diatoms), *Green Chem.* **11**(4), 580–588 (2009).

41 Luis, P., Ortiz, I., Aldaco, R., and Irabien, A., A novel group contribution method in the development of a QSAR for predicting the toxicity (Vibrio fischeri EC50) of ionic liquids, *Ecotox. Environ. Safe.* **67**(3), 423–429 (2007).

42 Torrecilla, J.S., Palomar, J., Lemus, J., and Rodriguez, F., A quantum-chemical-based guide to analyze/quantify the cytotoxicity of ionic liquids, *Green Chem.* **12**(1), 123–134 (2010).

43 Palomar, J., Torrecilla, J.S., Lemus, J., Ferro, V.R., and Rodríguez, F., A COSMO-RS based guide to analyze/quantify the polarity of ionic liquids and their mixtures with organic cosolvents, *Phys. Chem. Chem. Phys.* **12**(8), 1991–2000 (2010).

44 Baumann, M.D., Daugulis, A.J., and Jessop, P.G., Phosphonium ionic liquids for degradation of phenol in a two-phase partitioning bioreactor, *Appl. Microbiol. Biotechnol.* **67**(1), 131–137 (2005).

45 Bernot, R.J., Brueseke, M.A., Evans-White, M.A., and Lamberti, G.A., Acute and chronic toxicity of imidazolium-based ionic liquids on *Daphnia magna*, *Environ. Toxicol. Chem.* **24**(1), 87–92 (2005).

46 Cho, C.W., Jeon, Y.C., Pham, T.P.T., Vijayaraghavan, K., and Yun, Y.S., The ecotoxicity of ionic liquids and traditional organic solvents on microalga *Selenastrum capricornutum*, *Ecotox. Environ. Safe.* **71**(1), 166–171 (2008).

47 Clare, B., Sirwardana, A., and MacFarlane, D.R., Synthesis, purification and characterisation of ionic liquids, in: *Ionic liquids (Topics in current chemistry)*, ed. B. Kirchner, Springer, Berlin, Heidelberg, New York (2009), pp. 1–40.

48 Petkovic, M., Hartmann, D.O., Adamová, G., Seddon, K.R., Rebelo, L.P.N, and Silva Pereira, C., Unravelling the mechanism of toxicity of alkyltributylphosphonium chlorides in Aspergillus nidulans conidia, *New J. Chem.*, **36**(1), 56–63 (2012).

49 Cai, L., Jeewon, R., and Hyde, K.D., Phylogenetic investigations of *Sordariaceae* based on multiple gene sequences and morphology, *Mycol. Res.* **110**, 137–150 (2006).

50 Pernak, J., and Branicka, M., The properties of 1-alkoxymethyl-3-hydroxypyridinium and 1-alkoxymethyl-3-dimethylaminopyridinium chlorides, *J. Surfactants Deterg.* **6**(2), 119–123 (2003).

51 Schaffran, T., Justus, E., Elfert, M., Chen, T., and Gabel, D., Toxicity of N,N,N-trialkylammoniododecaborates as new anions of ionic liquids in cellular, liposomal and enzymatic test systems, *Green Chem.* **11**(9), 1458–1464 (2009).

52 Cornmell, R.J., Winder, C.L., Tiddy, G.J.T., Goodacre, R., and Stephens, G., Accumulation of ionic liquids in *Escherichia coli* cells, *Green Chem.* **10**(8), 836–841 (2008).

53 Stolte, S., Arning, J., Bottin-Weber, U., Matzke, M., Stock, F., Thiele, K., Uerdingen, M., Welz-Biermann, U., Jastorff, B., and Ranke, J., Anion effects on the cytotoxicity of ionic liquids, *Green Chem.* **8**(7), 621–629 (2006).

54 Pernak, J., Syguda, A., Mirska, I., Pernak, A., Nawrot, J., Prądzyńska, A., Griffin, S.T., and Rogers, R.D., Choline-derivative-based ionic liquids, *Chem. Eur. J.* **13**(24), 6817–6827 (2007).

55 Fukaya, Y., Iizuka, Y., Sekikawa, K., and Ohno, H., Bio ionic liquids: room temperature ionic liquids composed wholly of biomaterials, *Green Chem.* **9**(11), 1155–1157 (2007).

56 Boethling, R.S., Sommer, E., and DiFiore, D., Designing small molecules for biodegradability, *Chem. Rev.* **107**(6), 2207–2227 (2007).

57 Nockemann, P., Thijs, B., Driesen, K., Janssen, C.R., Van Hecke, K., Van Meervelt, L., Kossmann, S., Kirchner, B., and Binnemans, K., Choline saccharinate and choline acesulfamate: ionic liquids with low toxicities, *J. Phys. Chem. B* **111**(19), 5254–5263 (2007).

58 Dipeolu, O., Green, E., and Stephens, G., Effects of water-miscible ionic liquids on cell growth and nitro reduction using *Clostridium sporogenes*, *Green Chem.* **11**(3), 397–401 (2009).

59 Pham, T.P.T., Cho, C.W., Min, J., and Yun, Y.S., Alkyl-chain length effects of imidazolium and pyridinium ionic liquids on photosynthetic response of *Pseudokirchneriella subcapitata*, *J. Biosci. Bioeng.* **105**(4), 425–428 (2008).

60   Ranke, J., Müller, A., Bottin-Weber, U., Stock, F., Stolte, S., Arning, J., Stormann, R., and Jastorff, B., Lipophilicity parameters for ionic liquid cations and their correlation to *in vitro* cytotoxicity, *Ecotox. Environ. Safe.* **67**(3), 430–438 (2007).

61   Garcia, H., Ferreira, R., Petkovic, M., Ferguson, J.L., Leitão, M.C., Gunaratne, N., Seddon, K.R., Rebelo, L.P.N., and Silva Pereira, C., Dissolution of cork biopolymers by biocompatible ionic liquids, *Green Chem.* **12**(3), 367–369 (2010).

62   Klein, R., Kellermeier, M., Drechsler, M., Touraud, D., and Kunz, W., Solubilisation of stearic acid by the organic base choline hydroxide, *Colloid Surf. A Physicochem. Eng. Asp.* **338**(1–3), 129–134 (2009).

63   Stolte, S., Abdulkarim, S., Arning, J., Blomeyer-Nienstedt, A.K., Bottin-Weber, U., Matzke, M., Ranke, J., Jastorff, B., and Thoming, J., Primary biodegradation of ionic liquid cations, identification of degradation products of 1-methyl-3-octylimidazolium chloride and electrochemical wastewater treatment of poorly biodegradable compounds, *Green Chem.* **10**(2), 214–224 (2008).

64   Dechema, BATIL-2 Meeting, http://events.dechema.de/batil2.html (2009).

65   Romero, A., Santos, A., Tojo, J., and Rodríguez, A., Toxicity and biodegradability of imidazolium ionic liquids, *J. Hazard. Mater.* **151**(1), 268–273 (2008).

66   Arning, J., Stolte, S., Böschen, A., Stock, F., Pitner, W.R., Welz-Biermann, U., Jastorff, B., and Ranke, J., Qualitative and quantitative structure activity relationships for the inhibitory effects of cationic head groups, functionalised side chains and anions of ionic liquids on acetylcholinesterase, *Green Chem.* **10**(1), 47–58 (2008).

67   Docherty, K.M., Dixon, J.K., and Kulpa, C.F., Biodegradability of imidazolium and pyridinium ionic liquids by an activated sludge microbial community, *Biodegradation* **18**(4), 481–493 (2007).

68   Deive, F.J., Rodríguez, A., and Silva Pereira, C., Unpublished data (2010).

69   Gross, H., Strategies to unravel the function of orphan biosynthesis pathways: recent examples and future prospects, *Appl. Microbiol. Biotechnol.* **75**(2), 267–277 (2007).

70   Oh, D.C., Kauffman, C.A., Jensen, P.R., and Fenical, W., Induced production of emericellamides A and B from the marine-derived fungus *Emericella sp* in competing co-culture, *J. Nat. Prod.* **70**(4), 515–520 (2007).

71   Matzapetakis, M., Petkovic, M., Garcia, H., and Silva Pereira, C., Unpublished data (2010).

72   Larsen, T.O., Smedsgaard, J., Nielsen, K.F., Hansen, M.E., and Frisvad, J.C., Phenotypic taxonomy and metabolite profiling in microbial drug discovery, *Nat. Prod. Rep.* **22**(6), 672–695 (2005).

73   Newman, D.J., and Cragg, G.M., Natural products as sources of new drugs over the last 25 years, *J. Nat. Prod.* **70**(3), 461–477 (2007).

74   Peláez, F., Biological activities of fungal metabolites, in: *Handbook of industrial mycology*, ed. Z. An, Marcel Dekker, New York (2005), pp. 49–92.

75   Fox, E.M., and Howlett, B.J., Secondary metabolism: regulation and role in fungal biology, *Curr. Opin. Microbiol.* **11**(6), 481–487 (2008).

76   Shwab, E.K., and Keller, N.P., Regulation of secondary metabolite production in filamentous ascomycetes, *Mycol. Res.* **112**, 225–230 (2008).

77   Hertweck, C., Hidden biosynthetic treasures brought to light, *Nat. Chem. Biol.* **5**(7), 450–452 (2009).

78 Alves, P.C., Martins, I., Becker, J., and Silva Pereira, C., unpublished data (2010).

79 de Gonzalo, G., Lavandera, I., Durchschein, K., Wurm, D., Faber, K., and Kroutil, W., Asymmetric biocatalytic reduction of ketones using hydroxy-functionalised water-miscible ionic liquids as solvents, *Tetrahedron: Asymmetry* **18**(21), 2541–2546 (2007).

80 Das, D., Dasgupta, A., and Das, P.K., Improved activity of horseradish peroxidase (HRP) in "specifically designed" ionic liquid, *Tetrahedron Lett.* **48**(32), 5635–5639 (2007).

81 Lozano, P., De Diego, T., Gmouh, S., Vaultier, M., and Iborra, J.L., Dynamic structure-function relationships in enzyme stabilization by ionic liquids, *Biocatal. Biotransform.* **23**(3–4), 169–176 (2005).

82 Fujita, K., Forsyth, M., MacFarlane, D.R., Reid, R.W., and Elliott, G.D., Unexpected improvement in stability and utility of cytochrome c by solution in biocompatible ionic liquids, *Biotechnol. Bioeng.* **94**(6), 1209–1213 (2006).

83 Friedrich, M., Grosser, R.J., Kern, E.A., Inskeep, W.P., and Ward, D.M., Effect of model sorptive phases on phenanthrene biodegradation: molecular analysis of enrichments and isolates suggests selection based on bioavailability, *Appl. Environ. Microbiol.* **66**(7), 2703–2710 (2000).

84 Subramanian, B., Yang, Q.L., Yang, Q.J., Khodadoust, A.P., and Dionysiou, D.D., Photodegradation of pentachlorophenol in room temperature ionic liquids, *J. Photochem. Photobiol. A* **192**(2–3), 114–121 (2007).

10 Use of Ionic Liquids in Dye-Sensitised Solar Cells

JENNIFER M. PRINGLE

ARC Centre of Excellence for Electromaterials Science, Department of Materials Engineering, Monash University, Clayton, Victoria, Australia

ABSTRACT

Dye-sensitised solar cells (DSSCs) are increasingly being recognised as a viable alternative to conventional silicon-based solar cells; these devices have significant potential advantages in terms of material costs and toxicity, as well as the potential to manufacture ultra-thin and flexible cells. One of the key components of an efficient and stable DSSC is the electrolyte, and while ethanenitrile-based electrolytes currently provide the best efficiencies, the use of ionic liquids is extremely beneficial in terms of device stability. This chapter discusses the different aspects of research into the use of ionic liquids in DSSCs. A range of different anions and cations have been utilised in combination with a wide variety of different photosensitisers, including a number of organic dyes. The development of quasi-solid state electrolytes that utilise ionic liquids is a very promising area of research, while the development of ionic liquid-based devices with different redox couples is very important, but has thus far had only limited success. The use of alternative counter electrodes and different cell structures, in combination with ionic liquids, is also discussed.

10.1 INTRODUCTION

Humanity is facing an unprecedented energy crisis. Clearly, our reliance on rapidly diminishing and environmentally disastrous fossil fuels cannot continue; the global atmosphere does not have the capacity to indefinitely accept

Ionic Liquids UnCOILed: Critical Expert Overviews, First Edition. Edited by Natalia V. Plechkova and Kenneth R. Seddon.
© 2013 John Wiley & Sons, Inc. Published 2013 by John Wiley & Sons, Inc.

greenhouse gas emissions without dire environmental consequences. Thus, the search for alternative energy sources, with significantly reduced carbon emissions, becomes increasingly urgent.

The earth receives more energy from the sun in only a few days than was produced from all of the fossil fuel consumption over the whole of human history [1]. In fact, three weeks of sunshine offsets all known fossil fuel reserves [1]. Clearly, then, being able to effectively harness and store this energy could efficiently address the world energy crisis while avoiding the potentially catastrophic effects of continued carbon dioxide emissions. It has been calculated that only 1% of the world's marginal areas (scrubland, etc.) would need to be covered by 15% efficient photovoltaics to supply the entire world's forecast energy requirements [1].

In addition to the environmental considerations, solar energy is extremely advantageous as it allows electricity to be generated right where it is needed rather than having to be transported thousands of kilometres from the nearest power plant. Nearly half of the people in developing countries live in rural areas; solar power allows the supply of electricity to rural communities, not only for homes, but also to power water supplies, communications, street lighting, and other essential services. However, for this technology to become more cost-effective, and thus enable its widespread use, an improvement in the efficiency of these devices is paramount.

The majority of research into solar cells and their subsequent commercialisation has focussed on silicon cells, which utilise modified silicon as the light absorber and the charge transporter. However, the high manufacturing costs and the toxicity of the chemicals used in their preparation can be a significant deterrent to their widespread use. While research continues into addressing these concerns, attention is rapidly turning to the development of other types of solar cells, namely those based on organic compounds, such as dye-sensitised solar cells (DSSCs). These devices have advantages over the traditional silicon cells in terms of material cost and toxicity, as well as the potential to manufacture ultra-thin and flexible cells.

The dye-sensitised nanocrystalline TiO_2 solar cell was first reported by Grätzel in 1991 [2], and this group has subsequently led the research in the field. At the heart of this type of solar cell is a high surface area, nanocrystalline TiO_2 film with a monolayer of photoactive dye attached to the surface (Fig. 10.1). The surface of this semiconductor (the photoanode) is in contact with a redox mediator—normally I^-/I_3^- in a suitable electrolyte. When the cell is exposed to sunlight, photo-excitation of the dye results in electron insertion into the conduction band of the TiO_2, and the dye is subsequently regenerated by electron donation from the redox electrolyte. The mediator in the electrolyte is itself regenerated by electrons from the external circuit (through the cathode) and hence the electrical circuit is completed [3, 4]. For photovoltaic activity to occur, regeneration of the dye by electron donation from the redox couple must occur before the dye can recapture the original electron. An essential difference between this type of solar cell and the original silicon cells

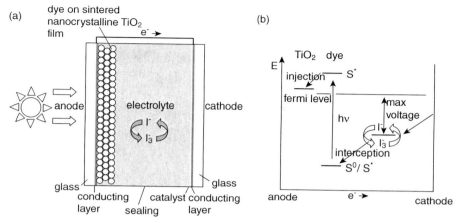

Figure 10.1   A schematic representation of (a) the structure of a dye-sensitised solar cell (not to scale), and (b) the energy levels within a DSSC.

is that in this system the light-absorbing material (the dye) is distinct from the charge carrier transport materials. This distinction significantly reduces the manufacturing cost of these cells as lower purity materials can be used, and the assembly processes need not be as rigorous as those required for the manufacture of silicon cells.

The efficiency of a solar cell is measured by recording the current–voltage characteristics of the cell under simulated solar irradiation. This irradiance is normally 100 mW cm^{-2}, standard reference AM 1.5 spectrum, at 25 °C, although these details should be specified in the experimental details of any report.

The efficiency of the cell, η, is given by:

$$\eta = J_{SC}V_{OC}FF/I_S$$

where I_S is the light intensity, J_{sc} is the short circuit current, FF is the fill factor, and V_{OC} is the open circuit voltage.

The short-circuit current density (J_{sc}) is the maximum current through the load under short-circuit conditions (when the potential difference is zero). The photocurrent of a cell depends on factors such as the difference in energy levels between the highest occupied molecular orbital (HOMO) and the lowest unoccupied molecular orbital (LUMO) of the dye, the energy gap between the LUMO of the dye and the conduction band of the TiO$_2$, and the diffusion of the redox couple through the electrolyte.

The open-circuit voltage of a cell is the maximum voltage that can be obtained at the load under open-circuit conditions—the difference in potential between the cathode and anode when there is no external load, that is, when the current is zero. In theory, the maximum potential output is the difference between the Fermi level of the TiO_2 and the redox potential of the electrolyte, with a maximum of about 0.9 V for the I^-/I_3^- redox couple (note that this latter value will depend on the composition of the electrolyte, which is important to keep in mind when comparing different solvent systems). A negative shift of the conduction band edge of the TiO_2 (as occurs when additives such as 4-t-butylpyridine are used) results in a higher V_{OC} but lower short-circuit currents. The quantum yield of photocurrent generation is normally evaluated by measurement of the incident photon to current conversion efficiency (IPCE); this will be over 80% for a good DSSC and can be as high as 90%.

The DSSC is made up of many components, each of which can have a significant influence on the efficiency of the cell. Clearly, the nature of the dye used will directly impact the efficiencies of the solar cell, and there has been significant work focussed in this area [5–7]. The majority of the dyes used in the development of DSSCs are ruthenium based, although, as discussed later, organic dyes have shown significant promise in recent years.

There are a number of additives that are commonly used to improve cell performance by various means. Additives such as 4-t-butylpyridine (TBP) [8], 1-methylbenzimidazole (MBI), or 1-butylbenzimidazole (NBB) [9] can increase the open-circuit voltage of the DSSC as a result of inhibition of the unwanted back electron transfer. These additives are believed to suppress the dark current at the TiO_2/electrolyte interface by blocking the surface states that are taking part in the charge transfer (TBP absorbs onto the surface of the TiO_2 because it is a Lewis base). This adsorption onto the TiO_2 surface may also cause the Fermi energy to increase, and its effect is concentration dependent [10]. Further, Li^+ or Mg^{2+} additives can absorb onto the TiO_2 surface and thereby positively charge the surface layer, which inhibits the undesirable electron loss (back-transfer into the dye or the redox couple) and drives the electron injection in the desired direction [11, 12]. Guanidinium thiocyanate (GNCS) is also commonly used to enhance the photovoltage and stability, but the exact mechanism is the subject of some debate [13–15]. Copper(I) iodide can also be used to enhance the performance of a DSSC with an ionic liquid, and again this is believed to be the result of adsorption of Cu^+ onto the TiO_2 surface [16]. The use of a co-absorbent, such as a phosphinic acid, applied to the semiconductor surface at the same time as the dye, can also improve the stability of the V_{OC} and photovoltaic efficiency of the DSSC [17].

The solvents typically used in DSSCs are volatile organic solvents such as ethanenitrile and propanenitrile, and the best efficiencies of DSSCs utilising molecular solvents are presently just over 11% [18, 19]. However, there are a number of problems associated with these solvents, particularly the high vapour pressure, which is exacerbated when the cell is exposed to considerable heat from the sun and can result in solvent evaporation. Reaction of water or dioxygen with these molecular solvents (if the solar cell is not perfectly sealed) will also significantly reduce their performance and lifetime. Further, there is strong drive towards the development of lighter, more flexible DSSCs using plastic substrates such as PEN (poly(ethylene-2,6-naphthalene dicarboxylate) or, more trivially, polyethylene naphthalate), which are also potentially much cheaper if they can be made using reel-to-reel processes. However, volatile solvents can permeate through this plastic film. As a result of these significant problems, there has now been considerable research into the applicability of ionic liquids as electrolytes for solar cells. Initially, the efficiency of these cells lagged considerably behind their molecular solvent-based counterparts, but they have improved rapidly in recent years, with efficiencies now of over 9% [20]. The negligible vapour pressure, good ionic conductivity, electrochemical and thermal stability, and non-flammability of ionic liquids are all advantageous properties for their use in solar cells and impart much better stability to the devices. Ionic liquid-based electrolytes are now commonly composed of more than one type of ionic liquid, where one is predominantly the solvent and one (or more) is the iodide source (e.g., the low viscosity 1-propyl-3-methylimidazoium iodide) as needed when the I^-/I_3^- redox couple is used. Indeed, using a mixture of two or more salts can result in highly conductive and fluid eutectic melts [21]. The use of an ionic liquid may also reduce or eliminate the problems of crystallisation of additives out of solution; 1-methylbenzimidazole (MBI) has been shown to crystallise out of a glutaronitrile-based electrolyte solution as $(MBI)_6(MBIH^+)_2(I^-)(I_3^-)$, and crystals were also observed to form in an analogous 3-methoxypropanenitrile solution [22]. The crystallisation of additives could lead to significant decreases in efficiencies of the DSSCs; not only would the beneficial effect of the additives be reduced or eliminated but it would also alter the composition of the electrolyte (e.g., reducing the concentration of I_3^-), and this problem may be exacerbated when the DSSC is exposed to lower temperatures during use. However, it has not yet been demonstrated that crystallisation of additives over time from ionic liquid-based electrolytes does not occur, and this is just one of the many aspects of the long-term stability of DSSCs that need to be considered.

It is important to bear in mind that DSSCs are extremely complex systems, with many different factors that can influence cell performance, and direct comparison of efficiency results from different research groups should be undertaken with caution. For example, in comparing different electrolyte systems, not only are there likely to be compositional differences (e.g., different concentrations and types of additives, use of pretreatments, different dyes, etc.), but there are also likely to be significant engineering differences. The surface area of the DSSCs tested can have a strong influence on efficiency, as can the quality of the materials used (the resistance of the conducting glass, purity of dye, screen-printing of the working electrodes, etc.), the testing conditions (light filtering and masking, simulator calibration, etc.), and, finally, the expertise of the researchers. As a result of these variables, it is an unfortunate fact that every group (or even different researchers within the same group) is likely to achieve different efficiencies for their "standard" DSSCs, which makes assessment of any new development rather subjective. In an ideal world, all published efficiencies would be independently verified in certified testing facilities, and every group would report the efficiencies of their latest DSSC development with reference to their best "standard" cell (e.g., with a pre-defined ethanenitrile-based electrolyte composition, cell size, etc.). In the meantime, authors reporting the use of new solvent systems, as is the focus of this chapter, are encouraged to report all efficiencies against the efficiency of their standard ethanenitrile-based cells.

It is also important to note that for all the different factors that affect the efficiency of a DSSC, as discussed above, the vast majority of past research into the optimisation of these parameters was focussed on molecular solvent systems rather than ionic liquids. Not only do we not yet fully understand the mode of action or efficacy of, for example, additives in ionic liquids, it is likely that more detailed investigations will allow further optimisation of dye structure, additives, pretreatments, semiconductor morphologies, etc, specifically for use with ionic liquid systems, and this could yield significant further efficiency enhancements.

The following discussion is not intended to be a comprehensive review of all of the research into the use of ionic liquids in DSSCs; for this purpose there are a number of excellent reviews available [5, 23, 24]. Rather, it is intended to be a critical overview of the field, primarily from the perspective of an ionic liquid chemist, in an attempt to highlight the most promising or critical areas of past and future research in this exciting area.

10.1.1   Room Temperature Ionic Liquid Electrolytes

There is a steadily increasing range of ionic liquids that have been investigated as electrolytes for photovoltaics (Fig. 10.2), although this is still only a fraction of those available. It is important to note that while this is a not a new field, rapid progress has been made recently, and the use of advanced analytical techniques can now provide a more in-depth insight into the different mechanisms at play in these complex devices, and the factors that limit the efficiencies, than was available in the initial reports of ionic liquid utilisation.

Papageorgiou et al. [39] were the first to recognise the potential for using ionic liquids in DSSCs. They reported that use of 1-hexyl-3-methylimidazolium iodide, [C_6mim]I, with an iodide/triiodide redox couple, can produce an extremely stable photovoltaic cell. The ratio of electrons generated to the quantity of dye present (the sensitiser turnover) was very high, at around 50 million electrons per dye molecule. At 1 sun, the performance of cells with the pure ionic liquid (with 7 mM iodine) was quite satisfactory, but at higher light intensity the current was found to be limited by the diffusion of the tri-iodide through the electrolyte. Measurement of the diffusion coefficient of triiodide in this ionic liquid, and in mixtures of [C_6mim]I and [C_4mim][OTf], indicated a change in the mechanism of mass- or charge transfer, to a chemical (electron) exchange charge carrier transfer mechanism like the Grotthuss mechanism [39]. This is an important consideration for photovoltaic systems, as it contradicts the assumption that an increase in iodide concentration induces more ion pairing and would therefore reduce diffusion. In fact, the

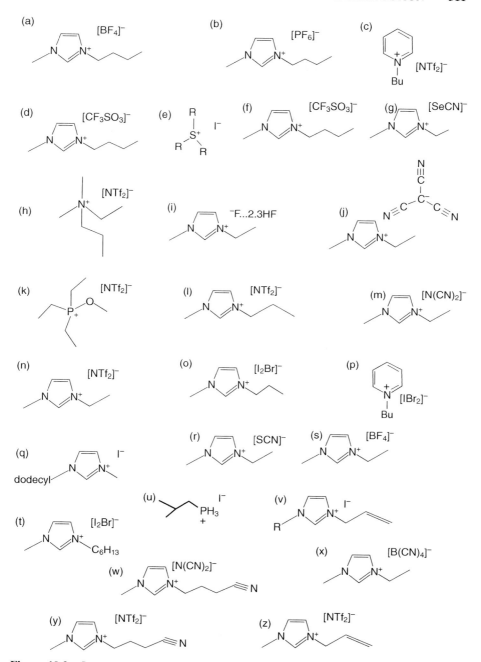

Figure 10.2 Some of the ionic liquids that have been used as electrolytes for DSSCs, with selected references; (a) [25], (b) [12], (c) [12], (d) [25], (e) [26], (f) [25], (g) [27], (h) [28], (i) [29], (j) [12, 30, 31], (k) [28], (l) [32], (m) [28, 31, 33, 34], (n) [12, 28], (o) [25], (p) [25], (q) [35], (r) [28, 31], (s) [12, 28], (t) [25], (u) [36], (v) [21, 37], (w) [32], (x) [21, 28, 38], (y) [32], and (z) [37].

triiodide diffusion rate in these systems does not obey the fundamental Stokes–Einstein equation because the product of viscosity and diffusion coefficient (the Walden product) is not constant at constant temperature but changes as a function of electrolyte composition. While the experimental Stokes–Einstein ratio remained almost constant for varying triiodide concentrations in the pure [C_6mim]I ionic liquid (almost constant diffusion coefficient and viscosity), a 4 M addition of iodine produced a massive drop in viscosity (from 1800 mP to 125 mP) [39]. It is also possible to decrease the viscosity of the electrolyte by the addition of a little molecular solvent, and this strategy is still used by some researchers to give better cell performance, but clearly this reduces the benefits of using a non-volatile ionic liquid. Thus, much of the subsequent research into the use of ionic liquids in DSSCs has focused on the development and utilisation of low viscosity, high-conductivity ionic liquid electrolytes, as discussed below. There has also been significant research into the transport of the I^-/I_3^- redox couple in different ionic liquids [40–44], and attempts to correlate the structure and physical properties of ionic liquids with their photovoltaic performance [45–50], in an attempt to identify the most dominant factors determining the efficiencies of these devices.

The effect of the iodide/triiodide concentration in a [C_2mim][NTf_2] ionic liquid was subsequently studied by Kawano et al. [12]. When this system contained a relatively high concentration of the redox couple ($[I^-]+[I_3^-] = 1.5$ M), the photocurrents of the cell were proportional to the photointensity, up to 100 mW cm^{-1}, that is, it did not become limited by diffusion or other processes. However, with a lower concentration of redox couple (0.5 M), the photocurrent saturated at 20 mW cm^{-2} photointensity. Thus, for good cell performance at high light intensities, it is essential for these viscous systems to have a high concentration of redox couple present to ensure that the flux of electrons towards the electrodes remains high. This flux is proportional to the concentration and diffusivity of the redox couple, which is normally inversely proportional to the viscosity of the solvent. However, the short-circuit currents of the ionic liquid-containing cells can be over 80% of that achieved for molecular solvents, despite the viscosity being over 10 times higher. Thus, the exchange-reaction-based diffusion rate dominates the charge transport processes at high concentrations of redox couple, but at low concentrations the diffusion processes dominate.

However, while the performance of an ionic liquid electrolyte will be enhanced by using higher concentrations of I^-/I_3^- than is necessary in a molecular solvent, it is important to note that there will be an optimum redox concentration, beyond which the cell performance will decrease; I_3^- is a strong absorber of visible light, which competes with the absorption by the dye, and it also enhances the unwanted back electron transfer from the conduction band of the TiO$_2$.

It is also important to note that the open-circuit voltage of the cell is directly related to the Nernst potential of the electrolyte and that this, in turn, is also related to the composition of the electrolyte.

The exchange reaction is considered to proceed via the following process [40]:

$$I^- + I_3^- \rightarrow I^- \ldots I_2 \ldots I^- \rightarrow I_3^- + I^-.$$

Four different methods for measuring the diffusion of triiodide in an ionic liquid mixture have recently been compared: impedance spectroscopy or polarisation measurements using thin layer cells, and cyclic voltammetry or chronoamperometry at microelectrodes [51]. Impedance spectroscopy was concluded to be the least accurate and takes the longest time, although all of the methods gave largely reproducible and concordant results, with a standard deviation of only about 2%. The measured triiodide diffusion coefficients in the mixture ([C$_3$mim]I and [C$_4$mim][BF$_4$]), with 10–100 wt% [C$_3$mim]I at 25 °C, were measured to be between 1.79×10^{-7} and 1.35×10^{-7} cm^2 s^{-1}. The non-Stokesian behaviour of the ionic liquid mixture was clearly demonstrated; although the viscosity increased by nearly 1000% when increasing the amount of [C$_3$mim]I from 10 to 100 wt%, the triiodide diffusion coefficients decreased by only about 30% [51]. As explained earlier, this is the result of the Grotthuss exchange mechanism, which contributes to the triiodide diffusion at high iodide concentrations.

Since the initial reports on the use of ionic liquids in DSSCs, research has focussed primarily on a few different areas, as discussed later. There has been a range of new ionic liquid systems investigated, primarily focussing on the development of the least viscous electrolyte systems but also, importantly, the development of quasi-solid state electrolytes using ionic liquids. The development of new redox couples is a very important area of research, while the design and synthesis of new sensitisers is also key, as these materials lie at the very heart of the device.

10.1.2 Anion Variations

As is well known in the wider ionic liquid field, variations in the anion can result in significant changes in the viscosity of an ionic liquid. However, changes to the anion can also alter the hydrophobicity, basicity, ionicity, etc. of the ionic liquid, and all of these parameters may influence the efficiency of a DSSC in an as yet unquantified manner. Kawano et al. [12] compared the performance of a number of different ionic liquids and found that efficiencies increased in the order:

$$[C_4mim][PF_6] < [C_4py][NTf_2] < [C_2mim][NTf_2] < [C_2mim][BF_4]$$
$$< [C_2mim][N(CN)_2]$$

up to a maximum efficiency of 3.8% (at 1 sun, N3 dye [defined in Fig. 10.3] [12]. With the exception of the dicyanamide ionic liquid, the J_{sc} changed depending on the ionic conductivity of the ionic liquid (higher conductivity

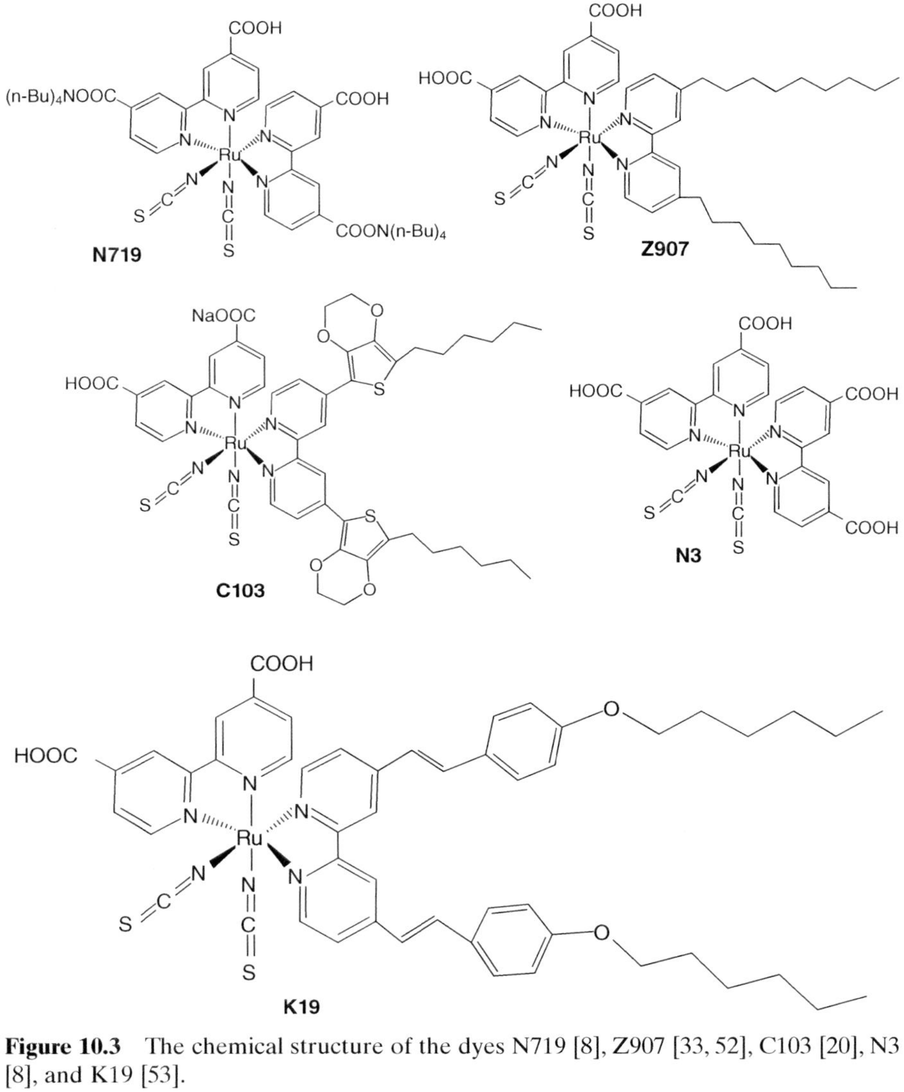

Figure 10.3   The chemical structure of the dyes N719 [8], Z907 [33, 52], C103 [20], N3 [8], and K19 [53].

gives higher photocurrent). The J_{SC} of ionic liquid DSSCs can be over 80% of those of the standard ethanenitrile-based DSSCs, despite the viscosity of the ionic liquid electrolytes being over 10 times higher. The dicyanamide ionic liquid gave a cell with an open-circuit voltage 100 mV higher than the others, suggesting that this ionic liquid has an influence on either the Fermi energy of the TiO_2, or on the equilibrium redox potential of the iodide/triiodide couple [12], and it is also the least viscous of those tested. They also found a beneficial effect of using additives TBP and LiI, as is observed for DSSCs with molecular solvents.

In the drive to utilise low-viscosity ionic liquid systems, interesting work has also been performed on an imidazolium F/HF system [29], but safety concerns relating to this system are yet to be clarified.

Some of the most fluid ionic liquids developed to date are those utilising the anions containing cyano groups (i.e., dicyanamide, $[N(CN)_2]^-$; thiocyanate, $[SCN]^-$; tetracyanoborate, $[B(CN)_4]^-$; and tricyanomethanide, $[C(CN)_3]^-$) and the performance of imidazolium salts of these anions in DSSCs has been compared [28, 31]. The tricyanomethanide anion is considered to be particularly interesting for this application, as it is the least viscous (19.56 cP at 25 °C for $[C_2mim][C(CN)_3]$) and is expected to exhibit less ion pairing as a result of its planarity and increased charge delocalisation. It is also less hygroscopic than the dicyanamide, which is a significant benefit for handling and device preparation. Further, the use of the dicyanamide ionic liquid has been reported to induce instability in the solar cell under visible light soaking [34]. The window of electrochemical stability of the tricyanomethanide ionic liquid is the smallest of the three species but is still sufficient for photovoltaic applications. The lower viscosity of the $[C(CN)_3]^-$ ionic liquids resulted in a higher short circuit current and open circuit voltage than for the $[N(CN)_2]^-$ salt, while those containing $[SCN]^-$ exhibited promising results at low light intensities [31]. Subsequently, efficiencies of >7% were reported using $[C_2mim][SCN]$ as part of a binary ionic liquid system (with $[C_3mim]I$, Z-907 dye [Fig. 10.3] and 3-phenylpropanoic acid as a co-adsorbed spacer) [34]. However, it has also since been suggested that $[C_2mim]X$ (X = $[SCN]$, $[SeCN]$, or $[C(CN)_3]$) are not stable under prolonged heating and light soaking, although no results are given, and that only $[C_2mim][B(CN)_4]$ is stable [21].

$[C_2mim][B(CN)_4]$ is also quite fluid (19.8 cP at 20 °C) as a result of the charge delocalisation over the four cyano groups of the tetracyanoborate anion. This ionic liquid (with $[C_3mim]I$, 13:7 volume ratio, and Z907 dye, Fig. 10.3) can be used to produce photovoltaic cells with 7.0% efficiencies and good thermal stabilities [38]. With heating up to 60 °C, the J_{SC} of the DSSC increased, primarily as a result of the decreased viscosity of the electrolyte, before decreasing upon further heating to 80 °C, probably due to the increased recapture of electrons by the triiodide (which reduces the electron collection efficiency). However, the temperature effects observed in this system were minor and are reportedly better than those of conventional silicon solar cells [38]. The DSSCs retained 90% of their initial activity after light soaking at 60 °C for 1000 hours under AM 1.5 full sunlight. During this time, there was a 40- to 50-mV drop in V_{OC} and a 1-mA cm^{-2} drop in J_{SC}, but this was somewhat compensated for by a 5% increase in fill factor.

The $[C_2mim][C(CN)_3]$ ionic liquid, as a 1:1 binary system with $[C_3mim]I$, has subsequently been used to produce 7.4% efficiencies (with Z907 dye [Fig. 10.3] and 0.5 M MBI) [30]. The use of these binary systems, where some of the $[C_3mim]I$ is replaced by a low-viscosity ionic liquid with weakly coordinating anions, can be very advantageous as it significantly lowers the viscosity, but the exact composition is critical to the efficiency of the device. As the concentration of [C$_3$mim]I in this system increases from 20% to 50% the increase in photocurrent density and efficiency of the device is consistent with the change in iodide diffusion flux. However, increasing the [C$_3$mim]I concentration further, from 50% to 80%, decreases the photocurrent density and efficiency. Transient absorption measurements show that the iodide can interact with the excited sensitiser, creating another route to dye deactivation, and thus high concentrations of [C$_3$mim]I can enhance the rate of reductive quenching of the excited dye [30]. It would also be interesting to determine how this compares with the high-concentration triiodide systems, and how significant the Grotthuss mechanism is at these high iodide concentrations.

The concept of binary systems has recently been taken even further, with the use of ternary melts to further improve the efficiencies [21]. Although [C$_1$mim]I and [C$_2$mim]I are solid at room temperature, they are both more conductive than the commonly used [C$_3$mim]I above their melting point, and a 1:1 mixture of these smaller iodide salts yields a highly conductive melt with a eutectic temperature of 47.5 °C. Further, when 1-allyl-3-methylimidazolium iodide ([allylmim]I) is also used (in a 1:1:1 molar ratio), a ternary melt with a melting point below 0 °C and conductivities three times higher than [C$_3$mim]I is formed [21]. Use of this electrolyte, in combination with commonly used additives ([C$_1$mim]I/[C$_2$mim]I/[allylmim]I/I$_2$/NBB/GNCS 8:8:8:1:2:0.4 molar ratio), gives device efficiencies of 7.1%. In fact, simply replacing [C$_3$mim]I with the eutectic melt of [C$_1$mim]I and [C$_2$mim]I gives a 17% improvement in efficiencies.

However, this efficiency can be further improved by using [C$_2$mim][B(CN)$_4$] instead of the allyl salt ([C$_1$mim]I/[C$_2$mim]I/[C$_2$mim][B(CN)$_4$]/I$_2$/NBB/GNCS 12:12:16:1.67:3.33:0.67) [21]. This mixture gives a higher V_{OC} as a result of the more positive Nernst potential of the electrolyte (due to the lower iodide concentration), and the lower triiodide concentration also yields a lower rate of charge recombination. This electrolyte system gives photovoltaic efficiencies of 8.2% (Z907 dye [Fig. 10.3] and 3-phenylpropanoic acid), and the stability of these devices (without the 3-phenylpropanoic acid co-absorbent) was tested by 1000 hours light soaking at 60 °C, after which it retained 93% of its efficiency. The influence of the co-absorbent on this stability was not reported.

Interestingly, the authors also report that drying of these electrolytes at 80 °C under vacuum (3 torr) for 8 hours did not result in the release of iodine (as determined by the iodine–starch colorimetric method), indicating sufficient complexation with the electrolyte and thus implying that loss of iodine through the long-term use of these devices is not a cause for concern.

These investigations clearly indicate that the use of binary or tertiary mixtures of ionic liquids can lead to significant enhancements in photovoltaic performance and that this is an area of research that should be actively pursued in the future. Optimisation of the composition of these complex electrolytes can be very time-consuming, with the influence of the multiple components and additives intricately linked and influencing the various photovoltaic parameters in different ways, but recent results suggest that this is a potentially valuable avenue of investigation. Further investigations into the volatility of iodine from these and other ionic liquid electrolyte systems is also strongly encouraged, in a drive to further illustrate the potential for the widespread and long-term use of these devices.

10.1.3   Cation Variations

In addition to variations in the ionic liquid anion used, there is also considerable scope for utilising ionic liquids with cations other than the traditional imidazolium species. For example, the use of an allyl-functionalised imidazolium ring produces ionic liquids that are supercooled at room temperature and exhibit viscosities lower than the commonly used [C$_3$mim]I [37]. Utilisation of a mixture of 1-allyl-3-ethylimidazolium iodide and 1-allyl-3-ethylimidazolium bistriflamide (13:7 by volume, with 0.2 M I$_2$, 0.5 M MBI and 0.1 M GNCS, with the ruthenium-based K60 dye) gave DSSC efficiencies of up to 6.8% at 1 sun (8% at low sun intensities). Further, the devices retained 91% of their photovoltaic efficiency after 1000 hours of light soaking at 60 °C. As observed for other ionic liquid systems, this drop was primarily due to a decrease in the open circuit voltage of the device.

Ramirez and Sanchez have studied the application of various tetraalkylphosphonium iodide ionic liquids in solar cells [36]. As these salts were not considered conductive enough to perform as electrolytes on their own, they were mixed with molecular solvents for testing, giving 5.7% efficiency. However, alternative phosphonium ionic liquids, such as triethyl(methoxymethyl)phosphonium bis{(trifluoromethyl)sulfonyl}amide ([P$_{222(101)}$][NTf$_2$], Fig. 10.5c) are much more fluid and have shown good efficiencies when used with organic dyes (see later) [28].

In a similar vein, a range of trialkylsulfonium iodides has been developed and tested [26]. While a number of these salts are solid at room temperature, others ([SEt$_2$Me]I, [SBuMe$_2$]I and [SBu$_2$Me]I) are lower melting and exhibit conductivities high enough to enable their use as electrolytes for DSSCs. The conductivities of these salts increase with addition of iodine, but the photocurrent of the most viscous electrolytes was still limited by the diffusion of [I$_3$]$^-$. The use of iodine-doped [SBu$_2$Me]I as an electrolyte achieved efficiencies of 3.7% at 0.1 sun (N719 dye, Fig. 10.3). Doping of these trialkylsulfonium iodides and polyiodides with copper and silver iodides gave efficiencies of up to 3.1% at 0.1 sun (with [SBu$_2$Me]I:AgI:I$_2$, 1:0.03:0.05) [54].

The performance of a number of nitrile and vinyl-substituted imidazolium ionic liquids, with the [NTf$_2$]$^-$ or [N(CN)$_2$]$^-$ anions as electrolytes in solar cells, has been tested [32]. These new ionic liquids were combined with [C$_3$mim]I (13:7 v/v, with 0.2 M I$_2$ and 0.5 M MBI) and used with the ruthenium-based K-19 dye (Fig. 10.3). The best result was obtained with 1-methyl-3-(3-cyanopropyl)imidazolium bistriflamide, which gave 5.9% at 1 sun and 60 °C.

However, the introduction of the nitrile functionality did not make a significant difference to the DSSC performance, even though the viscosities of the four ionic liquids tested were markedly different. The device utilising the 1-methyl-3-vinylimidazolium bistriflamide ionic liquid (which gave 5.6% at 1 sun and 60 °C) retained 90% of its efficiency after 35 days light soaking at 60 °C. The stability of the other ionic liquid systems was not reported. DSSCs utilising 0.8 M 1-methyl-3-propylnitrileimidazolium iodide in a 3-methoxypropanenitrile solvent (with 0.15 M I_2, 0.1 M GNCS and 0.5 M MBI) gave 7.2% efficiency at 1 sun (8% at low sun intensities) and only a slight decrease in efficiency with light soaking at 60 °C for 1000 hours.

10.2   IONIC LIQUIDS FOR QUASI-SOLID STATE DSSCs

10.2.1   Polyelectrolytes

Although the use of ionic liquid electrolytes in photovoltaic devices is beneficial for combating problems of solvent evaporation, their liquid state means that there remains some concern over leakage from improperly sealed cells. This can be addressed by their use in quasi-solid state polymer electrolytes or in gels (with or without nanoparticles). The performance of solid state electrolytes is often low as a result of poor contact with the nanoporous TiO_2 layer, which is made worse by expansion of the cell at high temperatures as a result of solar heating. In contrast, gel electrolytes are more malleable and thus should have a better contact with the electrodes and, thereby, improved cell performance. There has been considerable research into the use of polymer electrolytes in lithium batteries, and this concept is now receiving increasing attention from the photovoltaic community. In the lithium battery field, the gel electrolytes are generally made by combination of an acrylic monomer (the gelator) with a radical initiator in the liquid electrolyte, followed by cross-linking. One procedure for fabrication of a DSSC containing a chemically cross-linked gel electrolyte is to assemble an unsealed cell (i.e., with the electrodes and the dye-sensitised TiO_2 layer in place), inject the gel electrolyte precursor into the cell, effect the gelation (e.g., by heating), and then encapsulate the cell. Thus, for photovoltaic applications, the gel electrolyte synthesis must be achievable below the decomposition temperature of the dye, in the presence of some impurities (water, dioxygen, etc.) and without releasing unwanted by-products that would inhibit cell performance [55]. Unfortunately, the presence of the iodine-based redox couple required for DSSC operation also makes *in situ* polymerisation extremely difficult because iodine is a potent free-radical inhibitor. An alternative approach is to solidify the electrolyte by mixing with a polymer, or inorganic materials, and injecting the gel into the DSSC. There have been extensive investigations into the synthesis and application of gel electrolytes utilising molecular solvents for photovoltaics [23, 56], but efficiencies of these devices are generally low, there is still a volatile component, and the UV stability can be poor as a result of photodegradation of the polymer. Thus, attention has more recently turned to the use of ionic liquids in these systems, and this is a rapidly expanding area of ionic liquid/DSSC research [52, 57–61]. Some studies still utilise molecular solvents within the polymer matrix [58, 62], but it is clearly more desirable to remove these volatile components altogether.

For example, an ionic liquid ([C_3mim]I) can be incorporated into a poly(vinylidenefluoride-co-hexafluoropropene) matrix, and this gel used to form a quasi-solid state DSSC with a 5.3% efficiency (with the Z907 dye, Fig. 10.3) [52]. This efficiency is almost identical to that obtained in the absence of the polymer matrix, and the diffusion coefficients of the iodide and triiodide species are also very similar in both systems, suggesting that liquid channels are formed in the polymer phase through which the iodide and triiodide can diffuse (again primarily via a Grotthuss-type mechanism) [52]. An ionic liquid polymer, poly(1-oligo(ethylene glycol) methacrylate-3-methylimidazolium chloride), can also be used to solidify ionic liquids ([C_6mim]I and [C_2mim][BF_4]) [63]. This material has the advantage of exhibiting high solubility with inorganic salts, due to the presence of the ether oxygen atoms, and also good miscibility with organic salts as a result of the imidazolium groups. Using this quasi-solid state electrolyte system, efficiencies of up to 6.1% were obtained [63].

The fabrication of the most efficient DSSCs with quasi-solid state electrolytes relies on the ability of the electrolyte to effectively fill the pores of the TiO_2, which is a problem commonly encountered with high-viscosity or pre-gelled electrolytes. This can be overcome by using a system that can be inserted into the cell as a liquid at elevated temperatures and then cooled to room temperature to form a quasi-solid state electrolyte. For example, tetra(bromomethyl)benzene and poly(vinylpyridine) can be used as chemical cross-linkers for the *in situ* solidification of an ionic liquid ([C_3mim]I) without loss of photovoltaic performance [59]. This procedure results in micro-phase separation during the gelation and minimises the retardation of the I_3^-, thus maintaining cell performance.

For optimum performance, the electrolyte should contain as low a concentration of gelator as possible. The efficacy of low molecular weight organogelators in molecular solvents depends strongly on the solvent polarity. Cyclohexanecarboxylic acid-[(4-(3-tetradecylureido)phenyl]amide, which is good for gelling polar solvents, can be used at only 2 wt% concentration to gel a mixture of [C_3mim]I and [C_2mim][SCN], resulting in sol–gel transition temperatures of 108–120 °C depending on the ionic liquid mixture composition [64]. Importantly, inclusion of the additives required for DSSC applications does not change the gelation behaviour, and the gelation of the ionic liquid does not impair the solar cell performance, which was 6.3% (with K19 dye, Fig. 10.3). Furthermore, the device retained over 95% of its efficiency after light soaking at 60 °C for 1000 hours.

The *in situ* polymerisation of an ionic liquid, within the DSSC, can also be effected by using two cross-linking agents [65]. The electrolyte (I_2, NMB, LiI and two constituents of the chemically crosslinked gelators in [C_6mim]I) is inserted into the DSSC before heating at 90 °C for 30 minutes, which then forms a gel upon cooling to room temperature. The first gelation constituent was a polypyridyl-pendant poly(amidoamine) dendritic derivative (PPDDs), and the second were multifunctional halogen derivatives, such as 1,4-dibromobutane, 1,6-diiodohexane, or poly(ethyleneoxide) (PEO) with iodide groups on the chain ends. Using this technique, the devices retain over 90% of the liquid electrolyte performance, at around 3% (with N3 dye, Fig. 10.3).

The ionic liquid electrolyte can also be gelled using *in situ* photopolymerisation [66]. For the large-scale reel-to-reel manufacture of flexible DSSCs, rapid polymerisation of the electrolyte into a cross-linked gel is highly desirable, and the need for heating to elevated temperatures, as required in the examples given earlier, is problematic. Thus, the ability to effect polymerisation at room temperature simply by application of ultraviolet (UV) light is of significant interest. This was reported for the first time using 2-hydroxyethyl methacrylate (HEMA) monomer and TiO_2 nanoparticles as both the initiator and a co-gelator [66]. Tetra(ethylene glycol) diacrylate (TEGDA) was used as a co-monomer and cross-linker, with no additional initiator, thereby avoiding cell degradation by side reactions with the initiator or its by-products. Further, as the polymerisation mechanism is via charge transfer, initiated by the TiO_2 nanoparticles, it is not inhibited by the presence of radical inhibitors such as iodine. Variations in the ionic liquid, the amount of TiO_2, and the amount of co-monomer all influence the extent of gelation. The DSSC efficiency, for the best device, decreased from *ca.* 4.0% before polymerisation to 3.5% after polymerisation (with N719 dye, Fig. 10.3). This is largely due to a 65-mV decrease in the V_{OC}, but as this may be due to evaporation of some of the TBP from the electrolyte during polymerisation, this could be overcome in the future by use of non-volatile additives. The photocurrent also decreased on polymerisation, from *ca.* 7.9 mA cm^{-2} to *ca.* 7.4 mA cm^{-2}, due to lower mobility in the gel. However, the device performed well under lower light conditions, giving *ca.* 5% efficiencies at 0.39 sun. These promising initial results clearly indicate that photopolymerisation is an area of research worth further investigation. If a system can be developed that is polymerisable on a reel-to-reel processing time scale (i.e., using a strong industrial UV light source), without the use of any volatile components and with negligible loss of efficiency upon polymerisation, then this would clearly be a significant step towards the large-scale production of cheap and efficient DSSCs.

10.2.2 Nanocomposites

An alternative to physically or chemically gelating an ionic liquid is to use a nanocomposite electrolyte, for example, where the ionic liquid is combined with inorganic nanoparticles such as silica or titania. It should be noted that inorganic nanoparticles, or "fillers," have been used to increase the conductivity of a wide range of other electrolyte materials, prior to investigations into their use in DSSCs, and considerable insight can be gained from these studies [67–72]. Nanocomposite gel electrolytes are composed of two sub-phases, organic and inorganic, which are mixed at the nanoscale. The organic phase contains the components that provide ionic conductivity, and the inorganic phase (e.g., the silica) gels the organic phase and also assists in holding the two electrodes of the solar cell together [10]. The organic phase can be either a molecular solvent or, for lower volatility and increased stability, an ionic liquid. This is a very effective way of forming quasi-solid state DSSCs without the drop in efficiency that commonly occurs when solid state electrolytes are used.

The use of fumed silica nanoparticles (5 wt%, 12 nm particle size) in an ionic liquid electrolyte composed of 0.5 M I_2 and 0.45 M NMB in [C_3mim]I allows the formation of DSSCs (with Z907 dye, Fig. 10.3) that exhibit a 6.0% efficiency at 1 sun [73]. There is no drop in the efficiency of these devices compared with the systems without nanoparticles, indicating that the diffusion of I^-/I_3^- is not impeded, which is very promising for the development of quasi-solid state electrolytes.

It has been shown that SiO_2 nanoparticles can also be used to form a composite with [C_2mim][NTf_2] (plus [C_2mimI] and additives) [74]. Again, this has no detrimental effect on the ionic conductivity of the material, even though it changes in appearance from a gel to a powdery solid with an increasing concentration of nanoparticles. Further, the limiting current for the iodide/triiodide couple in the material actually increases with nanoparticle addition. This is attributed to the formation of iodide/triiodide-rich regions as a result of absorption of the ionic liquid cations onto the surface of the SiO_2 nanoparticles, which allows an increase in the exchange reaction-based diffusion. The addition of the SiO_2 nanoparticles not only increases the charge transport of the iodide/triiodide through the bulk but also increases the interfacial charge transport rate. Thus, again, the efficiency of the DSSC utilising the powdered electrolyte (3.7% at 1 sun) was equivalent to that utilising the liquid electrolyte [74].

One of the obstacles that must be overcome when designing solid state electrolytes for photovoltaic devices is that the materials must remain solid even at the high temperatures, to allow for the significant effects of solar heating on the cells. However, the physical properties of new quasi-solid state electrolytes are often not reported in detail, and this makes it difficult to assess whether the solidification is sufficient to prevent leakage at high temperatures. Many of the reported quasi-solid state DSSCs with efficiencies of over 4% have melting points below *ca.* 50°C, but the use of inorganic nanoparticles can address this problem. For example, the incorporation of the [BF_4]$^-$ anion (as [C_4mim][BF_4]) into a [C_4mim]I/silica nanoparticle system has been shown to significantly increase the solidification of the system and allow the synthesis of a high-temperature (>85°C) solid state nanocomposite material, which achieved efficiencies of 5.0% at 60°C in a solid state DSSC [75]. It is believed that this solidification behaviour is a result of the self-assembly of hydrogen bonds between the tetrafluoroborate anion (i.e., the strongly electronegative fluoride) and the hydroxyl groups on the silica [75]. It is also proposed that this interaction enhances the exclusion of water from the nanocomposite electrolyte and contributes to the significantly improved stability; the solid state DSSC showed no loss in performance after storage (but not light soaking) for 1000 hours at 60 °C, whereas the efficiency of the liquid DSSCs that was also tested dropped to 30% of its initial value. The stability of a similar system with the analogous hexafluorophosphate ionic liquid has also been reported [76].

The addition of TiO_2 nanoparticles, carbon nanotubes, carbon fibres, and carbon nanoparticles has also been shown to improve the performance of DSSCs with an $[C_2mim][NTf_2]$ electrolyte system [61], with the best results obtained using TiO_2 (giving an efficiency of 5.0% for the nanocomposite system, up from 4.21% for the ionic liquid), but again the mechanical properties of the gels, and the origins of the efficiency increases, were not reported in detail.

Layered α-zirconium phosphate has also been added to an ionic liquid electrolyte ($[C_3mim][H_2PO_4]$) to improve the photovoltaic performance of DSSCs, with the best results achieved for 6 wt% α-ZrP, although the device efficiencies in this report were very low [77]. Layered titanium phosphate is also effective; this material has larger inter-laminar cavities, which should be more suited to facilitation of the diffusion of the I_3^- in the electrolyte [78]. Indeed, X-ray powder diffraction (XRD) analysis suggests the intercalation of the $[C_3mim]^+$ cation from the electrolyte, while the diffusion of the I_3^- and the electron lifetime increased (using only 1 wt% α-TiP in a $[C_3mim]I/[C_2mim][BF_4]$ 13:7 molar ratio electrolyte). Thus, device efficiencies of 6.05% were achieved compared with 3.56% without the α-TiP. At higher concentrations of α-TiP, the diffusion of I_3^- decreases, reportedly due to an increase in the viscosity of the electrolyte. The viscosity and material properties of the 1 wt% composite were not given, but this filler material certainly might prove interesting for the further development of quasi-solid state electrolyte systems.

The concept of gelling ionic liquids with silica particles can be taken further via the use of functionalised silica particles [79], or ionic liquids derivatised with alkoxysilane groups (Fig. 10.4a) [80], and the use of these in the sol–gel process [81, 82]. Cells utilising the alkylsilane electrolyte, termed TMS-PMII, gave efficiencies of 3.2% (with N3 dye), [80],while using this material as a precursor for the sol–gel synthesis of a nanocomposite (Fig. 10.4b) for quasi-solid state devices gave efficiencies of 3.1% (1 sun, N3 dye, Fig. 10.3) [82].

Two alternative approaches have been explored by Cerneaux et al. [79], via either grafting of the aminopropyltriethoxysilane onto activated silica followed by quaternisation with an iodoalkane (Fig. 10.4c), or via the synthesis of new alkoxysilane precursors, with ethyl, heptyl, or dodecyl side chains, followed by grafting onto activated silica nanoparticles (Fig. 10.4d). This latter technique was found to give better product control. Photovoltaic efficiencies of up to 8.6% at 0.1 sun, or up to 6.6% at 1 sun, were achieved using these new quaternary ammonium iodide silica-based materials in solutions of

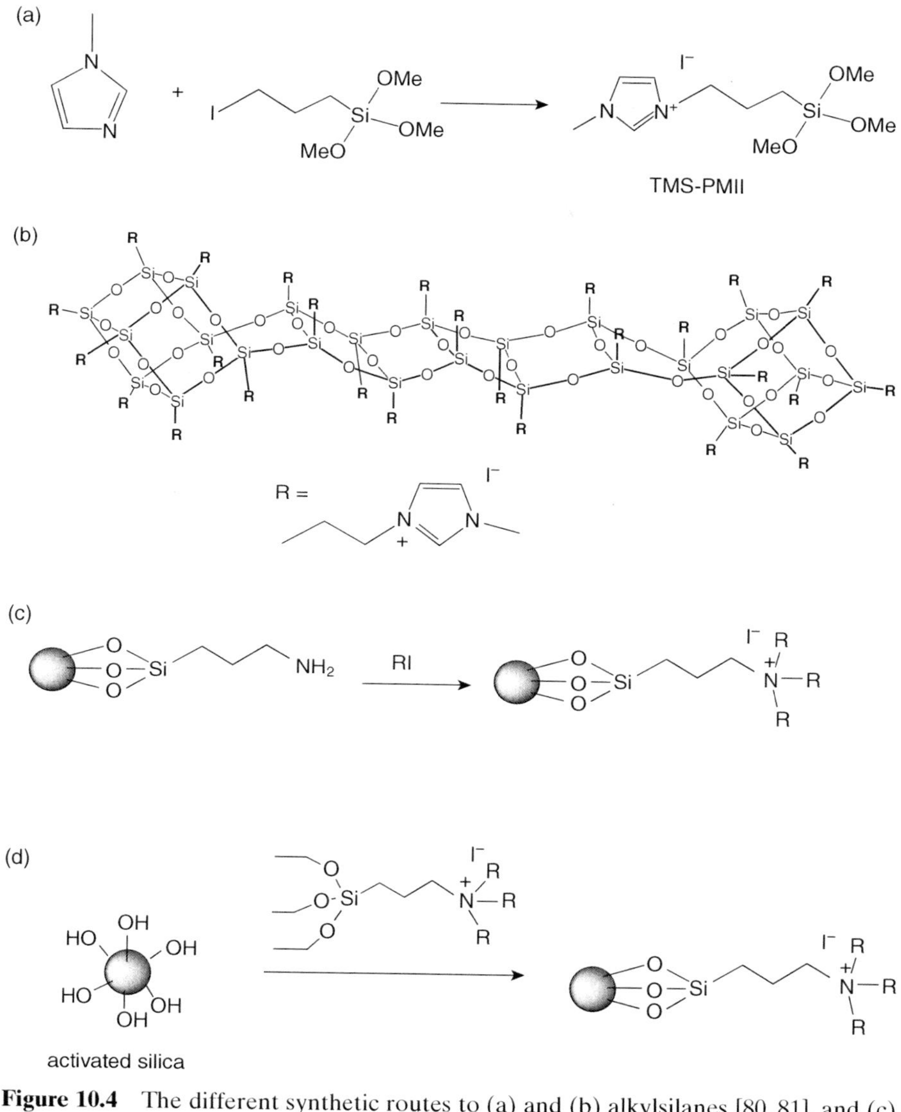

Figure 10.4 The different synthetic routes to (a) and (b) alkylsilanes [80, 81], and (c) and (d) functionalised silica materials [79], for use as iodide sources.

ethanenitrile or 3-methoxyethanenitrile (0.1 M, plus additives) [79], and investigations into the use of these new materials in ionic liquids are underway.

In summary, research towards the development of quasi-solid state electrolytes has made significant progress in recent years, and this is vital in the drive towards DSSCs with long-term stability. However, it should be noted that many of these systems still contain volatile components (most notably iodine, and in some cases TBP), and for the ultimate non-volatile electrolyte, future research needs to work towards alternatives to these species. Further, in reporting the development and characterisation of new quasi-solid state materials, authors are encouraged to give a clear indication of the physical properties of the gels/composites and the accessible operating temperatures. If a new electrolyte is synthesised that is only useful at low levels of gelation (i.e., before the transport of the redox couple through the electrolyte is significantly limited by diffusion) and is thus still relatively fluid, or that melts with only mild solar heating, then this represents only a small step towards the development of truly long-lasting, commercially viable devices.

10.2.3 Ionic Liquid Crystals and Plastic Crystals

The use of liquid crystals as electrolytes for solar cells utilises the strategy of promoting the reaction between iodide and triiodide by increasing the local concentrations of the species because the liquid crystal self-assembles into a layered structure [35]. 1-Dodecyl-3-methylimidazolium iodide forms into a bilayer structure (at temperatures between 27 and 45 °C), with the long alkyl chains interdigitated and the iodide species localised between the layers, and its use in DSSCs has been demonstrated [35]. Measurement of the diffusion limiting current in two different liquid crystal systems, at 40 °C using a microelectrode, showed that this is higher for the $[C_{12}mim]I/I_2$ (4.2×10^{-8} cm^2 s^{-1}) compared with the $[C_{11}mim]I/I_2$ mixture (3.2×10^{-8} cm^2 s^{-1}), despite the fact that the former electrolyte has a viscosity that is 2.5 times higher, suggesting that conductive pathways are formed in the material. This electrolyte can also be solidified using a gelator, which maintains the sub-micron ionic liquid crystal domains and results in an increase in the short-circuit current and conversion efficiency [83].

In contrast to liquid crystals, a plastic crystal consists of a regular three-dimensional crystalline lattice with the well-defined long-range position of species, but with molecules or ions that can exhibit orientational and/or rotational disorder [84, 85]. Such materials have a large number of degrees of freedom, and higher entropy than a fully ordered solid. As a result of this disorder, the significant amount of motion that is possible can lead to plasticity (i.e., ready deformation under an applied stress), high diffusivity, and, for ionic plastic crystals, phases with good ionic conductivity [86].

The plastic crystal succinonitrile can be doped with lithium salts and iodine to yield a DSSC electrolyte that displays high iodide and triiodide transport in the solid state [87, 88]. Succinonitrile can also be combined with $[C_4mim][BF_4]$ and silica nanoparticles to form a gel electrolyte that is quite thermally stable and conductive (6.6–18.2 mS cm^{-1} at 20–80 °C), and this electrolyte can give DSSC efficiencies of *ca.* 5% (with $[Bu_4N]I$ and I_2, Z907 dye, Fig. 10.3) [89]. The DSSC retained 93% of its initial efficiency after storage (but not light soaking) at 60 °C for 1000 hours. Importantly, this composite remains in a gelled state until over 80 °C, which allows for a wider DSSC operating temperature range than is suitable with an electrolyte of [C$_4$mim][BF$_4$] gelled only with silica, which becomes a liquid at 45 °C.

The disadvantage to molecular plastic crystals, however, is that they are still volatile and can evaporate at elevated temperatures. In contrast, organic ionic plastic crystals (which are structurally similar to ionic liquids but form plastic phases around room temperature) are non-volatile and thus suitable for use at higher temperatures [85]. We have recently demonstrated the use of a range of organic ionic plastic crystals as solid state electrolytes in DSSCs, with the best performance achieved using 1,1-dimethylpyrrolidinium dicyanamide, which gave efficiencies of *ca.* 5% at 1 sun [90].

10.2.4   Alternative Redox Couples

To further improve the efficiencies of photovoltaic cells containing ionic liquid electrolytes, it can be very beneficial to consider redox couples other than the traditional iodide/triiodide couple. There is a mismatch in the energy levels of the common ruthenium dyes and the iodide/triiodide couple, which limits the open-circuit potential of the DSSC, and there are continued concerns about the long-term stability of devices that contain iodine, which is both corrosive and volatile. Thus, there has been considerable effort expended in the search for new redox couples, with some success [91, 92], but little research has focussed on the development of alternative redox couples for ionic liquid electrolytes. The use of an ionic liquid may alter the energy levels of a redox couple compared with those values obtained in molecular solvent systems (i.e. the concentrations required may differ, to compensate for the lower diffusion rates, and hence the Nernst potential will change), and the solubility and stability may also differ, and thus this is clearly an area that warrants much more investigation.

A new [SeCN]$^-$/[(SeCN)$_3$]$^-$ redox couple, in the ionic liquid [C$_2$mim][SeCN], has shown very promising initial results [27]. This new ionic liquid is quite fluid (25 cP at 21 °C) and very conductive (14.1 mS cm^{-1}), and using this salt (with 0.15 M K[(SeCN)$_3$], 0.1 M GNCS, 0.5 M MBI, and the Z907 dye, Fig. 10.3) DSSC efficiencies of 7.5% at 1 sun (8.3% at lower sun intensities) were achieved. Electron donation from the [SeCN]$^-$ to the oxidised dye in this system is even faster than from the iodide ion to the oxidised dye in the usual [C$_3$mim]I systems. It is interesting to note that the good results obtained using the redox couple in this ionic liquid system are also considerably better that those previously obtained for this redox couple in ethanenitrile, emphasising the fact that the optimum redox couple is influenced by the type of electrolyte used, and that this factor should be considered when testing new ionic liquids or new redox systems [93]. The results of long-term stability tests under prolonged light soaking for this new redox couple/ionic liquid combination have not yet been reported and more research in this promising area is highly desirable.

It is also possible to substitute the triiodide-based system with one that utilises a mixed halide system, such as [IBr$_2$]$^-$ or [I$_2$Br]$^-$. It is predicted that, in these complex systems, there will be equilibria between the different anions and they may absorb less visible light than the triiodide, allowing higher concentrations to be used [94]. A range of interhalogen ionic liquids, such as $[C_3mim][[IBr_2], [C_6mim][IBr_2], [C_6mim][I_2Br], [C_4mim][IBr_2]$, and $[C_4py][IBr_2]$ (with varying amounts of GNCS and MBI, N719 dye) were tested in monolithic DSSCs (see later) [94]. Efficiencies of up to 2.4% were obtained for the undiluted ionic liquid electrolytes (higher when mixed with molecular solvents), which is primarily dictated by the viscosity of the different ionic liquids. The authors predict that cell optimisation would increase efficiencies, and it would also be interesting to assess the performance of these materials in more traditional "sandwich" DSSCs. The efficiencies of the cells decreased by between 9% and 14% after 1000 hours illumination at 350 W m^{-2}.

Recently, there has been a surge of interest in the use of cobalt-based redox couples, with efficiencies of over 12% demonstrated using an ethanenitrile-based electrolyte with a porphyrin dye [95, 96]. However, at the time of writing, the use of ionic liquid-based electrolytes with these new redox couples had not been explored and hence this is clearly an important future avenue of investigation.

10.2.5   Sensitiser Optimisation

There has been an immense amount of research undertaken to improve the performance of DSSCs through optimisation of the sensitiser [5–7]. For the solar cell to be efficient, the dye used must have a good spectral match to that of solar emission. Even subtle changes to the structure of the dye can impact the absorption spectra, but prediction of the effect of different structural changes is complex [97].

The Z907 dye (Fig. 10.3) has been frequently used for DSSCs utilising ionic liquids since it was reported that this dye gives better efficiencies than the N719 dye (Fig. 10.3) when used with a dicyanamide ionic liquid [33]. Using an ionic liquid electrolyte solution (0.1 M iodine, 0.1 M LiI, and 0.45 M MBI in $[C_3mim]I/[C_2mim][N(CN)_2]$ 13:7 by volume), the Z907 dye gave an efficiency of 6.6% at 1 sun, whereas the N719 dye gave only 5.0%. Dark current measurements indicate that this improvement is the result of the long hydrocarbon chains on the Z907 dye, which form a layer between the electrons trapped on the TiO$_2$ surface and the triiodide in the electrolyte and thus slow the rate of back transfer of electrons. The addition of LiI to this system significantly enhanced the efficiency, from 5.7% to the aforementioned 6.6%, as a result of increased dye regeneration rate and electron injection yield.

At the time of writing, the highest efficiency achieved using a ruthenium sensitiser and an ionic liquid electrolyte was 8.5% (at 1 sun, just over 9% at lower light intensities), using the C103 dye (Fig. 10.3) and a cholic acid-based co-adsorbent, with $[C_1mim]I/[C_2mim]I/[C_2mim][B(CN)_4]/I_2/NBB/GNCS$ (molar ratio 12/12/16/1.67/3.33/0.67) [20]. This new dye utilises electron-rich 3,4-ethylenedioxythiophene units to improve the molar extinction coefficient and the spectral match of the dye. It has been shown that, in ionic liquid-based DSSCs, the electron diffusion length is shorter than in ethanenitrile-based cells, as a result of faster charge recombination (primarily because of the need for higher triiodide concentrations). The C103 dye has a metal-to-ligand charge-transfer absorption band that is red shifted by 27 nm compared with the Z907 dye (giving better spectral matching), and a peak molar extinction coefficient (18.8×10^3 M^{-1} cm^{-1}) that is significantly better than the standard Z907 (12.2×10^3 M^{-1} cm^{-1}) and N719 (14.0×10^3 M^{-1} cm^{-1}) dyes [30, 98]. An improvement in extinction coefficient allows a thinner semiconductor layer to be used in the IL-based cells, which helps address the problem of shorter electron diffusion lengths. This new dye also performed exceptionally well in a low-volatility electrolyte utilising 3-methoxypropanenitrile; 9.6% at 1 sun (with 1 M [C_1mim]I, 0.15 M I_2, 0.5 M NBB, and 0.1 M GNCS).

In order to fully understand how the use of an ionic liquid in a DSSC influences the performance, it is crucial for detailed studies that directly compare parameters such as electron lifetime, diffusion lengths, and charge recombination rates to be undertaken. For example, in the DSSCs utilising the C103 dye, the primary cause of reduced efficiency in the DSSCs with the ionic liquid compared with the low-volatility electrolyte appears to be a decrease in the V_{OC} (the J_{SC}, V_{OC}, and FF are 17.51 mA cm^{-2}, 771 mV, and 0.709, respectively, for the DSSC with 3-methoxypropanenitrile, and 15.93 mA cm^{-2}, 710 mV, and 0.747, respectively, with the ionic liquid), and the authors suggest that this is due to a significant (45 mV) difference in the calculated electrolyte equilibrium potential [20]. It is interesting to note that the ionic liquid also gives a better fill factor, as is often observed, and this is postulated to be the result of effective screening of electric charges, with the high ion concentration enabling increased charge separation, but the exact mechanism is not fully understood.

Comparison of parameters such as the electron lifetime, diffusion coefficient, and diffusion length in different devices, using intensity-modulated photovoltage and photocurrent spectroscopy, can provide valuable information, and such analyses are very important for enhancing our understanding of these different systems. However, it should be noted that there is normally a significant difference in the electrolyte composition between the molecular solvent and ionic liquid-based electrolytes; for example, a significantly higher triiodide concentration is normally used in ionic liquids to overcome diffusion limitations, and this can result in a much higher charge recombination rate. This adds another level of complexity in the drive to clearly identify, and thus improve, the efficiency limitations in ionic liquid-based DSSCs. Clearly, it is most relevant to analyse, for example, the time scales of processes occurring in the most efficient ionic liquid-based DSSCs, but as optimisation of these devices can necessitate changes to the electrolyte composition (not to mention film thickness, particle size, device thickness, use of different co-adsorbents, and additives, etc.), to get the best performance out of a new dye, it is thus challenging to isolate these effects and gain the direct insight necessary to allow the target-specific design of new dyes for high-efficiency ionic liquid-based DSSCs.

There is also an increasing drive towards the development of dyes that do not contain ruthenium [7], partly due to the cost of this increasingly expensive and scarce metal, but also because there is limit to the extent to which the molar extinction coefficient of a ruthenium-based sensitiser can be increased. The molar extinction coefficient of organic dyes can be 2–3 times greater than ruthenium-based dyes, and this increase is particularly desirable for the development of ionic liquid-based DSSCs as it allows for the use of thinner photoactive layers, to counter the effect of the decreased electron diffusion lengths.

The molecular design of indoline dyes has achieved efficiencies of up to 7.2% with an ionic liquid [99]. The best-performing dye tested utilised two Rhodamine units and a longer octyl chain on the terminal Rhodamine unit (D205, Fig. 10.5), rather than those with one Rhodamine unit (D102) or with two units but a shorter ethyl group substituent (D149) [99]. The use of the octyl chain, on the D205 dye, gives a better V_{oc} and J_{sc} than that obtained with the ethyl-substituted dye, even though the chromophoric units are the same, which highlights how significant even subtle changes to the dye structure can be. The improved performance of the D205 dye is due to a negative shift in the conduction band edge and longer electron lifetimes, which suggests that absorption of the dye at the semiconductor surface inhibits the back reaction of the injected electron with the triiodide [99]. Unfortunately, these dyes also suffer from desorption from the TiO_2 surface over time, which is clearly detrimental to the stability of the DSSCs.

More stable organic dyes have been made utilising bithiophene [103], thienothiophene [100], and dithienothiophene [101] units, with a bisfluorenylaniline fragment, giving efficiencies of 5.8–7% with an ionic liquid electrolyte. These dyes have low free energies of solvation in polar solvents, thus reducing the dye desorption. The DSSCs incorporating the dye with the bithiophene unit (JK2) with an ionic liquid electrolyte (0.2 M I_2, 0.5 M MBI, and 0.1 M GNCS in a mixture of [C_3mim]I/[C_2mim][SCN] 13:7 by volume) gave over a 5.8% efficiency, which dropped to about 80% of its original value after light soaking at 60 °C for 1200 hours (primarily due to a drop in V_{oc} of about 120 mV) [103].

DSSCs incorporating a dye with fused dithienothiophene moieties (C203, Fig. 10.5) with an ionic liquid electrolyte ([C_1mim]I/[C_2mim]I/[C_2mim][B(CN)$_4$]/I_2/NBB/GNCS, molar ratio 12/12/16/1.67/3.33/0.67) give efficiencies of 7%, and a smaller drop in V_{OC} (47 mV) after light soaking under the same lifetime testing conditions as above [101]. However, it should be noted that use of this different ionic liquid electrolyte may influence both of these parameters. Nonetheless, it is interesting that this device showed only a very slight drop in photocurrent on moving from an ethanenitrile-based electrolyte to the ionic liquid. This indicates the success of the strategy of overcoming electron diffusion lengths via engineering a higher molecular extinction coefficient sensitiser for ionic liquid-based devices. The thienothiophene dye (C201, Fig. 10.5) gave similar efficiencies to the C203 device, with the same ionic liquid electrolyte composition, and similar stabilities [100, 101]. This dye also

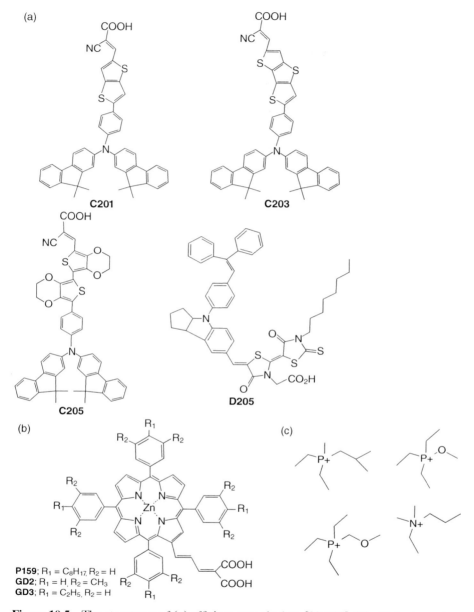

Figure 10.5 The structures of (a) efficient organic dyes [99–102], (b) efficient porphyrin dyes [28], and (c) the cations of some of the bistriflamide ionic liquids used in combination with the P159 porphyrin dye [28].

performed well in an all solid state device, with Spiro-OMeTAD as the hole transport material, giving efficiencies of 4.8%.

However, despite these promising results, the aforementioned organic dyes have narrower spectral responses than the best ruthenium-based dyes. Interestingly, the most efficient organic sensitiser reported to date also utilised a 3,4-ethylenedioxythiophene moiety (as does the most efficient ruthenium dye) [102]. Use of the 3,4-ethylenedioxythiophene unit in new donor–π–acceptor (D–π–A) organic dyes (Fig. 10.5) can achieve a significant red shift in the absorbance spectra and has allowed a new efficiency record of 7.6% to be obtained (at 1 sun, up to 8.5% at lower sun intensities, with the same tetracyanoborate ionic liquid electrolyte system as described earlier) [102]. This red shift is greater for the dye containing two 3,4-ethylenedioxythiophene (EDOT) units (C205, Fig. 10.5) than for the dye containing only one (C204) and is much larger than is observed for the shift between the bithiophene- and thiophene-based species [103]. The stability of these new dyes is also good, decreasing to 92% of their original value under the same lifetime testing conditions as used above.

The geometry of these new dyes appears to be advantageous, as the planarity of the biEDOT and the small torsion angle with respect to the neighbouring phenyl groups are likely to enhance the absorption onto the TiO$_2$ surface compared with the thiophene-based dyes, but without any decrease in molar absorption coefficient. This clearly suggests the potential benefit of incorporating electron-rich and planar conjugation units into future dye design. Once again, the authors observed a high charge recombination rate for the cells utilising the ionic liquid electrolyte compared with the ethanenitrile-based one and concluded that this is mainly a result of the high triiodide concentration [102]. The triiodide captures electrons from the conduction band of the TiO$_2$ at the semiconductor/electrolyte interface, and this depends on the electron occupancy of the trapped states. It is suggested that DSSCs utilising molecular solvent electrolytes appear to have less electron-trapping states below the conduction band edge [102] and that in this device the surfaces of the titania that are not fully coordinated are passivated by the NBB and GNCS additives. It is therefore suggested that, in the ionic liquid-based device, these additives do not have the same passivating action, possibly as a result of the high ionic concentration of the ionic liquid [102], in which case it would clearly be beneficial to develop new additives more suited to an ionic environment.

An alternative strategy to photosensitiser optimisation was recently demonstrated: co-sensitisation using two different organic dyes [104]. The use of two different organic dyes (SQ1 and JK2), with complimentary spectral responses, gives an increase in the photocurrent and efficiencies of the DSSC. Using an ionic liquid electrolyte (0.2 M I$_2$, 0.1 M GNCS, 0.5 M NBB in [C$_3$mim]I/[C$_2$mim][B(CN)$_4$] 65/35 by mol) with the two dyes together, efficiencies of 6.4% were achieved, which is higher than those obtainable using the individual dyes.

We have recently demonstrated the first use of a porphyrin sensitiser with an ionic liquid in a DSSC [28]. Three different dye structures were tested, GD1,

TABLE 10.1   The Physical Properties (at 25°C) of the Ionic Liquids Used in Combination with Porphyrin Dyes [28]

Ionic Liquid	Density/ g cm^{-3} ± 0.01	Viscosity/ mPa s ± 0.1	Conductivity/ mS cm^{-1} ± 1
[C$_2$mim][B(CN)$_4$]	1.04	21	16.1
[C$_2$mim][BF$_4$]	1.28	38	17.5
[C$_2$mim][NTf$_2$]	1.51	33	10.3
[C$_2$mim][N(CN)$_2$]	1.09	19	20.2
[C$_2$mim][SCN]	1.11	21	22.2
[N$_{1123}$][NTf$_2$]	1.41	101	2.30
[P$_{1224}$][NTf$_2$]^a	1.40	71	2.60
[P$_{222(201)}$][NTf$_2$] [105]	1.39	44	4.40
[P$_{222(101)}$][NTf$_2$] [105]	1.42	35	3.58

^{a} Measured at 30°C.

GD2, and P159 (Fig. 10.5b), with several different ionic liquids (Table 10.1). The general liquid electrolyte composition was [C$_1$mim]I/[C$_2$mim]I/IL/LiI/I$_2$/ NMB (12:12:16:1:1.67:4 by mol), and the ionic liquids tested were each chosen for their stability and fluidity, including the novel phosphonium ionic liquid diethylisobutylmethylphosphonium   bis{(trifluoromethyl)sulfonyl}amide (Figure 10.5c). Introduction of an ether side chain in the phosphonium cations significantly decreased the viscosity: [P$_{222(201)}$][NTf$_2$] is of comparable viscosity to the well-known [C$_2$mim][NTf$_2$], but with potentially better stability.

The P159 dye gave considerably improved photocurrents, when tested with the tetracyanoborate ionic liquid electrolyte, compared with the other two porphyrin dyes. It is postulated that this is due to the longer alkyl chains on the peripheral benzene rings, which form a barrier layer on the semiconductor surface and inhibit back electron transfer from the TiO$_2$, as previously postulated for the improved performance of the ruthenium-based sensitiser Z907 compared with N719 [33]. The efficiency of this device is even higher (5.2%) at lower (68%) sun intensities, which is significant as these devices are increasingly being regarded as ideal for the low light application market (where the silicon-based cells can suffer from a marked drop in performance).

Comparison of the performance of the P159 dye with a range of different ionic liquids is given in Table 10.2. Such comparisons are valuable in enhancing our understanding of how the nature of the ionic liquid influences the performance of DSSCs. For example, despite the fact that [C$_2$mim][SCN] and [C$_2$mim] [N(CN)$_2$] have similar viscosities and ionic conductivities to [C$_2$mim][B(CN)$_4$] (Table 10.1), they show relatively poor performance in the DSSCs. In fact, the performance of the DSSCs was not highly correlated with the physical properties of the ionic liquids, highlighting the importance of such studies over assumptions of a simple physical property/performance relationship. As [C$_2$mim][SCN] and [C$_2$mim][N(CN)$_2$] are highly hygroscopic and distinctly

TABLE 10.2   Photovoltaic Characteristics of DSSCs with Porphyrin P159 Dye and a Range of Different Ionic Liquid Electrolytes after Light Soaking for 30 Minutes [28]

Ionic Liquid	V_{oc}/mV	I_{sc}/mA cm^{-2}	Fill Factor	Efficiency/%
$[C_2mim][B(CN)_4]$	624 (±5)	11.9 (±0.2)	0.66 (±0.02)	4.9 (±0.1)
$[C_2mim][NTf_2]$	647 (±2)	8.3 (±0.6)	0.78 (±0.01)	4.2 (±0.1)
$[C_2mim][SCN]$	649 (±4)	5.3 (±0.4)	0.78 (±0.01)	2.7 (±0.2)
$[C_2mim][BF_4]$	613 (±3)	5.8 (±0.1)	0.74 (±0.01)	2.6 (±0.1)
$[C_2mim][N(CN)_2]$	652 (±10)	6.9 (±0.3)	0.73 (±0.04)	3.2 (±0.2)
$[P_{222(101)}][NTf_2]$	631 (±4)	8.9 (±0.3)	0.70 (±0.03)	3.9 (±0.2)
$[P_{222(201)}][NTf_2]$	649 (±5)	7.4 (±0.5)	0.74 (±0.02)	3.6 (±0.1)
$[P_{1224}][NTf_2]$	664 (±6)	7.0 (±0.3)	0.76 (±0.01)	3.5 (±0.1)
$[N_{2226}][NTf_2]$	636 (±7)	7.9 (±0.2)	0.75 (±0.03)	3.8 (±0.1)
$[N_{1123}][NTf_2]$	644 (±4)	7.8 (±0.1)	0.78 (±0.01)	3.9 (±0.1)
Standard ethanenitrile electrolyte				
Ethanenitrile	698 (±7)	12.3 (±0.1)	0.69 (±0.05)	6.0 (±0.1)

basic compared with $[C_2mim][B(CN)_4]$, it is postulated that the interaction between the electrolyte and the dye and/or titania will be different in these cases.

We also investigated the use of a new phosphonium ionic liquid with a small cation, plus two commercially available low-viscosity phosphonium ionic liquids (Table 10.2). The best efficiencies were achieved with the phosphonium ionic liquid containing the ether linkage, $[P_{222(101)}][NTf_2]$; 3.9% at 1 sun [28].

10.3   VARIATIONS FROM "TRADITIONAL" DSSC STRUCTURES

10.3.1   Monolithic Cells

Monolithic cells are an effective and relatively simple method of testing a range of components in dye-sensitised solar cells with good reproducibility. The traditional DSSC (Fig. 10.1) can be difficult to make; the electrode preparations, application of the dye, insertion of the electrolyte, and successful encapsulation and sealing against air and moisture can all cause manufacturing and reproducibility difficulties. This is a significant problem when the performance of cells utilising a range of different materials (electrolytes, dyes, additives, etc.) is being compared. Although the exact design may vary, a monolithic cell essentially consists of a layer of nanoporous TiO_2, an insulating ZrO_2 spacer layer, a carbon layer (the adhesion layer), and a platinised carbon on a conducting glass counter electrode (Fig. 10.6) [106, 107]. This design is likely to be much easier to produce in bulk by screen printing techniques. Further, monolithic multi-cells can be used for comparative testing of a range of different material components [107].

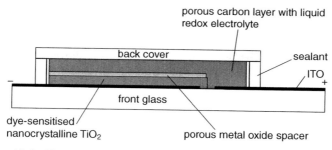

Figure 10.6 The basic structure of a monolithic dye-sensitised solar cell.

The influence of the anion component in binary mixtures of 1,3-dialkylimidazolium ionic liquids in monolithic DSSCs has been assessed [25]. Comparison of electrolytes containing [C₄mim]I and [C₄mim][BF₄], [C₄mim][SCN], [C₄mim][CF₃COO], and [C₄mim][CF₃SO₃] (with 0.1 M I₂ but no other additives, N719 dye) showed that the electron transport did not vary significantly with different anions, but they do influence the band edge position, recombination, and overall efficiencies. The charge-transfer resistance for the triiodide to iodide process was significantly affected by the type of ionic liquid mixture used but, surprisingly, the system with the highest charge-transfer resistance was also the one that gave the best solar cell performance (a mixture of [C₄mim]I and [C₄mim][SCN]). It is postulated that, in this system, [I₂(SCN)]⁻ formation may occur. The thiocyanate system gave a better performance at low light intensities (5.6% at 250 W/m²) than the other ionic liquids that are predicted to coordinate to a lesser extent to the I₂. The efficiencies of the cells were significantly lower at a higher light intensity (e.g., 2.2% for the thiocyanate system at 1000 W/m²), possibly due to charge transport limitations. The cells exhibited a *ca.* 10% decrease in efficiency after 1000 hours of continuous illumination at 350 W m⁻² [25]. However, it should be noted that although the efficiencies reported in this study are relatively low, and the exact reasons for this in the different ionic liquids are not clear, the electrolytes were not optimised and the potential benefits of this alternative geometry for ionic liquid-based DSSCs should not be overlooked.

10.3.2 Alternative Counter Electrodes

The platinum that is used as the catalytic layer on the counter electrode of a DSSC is a very expensive and increasingly scarce component, and the thermal treatment that is required for the synthesis of conventional platinum counter electrodes—typically *ca.* 400 °C for 15 minutes to produce the transparent platinum layer from H₂PtCl₆—is equally problematic to the wide-scale production of these devices. As the future of DSSCs is strongly tending towards the development of lightweight, cheaper, flexible devices utilising plastic substrates, any requirement for thermal processing is a significant problem. For inexpensive reel-to-reel type manufacture, it is preferable if the material can be applied by a standard printing method and rapidly dried/cured. Graphite, carbon black and carbon nanotubes have all been tested as replacements for platinum in traditional DSSCs, but efficiencies remain low [106, 108, 109].

Poly(3,4-ethylenedioxythiophene) (PEDOT) is a particularly interesting conducting polymer, as it can be highly conducting, with transparency to visible light and good stability at room temperature: it has been known for some time to be a potential alternative to platinum in the counter electrode catalyst rôle [110–112]. PEDOT can also be used as the hole transporter in DSSCs, either alone [113] or in combination with ionic liquids [114]. However, the widespread use of PEDOT as the electrocatalyst in DSSCs has so far been hampered by insufficient performance and processability. The majority of research into the development of PEDOT counter electrodes has focussed on the use of chemically synthesised PEDOT, which is dispersed into a solvent, coated onto a conducting substrate, dried at elevated temperatures, and then used in a DSSC containing a molecular solvent. The relative performance of PEDOT compared with platinum counter electrodes depends on factors such as the electrolyte [115], the interaction between the PEDOT and the substrate [116], and the nature of the PEDOT counter ion [110]. The charge-transfer resistance of PEDOT:poly(styrene sulfonate) (PSS) electrodes in an ethanenitrile-based electrolyte can be 30 times higher than that of a standard platinised fluorine-doped tin oxide (FTO) glass, which will be detrimental to the photovoltaic performance, although this resistance may be decreased when a gelled electrolyte is used [60]. The conductivity and performance of the commercially available PEDOT:PSS, which is sold as a water dispersion and can be deposited by, for example, spin coating then drying at elevated temperatures, can be increased by using solvents such as dimethyl sulfoxide with the addition of carbon black [117], or the addition of nanocrystalline TiO_2 [118]. The vast majority of the research undertaken to investigate the use of ionic liquids in DSSCs has utilised standard platinised FTO or ITO glass counter electrodes, with little attention being focussed on the counter electrode material.

Testing of DSSCs with chemically polymerised PEDOT-4-toluenesulfonate counter electrodes, with a [C_6mim]I-based ionic liquid electrolyte system, showed that a thicker (*ca.* 2 μm) PEDOT layer performed better than a thin (*ca.* 50 nm) layer, indicating the need for an increased surface area [119]. However, one possible drawback to PEDOT that has been chemically polymerised using an iron-based oxidant, such as iron(III) 4-toluenesulfonate, is that even with thorough washing the polymers may retain some Fe(II/III), and residual iron introduced into a DSSC can be very detrimental to the cell efficiency.

As an alternative to this, we have recently reported the use of PEDOT chemically polymerised using gold chloride in an ionic liquid medium, and the use of this material as the electrocatalyst in a DSSC with an ionic liquid electrolyte [120]. In addition to being iron-free, this polymer has the advantage that it can be solvent cast onto a conducting substrate from a trichloromethane dispersion and therefore does not require drying at elevated temperatures. Using the PEDOT-on-FTO glass counter electrodes, we achieved efficiencies of 4.8% (with [C_2mim][SCN] (35%) and [C_3mim]I (65%) plus I_2 (0.2 M) and TBP (0.5 M), N719 dye), compared with 5.1% with platinised FTO glass. With an ethanenitrile-base electrolyte, the efficiencies were 7.8% and 7.9%, respectively.

However, one of the problems that can arise with chemically polymerised PEDOT is that, depending on the method used to deposit the polymer on the glass (i.e., printing, spin coating, or solvent casting), the reproducibility of each electrode can be poor. In contrast, electrochemical deposition potentially allows much more control over the thickness and morphology of the PEDOT, plus it can be more cost-effective and suitable for large-scale production. The electrodeposition of PEDOT onto FTO glass from an ionic liquid ([C_4mim][NTf$_2$] with 0.1 M EDOT) was reported recently, and DSSCs with counter electrodes gave efficiencies of 7.9% with a molecular solvent electrolyte, compared with 8.7% with the DSSCs using the conventional platinised counter electrodes [121]. DSSCs with ionic liquids were not tested. Comparison of films grown for 30, 60, and 120 seconds showed better efficiencies with the thinner films, although the difference was very small.

We have recently demonstrated the use of electrochemical polymerisation for the synthesis of PEDOT on ITO-on-PEN plastic counter electrodes, which perform as well in DSSCs as platinised FTO glass using either molecular solvent or ionic liquid-based electrolytes (Table 10.3) [122].

The development of DSSCs on plastic substrates is highly desirable, as these allow the formation of light and flexible devices and potentially allow the solar cells to be produced on a large scale by reel-to-reel processes. Thus, it is very promising that using this technique, very short (5 seconds) electrodeposition times are sufficient to yield an electrocatalytically active PEDOT film, and therefore this process is potentially very cost-effective and suitable for scale up. Increasing the deposition time increases the roughness and colour of the film (Fig. 10.7), but we observed no statistically significant influence of the deposition time of the PEDOT layer (from 5 to 30 seconds) on the performance of the DSSC, indicating that the PEDOT-on-ITO-PEN cathodes are sufficiently electrocatalytic that this is not the limiting factor in determining the efficiency of the solar cell.

Finally, it has recently been demonstrated that cobalt sulfide (CoS) can also be electrodeposited onto ITO/PEN film (30 minutes deposition time, from an alkaline, pH = 10, aqueous solution of 5 mM cobalt(II) chloride hexahydrate and 150 mM thiourea) to form counter electrodes with electrocatalytic activity that exceeds that of platinum [124]. It is reported that the charge-transfer resistance of the CoS/PEN film is 1.8 Ω cm^2 at 20 °C, and that this is better than that of the Pt-coated-PEN (G24i Ltd, Cardiff, UK; 2.2 Ω cm^2) but not as low as thermally platinised FTO glass (1.3 Ω cm^2). The efficiency of a DSSC using the CoS/PEN film with a eutectic ionic liquid ([C_1mim]I/[C_2mim]I/

TABLE 10.3 The Photovoltaic Performance of DSSCs Utilising the New PEDOT-on-ITO-PEN Plastic Counter Electrodes, with Ethanenitrile or Ionic Liquid-Based Electrolytes (with Standard FTO Glass Working Electrodes), Compared with Platinised FTO Glass and the Commercially Available Pt/Ti Alloy on ITO-PEN [122, 123]

Electrolyte	Ethanenitrile Mixture			[C₂mim][SCN] Ionic Liquid Mixture			[C₂mim][B(CN)₄] Ionic Liquid Mixture		
Cathode	Pt on FTO Glass	Pt/Ti on PEN	PEDOT on ITO-PEN	Pt on FTO Glass	Pt/Ti on PEN	PEDOT on ITO-PEN	Pt on FTO Glass	Pt/Ti on PEN	PEDOT on ITO-PEN
I_{sc}/mA cm^{-2}	14.0	13.8	14.1	9.4	9.7	9.9	11.1	9.6	10.9
V_{oc}/mV	783.2	805.3	787.1	682.8	718.9	698.7	656.7	664.6	713.8
Fill factor	0.72	0.61	0.73	0.75	0.64	0.73	0.76	0.72	0.74
Efficiency/%	7.9	6.79	8.0	4.8	4.5	5.0	5.6	4.6	5.7

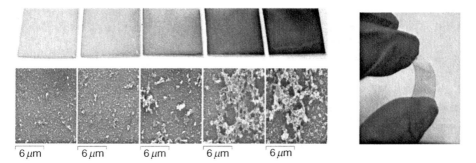

6 μm   6 μm   6 μm   6 μm   6 μm

Figure 10.7 Photographs (top) and scanning electron microscopy (SEM) images (lower) of PEDOT on ITO-PEN plastic (width 1.2 cm each) electrodeposited for different times (from left to right: 5, 10, 15, 30, and 45 seconds) showing the transparency, morphology, and flexibility of the films [122]. Reproduced with permission of The Royal Society of Chemistry.

[C₂mim][B(CN)₄]/I₂/NBB/GNCS 12:12:16:1.67:3.33:0.67, with Z907 dye and standard FTO glass working electrodes) was 6.5%, and the device retained 85% of its efficiency after 1000 hours light soaking at 60 °C. This drop in efficiency was primarily due to a decrease of *ca.* 120 mV in the V_{oc}.

10.4 CONCLUSIONS AND FUTURE DIRECTIONS

It is clear at this point, after a survey of the literature, that there is a significant range of room temperature ionic liquids that have proven effective as electrolytes for photovoltaic applications. The use of low-viscosity ionic liquids can be beneficial for cell performance, but the relationship between these parameters is not simple, with diffusion of the key triiodide species via a Grotthuss-type mechanism adding extra levels of complexity. Indeed, gelation of ionic liquids, by addition of polymers or nanoparticles, does not necessarily result in a decrease in efficiency, which is very promising in the drive towards efficient quasi-solid state DSSCs. Ionic liquids generally also appear to give a good fill factor, which is believed to be a result of effective screening of electric charges, with the high ion concentration enabling increased charge separation, but the exact mechanism is not fully understood. The influence of the nature of the ionic liquid on the energy levels of the redox couple, dye and TiO₂, and the efficacy of different additives, is also not entirely clear. Further elucidation of the influence of different ionic liquids on these parameters may be paramount in increasing the efficiency of ionic liquid-containing solar cells.

Optimisation of ionic liquid electrolytes is also paramount in the development of DSSCs on flexible substrates, as they do not permeate plastic in the way that molecular solvents do. The development of quasi-solid state electrolytes that are polymerised by UV light, in a very short time and without loss of efficiency, would also be extremely beneficial to their production on an industrial scale. The ability to produce efficient DSSCs on flexible plastic substrates entirely by reel-to-reel processing techniques will be a huge step forward in the development of commercially viable devices.

The development of novel dyes with optimised energy levels, or co-sensitisation using two dyes with complementary spectral responses, is clearly another promising area of research.

The physical properties of ionic liquids that are most pertinent to their choice for DSSCs are the conductivity and rate of diffusion of the redox species (which may not be linearly related to the viscosity), the electrochemical and thermal stability, resistance to UV degradation, and the viscosity of the ionic liquid (for both handling and conductivity reasons). A salt that is not hygroscopic is advantageous for handling and lifetime considerations, as these devices are very sensitive to oxygen and water, and parameters such as cost, toxicity, miscibility with additives, and large-scale availability must also be taken into consideration. Ultimately, of course, for commercialisation, these devices will have to be built and used on a large scale, and the development of large modules of DSSCs is still in its infancy. Finally, the long-term stability of these devices will continue to be an important consideration, and prolonged testing under accelerated conditions—with light soaking and elevated temperatures—is very important. Researchers are strongly encouraged to perform these tests as part of the development of any new system, and many of the results reported to date are reasonably encouraging, but it is important to note that for good lifetimes it is essential to have very good sealing of the device, and this may require specialist techniques and materials. Finally, for the ultimate in device lifetimes, the development of quasi-solid state solar cells is a priority. The polyelectrolyte, nanocomposite, and plastic crystal systems are all promising, each with their own particular advantages and disadvantages, and further research into each of these areas promises to ultimately yield truly long-lasting and efficient devices.

REFERENCES

1   Green, M., *Power to the people*, University of New South Wales Press Ltd, Sydney (2002).

2   O'Regan, B., and Grätzel, M., A low-cost, high-efficiency solar cell based on dye-sensitized colloidal titanium dioxide films, *Nature* **353**, 737–740 (1991).

3   Sorrell, C.C., Sugihara, S., and Nowotny, J., eds., *Materials for energy conversion devices*, Woodhead Pub., Cambridge; CRC Press, Boca Raton, FL (2005).

4   Archer, M.D., and Hill, R., eds., *Series on photoconversion of solar energy—volume 1: clean electricity from photovoltaics*, Imperial College Press, London (2001).

5   Zakeeruddin, S.M., and Grätzel, M., Solvent-free ionic liquid electrolytes for mesoscopic dye-sensitized solar cells, *Adv. Funct. Mater.* **19**, 2187–2202 (2009).

6   Durrant, J.R., Haque, S.A., and Palomares, E., Photochemical energy conversion: from molecular dyads to solar cells, *Chem. Commun.* **31**, 3279–3289 (2006).

7   Grätzel, M., Recent advances in sensitized mesoscopic solar cells, *Acc. Chem. Res.* **42**, 1788–1798 (2009).

8   Nazeeruddin, M.K., Kay, A., Rodicio, I., Humphry-Baker, R., Mueller, E., Liska, P., Vlachopoulos, N., and Grätzel, M., Conversion of light to electricity by cis-X_2bis(2,2'-bipyridyl-4,4'-dicarboxylate)ruthenium(II) charge-transfer sensitizers ($X = Cl^-$, Br^-, I^-, CN^-, and SCN^-) on nanocrystalline titanium dioxide electrodes, *J. Am. Chem. Soc.* **115**, 6382–6390 (1993).

9   Kusama, H., and Arakawa, H., Influence of benzimidazole additives in electrolytic solution on dye-sensitized solar cell performance, *J. Photochem. Photobiol. A* **162**, 441–448 (2004).

10 Stathatos, E., Lianos, P., Zakeeruddin, S.M., Liska, P., and Grätzel, M., A quasi-solid-state dye-sensitized solar cell based on a sol-gel nanocomposite electrolyte containing ionic liquid, *Chem. Mater.* **15**, 1825–1829 (2003).

11 Kambe, S., Nakade, S., Kitamura, T., Wada, Y., and Yanagida, S., Influence of the electrolytes on electron transport in mesoporous TiO_2-electrolyte systems, *J. Phys. Chem. B* **106**, 2967–2972 (2002).

12 Kawano, R., Matsui, H., Matsuyama, C., Sato, A., Susan, M.A.B.H., Tanabe, N., and Watanabe, M., High performance dye-sensitized solar cells using ionic liquids as their electrolytes, *J. Photochem. Photobiol. A* **164**, 87–92 (2004).

13 Zhang, C., Huang, Y., Huo, Z., Chen, S., and Dai, S., Photoelectrochemical effects of guanidinium thiocyanate on dye-sensitized solar cell performance and stability, *J. Phys. Chem. C* **113**, 21779–21783 (2009).

14 Kopidakis, N., Neale, N.R., and Frank, A.J., Effect of an adsorbent on recombination and band-edge movement in dye-sensitized TiO_2 solar cells: evidence for surface passivation, *J. Phys. Chem. B* **110**, 12485–12489 (2006).

15 Lee, K.-M., Suryanarayanan, V., Ho, K.-C., Justin Thomas, K.R., and Lin, J.T., Effects of co-adsorbate and additive on the performance of dye-sensitized solar cells: a photophysical study, *Sol. Energy Mater. Sol. Cells* **91**, 1426–1431 (2007).

16 Chen, L.-H., Xue, B.-F., Liu, X.-Z., Li, K.-X., Luo, Y.-H., Meng, Q.-B., Wang, R.-L., and Chen, L.-Q., Efficiency enhancement of dye-sensitized solar cells: using salt CuI as an additive in an ionic liquid, *Chin. Phys. Lett.* **24**, 555–558 (2007).

17 Wang, P., Zakeeruddin, S.M., Humphry-Baker, R., Moser, J.E., and Grätzel, M., Molecular-scale interface engineering of TiO_2 nanocrystals: improving the efficiency and stability of dye-sensitized solar cells, *Adv. Mater.* **15**, 2101–2104 (2003).

18 Chiba, Y., Islam, A., Watanabe, Y., Komiya, R., Koide, N., and Han, L., Dye-sensitized solar cells with conversion efficiency of 11.1%, *Jpn. J. Appl. Phys., Part 2: Lett. Exp. Lett.* **45**, L638–L640 (2006).

19 Gao, F., Wang, Y., Shi, D., Zhang, J., Wang, M., Jing, X., Humphry-Baker, R., Wang, P., Zakeeruddin, S.M., and Grätzel, M., Enhance the optical absorptivity of nanocrystalline TiO_2 film with high molar extinction coefficient ruthenium sensitizers for high performance dye-sensitized solar cells, *J. Am. Chem. Soc.* **130**, 10720–10728 (2008).

20 Shi, D., Pootrakulchote, N., Li, R., Guo, J., Wang, Y., Zakeeruddin, S.M., Grätzel, M., and Wang, P., New efficiency records for stable dye-sensitized solar cells with low-volatility and ionic liquid electrolytes, *J. Phys. Chem. C* **112**, 17046–17050 (2008).

21 Bai, Y., Cao, Y., Zhang, J., Wang, M., Li, R., Wang, P., Zakeeruddin, S.M., and Grätzel, M., High-performance dye-sensitized solar cells based on solvent-free electrolytes produced from eutectic melts, *Nat. Mater.* **7**, 626–630 (2008).

22 Fischer, A., Pettersson, H., Hagfeldt, A., Boschloo, G., Kloo, L., and Gorlov, M., Crystal formation involving 1-methylbenzimidazole in iodide/triiodide electrolytes for dye-sensitized solar cells, *Sol. Energy Mater. Sol. Cells* **91**, 1062–1065 (2007).

23 Li, B., Wang, L., Kang, B., Wang, P., and Qiu, Y., Review of recent progress in solid-state dye-sensitized solar cells, *Sol. Energy Mater. Sol. Cells* **90**, 549–573 (2006).

24 Gorlov, M., and Kloo, L., Ionic liquid electrolytes for dye-sensitized solar cells, *Dalton Trans.* **20**, 2655–2666 (2008).

25 Fredin, K., Gorlov, M., Pettersson, H., Hagfeldt, A., Kloo, L., and Boschloo, G., On the influence of anions in binary ionic liquid electrolytes for monolithic dye-sensitized solar cells, *J. Phys. Chem. C* **111**, 13261–13266 (2007).

26 Paulsson, H., Hagfeldt, A., and Kloo, L., Molten and solid trialkylsulfonium iodides and their polyiodides as electrolytes in dye-sensitized nanocrystalline solar cells, *J. Phys. Chem. B* **107**, 13665–13670 (2003).

27 Wang, P., Zakeeruddin, S.M., Moser, J.-E., Humphry-Baker, R., and Grätzel, M., A solvent-free, $SeCN^-/(SeCN)_3^-$ based ionic liquid electrolyte for high-efficiency dye-sensitized nanocrystalline solar cells, *J. Am. Chem. Soc.* **126**, 7164–7165 (2004).

28 Armel, V., Pringle, J.M., Forsyth, M., MacFarlane, D.R., Officer, D.L., and Wagner, P., Ionic liquid electrolyte porphyrin dye sensitized solar cells, *Chem. Commun.* **46**, 3146–3148 (2010).

29 Matsumoto, H., Matsuda, T., Tsuda, T., Hagiwara, R., Ito, Y., and Miyazaki, Y., The application of room temperature molten salt with low viscosity to the electrolyte for dye-sensitized solar cell, *Chem. Lett.* **30**, 26–27 (2001).

30 Wang, P., Wenger, B., Humphry-Baker, R., Moser, J.-E., Teuscher, J., Kantlehner, W., Mezger, J., Stoyanov, E.V., Zakeeruddin, S.M., and Grätzel, M., Charge separation and efficient light energy conversion in sensitized mesoscopic photoelectrochemical cells based on binary ionic liquids, *J. Am. Chem. Soc.* **127**, 6850–6856 (2005).

31 Dai, Q., Menzies, D.B., MacFarlane, D.R., Batten, S.R., Forsyth, S., Spiccia, L., Cheng, Y.-B., and Forsyth, M., Dye-sensitized nanocrystalline solar cells incorporating ethylmethylimidazolium-based ionic liquid electrolytes, *C. R. Chim.* **9**, 617–621 (2006).

32 Mazille, F., Fei, Z., Kuang, D., Zhao, D., Zakeeruddin, S.M., Grätzel, M., and Dyson, P.J., Influence of ionic liquids bearing functional groups in dye-sensitized solar cells, *Inorg. Chem.* **45**, 1585–1590 (2006).

33 Wang, P., Zakeeruddin, S.M., Moser, J.-E., and Grätzel, M., A new ionic liquid electrolyte enhances the conversion efficiency of dye-sensitized solar cells, *J. Phys. Chem. B* **107**, 13280–13285 (2003).

34 Wang, P., Zakeeruddin, S.M., Humphry-Baker, R., and Grätzel, M., A binary ionic liquid electrolyte to achieve >=7% power conversion efficiencies in dye-sensitized solar cells, *Chem. Mater.* **16**, 2694–2696 (2004).

35 Yamanaka, N., Kawano, R., Kubo, W., Kitamura, T., Wada, Y., Watanabe, M., and Yanagida, S., Ionic liquid crystal as a hole transport layer of dye-sensitized solar cells, *Chem. Commun.*, 740–742 (2005).

36 Ramirez, R.E., and Sanchez, E.M., Molten phosphonium iodides as electrolytes in dye-sensitized nanocrystalline solar cells, *Sol. Energy Mater. Sol. Cells* **90**, 2384–2390 (2006).

37 Fei, Z., Kuang, D., Zhao, D., Klein, C., Ang, W.H., Zakeeruddin, S.M., Grätzel, M., and Dyson, P.J., A supercooled imidazolium iodide ionic liquid as a low-viscosity electrolyte for dye-sensitized solar cells, *Inorg. Chem.* **45**, 10407–10409 (2006).

38 Kuang, D., Wang, P., Ito, S., Zakeeruddin, S.M., and Grätzel, M., Stable mesoscopic dye-sensitized solar cells based on tetracyanoborate ionic liquid electrolyte, *J. Am. Chem. Soc.* **128**, 7732–7733 (2006).

39 Papageorgiou, N., Athanassov, Y., Armand, M., Bonhote, P., Pettersson, H., Azam, A., and Grätzel, M., The performance and stability of ambient temperature molten salts for solar cell applications, *J. Electrochem. Soc.* **143**, 3099–3108 (1996).

40 Kawano, R., and Watanabe, M., Anomaly of charge transport of an iodide/tri-iodide redox couple in an ionic liquid and its importance in dye-sensitized solar cells, *Chem. Commun.* **16**, 2107–2109 (2005).

41 Cao, Y., Zhang, J., Bai, Y., Li, R., Zakeeruddin, S.M., Grätzel, M., and Wang, P., Dye-sensitized solar cells with solvent-free ionic liquid electrolytes, *J. Phys. Chem. C* **112**, 13775–13781 (2008).

42 Wachter, P., Schreiner, C., Zistler, M., Gerhard, D., Wasserscheid, P., and Gores, H.J., A microelectrode study of triiodide diffusion coefficients in mixtures of room temperature ionic liquids, useful for dye-sensitised solar cells, *Microchim. Acta* **160**, 125–133 (2008).

43 Zistler, M., Schreiner, C., Wachter, P., Wasserscheid, P., Gerhard, D., and Gores, H.J., Electrochemical characterization of 1-ethyl-3-methylimidazolium thiocyanate and measurement of triiodide diffusion coefficients in blends of two ionic liquids, *Int. J. Electrochem. Sci.* **3**, 236–245 (2008).

44 Yu, Z., Gorlov, M., Nissfolk, J., Boschloo, G., and Kloo, L., Investigation of iodine concentration effects in electrolytes for dye-sensitized solar cells, *J. Phys. Chem. C* **114**, 10612–10620 (2010).

45 Furube, A., Wang, Z.-S., Sunahara, K., Hara, K., Katoh, R., and Tachiya, M., Femtosecond diffuse reflectance transient absorption for dye-sensitized solar cells under operational conditions: effect of electrolyte on electron injection, *J. Am. Chem. Soc.* **132**, 6614–6615 (2010).

46 Chen, Y.H., Chen, W.H., and Hong, C.W., Molecular simulation and performance prediction of photoelectrochemical solar cells, *J. Electrochem. Soc.* **156**, P163–P168 (2009).

47 Fabregat-Santiago, F., Bisquert, J., Palomares, E., Otero, L., Kuang, D., Zakeeruddin, S.M., and Grätzel, M., Correlation between photovoltaic performance and impedance spectroscopy of dye-sensitized solar cells based on ionic liquids, *J. Phys. Chem. C* **111**, 6550–6560 (2007).

48 Oda, T., Tanaka, S., and Hayase, S., Analysis of dominant factors for increasing the efficiencies of dye-sensitized solar cells: comparison between ethanenitrile and ionic liquid based electrolytes, *Jpn. J. Appl. Phys. Part 1* **45**, 2780–2787 (2006).

49 Zistler, M., Wachter, P., Schreiner, C., Fleischmann, M., Gerhard, D., Wasserscheid, P., Hinsch, A., and Gores, H.J., Temperature dependent impedance analysis of binary ionic liquid electrolytes for dye-sensitized solar cells, *J. Electrochem. Soc.* **154**, B925–B930 (2007).

50 Wang, M., Chen, P., Humphry-Baker, R., Zakeeruddin, S.M., and Grätzel, M., The influence of charge transport and recombination on the performance of dye-sensitized solar cells, *Chem. Phys. Chem.* **10**, 290–299 (2009).

51 Zistler, M., Wachter, P., Wasserscheid, P., Gerhard, D., Hinsch, A., Sastrawan, R., and Gores, H.J., Comparison of electrochemical methods for triiodide diffusion coefficient measurements and observation of non-Stokesian diffusion behaviour in binary mixtures of two ionic liquids, *Electrochim. Acta* **52**, 161–169 (2006).

52 Wang, P., Zakeeruddin, S.M., Exnar, I., and Grätzel, M., High efficiency dye-sensitized nanocrystalline solar cells based on ionic liquid polymer gel electrolyte, *Chem. Commun.* **24**, 2972–2973 (2002).

53 Wang, P., Klein, C., Humphry-Baker, R., Zakeeruddin, S.M., and Grätzel, M., A high molar extinction coefficient sensitizer for stable dye-sensitized solar cells, *J. Am. Chem. Soc.* **127**, 808–809 (2005).

54 Paulsson, H., Berggrund, M., Svantesson, E., Hagfeldt, A., and Kloo, L., Molten and solid metal-iodide-doped trialkylsulphonium iodides and polyiodides as electrolytes in dye-sensitized nanocrystalline solar cells, *Sol. Energy Mater. Sol. Cells* **82**, 345–360 (2004).

55 Murai, S., Mikoshiba, S., Sumino, H., and Hayase, S., Quasi-solid dye-sensitized solar cells containing chemically crosslinked gel. How to make gels with a small amount of gelator, *J. Photochem. Photobiol. A* **148**, 33–39 (2002).

56 Nei De Freitas, J., Nogueira, A.F., and De Paoli, M.-A., New insights into dye-sensitized solar cells with polymer electrolytes, *J. Mater. Chem.* **19**, 5279–5294 (2009).

57 Kubo, W., Makimoto, Y., Kitamura, T., Wada, Y., and Yanagida, S., Quasi-solid-state dye-sensitized solar cell with ionic polymer electrolyte, *Chem. Lett.* **9**, 948–949 (2002).

58 Lee, H.-J., Lee, J.-K., Kim, M.-R., Shin, W.S., Jin, S.-H., Kim, K.-H., Park, D.-W., and Park, S.-W., Influence of ionic liquids in quasi-solid state electrolyte on dye-sensitized solar cell performance, *Mol. Cryst. Liq. Cryst.* **462**, 75–81 (2007).

59 Murai, S., Mikoshiba, S., Sumino, H., Kato, T., and Hayase, S., Quasi-solid dye sensitized solar cells filled with phase-separated chemically cross-linked ionic gels, *Chem. Commun.* **13**, 1534–1535 (2003).

60 Shibata, Y., Kato, T., Kado, T., Shiratuchi, R., Takashima, W., Kaneto, K., and Hayase, S., Quasi-solid dye sensitized solar cells filled with ionic liquid-increase in efficiencies by specific interaction between conductive polymers and gelators, *Chem. Commun.* **21**, 2730–2731 (2003).

61 Usui, H., Matsui, H., Tanabe, N., and Yanagida, S., Improved dye-sensitized solar cells using ionic nanocomposite gel electrolytes, *J. Photochem. Photobiol. A* **164**, 97–101 (2004).

62 Suryanarayanan, V., Lee, K.-M., Ho, W.-H., Chen, H.-C., and Ho, K.-C., A comparative study of gel polymer electrolytes based on PVDF-HFP and liquid electrolytes, containing imidazolinium ionic liquids of different carbon chain lengths in DSSCs, *Sol. Energy Mater. Sol. Cells* **91**, 1467–1471 (2007).

63 Wang, M., Yin, X., Xiao, X.R., Zhou, X., Yang, Z.Z., Li, X.P., and Lin, Y., A new ionic liquid based quasi-solid state electrolyte for dye-sensitized solar cells, *J. Photochem. Photobiol. A* **194**, 20–26 (2008).

64 Mohmeyer, N., Kuang, D., Wang, P., Schmidt, H.-W., Zakeeruddin, S.M., and Grätzel, M., An efficient organogelator for ionic liquids to prepare stable quasi-solid-state dye-sensitized solar cells, *J. Mater. Chem.* **16**, 2978–2983 (2006).

65 Wang, L., Fang, S.-B., and Lin, Y., Novel polymer electrolytes containing chemically crosslinked gelators for dye-sensitized solar cells, *Polym. Adv. Technol.* **17**, 512–517 (2006).

66 Winther-Jensen, O., Armel, V., Forsyth, M., and MacFarlane, D.R., In situ photopolymerization of a gel ionic liquid electrolyte in the presence of iodine and its use in dye sensitized solar cells, *Macromol. Rapid Commun.* **31**, 479–483 (2010).

67 Bhattacharyya, A.J., and Maier, J., Second phase effects on the conductivity of non-aqueous salt solutions: "Soggy sand electrolytes", *Adv. Mater.* **16**, 811–814 (2004).

68 Bhattacharyya, A.J., Maier, J., Bock, R., and Lange, F.F., New class of soft matter electrolytes obtained via heterogeneous doping: percolation effects in "soggy sand" electrolytes, *Solid State Ionics* **177**, 2565–2568 (2006).

69 Chung, S.H., Wang, Y., Persi, L., Croce, F., Greenbaum, S.G., Scrosati, B., and Plichta, E., Enhancement of ion transport in polymer electrolytes by addition of nanoscale inorganic oxides, *J. Power Sources* **97–98**, 644–648 (2001).

70 Liang, C.C., Conduction characteristics of the lithium iodide-aluminum oxide solid electrolytes, *J. Electrochem. Soc.* **120**, 1289–1292 (1973).

71 Maier, J., Ionic conduction in space charge regions, *Prog. Solid State Chem.* **23**, 171–263 (1995).

72 Maier, J., Nanoionics: ionic charge carriers in small systems, *Phys. Chem. Chem. Phys.* **11**, 3011–3022 (2009).

73 Wang, P., Zakeeruddin, S.M., Comte, P., Exnar, I., and Grätzel, M., Gelation of ionic liquid-based electrolytes with silica nanoparticles for quasi-solid-state dye-sensitized solar cells, *J. Am. Chem. Soc.* **125**, 1166–1167 (2003).

74 Katakabe, T., Kawano, R., and Watanabe, M., Acceleration of redox diffusion and charge-transfer rates in an ionic liquid with nanoparticle addition, *Electrochem. Solid-State Lett.* **10**, F23–F25 (2007).

75 Yang, H., Yu, C., Song, Q., Xia, Y., Li, F., Chen, Z., Li, X., Yi, T., and Huang, C., High-temperature and long-term stable solid-state electrolyte for dye-sensitized solar cells by self-assembly, *Chem. Mater.* **18**, 5173–5177 (2006).

76 Chen, Z., Li, F., Yang, H., Yi, T., and Huang, C., A thermostable and long-term-stable ionic-liquid-based gel electrolyte for efficient dye-sensitized solar cells, *Chem. Phys. Chem.* **8**, 1293–1297 (2007).

77 Wang, N., Lin, H., Li, J., and Li, X., Improved quasi-solid dye-sensitized solar cells by composite ionic liquid electrolyte including layered alpha -zirconium phosphate, *Appl. Phys. Lett.* **89**, 194104 (2006).

78 Cheng, P., Lan, T., Wang, W., Wu, H., Yang, H., Deng, C., Dai, X., and Guo, S., Improved dye-sensitized solar cells by composite ionic liquid electrolyte incorporating layered titanium phosphate, *Sol. Energy* **84**, 854–859 (2010).

79 Cerneaux, S., Zakeeruddin, S.M., Pringle, J.M., Cheng, Y.-B., Grätzel, M., and Spiccia, L., Novel nano-structured silica-based electrolytes containing quaternary ammonium iodide moieties, *Adv. Funct. Mater.* **17**, 3200–3206 (2007).

80 Jovanovski, V., Stathatos, E., Orel, B., and Lianos, P., Dye-sensitized solar cells with electrolyte based on a trimethoxysilane-derivatized ionic liquid, *Thin Solid Films* **511–512**, 634–637 (2006).

81 Jovanovski, V., Orel, B., Jese, R., Vuk, A.S., Mali, G., Hocevar, S.B., Grdadolnik, J., Stathatos, E., and Lianos, P., Novel polysilsesquioxane-I⁻/I₃⁻ ionic electrolyte for dye-sensitized photoelectrochemical cells, *J. Phys. Chem. B* **109**, 14387–14395 (2005).

82 Stathatos, E., Jovanovski, V., Orel, B., Jerman, I., and Lianos, P., Dye-sensitized solar cells made by using a polysilsesquioxane polymeric ionic fluid as redox electrolyte, *J. Phys. Chem. C* **111**, 6528–6532 (2007).

83 Yamanaka, N., Kawano, R., Kubo, W., Masaki, N., Kitamura, T., Wada, Y., Watanabe, M., and Yanagida, S., Dye-sensitized TiO₂ solar cells using imidazolium-type ionic liquid crystal systems as effective electrolytes, *J. Phys. Chem. B* **111**, 4763–4769 (2007).

84 Timmermans, J., Plastic crystals: a historical review, *Phys. Chem. Solids* **18**, 1–8 (1961).

85 Pringle, J.M., Howlett, P.C., MacFarlane, D.R., and Forsyth, M., Organic ionic plastic crystals: recent advances, *J. Mater. Chem.* **20**, 2056–2062 (2010).

86 MacFarlane, D.R., and Forsyth, M., Plastic crystal electrolyte materials: new perspectives on solid state ionics, *Adv. Mater.* **13**, 957–966 (2001).

87 Dai, Q., MacFarlane, D.R., and Forsyth, M., High mobility I⁻/I₃⁻ redox couple in a molecular plastic crystal: a potential new generation of electrolyte for solid-state photoelectrochemical cells, *Solid State Ionics* **177**, 395–401 (2006).

88 Dai, Q., MacFarlane, D.R., Howlett, P.C., and Forsyth, M., Rapid I⁻/I₃⁻ diffusion in a molecular-plastic-crystal electrolyte for potential application in solid-state photoelectrochemical cells, *Angew. Chem. Int. Ed.* **44**, 313–316 (2005).

89 Chen, Z., Yang, H., Li, X., Li, F., Yi, T., and Huang, C., Thermostable succinonitrile-based gel electrolyte for efficient, long-life dye-sensitized solar cells, *J. Mater. Chem.* **17**, 1602–1607 (2007)

90 Armel, V., Forsyth, M., MacFarlane, D.R., and Pringle, J.M., Organic ionic plastic crystal electrolytes; a new class of electrolyte for high efficiency solid state dye-sensitized solar cells, *Energy Environ. Sci.* **4**, 2234–2239 (2011).

91 Yanagida, S., Yu, Y., and Manseki, K., Iodine/iodide-free dye-sensitized solar cells, *Acc. Chem. Res.* **42**, 1827–1838 (2009).

92 Li, D., Li, H., Luo, Y., Li, K., Meng, Q., Armand, M., and Chen, L., Non-corrosive, non-absorbing organic redox couple for dye-sensitized solar cells, *Adv. Funct. Mater.* **20**, 3358–3365 (2010).

93 Oskam, G., Bergeron, B.V., Meyer, G.J., and Searson, P.C., Pseudohalogens for dye-sensitized TiO₂ photoelectrochemical cells, *J. Phys. Chem. B* **105**, 6867–6873 (2001).

94   Gorlov, M., Pettersson, H., Hagfeldt, A., and Kloo, L., Electrolytes for dye-sensitized solar cells based on interhalogen ionic salts and liquids. *Inorg. Chem.* **46**, 3566–3575 (2007).

95   Yella, A., Lee, H.-W., Tsao, H.N., Yi, C., Chandiran, A.K., Nazeeruddin, M.K., Diau, E.W.-G., Yeh, C.-Y., Zakeeruddin, S.M., and Grätzel, M., Porphyrin-sensitized solar cells with cobalt (II/III)-based redox electrolyte exceed 12% efficiency, *Science*, **334**, 629–634 (2011).

96   Hamann, T.W., The end of iodide? Cobalt complex redox shuttles in DSSCs, *Dalton Trans.* **41**, 3111–3115 (2012).

97   Lind, S.J., Gordon, K.C., Gambhir, S., and Officer, D.L., A spectroscopic and DFT study of thiophene-substituted metalloporphyrins as dye-sensitized solar cell dyes, *Phys. Chem. Chem. Phys.* **11**, 5598–5607 (2009).

98   Zakeeruddin, S.M., Nazeeruddin, M.K., Humphry-Baker, R., Pechy, P., Quagliotto, P., Barolo, C., Viscardi, G., and Grätzel, M., Design, synthesis, and application of amphiphilic ruthenium polypyridyl photosensitizers in solar cells based on nanocrystalline TiO_2 films, *Langmuir* **18**, 952–954 (2002).

99   Kuang, D., Uchida, S., Humphry-Baker, R., Zakeeruddin Shaik, M., and Grätzel, M., Organic dye-sensitized ionic liquid based solar cells: remarkable enhancement in performance through molecular design of indoline sensitizers, *Angew. Chem. Int. Ed.* **47**, 1923–1927 (2008).

100 Wang, M., Xu, M., Shi, D., Li, R., Gao, F., Zhang, G., Yi, Z., Humphry-Baker, R., Wang, P., Zakeeruddin, S.M., and Grätzel, M., High-performance liquid and solid dye-sensitized solar cells based on a novel metal-free organic sensitizer, *Adv. Mater.* **20**, 4460–4463 (2008).

101 Qin, H., Wenger, S., Xu, M., Gao, F., Jing, X., Wang, P., Zakeeruddin Shaik, M., and Grätzel, M., An organic sensitizer with a fused dithienothiophene unit for efficient and stable dye-sensitized solar cells, *J. Am. Chem. Soc.* **130**, 9202–9203 (2008).

102 Xu, M., Wenger, S., Bala, H., Shi, D., Li, R., Zhou, Y., Zakeeruddin, S.M., Grätzel, M., and Wang, P., Tuning the energy level of organic sensitizers for high-performance dye-sensitized solar cells, *J. Phys. Chem. C* **113**, 2966–2973 (2009).

103 Kim, S., Lee, J.K., Kang, S.O., Ko, J., Yum, J.H., Fantacci, S., De Angelis, F., Di Censo, D., Nazeeruddin, M.K., Grätzel, M., Molecular engineering of organic sensitizers for solar cell applications, *J. Am. Chem. Soc.* **128**, 16701–16707 (2006).

104 Kuang, D., Walter, P., Nüesch, F., Kim, S., Ko, J., Comte, P., Zakeeruddin, S.M., Nazeeruddin, M.K., and Grätzel, M., Co-sensitization of organic dyes for efficient ionic liquid electrolyte-based dye-sensitized solar cells, *Langmuir* **23**, 10906–10909 (2007).

105 Tsunashima, K., and Sugiya, M., Physical and electrochemical properties of low-viscosity phosphonium ionic liquids as potential electrolytes, *Electrochem. Commun.* **9**, 2353–2358 (2007).

106 Kay, A., and Grätzel, M., Low cost photovoltaic modules based on dye sensitized nanocrystalline titanium dioxide and carbon powder, *Sol. Energy Mater. Sol. Cells* **44**, 99–117 (1996).

107 Pettersson, H., Gruszecki, T., Bernhard, R., Haeggman, L., Gorlov, M., Boschloo, G., Edvinsson, T., Kloo, L., and Hagfeldt, A., The monolithic multicell: a tool for

testing material components in dye-sensitized solar cells, *Prog. Photovoltaics* **15**, 113–121 (2007).

108 Murakami, T.N., and Grätzel, M., Counter electrodes for DSC: application of functional materials as catalysts, *Inorg. Chim. Acta* **361**, 572–580 (2008).

109 Trancik, J.E., Calabrese Barton, S., and Hone, J., Transparent and catalytic carbon nanotube films, *Nano Lett.* **8**, 982–987 (2008).

110 Saito, Y., Kitamura, T., Wada, Y., and Yanagida, S., Application of poly(3,4-ethylenedioxythiophene) to counter electrode in dye-sensitized solar cells, *Chem. Lett.* **10**, 1060–1061 (2002).

111 Sirimanne, P.M., Winther-Jensen, B., Weerasinghe, H.C., and Cheng, Y.-B., Towards an all-polymer cathode for dye sensitized photovoltaic cells, *Thin Solid Films* **518**, 2871–2875 (2010).

112 Winther-Jensen, B. and MacFarlane, D.R., New generation, metal-free electrocatalysts for fuel cells, solar cells and water splitting, *Energy Environ. Sci.* **4**, 2790–2798 (2011).

113 Xia, J., Masaki, N., Lira-Cantu, M., Kim, Y., Jiang, K., and Yanagida, S., Influence of doped anions on poly(3,4-ethylenedioxythiophene) as hole conductors for iodine-free solid-state dye-sensitized solar cells, *J. Am. Chem. Soc.* **130**, 1258–1263 (2008).

114 Senadeera, R., Fukuri, N., Saito, Y., Kitamura, T., Wada, Y., and Yanagida, S., Volatile solvent-free solid-state polymer-sensitized TiO_2 solar cells with poly(3,4-ethylenedioxythiophene) as a hole-transporting medium, *Chem. Commun.* **17**, 2259–2261 (2005).

115 Shibata, Y., Kato, T., Kado, T., Shiratuchi, R., Takashima, W., Kaneto, K., and Hayase, S., Quasi-solid dye sensitised solar cells filled with ionic liquid—increase in efficiencies by specific interaction between conductive polymers and gelators, *Chem. Commun.* **21**, 2730–2731 (2003).

116 Xia, J., Masaki, N., Jiang, K., and Yanagida, S., The influence of doping ions on poly(3,4-ethylenedioxythiophene) as a counter electrode of a dye-sensitized solar cell, *J. Mater. Chem.* **17**, 2845–2850 (2007).

117 Chen, J.-G., Wei, H.-Y., and Ho, K.-C., Using modified poly(3,4-ethylene dioxythiophene): poly(styrene sulfonate) film as a counter electrode in dye-sensitized solar cells, *Sol. Energy Mater. Sol. Cells* **91**, 1472–1477 (2007).

118 Muto, T., Ikegami, M., Kobayashi, K., and Miyasaka, T., Conductive polymer-based mesoscopic counter electrodes for plastic dye-sensitized solar cells, *Chem. Lett.* **36**, 804–805 (2007).

119 Saito, Y., Kubo, W., Kitamura, T., Wada, Y., and Yanagida, S., I-/I3- redox reaction behavior on poly(3,4-ethylenedioxythiophene) counterelectrode in dye-sensitized solar cells, *J. Photochem. Photobiol. A* **164**, 153–157 (2004).

120 Pringle, J.M., Armel, V., Forsyth, M., and MacFarlane, D.R., PEDOT-coated counter electrodes for dye-sensitized solar cells, *Aust. J. Chem.* **62**, 348–352 (2009).

121 Ahmad, S., Yum, J.-H., Xianxi, Z., Grätzel, M., Butt, H.-J., and Nazeeruddin, M.K., Dye-sensitized solar cells based on poly(3,4-ethylenedioxythiophene) counter electrode derived from ionic liquids, *J. Mater. Chem.* **20**, 1654–1658 (2010).

122 Pringle, J.M., Armel, V., and MacFarlane, D.R., Electrodeposited PEDOT-on-plastic cathodes for dye-sensitized solar cells, *Chem. Commun.* **46**, 5367–5369 (2010).

123 Ikegami, M., Miyoshi, K., Miyasaka, T., Teshima, K., Wei, T.C., Wan, C.C., and Wang, Y.Y., Platinum/titanium bilayer deposited on polymer film as efficient counter electrodes for plastic dye-sensitized solar cells, *Appl. Phys. Lett.* **90**, 153122 (3pp.) (2007).

124 Wang, M., Anghel, A.M., Marsan, B., Cevey Ha, N.-L., Pootrakulchote, N., Zakeeruddin, S.M., and Grätzel, M., CoS supersedes Pt as efficient electrocatalyst for triiodide reduction in dye-sensitized solar cells, *J. Am. Chem. Soc.* **131**, 15976–15977 (2009).

11 Phase Behaviour of Gases in Ionic Liquids

MARK B. SHIFLETT

DuPont Central Research and Development, Experimental Station, Wilmington, Delaware, USA

AKIMICHI YOKOZEKI

32 Kingsford Lane, Spencerport, New York, USA

ABSTRACT

The main purpose of this chapter is to emphasise the importance of characterising the global phase behaviour of gases in ionic liquids, and how this can provide insight into new applications. Solubility measurements of several gases in ionic liquids are discussed and important experimental details are highlighted when measuring the vapour–liquid equilibria using a gravimetric microbalance and vapour–liquid–liquid equilibria using a mass–volume technique. The global phase behaviour has been well correlated with a modified cubic equation, and many of the systems described exhibit type-III and type-V phase behaviour according to the van Konynenburg–Scott classification. We have found that gases can exhibit different solubility behaviours in ionic liquids (i.e., physical and chemical absorption) and that these behaviours can be analysed with the present equation of state using a simple association model and excess thermodynamic functions. Gas separation using ionic liquids is discussed, which includes examples of ternary phase behaviour (experiments and model calculations) and what will be required to develop future commercial processes.

11.1 INTRODUCTION

Ionic liquids, a relatively new class of materials, have been defined as molten salts with a melting temperature below about 373 K [1, 2]. A variety

Ionic Liquids UnCOILed: Critical Expert Overviews, First Edition. Edited by Natalia V. Plechkova and Kenneth R. Seddon.

of thermophysical property data has been published for ionic liquids, and many new applications have been proposed during the past 10 years [1–9]. Several reviews have been written on the solubility of gases (particularly noncondensable gases) in ionic liquids [10–13], with the results reported as the Henry's Law constant. Here, we broaden the meaning of gas solubility in ionic liquids to include the global phase behaviour (i.e., vapour–liquid equilibria [VLE], vapour–liquid–liquid equilibria [VLLE], and liquid–liquid equilibria [LLE]).

The main focus of this review is to emphasise the importance of characterising the global phase behaviour (i.e., gas composition from $0 \leq x \leq 1$) over a wide temperature and pressure range. By studying ionic liquids over a wide range of conditions, we have discovered several systems which exhibit an immiscible region (VLLE and LLE) that can be classified according to the type of high-pressure phase behaviour defined by van Konynenburg and Scott [14, 15]. In order to understand such a wide range of phase behaviour, we have developed an equation of state (EOS) method based on our original refrigerant–oil correlation [16].

Initially, we had some doubt about whether a non-electrolyte EOS could work for ionic liquid mixtures. However, we discovered that the EOS method works well for $PVTx$ phase calculations. This should not be surprising since the thermodynamic equilibrium for VLE does not depend on knowing the "detailed liquid structure." Other well-known examples that have also been well correlated with solution models and ordinary non-electrolyte EOS methods include hydrogen-bonded liquids (e.g., [alcohol + water]) [17] or electrolyte solutions (e.g., [HCl + water] [18, 19]; [NH₃ + water] [20]).

In this chapter, we focus on our major interest (global phase behaviour) and discuss our experimental methods and EOS modelling.

We have intensively studied the phase behaviour of several important industrial gases such as carbon dioxide [21–25], hydrofluorocarbons [26–41], ammonia [42, 43], hydrogen [44], sulfur dioxide [45–47], hydrogen sulfide [48, 49], and nitrous oxide [50], as well as other liquid solvents (alcohols, benzene derivatives) [51–55] that are not reviewed in this chapter. We have found that these gases exhibit different solubility behaviours in ionic liquids (i.e., "physical" and "chemical" absorption). These behaviours were analysed with the present EOS using a simple association model with the use of excess thermodynamic functions (excess Gibbs free energy, G^E, excess enthalpy, H^E, and excess entropy, S^E).

As mentioned earlier, the thermodynamic analyses alone cannot determine the detailed liquid structure; therefore, we often state in our publications that theoretical modellers should use our results to provide such information. As an example, we show an encouraging case that compares our (NH₃ + ionic liquid) measurements with molecular simulations from Prof. Ed Maginn's group. Finally, we explore gas separation using ionic liquids and provide some examples of ternary phase behaviour with experiments and model calculations. We conclude, for future directions, that there will be a need for a more thorough understanding of (gas + ionic liquid) interactions that will be required to develop commercial applications.

11.2   EXPERIMENTAL TECHNIQUES

In this section, we discuss our experimental technique using a gravimetric microbalance to measure gas solubility, and point out important details for properly correcting for gravitational forces. We also discuss our mass–volume method for measuring (VLLE and LLE), and emphasise the importance of careful error analysis. This section is concluded with a review of other gas solubility techniques and emphasises the importance of safety procedures.

The determination of the solubility of a gas in an ionic liquid requires several steps:

(i)   Purification and characterisation of the solute gas and ionic liquid.

(ii)   Thorough drying and degassing of the ionic liquid.

(iii) Equilibration of the gas and liquid phases under the conditions of known constant temperature and pressure.

(iv) Measurements that determine the composition of the gas in the liquid phase.

(v)   Proper error analysis to estimate the uncertainty.

(vi) Application of a thermodynamic model to validate results.

Any paper reporting data on the solubility of a gas in an ionic liquid should include an adequate description of these steps, and comparison measurements on a standard system to allow the user to judge the reliability of the data [56]. For this purpose, the International Union of Pure and Applied Chemistry (IUPAC) sponsored a project to make physical property measurements available for comparing the gas solubility of carbon dioxide in a reference ionic liquid,   1-hexyl-3-methylimidazolium   bis((trifluoromethyl)sulfonyl)amide, abbreviated $[C_6mim][NTf_2]$ [57, 58]. The report provides both recommended gas absorption methods and reference values for VLE and LLE of carbon dioxide in $[C_6mim][NTf_2]$ [58].

11.2.1   Gravimetric Methods

A common technique used to measure the solubility of gases in ionic liquids is the gravimetric method. Although this method was originally developed to measure gas adsorption on solids, it can be applied to ionic liquids because the lack of vapour pressure prevents evaporation of the sample. This technique was first used by Brennecke and coworkers [59–61].

We have extensively used a Hiden gravimetric microbalance (IGA003) [62] to measure the solubility and diffusivity of gases in ionic liquids over a range of temperatures (283–348 K) and pressures (0–2 MPa). The advantages of a microbalance include the minimal sample size of the ionic liquid (<100 mg) required, the ability to automate the measurement process to take several PTx data, and the flexibility to measure both absorption and desorption isotherms. When done properly, the gravimetric analysis provides a direct and powerful method for assessing both gas solubility and diffusivity. The two critical factors that must be considered include a proper correction for the buoyancy effects of the system, and allowance of sufficient time to reach equilibrium (i.e., no mixing is possible). Particularly for certain ionic liquids at low temperature, the viscosity can be large and the diffusion-controlled gas absorption process can take several hours or days to reach equilibrium at a given T and P.

Here, we provide critical details for properly correcting for a number of balance forces. Balance forces include:

1. Changes in the buoyant forces due to changes in pressure and temperature.
2. Aerodynamic drag forces created by the flow of gases.
3. Changes in the balance sensitivity due to changes in temperature and pressure.
4. Volumetric changes in the samples due to expansivity.

The sum of these forces can be quite large and can lead to large errors if not carefully accounted for during the experimental measurements and data reduction. The buoyancy correction follows from Archimedes' principle, that there is an upward force exerted on an object equivalent to the mass of fluid displaced. The upward force (C_b) due to buoyancy is calculated using Equation (11.1),

$$C_b = Buoyancy = gV_i\rho_g(T, P) = g\frac{m_i}{\rho_i}\rho_g(T, P) \tag{11.1}$$

where the mass of the gas displaced is equivalent to the volume of the submersed object (V_i) multiplied by the density (ρ_g) of the gas at a given (T,P) and the gravitational acceleration (g).

If the volume of the object remains constant, V_i can simply be calculated from knowledge of the mass (m_i) and density (ρ_i) of the object. To improve the accuracy of the buoyancy correction, we calculate the gas densities using a computer program (REFPROP, v.7) [63] instead of using the gas densities provided in the Hiden Isochema software.

The buoyancy correction using the IGA003 system involves several objects for weighing the sample. Table 11.1 provides a list of each critical component along with the object's weight, material, density, and temperature. The component arrangement in Figure 11.1 leads to a mass balance as shown by Equation (11.2): this accounts for the summation of all components, the absorbed gas

TABLE 11.1   Microbalance Components Contributing to Buoyancy Calculation

Subscript	Item	Weight/g	Material	Density/g cm^{-3}	Temperature/K
S	Dry sample	m_s	Ionic liquid	ρ_s	Sample T (T_s)
A	Interacted gas	m_a	Gas	ρ_a	(T_s)
i_1	Sample container	0.5986	Pyrex	2.23	(T_s)
i_2	Wire	0.051	Tungsten	21.0	(T_s)
i_3	Chain	0.3205	Gold	19.3	303.15
j_1	Counter-weight	0.8054	Stainless steel	7.9	298.15
j_2	Hook	0.00582	Tungsten	21.0	298.15
j_3	Chain	0.2407	Gold	19.3	303.15

mass (m_a) and a correction factor (C_f), which is the result of the balance sensitivity to T, P. To be accurate, the density of air (ρ_{air}) at ambient temperature and pressure should be subtracted from ρ_i and ρ_j because the components were initially weighed in air.

$$\sum_{i=1} m_i - \sum_{j=1} m_j - \sum_{i=1} \frac{m_i}{\rho_i} \rho_g(T_i, P) + \sum_{j=1} \frac{m_j}{\rho_j} \rho_g(T_j, P) + m_s + m_a$$
$$- \frac{m_s}{\rho_s(T_s)} \rho_g(T_s, P) - \frac{m_a}{\rho_a(T_s)} \rho_g(T_s, P) - C_f(T_s, P) = reading \tag{11.2}$$

The largest contributions in Equation (11.2) are typically those of the sample container, sample, and counter weight. The other referenced objects in Table 11.1 contribute less because of their larger densities (denominators in Eq. 11.2). Accurate physical densities as a function of temperature for the ionic liquid must also be known or measured.

The system was operated in static mode, which eliminates any aerodynamic drag forces due to flowing gases. The microbalance is also designed with internal baffles to minimise convective flows. Electrobalances are sensitive to T and P fluctuations on the beam arm and internal electronics. To minimise this effect, the balance electronics are heated externally with a band heater to a temperature of 318.15 ± 0.1 K. In addition, the sample temperature (T_s) provided in Table 11.1 is measured, while other temperatures are estimated. The correction factor (C_f) for the balance sensitivity was determined as a function of T and P by measuring the buoyancy effect without sample load, and calculating a least-squares fit to tare the balance.

It is important that all contributions to the buoyancy correction be included in the gas solubility calculation. A common mistake is neglecting the expansion

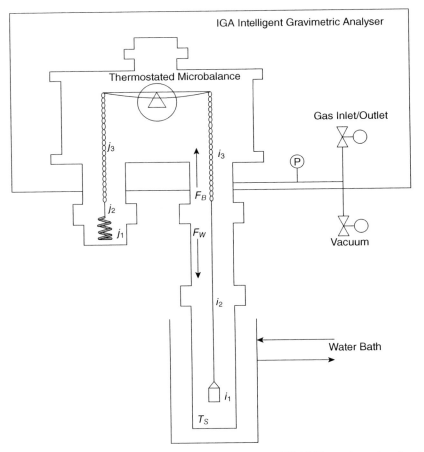

Figure 11.1   Schematic diagram of the Hiden Isochema IGA003 gravimetric microbalance. Symbols: arrow F_B indicates direction due to buoyancy force on sample side of balance; arrow F_W indicates direction of weight due to gravity on sample side of balance; additional symbols are described in Table 11.1, see Section 11.2.1 for details.

of the sample. In order to make a proper buoyancy correction due to the liquid volume change, a simple mole fraction average for the molar volume, $\tilde{V}_m$, may be used.

$$\tilde{V}_m(T, P) = \tilde{V}_{IL}(1-x) + \tilde{V}_g x, \tag{11.3}$$

where $\tilde{V}_i = MW_i/\rho_i$ and x represents the molar fraction of the gas in the solution.

$$V_m(T, P) = \tilde{V}_m(T, P)\left[\left(\frac{m_{IL}}{MW_{IL}}\right) + \left(\frac{m_g}{MW_g}\right)\right] \tag{11.4}$$

$$\frac{m_s}{\rho_s(T_s)}\rho_g(T_s, P) + \frac{m_a}{\rho_a(T_s)}\rho_g(T_s, P) = V_m(T, P)\rho_g(T, P).\qquad(11.5)$$

As a first approximation, Equations (11.3) and (11.4) were used to estimate the change in the liquid sample volume, V_m, at the measured T, P conditions. Equation (11.5) can be substituted into Equation (11.2) to account for the buoyancy change with respect to sample expansion. We have also discovered that an approximate $\tilde{V}_g$ can be defined for temperatures above the T_c using Equation (11.6),

$$\tilde{V}_g = (1 - \alpha)\tilde{V}_{IL},\qquad(11.6)$$

where $\tilde{V}_{IL}$ is the molar liquid volume of the ionic liquid at T and α is a unique temperature-independent constant for each binary system (see Reference [26] for details). It should be mentioned here that the present approximate $\tilde{V}_g$ in Equation (11.6) is sufficient for the present data analysis in Equation (11.3), since it is used merely in the correction term. In the case of CO_2 absorbing in [C_4mim][PF_6] or [C_4mim][BF_4], the change in molar liquid volume at the measured T, P conditions is from 0 to -30% [21], with a measurable effect on the final solubility measurements of 0 to +3 mole %.

One weak point to using a gravimetric microbalance is a limitation in pressure (IGA003 maximum pressure is 2.0 MPa); however, Hiden is developing a high-pressure microbalance to overcome this issue. Another option is to use a magnetic suspension balance that can operate at higher pressures [10]. This type of a balance is magnetically coupled to the sorption vessel, so the balance is not in contact with the gas. Therefore, absorption of corrosive gases such as SO_2, H_2S, and NH_3 can be measured [64]; however obtaining accurate results requires larger sample sizes ($\sim$1 to 2 g), which for some systems can translate to a long equilibration time.

11.2.2 Liquid–Liquid Equilibria Measurements

Liquid–liquid equilibria measurements were conducted using a simple mass–volume technique [39, 40]. When a binary system exhibits a liquid–liquid separation (or VLLE), it is a *univariant* state according to the Gibbs phase rule. This means that at a given intensive variable, for example, temperature, there is no freedom for other intensive variables. All other variables such as compositions, pressure, and densities of the system are uniquely determined regardless of any different extensive variables (volume of each phase and total mass of the system). The overall feed composition merely changes the physical volume in each phase, but the composition and density in each phase remains constant as long as the three phases exist at the fixed T.

This unique VLLE state of a binary system can be determined experimentally using a simple apparatus by mass and volume measurements *alone* without using any analytical method for the composition analysis. A set of mass and volume measurements of two sample containers at a constant temperature is sufficient to determine the required thermodynamic properties using the following equations [39, 40].

The total moles of compound-I (e.g., ionic liquid) in sample-0 (n_{IT0}) and sample-1 (n_{IT1}) are known from the measured mass and molecular weight of compound-I in each container and the following relation holds:

$$n_{I0} + n'_{I0} = n_{IT0} \tag{11.7}$$

$$n_{I1} + n'_{I1} = n_{IT1} \tag{11.8}$$

where the subscripts $I0$ and $I1$ represent the moles of compound-I in the upper liquid phase L of the sample-0 and sample-1 containers, respectively, and the superscript $'$ means the moles in the lower liquid phase L′. Molar volumes of the two liquids are unchanged (in the sample-0 and sample-1 containers) because VLLE of a binary system is a *univariant* state:

$$\frac{v_1}{v_0} = \frac{n_{R1} + n_{I1}}{n_{R0} + n_{I0}} \tag{11.9}$$

$$\frac{v'_1}{v'_0} = \frac{n'_{R1} + n'_{I1}}{n'_{R0} + n'_{I0}}, \tag{11.10}$$

where the subscripts $R0$ and $R1$ represent the moles of compound-R (e.g., refrigerant or other gas) in the upper liquid phase L of the sample-0 and sample-1, respectively, and the superscript $'$ means the moles in the lower liquid phase L′. Subscripts (1 and 0) of the physical volume, v, correspond to the sample-0 and sample-1 containers for upper liquid phase L, respectively, and the superscript $'$ means the volumes in the lower liquid phase L′.

Next, we observe that the mole fractions of the two liquids are unchanged in both containers because VLLE of a binary system is a *univariant* state:

$$\frac{n_{R0}}{n_{R0} + n_{I0}} = \frac{n_{R1}}{n_{R1} + n_{I1}} \tag{11.11}$$

$$\frac{n'_{R0}}{n'_{R0} + n'_{I0}} = \frac{n'_{R1}}{n'_{R1} + n'_{I1}}. \tag{11.12}$$

Compound-R total moles (n_{RT0} and n_{RT1}) in both containers are also known from the measured mass and molecular weight of compound-R in each container and the following relation holds:

$$n_{R0} + n'_{R0} + n_{g0} = n_{RT0} \tag{11.13}$$

$$n_{R1} + n'_{R1} + n_{g1} = n_{RT1} \tag{11.14}$$

The vapour pressure for many ionic liquids is negligible, so the gas phase contains only compound-R. Therefore, gas phase moles n_{g0} and n_{g1} for sample-0 and sample-1, respectively, can be obtained from the measured gas volume, v_g, and using an EOS [63] to calculate the vapour density. In most cases, the vapour can be assumed to be at the saturation condition; therefore, knowing only T, the vapour density can be calculated. For low-pressure fluids such as alcohols and water the contributions of n_{g0} and n_{g1} can be safely ignored; however, for high-pressure refrigerant gases the vapour-phase moles must be included in the analysis.

The experimentally known quantities are n_{IT0}, n_{IT1}, n_{RT0}, n_{RT1}, n_{g0}, n_{g1}, v_1, v_0, v_1', and there are eight unknown parameters (n_{R0}, n_{I0}, n_{R1}, n_{I1}, n_{R0}', n_{I0}', n_{R1}', n_{I1}'). These eight unknown parameters can be obtained by solving the above eight independent equations (see Refs [39] and [40] for details).

The most difficult problem with the experiments was establishing the equilibrium state. Monitoring the change in the height of each phase until no further change occurred assured that equilibrium had been reached. Mixing time to reach equilibrium can take several days, and is one of the most critical properties for properly measuring VLLE whether using this method or the more common cloud-point method [65–70]. The time required to effectively mix the (gas + ionic liquid) systems and reach equilibrium depends on the choice of ionic liquid and the fluid viscosity.

The gas solubility or LLE measurements are of little use without specifying the appropriate error. Experimental errors are of two types: random (indeterminate) and systematic (determinate). Random errors are present in all experimental measurements and lead to different values when the measurement is repeated many times (i.e., indeterminate) assuming all other conditions are held constant. Random errors can have many causes, including operator errors, fluctuating experimental conditions, and variability of measuring instruments. Systematic errors have the same magnitude and direction when the measurement is repeated several times, and if identified (i.e., determinate) can be corrected. A common systematic error is a miscalibrated instrument.

Total errors ($\delta x_{TE} = \sqrt{\delta x_{RE}^2 + \delta x_{SE}^2}$) were estimated by calculating both the overall random (δx_{RE}) and systematic error (δx_{SE}). The overall random error was estimated using the following uncertainty propagation method:

$$\delta x_{RE} = \sqrt{\sum_{i=1}^{n} \left[\delta p_i \left(\frac{\partial x}{\partial p_i} \right) \right]^2}, \qquad (11.15)$$

where δx_{RE} is the mole fraction or molar volume random uncertainty of the liquid composition, $\partial x / \partial p_i$ is the partial derivative of x with respect to the i-th experimental parameter p_i, which is calculated from the sensitivity analysis of each parameter, and δp_i is the estimated uncertainty of the experimental parameter p_i.

In the case of the LLE measurements, the following 12 experimental parameters were considered to have an effect on the random errors: a, b, n_{g0}, n_{g1}, h', and h for both sample containers (sample-0 and sample-1). The heights were found to have the largest overall effect on the random error. The systematic error (δx_{SE}) includes properly correcting for the area expansion, meniscus, and vapour-phase moles.

A good experimentalist must use proper error analysis throughout the entire measurement process from the selection of the analytical method to the final data analysis and reporting of results. Our opinion is that poor-quality data are worse than no data, and high-quality data will survive forever.

11.2.3 Other Gas Solubility Techniques

11.2.3.1 Synthetic Methods Several variations of the synthetic (or stoicheiometric) method exist that involve adding a precise amount of gas and ionic liquid into a high-pressure view cell with a known volume. One method used by several researchers involves increasing the pressure (at constant temperature) until all the gas dissolves in the liquid and the last bubble disappears [71–79]. A variation of this method by Lim and coworkers involves continuing to increase the pressure using a variable–volume view cell and then slowly decreasing the pressure until the first bubble appears [80]. Maurer and coworkers used a similar method, but added and withdrew known amounts of the ionic liquid to pressurise or depressurise the mixture, and observed the phase change [12, 81–83]. Another variation by Ren and Scurto measures the vapour and liquid phase volumes and is capable of determining the solubility, molar volume, volume expansion, and density of the liquid solution simultaneously [84–86].

A pressure drop technique is also used. In this method, a gas is added to a calibrated volume (V_1) at a given T and P. A known amount of ionic liquid is added to a second calibrated volume (V_2), which is connected by a valve to V_1. When the valve is opened, the gas fills both volumes and dissolves into the ionic liquid. Measurement of the pressure drop at equilibrium allows the number of moles of gas to be calculated in the vapour and liquid phase (by difference). This technique has been used by Costa Gomes and coworkers in glassware at low pressures and is ideal for measuring dilute solutions to obtain Henry's Law constants [87–91].

11.2.3.2 Chromatography Methods Gas chromatography has been used to calculate infinite dilution activity coefficients (and the equivalent Henry's Law constants). A chromatography column is coated with the ionic liquid and the solute (gas or liquid) is introduced with a carrier gas. The retention time of the solute is measured at steady state and the strength of the interaction of the solute in the ionic liquid determines the infinite dilution activity coefficient. This technique has been used by Heintz and coworkers to measure the infinite dilution activity coefficients of liquids in ionic liquids [92–94].

11.2.3.3 Spectroscopic Methods Infrared and proton nuclear magnetic resonance (^{1}H NMR) spectroscopies have been used to measure the solubility of gases in ionic liquids. Welton and coworkers used attenuated total reflectance infrared (ATR-IR) and ^{1}H NMR spectroscopies to measure the solubility of carbon dioxide and dihydrogen in ionic liquids, respectively [95, 96].

11.2.4 Safety

Many of these experimental methods involve handling gases under elevated T and P; therefore, all equipment should have the proper T and P ratings, relief devices, and safety factors. In some cases, the gases may be flammable and/or toxic and proper handling, ventilation, and monitoring is required. In addition, there is still little known about the toxicity of many ionic liquids. Appropriate personal protective equipment (e.g., impervious gloves and eye protection) should always be worn. Proper disposal of ionic liquid waste must be according to environmental guidelines.

11.3 MODELLING GAS SOLUBILITY

Simply measuring and reporting the solubility of a gas in an ionic liquid does not ensure that the results are thermodynamically consistent. A thermodynamic model can be applied not only to check the data quality, but also for making predictions. In fact, we have shown that thermodynamic phase behaviours of gases with ionic liquids can be well modelled with activity models and an ordinary EOS method.

Several activity models are available in the literature [97–99] and we have successfully analysed our experimental (PTx) results using the non-random two liquid (NRTL) model [98, 99]. As mentioned in the introduction, treating the ionic liquid as an undissociated species (i.e., non-electrolytes) may not be a bad assumption. Most of the electrolyte-solution models assume the complete dissociation, which is not always a valid assumption. As far as the phase behaviour correlation is concerned, it seems well known that non-electrolyte-solution models work well even for electrolyte solutions [100–103].

However, activity models (or any solution models) are inaccurate (or undefined) at high temperatures, particularly near and above the T_c of the gaseous species. Therefore, extrapolations for phase behaviours over wide T and P ranges must be treated with caution. Also, the prediction of LLE based on only VLE data, or vice versa, is not numerically accurate with any conventional activity models [98]. More reliable predictions of large-scale (global) phase behaviours may be made using a proper EOS, which is discussed in the next section.

11.3.1 EOS Modelling

In order to understand the global phase behaviour, we have employed a generic Redlich–Kwong (RK) type of cubic EOS [16, 104], which is written in the following form (where R: universal gas constant):

$$P = \frac{RT}{V-b} - \frac{a(T)}{V(V+b)} \tag{11.16}$$

$$a(T) = 0.427480 \frac{R^2 T_c^2}{P_c} \alpha(T) \tag{11.17}$$

$$b = 0.08664 \frac{RT_c}{P_c}. \tag{11.18}$$

The temperature-dependent part of the a parameter in the EOS for pure compounds is modelled by the following empirical form [104]:

$$\alpha(T) = \sum_{k=0}^{\leq 3} \beta_k (1/T_r - T_r)^k, \quad \text{for } T_r \equiv T/T_c \leq 1, \tag{11.19a}$$

$$\alpha(T) = \beta_0 + \beta_1 [\exp\{2(1-T_r)\} - 1], \quad \text{for } T_r \geq 1. \tag{11.19b}$$

The coefficients, β_k, are determined to reproduce the vapour pressure of each pure compound. It should be noted that Equation (11.19b) is implemented for $T_r \geq 1$ in order for $\alpha(T)$ to be physically meaningful for gases at even very high temperatures: always $\alpha(T) > 0$ and a decreasing function with T. At $T_r = 1$, Equations (11.19a) and (11.19b) are set to be analytically continuous.

For ionic liquids, however, usually no vapour pressure data are available (as they are practically non-volatile); furthermore, only estimated data for the critical parameters (T_c and P_c) exist [105–107]. The critical parameters can be estimated in various ways. As discussed in Reference [16], rough estimates for the critical parameters of non-volatile compounds are sufficient for the present purpose. On the other hand, the temperature-dependent part of the a parameter of ionic liquids (Eq. 11.19) is significantly important when we try to correlate the solubility (pressure–temperature–composition: PTx) data, although the vapour pressure of ionic liquids is essentially zero at the temperature of interest. Therefore, the coefficient β_1 for ionic liquids in Equation (11.19) is usually treated as an adjustable fitting parameter using $\beta_0 = 1$ and $\beta_2 = \beta_3 = 0$ in the solubility data analysis, together with the binary interaction parameters discussed in Equations (11.20)–(11.23). However, once β_1 was determined for a particular ionic liquid using a certain binary system, it will be used as a fixed constant for any other binary systems containing that ionic liquid.

Then, the a and b parameters for a general N-component mixture are modelled in terms of binary interaction parameters [16]:

$$a = \sum_{i,j=1}^{N} \sqrt{a_i a_j} f_{ij}(T)(1-k_{ij}) x_i x_j, \quad a_i = 0.427480 \frac{R^2 T_{ci}^2}{P_{ci}} \alpha_i(T) \tag{11.20}$$

$$f_{ij}(T) = 1 + \tau_{ij}/T, \quad \text{where} \quad \tau_{ij} = \tau_{ji}, \quad \text{and} \quad \tau_{ii} = 0. \tag{11.21}$$

$$k_{ij} = \frac{l_{ij} l_{ji} (x_i + x_j)}{l_{ji} x_i + l_{ij} x_j}, \quad \text{where} \quad k_{ii} = 0 \tag{11.22}$$

$$b = \frac{1}{2}\sum_{i,j=1}^{N}(b_i + b_j)(1-k_{ij})(1-m_{ij})x_i x_j, \quad b_i = 0.08664\frac{RT_{ci}}{P_{ci}}, \quad (11.23)$$

where $m_{ij} = m_{ji}$, and $m_{ii} = 0$; T_{ci}: critical temperature of i-th species; P_{ci}: critical pressure of i-th species; and x_i: mole fraction of i-th species.

In the above model, there are a maximum of four binary interaction parameters: l_{ij}, l_{ji}, m_{ij}, and τ_{ij} for each binary pair. It should be noted that when $l_{ij} = l_{ji}$ in Equation (11.20) and $f_{ij} = 1$ in Equation (11.21), Equation (11.20) becomes the ordinary quadratic-mixing rule for the a parameter. The present EOS model has been successfully applied for highly asymmetric (with respect to the polarity and size) mixtures such as various refrigerant/oil mixtures [16, 104]. The fugacity coefficient ϕ_i of i-th species for the present EOS model, which is needed for the phase equilibrium calculation, is given by:

$$\ln\phi_i = \ln\frac{RT}{P(V-b)} + b_i'\left(\frac{1}{V-b} - \frac{a}{RTb(V+b)}\right) + \frac{a}{RTb}\left(\frac{a_i'}{a} - \frac{b_i'}{b} + 1\right)\ln\frac{V}{V+b}, \quad (11.24)$$

where

$$a_i' \equiv \left(\frac{\partial na}{\partial n_i}\right)_{n_{j\neq i}} \quad \text{and} \quad b_i' \equiv \left(\frac{\partial nb}{\partial n_i}\right)_{n_{j\neq i}}$$

where n = total mole number and n_i = mole number of i-th species (or $x_i = n_i/n$). The explicit forms of a_i' and b_i' may be useful for readers and are given by:

$$a_i' = 2\sum_{j=1}^{N}\sqrt{a_i a_j}\,f_{ij}x_j\left\{1-k_{ij} - \frac{l_{ij}l_{ji}(l_{ij}-l_{ji})x_i x_j}{(l_{ji}x_i + l_{ij}x_j)^2}\right\} - a \quad (11.25)$$

$$b_i' = \sum_{j=1}^{N}(b_i + b_j)(1-m_{ij})x_j\left\{1-k_{ij} - \frac{l_{ij}l_{ji}(l_{ij}-l_{ji})x_i x_j}{(l_{ji}x_i + l_{ij}x_j)^2}\right\} - b. \quad (11.26)$$

Phase equilibria ($\alpha, \beta, \gamma,...$, coexisting phases) for an N-component system can be obtained by solving the following equilibrium conditions:

$$x_i^\alpha \phi_i^\alpha = x_i^\beta \phi_i^\beta = x_i^\gamma \phi_i^\gamma = \ldots\ldots, (i = 1,\ldots, N) \quad (11.27)$$

where

$x_i^\alpha, x_i^\beta, x_i^\gamma, \ldots$: mole fractions of $\alpha, \beta, \gamma, \ldots$ phases for the i-th species
$\phi_i^\alpha, \phi_i^\beta, \phi_i^\gamma, \ldots$: fugacity coefficients of $\alpha, \beta, \gamma, \ldots$ phases for the i-th species.

In order to develop a reliable EOS model over a wide range of T and P for the IUPAC reference system (CO_2 + [C$_6$mim][NTf$_2$]), we have combined our low-pressure solubility data [22] with the high-pressure data by Kumełan et al. [108]. Both groups used the same sample of [C$_6$mim][NTf$_2$] supplied by the

TABLE 11.2   Pure Component EOS Constants for the (CO_2 + [C_6mim][NTf_2]) System[a]

Compound	Molar Mass/g mol^{-1}	T_c /K	P_c /kPa	β_0 –	β_1 –	β_2 –	β_3 –
CO_2	44.01	304.13	7385	1.0005	0.43866	−0.10498	0.06250
[C_6mim][NTf_2]	447.42	815.0	1611	1.0	0.50036	0	0

[a] The critical parameters for [C_6mim][NTf_2] were estimated with the method proposed by Vetere [109], using two-liquid density data and an assumed critical compressibility factor of 0.253.

TABLE 11.3   Optimal Binary Interaction Parameters for the (CO_2 + [C_6mim][NTf_2]) System[a]

System (1)/(2)	l_{12}	l_{21}	$m_{12} = m_{21}$	$\tau_{12} = \tau_{21}$/K
CO_2/[C_6mim][NTf_2]	0.34880	0.32217	−0.40939	95.073

[a] Determined by non-linear least squares analyses using the combined VLE data from [22] and [108]. The standard deviation of the pressure fit is 0.034 MPa.

IUPAC project [57, 58]. The pure component EOS constants are provided in Table 11.2. Optimal binary interaction parameters (l_{ij}, l_{ji}, m_{ij}, and τ_{ij}) determined using a non-linear least-squares method for the solubility data are shown in Table 11.3. The combined experimental PTx data are compared with the present model calculations in Figure 11.2. The standard deviation for the P versus x_1 fit is 0.034 MPa for the present analysis.

Now that we have developed an EOS model for the (CO_2 + [C_6mim][NTf_2]) system, other data in the literature (which were not used in the EOS-model development) can be compared with the present EOS predictions. Such comparisons serve two purposes: first, to see whether our model can predict other observed solubility data, and second, to judge whether other reported data are thermodynamically consistent (or can be explained by a thermodynamically consistent EOS model) [110].

In Figure 11.3, three sets of solubility data are compared with the EOS model prediction at 333.2 K. Data by Ren et al. [84] and Kumełan et al. [108] are consistent with the EOS model, but data by Aki et al. [111] scatter around the EOS prediction. EOS model predictions are also in excellent agreement with data by Kim et al. at 298.2 K [112].

Inconsistent solubility data in the literature may be due to the sample differences used in the experiment. In fact, samples of [C_6mim][NTf_2] used in References [84, 108, 111, 112] are all different; References [22] and [108] used the same IUPAC sample, while all others are home-made samples. The data of Ren et al. [84] and Kim et al. [112] are consistent with those based on the IUPAC sample [57, 58]. Therefore, it is difficult to imagine that the data inconsistency of Aki et al. [111] is solely due to the sample difference (or purity problem). This type of analysis using an EOS model is ideal for determining which data in the literature are of the highest quality.

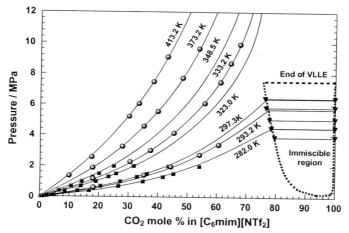

Figure 11.2   *PTx* phase diagram for (CO_2 + [C_6mim][NTf_2]) mixtures. Solid lines: the present EOS model calculations. Dotted lines: predicted liquid–liquid separation gap by the present EOS. Symbols: experimental VLE data: squares [22] and circles [108], which were used to construct the present EOS model; experimental VLLE data: triangles [22]. Adapted with permission from Reference [22]; Copyright (2007) American Chemical Society.

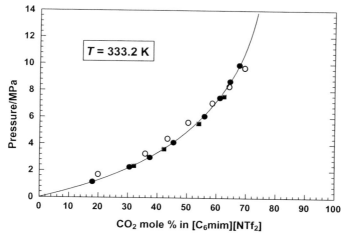

Figure 11.3   Isothermal *Px* phase diagram for (CO_2 + [C_6mim][NTf_2]) mixtures at about *T* = 333.2 K. Line: the present EOS model. Symbols: experimental data; squares [84]; solid circles [108]; open circles [111].

TABLE 11.4   Experimental VLLE for (CO_2 (1) + [C_6mim][NTf_2] (2))

Temperature/K	x'_1 / mol %	x_1/mol %	$\bar{V}'^{\,a}$ / cm³ mol⁻¹	$\bar{V}^{\,a}$/ cm³ mol⁻¹	$\bar{V}^{ex'\,b}$/ cm³ mol⁻¹	$\bar{V}^{ex\,b}$/ cm³ mol⁻¹
277.2	79.5 ± 0.4	99.9 ± 0.1	97.5 ± 1.8	48.9 ± 0.4	−7.4 ± 1.8	−0.1 ± 0.4
282.8	78.3 ± 0.4	99.9 ± 0.1	102.2 ± 1.8	51.1 ± 0.4	−7.8 ± 1.8	0.0 ± 0.4
287.7	77.7 ± 0.4	99.9 ± 0.1	103.8 ± 1.8	53.6 ± 0.4	−9.7 ± 1.8	0.0 ± 0.4
293.8	76.5 ± 0.4	99.9 ± 0.1	108.2 ± 1.8	57.4 ± 0.4	−12.1 ± 1.8	0.0 ± 0.4
298.1	76.2 ± 0.4	99.9 ± 0.1	109.5 ± 1.8	61.6 ± 0.4	−15.0 ± 1.8	−0.4 ± 0.4

[a] Observed molar volume.
[b] Excess molar volume.

Using only the VLE data, the present EOS model predicted a partial immiscibility with CO_2-rich compositions, as shown in Figure 11.2. Therefore, we have conducted VLLE experiments using our mass–volume technique (see Section 11.2.2), and the present prediction has been well confirmed as presented in Figure 11.2. The VLLE experiments have also provided the excess molar volume of this binary system, and show large negative values at the ionic liquid-rich compositions, as shown in Table 11.4. The excess molar volume of each liquid solution ($V^{ex'}$ and V^{ex}) can be obtained, by use of the pure component molar volumes V_1^0 (CO_2) and V_2^0 (ionic liquid) using:

$$V^{ex'} = V_m - x'_1 V_1^0 - x'_2 V_2^0 \quad \text{or} \quad V^{ex} = V_m - x_1 V_1^0 - x_2 V_2^0 \tag{11.28}$$

where V_m is the measured molar volume of the mixture ($V_m = V'$ for the lower phase L′ or $V_m = V$ for the upper phase L), and (x'_1, x'_2 or x_1, x_2) are mole fractions of CO_2 (1) and ionic liquid (2) in phase L′ and L, respectively. Saturated liquid molar volumes for CO_2 (as well as many other gases) were calculated using an EOS program (REFPROP) [63]. Molar volumes for ionic liquids were calculated from known or measured liquid density data [108]. The present phase behaviours and the large negative excess molar volumes shown for the (CO_2 + [C_6mim][NTf_2]) system have also been found for other (CO_2 + ionic liquids) mixtures, for example, (CO_2 + [C_4mim][PF_6]), (CO_2 + [C_4mim][BF_4]), and (CO_2 + [C_4mim][C_1CO_2]) [21, 23], and other (gas + ionic liquid) systems [30, 36, 39, 40, 45–48]. A size effect, polarity (dipole and quadrupole moments), and hydrogen bonding (i.e., HFCs) may all be involved.

11.3.2   Types of Phase Behaviour

With the present EOS model, many PTx diagrams have been calculated in order to understand the global phase behaviour of gases including: CO_2, HFCs, NH_3, SO_2, H_2S, and N_2O in a variety of ionic liquids. Figure 11.4 shows the PTx diagram for (trifluoromethane (HFC-23) and [C_4mim][PF_6]) [30]. Using only low-pressure VLE measurements, the EOS model predicts liquid–liquid phase separations with a lower critical solution temperature (LCST) in the HFC-

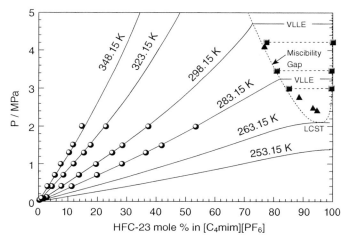

Figure 11.4   *PTx* phase diagram for (HFC-23 (CHF$_3$) + [C$_4$mim][PF$_6$]) mixtures. Solid lines: the present EOS model calculations. Dotted lines: predicted liquid–liquid separation gap by the present EOS. Symbols: experimental VLE data: circles [26], which were used to construct the present EOS model; experimental VLLE data: squares [30]; experimental cloud-point data: triangles [30].

23-rich side solutions. LLE and cloud point measurements [30] confirm the existence of the LCST, which supports the type-V mixture behaviour for this system, according to the classification of van Konynenburg and Scott [14, 15]. Other HFC binary systems that indicate type-V mixture behaviour have also been measured [32, 33, 36, 39, 40]. In some systems, an LCST did not exist; therefore, these systems belong to the type-III mixture behaviour [36, 40]. Figure 11.5 shows the *PTx* diagram for (fluoromethane (HFC-41) and [C$_4$mim][PF$_6$]), which belongs to the type-III mixture behaviour [36, 40]. The (CO$_2$ + [C$_6$mim][NTf$_2$]) system shown in Figure 11.2 is also likely to be exhibiting type-III mixture behaviour. In this case, the LLE will intersect with the solid–liquid equilibria such that no LCST can exist.

11.3.3   Physical and Chemical Absorption

It is well known that CO$_2$ possesses relatively high solubility in ionic liquids; however most cases are for physisorption [13, 21, 22, 60, 61, 72, 73, 76, 80, 84, 86, 87, 89, 90, 95, 111–117]. Observed isothermal *Px* plots show essentially three types of absorption, which are illustrated in Figure 11.6 using the three typical binary systems: (CO$_2$ + 1-ethyl-3-methylimidazolium trifluoroethanoate, [C$_2$mim][CF$_3$CO$_2$]) (Case A); (CO$_2$ + 1-hexyl-3-methylimidazolium tris(pentafluoroethyl)trifluorophosphate, [C$_6$mim][(C$_2$F$_5$)$_3$PF$_3$]) (Case B); and (CO$_2$ + 1-ethyl-3-methylimidazolium ethanoate, [C$_2$mim][C$_1$CO$_2$]) (Case C) binary systems [21].

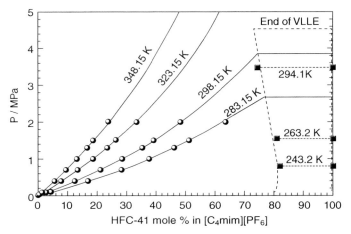

Figure 11.5 *PTx* phase diagram for (HFC-41 (CH$_3$F) + [C$_4$mim][PF$_6$]) mixtures. Solid lines: the present EOS model predictions. Dotted lines: predicted liquid–liquid separation gap by the present EOS. Symbols: experimental VLE data: circles [29], which were used to construct the present EOS model; experimental VLLE data: squares [40]. Adapted with permission from Reference [40]. Copyright (2006) American Chemical Society.

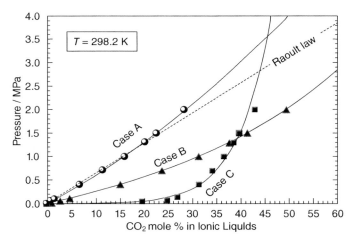

Figure 11.6 Isothermal *Px* phase diagram for CO$_2$ + ionic liquid mixtures at about *T* = 298.2 K. Three typical cases (Cases A, B, and C) are shown. Solid lines: calculated from the present EOS model. Symbols: the present experimental data, circles = [C$_2$mim][CF$_3$CO$_2$], triangles = [C$_6$mim][(C$_2$F$_5$)$_3$PF$_3$], and squares = [C$_2$mim][C$_1$CO$_2$]. Dotted line: the Raoult's law solubility; see Reference [24] for details. Adapted with permission from Reference [24]. Copyright (2008) American Chemical Society.

The solubility behaviour of Case A is close to the case of Raoult's law (or positive deviations from Raoult's law). Cases B and C show large negative deviations from Raoult's law, and particularly in Case C, the negative deviation is so large that the high pressure of CO_2 gas becomes practically zero in the CO_2 mole fraction range of less than about 0.3. Highly negative deviations from Raoult's law suggest chemical absorption instead of ordinary physical absorption, that is, chemical complex formation [24].

These solubility behaviours are more clearly understood in terms of the thermodynamic excess functions (excess Gibbs G^E, excess enthalpy H^E, and excess entropy (multiplied by T) TS^E energies). The residual properties ($\Delta M' \equiv M^{\text{ideal gas}} - M$) are useful for calculations of the excess properties and the property changes. Their explicit forms for the present EOS are:

$$\Delta H' = \left(\frac{T}{b}\frac{da}{dT} - \frac{a}{b}\right)\ln\frac{V}{V+b} + RT(1-Z), \tag{11.29}$$

$$\Delta S' = \frac{1}{b}\frac{da}{dT}\ln\frac{V}{V+b} + R\ln\frac{RT}{P(V-b)}. \tag{11.30}$$

The temperature derivative of the a parameter can be written for the present EOS as,

$$\frac{da}{dT} = \sum_{i=1}^{N}\sum_{j=1}^{N}(0.5A_1A_2A_3 + A_4A_3')(1-k_{ij})x_ix_j, \tag{11.31}$$

with the following definition of symbols:

$$A_1 \equiv (a_ia_j)^{-1/2} \tag{11.32}$$

$$A_2 \equiv a_j\frac{da_i}{dT} + a_i\frac{da_j}{dT} \tag{11.33}$$

$$A_3 \equiv f_{ij} \tag{11.34}$$

$$A_4 \equiv 1/A_1 \tag{11.35}$$

$$A_3' = f_{ij}' = -\tau_{ij}/T^2. \tag{11.36}$$

Here, da_i/dT ($i = i$ or j) above is given by:

$$\frac{da_i}{dT} = A_0\sum_{k=0}^{3}\beta_{ki}k(T_{ci}/T - T/T_{ci})^{k-1}(-T_{ci}/T^2 - 1/T_{ci}), \tag{11.37}$$

where

$$A_0 \equiv 0.427480\frac{R^2T_{ci}^2}{P_{ci}}. \tag{11.38}$$

The excess properties can be written using the residual properties obtained above, with the dimensionless forms (R = universal gas constant):

$$\frac{H^E}{RT} = \sum_{i=1}^{N} x_i \frac{\Delta H_i'(\text{pure compound})}{RT} - \frac{\Delta H'(\text{mixture})}{RT} \tag{11.39}$$

$$\frac{S^E}{R} = \sum_{i=1}^{N} x_i \frac{\Delta S_i'(\text{pure compound})}{R} - \frac{\Delta S'(\text{mixture})}{R} \tag{11.40}$$

$$\frac{G^E}{RT} = \frac{H^E}{RT} - \frac{S^E}{R} \tag{11.41}$$

or, equivalently,

$$\frac{G^E}{RT} = \sum_{i=1}^{N} x_i \ln \frac{\phi_i(\text{species } i \text{ in mixture})}{\phi_i(\text{pure species } i)}. \tag{11.42}$$

Sufficiently negative values in G^E usually indicate some chemical complex formations; the heat of mixing (or H^E) is more negative than TS^E. A minimum in G^E occurs around 50 mole % for the (CO_2 + [C_6mim][(C_2F_5)$_3$PF$_3$]) system, suggesting the 1:1 complex formation, while a G^E minimum around 33 mole % for CO_2 in [C_2mim][C_1CO_2] indicates the 1:2 (CO_2:[C_2mim][C_1CO_2]) complex formation [24].

11.3.4   Ideal Association Modelling

In the previous section, highly non-ideal phase behaviours for some of the binary systems were observed, and those behaviours have been successfully correlated with our EOS model. Here, in order to understand the physical meaning of such highly non-ideal phase behaviours (for the binary systems of Cases B and C in the previous section), we interpret them as a consequence of chemical absorptions [118–123]. Now, consider the following two types of chemical associations (or complex formations) in a liquid solution of species A and B:

$$A + B \rightleftharpoons AB, \text{ with an equilibrium constant } K_1 \tag{11.43}$$

$$A + 2B \rightleftharpoons AB_2, \text{ with an equilibrium constant } K_2 \tag{11.44}$$

Thus, in this solution there exist four species A, B, AB, and AB$_2$. If it is assumed that these species form an *ideal solution*, the following thermodynamic excess functions can be derived [118, 119]:

$$\frac{G^E}{RT} = (1 - x_B) \ln \frac{1 - z_B}{(1 - x_B)(1 + K_1 z_B + K_2 z_B^2)} + x_B \ln \frac{z_B}{x_B}, \tag{11.45}$$

$$\frac{H^E}{RT} = \frac{(1-x_B)z_B\left(K_1\Delta H_1 + K_2\Delta H_2 z_B\right)}{RT\left(1 + K_1 z_B + K_2 z_B^2\right)}, \tag{11.46}$$

where x_B = the stoicheiometric (or feed) mole fraction of species B, z_B = the true (or actually existing) mole fraction of B in the solution, and ΔH_1 and ΔH_2 are the heats of complex formation for the association reactions of Equations (11.43) and (11.44), respectively. x_B is related to z_B by [118, 119]:

$$x_B = \frac{(1+K_1)z_B + K_2 z_B^2(1-z_B)}{1 + K_1 z_B(2 - z_B) + K_2 z_B^2(3 - 2z_B)}. \tag{11.47}$$

Using the present EOS, the excess functions, G^E and H^E (Eq. 11.41 or Eq. 11.42, and Eq. 11.39), can be calculated at given T, P, and compositions, where A is CO_2 and B is an ionic liquid. The temperature is taken as 298.15 K, and $P = 6.5$ MPa, which is close to the vapour pressure of CO_2 at 298.15 K. Then, G^E and H^E from the EOS correlation can be calculated as a function of x_A ($= 1-x_B$).[†]

On the other hand, G^E and H^E from the association model can be evaluated similarly as a function of x_B ($= 1-x_A$) with four unknown parameters: K_1, K_2, ΔH_1, and ΔH_2 (see Eqs. 11.43–11.46). These unknown parameters have been determined using a non-linear least squares method by minimising the differences of both G^E and H^E functions between the EOS and the association models (see Fig. 11.7 for example). The mole fraction range used in the analysis

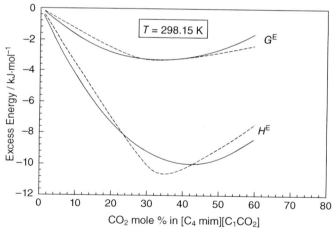

Figure 11.7 Analysis of the ideal association model using an example of the (CO_2 + [C_4mim][C_1CO_2]) system. Solid lines: the EOS model calculation. Dotted lines: results from the association model by the least-squares analysis of G^E and H^E; see Reference [24] for details. Adapted with permission from Reference [24]. Copyright (2008) American Chemical Society.

[†] See Reference [24] for additional details with figures.

is $0 < x_A < 0.6$, since solutions with higher CO_2 mole fractions (about >0.7) become immiscible liquids, as mentioned earlier. After some trial-and-error analyses, we have found that only one type of the complex (AB or AB_2) is dominating in the present liquid solution for Cases B and C. This fact is consistent with the observation of the minimum G^E location of about $x_A = 0.5$ and $x_A = 0.33$, mentioned in the previous Section 11.3.4. Results of the present analyses are shown in Table 11.5.

11.3.5   Henry's Law Constant Modelling

Although Henry's Law constants do not tell the whole story about the solubility behaviour of gases, they are often reported in literature as the limiting solubility at the infinite dilution. The Henry's Law constant, k_H, for a solute (here species 1) is defined as:

$$f_1^V \equiv P\phi_1^V y_1 = k_H x_1, (x_1 \rightarrow 0),$$ (11.48)

where the superscript V means a vapour-phase property, and x and y are liquid- and vapour-phase mole fractions of species 1, respectively. The vapour-phase fugacity, f_1^V, of species 1 must be equal to the liquid-phase fugacity, f_1^L, at the VLE (or solubility measurements):

$$f_1^L \equiv P\phi_1^L x_1.$$ (11.49)

Then, Equations (11.48) and (11.49) lead to:

$$k_H = P\phi_1^L (x_1 \rightarrow 0).$$ (11.50)

TABLE 11.5   Equilibrium Constant (K_i), Enthalpy (ΔH_i) of Complex Formation, and Henry's Law Constant (k_H) for the Binary CO_2 System with [C_2mim][CF_3CO_2], [C_6mim][(C_2F_5)$_3$PF$_3$] or [C_2mim][C_1CO_2] at 298.15 K

	System with Room Temperature Ionic Liquid (RTIL)		
	[C_2mim][CF_3CO_2]	[C_6mim][(C_2F_5)$_3$PF$_3$]	[C_2mim][C_1CO_2]
K_1	Physical absorption	1.007 ± 0.042	0
K_2	Physical absorption	0	220.3 ± 13.4
ΔH_1/kJ mol^{-1}	–	-10.7 ± 0.75	–
ΔH_2/kJ mol^{-1}	–	–	-30.81 ± 0.46
k_H/MPa	5.20	2.28	5.12×10^{-3}

The physical meaning of k_{II} (normalised by a system unit pressure) may be obtained by the direct relationship to the excess chemical potential of solute at infinite dilution [118, 124]:

$$k_{II}(T, P) = \exp\frac{\mu_1^\infty - \mu_1^0}{RT}, \tag{11.51}$$

where μ_1^∞ is the chemical potential of the solute at the infinitely dilute solution state (in the present case, at the system T and $P \to 0$), and μ_1^0 is the chemical potential referring to the pure gas (species 1) at the system T and at a pressure of 1 atm. Thus, the Henry's Law constant is an important quantity for theoretical works that evaluate the intermolecular potential between a solute molecule and a solvent molecule.

The Henry's Law constant can be obtained *conventionally* from the experimental solubility (PTx) data using the following relation, Equation (11.52) [97]:

$$k_{II} = \lim_{x_1 \to 0} \frac{f_1^V(T, P, y_1)}{x_1} \approx \left(\frac{df_1^V}{dx_1}\right)_{x_1=0}, \tag{11.52}$$

where f_1^V is the vapour-phase fugacity of the pure gas (the present case: CO_2 with $y_1 = 1$) and can be calculated by a proper EOS model [63] at a given experimental (T,P). The fugacity for $x_1 \ll 1$ can be fitted to a proper-order polynomial of x_1 in order to use Equation (11.52): for example, $f_1^V = a_0 + a_1 x_1 + a_2 x_1^2$, and then $k_{II} = a_1$.

Another (or often more rigorous) method to obtain the Henry's Law constant is to use an EOS correlation for the entire experimental PTx data, using the relationship of Equation (11.50). In the present case, $P \approx 0$ when x_1 (solute CO_2) $\to 0$, since the solvent (ionic liquid, species 2) is practically non-volatile. When $P = 0$, the present EOS (Eqs. 11.16–11.23) provides an explicit form of the EOS volume parameter, V_0 as:

$$V_0 = \frac{1}{2}\left(\frac{a_2}{RT} - b_2\right)\left(1 - \sqrt{1 - \frac{4b_2 a_2 / RT}{(b_2 - a_2 / RT)^2}}\right). \tag{11.53}$$

Then, Equation (11.50) can be rewritten as:

$$\ln k_{II} = \ln P\phi_1^l, \tag{11.54}$$

where $\ln P\phi_1^l$ is calculated by Equation (11.24) with $V = V_0, a = a_2, b = b_2, P = 1$ (with a system pressure unit, since P is eliminated out in $\ln P\phi_1^l$) and:

$$a_1' = 2\sqrt{a_1 a_2}\, f_{12}(1 - k_{12}) - a_2, \tag{11.55}$$

$$b_1' = (b_1 + b_2)(1 - k_{12})(1 - m_{12}) - b_2. \tag{11.56}$$

Some comments on Henry's Law constant, k_H, should be made here. In most of the typical physical absorption cases, such as Case A (CO_2 + [C_2mim][CF_3CO_2]), the *conventional method* to determine k_H, as described above, works very well, if one chooses a *proper range* of compositions (not including *too large amounts* of the solute) and a *proper-order* polynomial in order to fit the fugacity data. In some cases, the *proper composition range* is not obvious, and one needs some trial-and-error analyses to get a reliable k_H. In the case of a small k_H (typically less than about 3 MPa at 298 K), the conventional method becomes unreliable [24]; large uncertainties may result, depending strongly on the degree of the fitting polynomials, range of compositions, and whether the origin (zero pressure and zero composition) is included in the data set. On the other hand, the *EOS method* works well for any case and is quite reliable using all experimental solubility data. Finally, it is useful to say that a k_H value less than about 2.3 MPa at 298 K [24] would be the case of chemical absorption for CO_2 in ionic liquids, as can be seen in Table 11.5.

11.3.6 Molecular Modelling

Recently, a tremendous amount of work has been carried out to predict gas solubility (and other phase equilibria properties) in ionic liquids. Conductor-like screening models (COSMO-RS and COSMO-SAC) [125, 126] have been developed to calculate the chemical potential of a species in a mixture using quantum-mechanical calculations. These models have been used to predict the Henry's Law constant as a screening method for the molecular design of ionic liquids to capture CO_2 [127]. However, these models still require further development to accurately predict the global phase behaviour of gases in ionic liquids (i.e., particularly hydrofluorocarbon and chemical complexing systems).

Monte Carlo and molecular dynamic simulations have also been used to compute complete isotherms for gases in ionic liquids. Prof. Ed Maginn's group at the University of Notre Dame recently calculated the isotherms for (NH_3 + [C_2mim][NTf_2]) at 298, 333, and 348 K [128]. The isotherms show reasonable agreement with measured VLE data [42, 43]. Most of the difference is attributed to the model used for NH_3. When the simulated isotherms were normalised using the NH_3 saturation pressure, the agreement with experimental data was excellent. Additional properties, such as enthalpy of mixing, partial molar volumes, and total liquid volume expansion were also calculated. The power of the simulations is not only to predict thermophysical mixture properties, as mentioned earlier, but also to give insight into the detailed liquid structure. The simulations showed that the NH_3 interacts more strongly with the cation than the anion, due to hydrogen-bonding interactions between the basic nitrogen atom of NH_3 and the acidic hydrogen atoms on the cation ring [128]. This is in contrast to other observations for gases in ionic liquids, where interactions with the anion can be stronger. Understanding these interactions is an important step in the design of ionic liquids for applications such as gas separation and capture. We have also proposed a new application in our previous work [42, 43] that ammonia in ionic liquids can be used for absorption refrigeration. Our original motivation to study the solubility of ammonia in ionic liquids was to evaluate the feasibility of replacing the (NH_3 + water) absorption refrigeration cycle with (NH_3 + ionic liquids). The minimal vapour pressure of ionic liquids would eliminate the need for a rectifier in these systems and potentially reduce equipment costs. Coefficient of performance calculations with a variety of ionic liquids indicates that the energy efficiency is similar to the (NH_3 + water) system [42, 43] and further studies with (water + ionic liquid) systems have been compared with the (water + LiBr) system [129].

11.4   TERNARY PHASE BEHAVIOUR

In order to study the phase behaviour of a ternary system, we have developed models based on our EOS [44–46, 48]. Pure component and binary interaction parameters are developed using the same procedures as outlined in Section 11.3.1. Even if the solubility behaviour of each binary system is well correlated with the present EOS model, the phase behaviour prediction of the *ternary system* may not always be guaranteed based on the binary interaction parameters alone [48]. Particularly for systems containing supercritical fluids and/or non-volatile compounds such as ionic liquids, the validity of an EOS model for ternary mixtures must be checked experimentally. Ternary VLE experiments are not widely available in the literature and are a very sensitive check of the EOS model.

We have confirmed the validity of our ternary EOS model for several systems [44–46, 48] and provide an example for CO_2/SO_2/1-butyl-3-methylimidazolium methyl sulfate [C_4mim][C_1SO_4]) [46]. Ternary VLE measurements were performed using a chromatography method to analyse the vapour composition, which confirmed the validity of the ternary EOS model [46]. The isothermal ternary phase diagram predicted by the present EOS at 298.15 K is shown in Figure 11.8. A large portion of the ternary composition exhibits the liquid–liquid separation (LLE) that reflects the immiscibility gap in the binary (CO_2 + [C_4mim][C_1SO_4]) system [46].

Now that the present EOS model has been verified, we can predict with confidence the solubility behaviour of the present ternary system. In order to assess the feasibility of the gas separation by the extractive distillation or selective absorption method, gaseous selectivity $\alpha_{A/B}$, ability to separate gases A and B in the gas phase, or gaseous absorption selectivity $S_{B/A}$ in the liquid phase are commonly defined:

$$\alpha_{A/B} = S_{B/A} = \left(\frac{y_A}{x_A}\right)\bigg/\left(\frac{y_B}{x_B}\right), \tag{11.57}$$

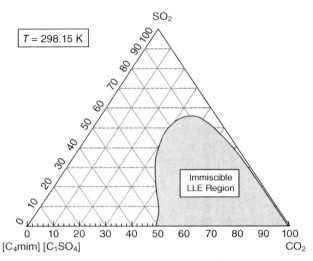

Figure 11.8   Isothermal ternary (CO_2 + SO_2 + [C_4mim][C_1SO_4]) phase diagram calculated by the present EOS model at $T = 298.15$ K; see Reference [46] for details. Adapted with permission from Reference [46]. Copyright (2010) American Chemical Society.

where x_A (or x_B) and y_A (or y_B) are the mole fractions of A (or B) in the ionic liquid solution phase and vapour phase, respectively [44–46, 48, 52, 129, 130]. Here, CO_2 is denoted as A and SO_2 as B. The CO_2/SO_2 selectivity ($\alpha_{A/B}$) in the gas phase has been examined using the present EOS model at various T, P, and feed compositions. In Figure 11.9a, the CO_2/SO_2 selectivity ($\alpha_{A/B}$) is plotted as a function of the ionic liquid [C_4mim][C_1SO_4] concentration for ternary mixtures with three CO_2/SO_2 mole ratios (1/9, 1/1, and 9/1) at $T = 298.15$ K and $P = 1$ bar. The ternary mixture with the lowest CO_2/SO_2 mole ratio ($CO_2/SO_2 = 1/9$) or highest SO_2 concentration shows a significant increase in selectivity ($\alpha_{A/B}$) from 42 to 346 with increasing ionic liquid addition. The equimolar CO_2/SO_2 case ($CO_2/SO_2 = 1/1$) shows a similar increase in selectivity from 90 to 310. For the highest CO_2/SO_2 mole ratio ($CO_2/SO_2 = 9/1$) or lowest SO_2 concentration, the selectivity was initially high (258) for even a small addition (1 mole %) of [C_4mim][C_1SO_4] and increased to a maximum selectivity of about 348 at about 25 mole % ionic liquid addition. This characteristic behaviour is due to the fact that SO_2 has much higher solubility in [C_4mim][C_1SO_4] than CO_2, and the large SO_2 concentration is absorbed in the liquid phase, leading to the high CO_2/SO_2 selectivity in the gas phase. The degree of selectivity is largely dependent on the CO_2/SO_2 feed composition and amount of ionic liquid as shown in Figure 11.9a [46].

In order to provide clear insights into the change in selectivity due to the ionic liquid addition, Figure 11.9b shows the selectivity *without* [C_4mim][C_1SO_4] at 298.15 K plotted as a function of pressure for the same CO_2/SO_2

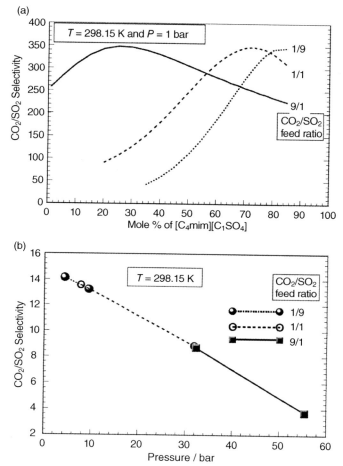

Figure 11.9   (a) Plots of calculated selectivity defined by Equation (11.57) versus [C₄mim][C₁SO₄] mole % with three different CO₂/SO₂ feed ratios at T = 298.15 K and P = 1 bar. (b) Selectivity plots without ionic liquid [C₄mim][C₁SO₄] as a function of total pressure at T = 298.15 K; lines: dotted line = 1/9 CO₂/SO₂ feed mole ratio; broken line = 1/1 CO₂/SO₂ feed mole ratio; solid line = 9/1 CO₂/SO₂ feed mole ratio; see Reference [46] for details. Adapted with permission from Reference [46]. Copyright (2010) American Chemical Society.

feed ratios (1/9, 1/1, and 9/1). The selectivity enhancement due to the ionic liquid addition can be well observed from the comparison between Figure 11.9a and Figure 11.9b. For example, the feed ratio of 9/1 (CO₂/SO₂) with the ionic liquid has a selectivity range of 226–348, while the corresponding case *without* the ionic liquid shows a selectivity range of 3–9. Similar trends are seen for other feed ratio cases [46].

11.5   NEW DIRECTIONS

A deeper understanding of molecular interactions between gases and ionic liquids is still required. The focus on the solubility of important industrial gases in ionic liquids must now include transport and calorimetry measurements. Only a few articles have been published on diffusion measurements [21, 27, 28, 131] and enthalpies of dissolution [132] in ionic liquids. Typical diffusion coefficients for CO_2 in imidazolium-based ionic liquids range from 1×10^{-5} to 1×10^{-6} cm^2 s^{-1} [21, 131]. Heat of mixing calculations using EOS models are only semi-quantitative; therefore, experimental measurements are needed to develop process simulations and accurate economic models.

Molecular modelling such as COSMO-RS is appealing, but the accuracy in the VLE predictions is still qualitative [127]. Modelling by Prof. Ed Maginn's group is promising as discussed in Section 11.3.6, but calculations are time-consuming and still untested for VLLE calculation. Continuing work to improve the accuracy and speed of these calculations is still needed. The use of simulations to guide the choice of ionic liquids to synthesise and test, combined with accurate experimental measurements to refine the model predictions, will improve molecular understanding of gas–ionic liquid interactions and accelerate new process development.

New materials are also being developed for gas separation using supported ionic liquid membranes (SILMs), where the ionic liquid is confined in the pores of the membrane [133–139]. Noble and coworkers are developing SILMs for studying the separation of gas mixtures such as CO_2, CH_4, and N_2 [133, 138, 139].

In addition to the basic physical properties, knowledge directly related to applications is required. Although the solubilities of several gases have been measured in a variety of ionic liquids, only a few commercial examples have been reported. One example developed by Air Products and Chemicals, Inc., uses ionic liquids for storing, transporting, and delivering toxic, flammable, or reactive gases. They have developed ionic liquids such as 1-butyl-3-methylimidazolium trichlorodicuprate(I) [C$_4$mim][Cu$_2$Cl$_3$] and 1-butyl-3-methylimidazolium tetrafluoroborate [C$_4$mim][BF$_4$], which can reversibly store the toxic gases, phosphine (PH$_3$) and boron trifluoride (BF$_3$), respectively [140]. In order for additional applications for the gas processing industry to be implemented, technoeconomic evaluations must be prepared, and the ionic liquid-based process must show a significant financial benefit when compared with existing processes. We have prepared such an analysis for the first time to compare the separation of CO_2 from flue gas using monoethanolamine with the ionic liquid [C$_4$mim][C$_1$CO$_2$] [141].

11.6   CONCLUDING REMARKS

We have discussed the solubility of several gases in ionic liquids and highlighted important experimental details when measuring VLE using a gravimetric microbalance and VLLE using a mass–volume technique. The global phase behaviour has been well correlated with a modified cubic EOS, which can reliably predict VLLE behaviour based on only low-pressure VLE data. Many of the systems shown demonstrate type-III and type-V phase behaviour according to the van Konynenburg–Scott classification. Large negative excess molar volumes indicate that the molecular size effect, dipole/quadrupole interactions, and hydrogen-bonding capability are important. Our EOS model has also been used to calculate thermodynamic excess functions and Henry's Law constants. Based on these excess functions, we have developed an association model to evaluate physical versus chemical absorption and understand chemical complexation.

Molecular-level understanding of the gas–ionic liquid interactions is still in progress and should provide further insight into the detailed liquid structure that can guide future experiments. Once an application and ionic liquid are identified, several other properties must still be evaluated. First, the cost of the ionic liquid and economic benefit must be calculated and clearly show a financial benefit compared with the incumbent technology. The physical and chemical stability of the ionic liquid must be demonstrated in the presence of the gases (and impurities) at operating conditions and required time scales. Materials of construction, corrosion, transport limitations, foaming, toxicity, waste handling, to name a few, are additional details that must also be considered. Traditional unit operations (i.e., columns and tanks) may be employed, but new materials such as supported ionic liquid membranes should also be considered. Although ionic liquid-based processes share the same hurdles to development and implementation as any new chemical process, it is our belief that ionic liquids still hold the possibility for future absorption-based gas separations.

ACKNOWLEDGEMENTS

The authors thank Mr. Brian L. Wells, Mrs. Anne Marie S. Niehaus, and Mr. Joe Nestlerode at the DuPont Experimental Station for their assistance with the gas solubility measurements, and DuPont Central Research and Development for supporting this work.

REFERENCES

1 Wasserscheid, P., and Welton, T., eds., *Ionic liquids in synthesis*, 1st ed., Wiley-VCH Verlag GmbH & Co. KGaA, Weinheim (2003).

2 Wasserscheid, P., and Welton, T., eds., *Ionic liquids in synthesis*, 2nd ed., Wiley-VCH Verlag GmbH & Co. KGaA, Weinheim (2008).

3 Rogers, R.D., Seddon, K.R., and Volkov, S., eds., *Green industrial applications of ionic liquids, NATO science series II: mathematics, physics and chemistry*, Vol. 92, Kluwer, Dordrecht (2002).

4   Plechkova, N.V., Rogers, R.D., and Seddon, K.R., eds., *Ionic liquids: from knowledge to application*, ACS Symp. Ser., Vol. 1030, American Chemical Society, Washington, D.C. (2009).

5   Brazel, C.S., and Rogers, R.D., eds., *Ionic liquids in polymer systems: solvents, additives, and novel applications*, ACS Symp. Ser., Vol. 913, American Chemical Society, Washington, D.C. (2005).

6   Rogers, R.D., and Seddon, K.R., eds., *Ionic liquids IIIB: fundamentals, progress, challenges, and opportunities: transformations and processes*, ACS Symp. Ser., Vol. 902, American Chemical Society, Washington, D.C. (2005).

7   Rogers, R.D., and Seddon, K.R., eds., *Ionic liquids IIIA: Fundamentals, progress, challenges, and opportunities: Properties and structure*, ACS Symp. Ser., Vol. 901, American Chemical Society, Washington, D.C. (2005).

8   Rogers, R.D., and Seddon, K.R., eds., *Ionic liquids as green solvents: progress and prospects*, ACS Symp. Ser., Vol. 856, American Chemical Society, Washington, D.C. (2003).

9   Rogers, R.D., and Seddon, K.R., eds., *Ionic liquids: industrial applications to green chemistry*, ACS Symp. Ser., Vol. 818, American Chemical Society, Washington, D.C. (2002).

10 Brennecke, J.F., Lopez-Castillo, Z.K., and Mellein, B.R., Gas solubilities in ionic liquids and related measurement techniques, in: *Ionic liquids in chemical analysis*, ed. M. Koel, CRC Press, Boca Raton, FL (2009), pp. 229–241.

11 Anderson, J.L., Anthony, J.L., Brennecke, J.F., and Maginn, E.J., Gas solubilities in ionic liquids, in: *Ionic liquids in synthesis*, Vol. 1, 2nd ed., ed. P. Wasserscheid, and T. Welton, Wiley-VCH Verlag GmbH & Co. KGaA, Weinheim (2008), pp. 103–129.

12 Maurer, G., and Pérez-Salado Kamps, Á., Solubility of gases in ionic liquids, aqueous solutions, and mixed solvents, in: *Developments and applications in solubility*, ed. T.M. Letcher, Royal Society of Chemistry, Cambridge (2007), pp. 41–58.

13 Brennecke, J.F., Anthony, J.L., and Maginn, E.J., Gas solubilities in ionic liquids, in: *Ionic liquids in synthesis*, 1st ed., ed. P. Wasserscheid, and T. Welton, Wiley-VCH Verlag GmbH & Co. KGaA, Weinheim (2003), pp. 81–93.

14 Scott, R.L., and van Konynenburg, P.H., Static problems of solutions. Van der Waals and related models for hydrocarbon mixtures, *Discuss Faraday Soc.* **49**, 87–97 (1970).

15 van Konynenburg, P.H., and Scott, R.L., Critical lines and phase equilibria in binary van der Waals mixtures, *Phil. Trans.* **A298**, 495–540 (1980).

16 Yokozeki, A., Solubility of refrigerants in various lubricants, *Int. J. Thermophys.* **22**, 1057–1071 (2001).

17 Gmehling, J., Onken, U., Arlt, W., Grenzheuser, P., Weidlich, U., Kolbe, B., and Rarey, J., Vapour-liquid equilibrium data collection, Part 2a: Alcohols, *Dechema Chemistry Data Series*, 1.2a 6, 1–750 (1986).

18 Gibbons, R.M., and Langhton, A.P., An Equation of State for hydrochloric acid solutions, *Fluid Phase Equilibr.* **18**, 61–68 (1984).

19 Stryjek, R., and Vera, J.H., Vapour-liquid equilibrium of hydrochloric acid solutions with the PRSV equation of state, *Fluid Phase Equilibr.* **25**, 279–290 (1986).

20 Tillner-Roth, R., and Friend, D.G., A Helmholtz free energy formulation of the thermodynamic properties of the mixture (water + ammonia), *J. Phys. Chem. Ref. Data* **27**, 63–96 (1998).

21 Shiflett, M.B., and Yokozeki, A., Solubilities and diffusivities of carbon dioxide in ionic liquids: [bmim][PF_6] and [bmim][BF_4], *Ind. Eng. Chem. Res.* **44**, 4453–4464 (2005).

22 Shiflett, M.B., and Yokozeki, A., Solubility of CO_2 in room-temperature ionic liquid [hmim][Tf_2N], *J. Phys. Chem. B* **111**, 2070–2074 (2007).

23 Shiflett, M.B., Kasprzak, D.J., Junk, C.P., and Yokozeki, A., Phase behaviour of carbon dioxide + [bmim][Ac] mixtures, *J. Chem. Thermodyn.* **40**, 25–31 (2008).

24 Yokozeki, A., Shiflett, M.B., Junk, C.P., Grieco, L.M., and Foo, T., Physical and chemical absorptions of carbon dioxide in room-temperature ionic liquids, *J. Phys. Chem. B* **112**, 16654–16663 (2008).

25 Shiflett, M.B., and Yokozeki, A., Phase behaviour of carbon dioxide in ionic liquids: [emim][acetate], [emim][trifluoroacetate], and [emim][acetate] + [emim][trifluoroacetate] mixtures, *J. Chem. Eng. Data* **54**, 108–114 (2009).

26 Shiflett, M.B., and Yokozeki, A., Solubility and diffusivity of hydrofluorocarbons in room-temperature ionic liquids, *AIChE J.* **52**, 1205–1219 (2006).

27 Shiflett, M.B., Junk, C.P., Harmer, M.A., and Yokozeki, A., Solubility and diffusivity of difluoromethane in room-temperature ionic liquids, *J. Chem. Eng. Data* **51**, 483–495 (2006).

28 Shiflett, M.B., Junk, C.P., Harmer, M.A., and Yokozeki, A., Solubility and diffusivity of 1,1,1-tetrafluoroethane in room-temperature ionic liquids, *Fluid Phase Equilibr.* **242**, 220–232 (2006).

29 Shiflett, M.B., and Yokozeki, A., Gaseous absorption of fluoromethane, fluoroethane, and 1,1,2,2-tetrafluoroethane in 1-butyl-3-methylimidazolium hexafluorophosphate, *Ind. Chem. Eng. Res.* **45**, 6375–6382 (2006).

30 Yokozeki, A., and Shiflett, M.B., Global phase behaviours of trifluoromethane in room-temperature ionic liquid [bmim][PF_6], *AIChE J.* **52**, 3952–3957 (2006).

31 Shiflett, M.B., and Yokozeki, A., Phase equilibria of hydrofluorocarbon-4310mee mixtures with ionic liquids: miscibility of threo- and erythro-diasteromers in ionic liquids, *Ind. Eng. Chem. Res.* **47**, 926–934 (2008).

32 Shiflett, M.B., and Yokozeki, A., Hydrogen substitution effect on the solubility of perhalogenated compounds in ionic liquid [bmim][PF_6], *Fluid Phase Equilibr.* **259**, 210–217 (2007).

33 Shiflett, M.B., and Yokozeki, A., Solubility differences of halocarbon isomers in ionic liquid [emim][Tf_2N], *J. Chem. Eng. Data* **52**, 2007–2015 (2007).

34 Kumełan, J., Pérez-Salado Kamps, Á., Tuma, D., Yokozeki, A., Shiflett, M.B., and Maurer, G., Solubility of tetrafluoromethane in the ionic liquid [hmim][Tf_2N], *J. Phys. Chem. B* **112**, 3040–3047 (2008).

35 Shiflett, M.B., and Yokozeki, A., Binary vapour-liquid and vapour-liquid-liquid equilibria of hydrofluorcarbons (HFC-125 and HFC-143a) and hydrofluorethers (HFE-125 and HFE-143a) with ionic liquid [emim][Tf_2N], *J. Chem. Eng. Data* **53**, 492–497 (2008).

36 Shiflett, M.B., and Yokozeki, A., Solubility of fluorocarbons in room temperature ionic liquids, in: *Ionic liquids: from knowledge to applications*, ACS Symp. Ser.,

Vol. 1030, ed. N.V. Plechkova, R.D. Rogers, and K.R. Seddon, American Chemical Society, Washington, D.C. (2009), pp. 21–42.

37 Ren, W., Scurto, A.M., Shiflett, M.B., and Yokozeki, A., Phase behaviour and equilibria of ionic liquids and refrigerants: 1-Ethyl-3-methyl-imidazolium Bis(trifluoromethylsulfonyl)imide ([EMIm][Tf$_2$N]) and R-134a, in: *Gas-expanded liquids and near-critical media: green chemistry and engineering*, ACS Symp. Ser., Vol. 1006, ed. K.W. Hutchenson, A.M. Scurto, and B. Subramaniam, American Chemical Society, Washington, D.C. (2009), pp. 112–128.

38 Shiflett, M.B., and Yokozeki, A., Separation of difluoromethane and pentafluoroethane by extractive distillation using ionic liquid, *Chem. Today* **24**, 28–30 (2006).

39 Shiflett, M.B., and Yokozeki, A., Vapour-liquid-liquid equilibria of pentafluoroethane and ionic liquid [bmim][PF$_6$] mixtures studied with the volumetric method, *J. Phys. Chem. B* **110**, 14436–14443 (2006).

40 Shiflett, M.B., and Yokozeki, A., Vapour-liquid-liquid equilibria of hydrofluorocarbons and 1-butyl-3-methylimidazolium hexafluorophosphate, *J. Chem. Eng. Data* **51**, 1931–1939 (2006).

41 Shiflett, M.B., and Yokozeki, A., Liquid-liquid equilibria of hydrofluoroethers and ionic liquid [emim][Tf$_2$N], *J. Chem. Eng. Data* **52**, 2413–2418 (2007).

42 Yokozeki, A., and Shiflett, M.B., Ammonia solubilities in room-temperature ionic liquids, *Ind. Eng. Chem. Res.* **46**, 1605–1610 (2007).

43 Yokozeki, A., and Shiflett, M.B., Vapour-liquid equilibria of ammonia + ionic liquid mixtures, *Appl. Energy* **84**, 1258–1273 (2007).

44 Yokozeki, A., and Shiflett, M.B., Hydrogen purification using room-temperature ionic liquids, *Appl. Energy* **84**, 351–361 (2007).

45 Yokozeki, A., and Shiflett, M.B., Separation of carbon dioxide and sulfur dioxide gases using room-temperature ionic liquid [hmim][Tf$_2$N], *Energy Fuels* **23**, 4701–4708 (2009).

46 Shiflett, M.B., and Yokozeki, A., Separation of carbon dioxide and sulfur dioxide using room-temperature ionic liquid [bmim][MeSO$_4$], *Energy Fuels* **24**, 1001–1008 (2010).

47 Shiflett, M.B., and Yokozeki, A., Chemical absorption of sulfur dioxide in room-temperature ionic liquids, *Ind. Eng. Chem. Res.* **49**, 1370–1377 (2010).

48 Shiflett, M.B., and Yokozeki, A., Separation of CO$_2$ and H$_2$S using room-temperature ionic liquid [bmim][PF$_6$], *Fluid Phase Equilibr.* **294**, 105–113 (2010).

49 Shiflett, M.B., Niehaus, A.M.S., and Yokozeki, A., Separation of CO$_2$ and H$_2$S using room-temperature ionic liquid [bmim][MeSO$_4$], *J. Chem. Eng. Data* **55**, 4785–4793 (2010).

50 Shiflett, M.B., Niehaus, A.M.S., and Yokozeki, A., Separation of N$_2$O and CO$_2$ using room-temperature ionic liquid [bmim][BF$_4$], *J. Phys. Chem. B* **115**, 3478–3487 (2010).

51 Shiflett, M.B., and Yokozeki, A., Liquid liquid equilibria in binary mixtures of 1,3-propanediol + ionic liquids [bmim][PF$_6$], [bmim][BF$_4$], and [emim][BF$_4$], *J. Chem. Eng. Data* **52**, 1302–1306 (2007).

52 Yokozeki, A., and Shiflett, M.B., Binary and ternary phase diagrams of benzene, hexafluorobenzene, and ionic liquid [emim][Tf$_2$N] using equations of state, *Ind. Eng. Chem. Res.* **47**, 8389–8395 (2008).

53 Shiflett, M.B., and Yokozeki, A., Liquid-liquid equilibria in binary mixtures containing fluorinated benzenes and ionic liquid 1-ethyl-3-methylimidazolium bis(trifluoromethylsulfonyl)imide, *J. Chem. Eng. Data* **53**, 2683–2691 (2008).

54 Shiflett, M.B., Niehaus, A.M.S., and Yokozeki, A., Liquid-liquid equilibria in binary mixtures containing chlorobenzene, bromobenzene, and iodobenzene with ionic liquid 1-ethyl-3-methylimidazolium bis(trifluoromethylsulfonyl)imide, *J. Chem. Eng. Data* **54**, 2090–2094 (2009).

55 Shiflett, M.B., and Niehaus, A.M.S., Liquid-liquid equilibria in binary mixtures containing substituted benzenes with ionic liquid 1-ethyl-3-methylimidazolium bis(trifluoromethylsulfonyl)imide, *J. Chem. Eng. Data* **1**, 346–353 (2010).

56 Battino, R., and Clever, H.L., The solubility of gases in water and seawater, in: *Developments and applications in solubility*, ed. T.M. Letcher, Royal Society of Chemistry Publishing, Cambridge (2007), pp. 66–78.

57 Marsh, K.N., Brennecke, J.F., Chirico, R.D., Frenkel, M., Heintz, A., Magee, J.W., Peters, C.J., Rebelo, L.P.N., and Seddon, K.R., Thermodynamic and thermophysical properties of the reference ionic liquid: 1-hexyl-3-methylimidazolium bis[(trifluoromethyl)sulfonyl]amide (including mixtures) part 1. Experimental methods and results, *Pure Appl. Chem.* **81**, 781–790 (2009).

58 Chirico, R.D., Diky, V., Magee, J.W., Frenkel, M., and Marsh, K.N., Thermodynamic and thermophysical properties of the reference ionic liquid: 1-hexyl-3-methylimidazolium bis[(trifluoromethyl)sulfonyl]amide (including mixtures) part 2. Critical evaluation and recommended property values, *Pure Appl. Chem.* **81**, 791–828 (2009).

59 Anthony, J.L., Maginn, E.J., and Brennecke, J.F., Solution thermodynamics of imidazolium-based ionic liquids and water, *J. Phys. Chem. B* **105**, 10942–10949 (2001).

60 Anthony, J.L., Maginn, E.J., and Brennecke, J.F., Solubilities and thermodynamic properties of gases in the ionic liquid 1-n-butyl-3-methylimidazolium hexafluorophosphate, *J. Phys. Chem. B* **106**, 7315–7320 (2002).

61 Cadena, C., Anthony, J.L., Shah, J.K., Morrow, T.I., Brennecke, J.F., and Maginn, E.J., Why is CO_2 so soluble in imidazolium-based ionic liquids? *J. Am. Chem. Soc.* **126**, 5300–5308 (2004).

62 Hiden Isochema, Ltd., Private communication, Warrington, UK, http://www.hidenisochema.com (2010).

63 Lemmon, E.W., McLinden, M.O., and Huber, M.L., A computer program, REFPROP, Reference Fluid Thermodynamic and Transport Properties, ver. 7, National Institute of Standards and Technology, Gaithersburg, MD (2002).

64 Anderson, J.L., Dixon, J.K., Maginn, E.J., and Brennecke, J.F., Measurement of SO_2 solubility in ionic liquids, *J. Phys. Chem. B* **110**, 15059–15062 (2006).

65 Wu, C.-T., Marsh, K.N., Deev, A.V., and Boxall, J.A., Liquid-liquid equilibria of room-temperature ionic liquids and butan-1-ol, *J. Chem. Eng. Data* **48**, 486–491 (2003).

66 Heintz, A., Lehmann, J.K., and Wertz, C., Thermodynamic properties of mixtures containing ionic liquids. 3. Liquid-liquid equilibria of binary mixtures of 1-ethyl-3-methylidazolium bis(trifluoromethylsulfonyl)imide with propan-1-ol, butan-1-ol, and pentan-1-ol, *J. Chem. Eng. Data* **48**, 472–474 (2003).

67   Crosthwaite, J.M., Aki, S.N.V.K., Maginn, E.J., and Brennecke, J.F., Liquid-phase behaviour of imidazolium-based ionic liquids with alcohols: effect of hydrogen bonding and non-polar interactions, *Fluid Phase Equilib.* **228**, 303–309 (2005).

68   Crosthwaite, J.M., Aki, S.N.V.K., Maginn, E.J., and Brennecke, J.F., Liquid-phase behaviour of imidazolium-based ionic liquids with alcohols, *J. Phys. Chem. B* **108**, 5113–5119 (2004).

69   Wagner, M., Stanga, O., and Schröer, W., The liquid-liquid coexistence of binary mixtures of the room-temperature ionic liquid 1-methyl-3-hexylimidazolium tetrafluoroborate with alcohols, *Phys. Chem. Chem. Phys.* **6**, 4421–4431 (2004).

70   Wagner, M., Stanga, O., and Schröer, W., Corresponding states analysis of the critical points in binary solutions of room-temperature ionic liquids, *Phys. Chem. Chem. Phys.* **5**, 3943–3950 (2003).

71   Carvalho, P.J., Alvarez, V.H., Schröder, B., Gil, A.M., Marrucho, I.M., Aznar, M., Santos, L.M.N.B.F., and Coutinho, J.A.P., Specific solvation interactions of CO_2 on acetate and trifluoroacetate imidazolium based ionic liquids at high pressures, *J. Phys. Chem. B* **113**, 6803–6812 (2009).

72   Carvalho, P.J., Álvarez, V.H., Machado, J.J.B., Pauly, J., Daridon, J., Marrucho, I.M., Aznar, M., and Coutinho, J.A.P., High pressure phase behaviour of carbon dioxide in 1-alkyl-3-methylimidazolium bis(trifluoromethylsulfonyl)imide ionic liquids, *J. Supercrit. Fluids* **48**, 99–107 (2009).

73   Carvalho, P.J., Alvarez, V.H., Marrucho, I.M., Aznar, M., and Coutinho, J.A.P., High pressure phase behaviour of carbon dioxide in 1-butyl-3-methylimidazolium bis(trifluoromethylsulfonyl)imide and 1-butyl-3-methylimidazolium dicyanamide ionic liquids, *J. Supercrit. Fluids* **50**, 105–111 (2009).

74   Carvalho, P.J., Alvarez, V.H., Marrucho, I.M., Aznar, M., and Coutinho, J.A.P., High carbon dioxide solubilities in trihexyltetradecylphosphonium-based ionic liquids, *J. Supercrit. Fluids* **52**, 258–265 (2010).

75   Shariati, A., and Peters, C.J., High-pressure phase behaviour of systems with ionic liquids: measurements and modeling of the binary system fluoroform + 1-ethyl-3-methylimidazolium hexafluorophosphate, *J. Supercrit. Fluids* **25**, 109–117 (2003).

76   Shariati, A., and Peters, C.J., High-pressure phase behaviour of systems with ionic liquids Part II. The binary system carbon dioxide + 1-ethyl-3-methyllimidazolium hexafluorophosphate, *J. Supercrit. Fluids* **29**, 43–48 (2004).

77   Shariati, A., and Peters, C.J., High-pressure phase behaviour of systems with ionic liquids Part III. The binary system carbon dioxide + 1-hexyl-3-methylimidazolium hexafluorophosphate, *J. Supercrit. Fluids* **30**, 139–144 (2004).

78   Costantini, M., Toussaint, V.A., Shariati, A., Peters, C.J., and Kikic, I., High-pressure phase behaviour of systems with ionic liquids Part IV. The binary system carbon dioxide + 1-hexyl-3-methylimidazolium tetrafluoroborate, *J. Chem. Eng. Data* **50**, 52–55 (2005).

79   Kroon, M.C., Shariati, A., Costantini, M., Van Spronsen, J., Witkamp, G.-J., Sheldon, R.A., and Peters, C.J., High-pressure phase behaviour of systems with ionic liquids Part V. The binary system carbon dioxide + 1-butyl-3-methylimidazolium tetrafluoroborate, *J. Chem. Eng. Data* **50**, 173–176 (2005).

80   Song, H.N., Byung-Chul, L., and Lim, J.S., Measurement of CO_2 solubility in ionic liquids: [BMP][TfO] and [P14,6,6,6][Tf$_2$N] by measuring bubble-point pressure, *J. Chem. Eng. Data* **55**, 891–896 (2010).

81 Kumełan, J., Tuma, D., Pérez-Salado Kamps, Á., and Maurer, G., Solubility of the single gases carbon dioxide and hydrogen in the ionic liquid [bmpy][Tf₂N], *J. Chem. Eng. Data* **55**, 165–172 (2010).

82 Kumełan, J., Tuma, D., Pérez-Salado Kamps, Á., and Maurer, G., Solubility of the single gases carbon monoxide and oxygen the ionic liquid [hmim][Tf₂N], *J. Chem. Eng. Data* **54**, 966–971 (2009).

83 Kumełan, J., Tuma, D., Pérez-Salado Kamps, Á., and Maurer, G., Solubility of the H₂ in the ionic liquid [hmim][Tf₂N], *J. Chem. Eng. Data* **51**, 1364–1367 (2006).

84 Ren, W., and Scurto, A.M., High-pressure phase equilibria with compressed gases, *Rev. Sci. Instrum.* **78**, 125104 (2007).

85 Ren, W., and Scurto, A.M., Phase equilibria of imidazolium ionic liquids and the refrigerant gas 1,1,1,2-tetrafluoroethane (R-134a), *Fluid Phase Equilib.* **286**, 1–7 (2009).

86 Ren, W., Sensenich, B., and Scurto, A.M., High-pressure phase equilibria of {carbon dioxide (CO₂) + n-alkyl-imidazolium bis(trifluoromethylsulfonyl)amide} ionic liquids, *J. Chem. Thermodyn.* **42**, 305–311 (2010).

87 Husson-Borg, P., Majer, V., and Costa Gomes, M.F., Solubilities of oxygen and carbon dioxide in butyl methyl imidazolium tetrafluoroborate as a function of temperature at pressures close to atmospheric pressure, *J. Chem. Eng. Data* **48**, 480–485 (2003).

88 Jacquemin, J., Costa Gomes, M.F., Husson, P., and Majer, V., Solubility of carbon dioxide, ethane, methane, oxygen, nitrogen, hydrogen, argon, and carbon monoxide in 1-butyl-3-methylimidazolium tetrafluoroborate between temperatures 283 K and 343 K and at pressures close to atmospheric, *J. Chem. Thermodyn.* **38**, 490–502 (2006).

89 Jacquemin, J., Husson, P., Majer, V., and Costa Gomes, M.F., Low-pressure solubilities and thermodynamics of solvation of eight gases in 1-butyl-3-methylimidazolium hexafluorophosphate, *Fluid Phase Equilib.* **240**, 87–95 (2006).

90 Costa Gomes, M.F., Low-pressure solubility and thermodynamic of salvation of carbon dioxide, ethane, and hydrogen in 1-hexyl-3-methylimidazolium bis(trifluoromethylsulfonyl)amide between temperatures of 283 K and 343 K, *J. Chem. Eng. Data* **52**, 472–475 (2007).

91 Pison, L., Canongia Lopes, J.N., Rebelo, L.P.N., Padua, A.A.H., and Costa Comes, M.F., Interactions of fluorinated gases with ionic liquids: solubility of CF₄, C₂F₆, and C₃F₈ in trihexyltetradecylphosphonium bis(trifluoromethylsulfonyl)amide, *J. Phys. Chem. B* **112**, 12394–12400 (2008).

92 Heintz, A., Kulikov, D.V., and Verevkin, S.P., Thermodynamic properties of mixtures containing ionic liquids. 1. Activity coefficients at infinite dilution of alkanes, alkenes, and alkylbenzenes in 4-methyl-n-butylpyridinium tetrafluoroborate using gas-liquid chromatography, *J. Chem. Eng. Data* **46**, 1526–1529 (2001).

93 Heintz, A., Kulikov, D.V., and Verevkin, S.P., Thermodynamic properties of mixtures containing ionic liquids. Activity coefficients at infinite dilution of polar solutes in 4-methyl-n-butyl-pyridinium tetrafluoroborate using gas-liquid chromatography, *J. Chem. Thermodyn.* **34**, 1341–1347 (2002).

94 Heintz, A., Kulikov, D.V., and Verevkin, S.P., Thermodynamic properties of mixtures containing ionic liquids. 2. Activity coefficients at infinite dilution of hydrocarbons and polar solutes in 1-methyl-3-ethyl-imidazolium bis

(trifluoromethyl-sulfonyl) amide and in 1,2-dimethyl-3-ethyl-imidazolium bis(trifluoromethyl-sulfonyl) amide using gas-liquid chromatography, *J. Chem. Eng. Data* **47**, 894–899 (2002).

95   Kazarian, S.G., Briscoe, B.J., and Welton, T., Combining ionic liquids and supercritical fluids: in situ ATR-IR study of CO_2 dissolved in two ionic liquids at high pressures, *Chem. Commun.*, 2047–2048 (2000).

96   Dyson, P.J., Laurenczy, G., Ohlin, A., Vallance, J., and Welton, T., Determination of hydrogen concentration in ionic liquids and the effect (or lack of) on rates of hydrogenation, *Chem. Commun.*, 2418–2419 (2003).

97   Van Ness, C.H., and Abbott, M.M., *Classical thermodynamics of nonelectrolyte solutions*, McGraw-Hill, New York (1982).

98   Walas, S.M., *Phase equilibria in chemical engineering*, Butterworth, Boston (1985), pp. 178–183.

99   Reid, R.C., Prausnitz, J.M., and Poling, R.E., *The properties of gases and liquids*, 4th ed., McGraw-Hill, New York (1987).

100   Seiler, M., Jork, C., Kavarou, A., Arlt, W., and Hirsch, R., Separation of azeotropic mixtures using hyperbranched polymers or ionic liquids, *AIChE J.* **50**, 2439–2454 (2004).

101   Kato, R., Krummen, M., and Gmehling, J., Measurement and correlation of vapour-liquid equilibria and excess enthalpies of binary systems containing ionic liquids and hydrocarbons, *Fluid Phase Equilib.* **224**, 47–54 (2004).

102   Döker, M., and Gmehling, J., Measurement and prediction of vapour-liquid equilibria of ternary systems containing ionic liquids, *Fluid Phase Equilib.* **227**, 255–266 (2005).

103   Anderko, A., Wang, P., and Rafal, M., Electrolyte solutions: from thermodynamics and transport property models to simulation of industrial processes, *Fluid Phase Equilib.* **194–197**, 123–142 (2002).

104   Yokozeki, A., Solubility correlation and phase behaviours of carbon dioxide + lubricant oil mixtures, *Appl. Energy* **84**, 159–175 (2007).

105   Valderrama, J.O., and Robles, P.A., Critical properties, normal boiling temperatures, and acentric factors of fifty ionic liquids, *Ind. Eng. Chem. Res.* **46**, 1338–1344 (2007).

106   Valderrama, J.O., Sanga, W.W., and Lazzś, J.A., Critical properties, normal boiling temperatures, and acentric factors of another 200 ionic liquids, *Ind. Eng. Chem. Res.* **47**, 1318–1330 (2008).

107   Valderrama, J.O., and Rojas, R.E., Critical properties of ionic liquids. Revisited, *Ind. Eng. Chem. Res.* **48**, 6890–6900 (2009).

108   Kumełan, J., Pérez-Salado Kamps, Á., Tuma, D., and Maurer, G., Solubility of CO_2 in the ionic liquid [hmim][Tf$_2$N], *J. Chem. Thermodyn.* **38**, 1396–1401 (2006).

109   Vetere, A., Again the Rackett equation, *Chem. Eng. J.* **49**, 27–33 (1992).

110   Valderrama, J.O., Reátegui, A., and Sanga, W.W., Thermodynamic consistency test of vapour-liquid equilibrium data for mixtures containing ionic liquids, *Ind. Eng. Chem. Res.* **47**, 8416–8422 (2008).

111   Aki, S.N.V.K., Mellein, B.R., Saurer, E.M., and Brennecke, J.F., High-pressure phase behaviour of carbon dioxide with imidazolium-based ionic liquids, *J. Phys. Chem. B* **108**, 20355–20365 (2004).

112   Kim, Y.S., Choi, W.Y., Jang, J.H., Yoo, K., and Lee, C.S., Solubility measurement and prediction of carbon dioxide in ionic liquids, *Fluid Phase Equilib.* **228**, 439–455 (2005).

113   Baltus, R.E., Culbertson, B.H., Dai, S., Luo, H., and DePaoli, D.W., Low-pressure solubility of carbon dioxide in room-temperature ionic liquids measured with a quartz crystal microbalance, *J. Phys. Chem. B* **108**, 721–727 (2004).

114   Kumełan, J., Pérez-Salado Kamps, Á., Tuma, D., and Maurer, G., Solubility of CO_2 in the ionic liquids [bmim][CH_3SO_4] and [bmim][PF_6], *J. Chem. Eng. Data* **51**, 1802–1807 (2006).

115   Muldoon, M.J., Aki, S.N.V.K., Anderson, J.L., Dixon, J.K., and Brennecke, J.F., Improving carbon dioxide solubility in ionic liquids, *J. Phys. Chem. B* **111**, 9001–9009 (2007).

116   Kim, Y.S., Jang, J.H., Lim, B.D., Kang, J.W., and Lee, C.S., Solubility of mixed gases containing carbon dioxide in ionic liquids: measurements and predictions, *Fluid Phase Equilib.* **256**, 70–74 (2007).

117   Lee, B.-C., and Outcalt, S.L., Solubilities of gases in ionic liquid 1-*n*-butyl-3-methylimidazolium Bis(trifluoromethylsulfonyl)imide, *J. Chem. Eng. Data* **51**, 892–897 (2006).

118   Acree, W.E., Jr., *Thermodynamic properties of nonelectrolyte solutions*, Academic Press, New York (1984).

119   McGlashan, M.L., and Rastogi, R.P., The thermodynamics of associated mixtures. Part 1.-Dioxan + chloroform, *Trans. Faraday Soc.* **54**, 496–501 (1958).

120   Ohta, T., Asano, H., and Nagata, I., Thermodynamic study of complex formation in four binary liquid mixtures containing chloroform, *Fluid Phase Equilib.* **4**, 105–114 (1980).

121   Matusi, T., Hepler, L.G., and Fenby, D.V., Thermodynamic investigation of complex formation by hydrogen bonding in binary liquid systems. Chloroform with triethylamine, dimethyl sulfoxide, and acetone, *J. Phys. Chem.* **77**, 2397–2400 (1973).

122   Handa, Y.P., Fenby, D.V., and Jones, D.E., Vapour pressures of triethylamine + chloroform and of triethylamine + dichloromethane, *J. Chem. Thermodyn.* **7**, 337–343 (1975).

123   Hepler, L.G., and Fenby, D.V., Thermodynamic study of complex formation between triethylamine and chloroform, *J. Chem. Thermodyn.* **5**, 471–475 (1973).

124   Denbigh, K., *The principles of chemical equilibrium*, 3rd ed., Cambridge University Press, London (1971).

125   Klamt, A., and Eckert, F., Fast solvent screening via quantum chemistry: COSMO-RS approach, *AIChE J.* **48**, 369–385 (2002).

126   Lin, S.-T., and Sandler, S.I., A priori phase equilibrium prediction from a segment contribution solvation model, *Ind. Eng. Chem. Res.* **41**, 899–913 (2002).

127   Zhang, X., Liu, Z., and Wang, W., Screening of ionic liquids to capture CO_2 by COSMO-RS and experiments, *AIChE J.* **54**, 2717–2728 (2008).

128   Shi, W., and Maginn, E.J., Molecular simulation of ammonia absorption in the ionic liquid 1-ethyl-3-methylimidazolium bis(trifluoromethylsulfonyl)imide ([emim][Tf_2N]), *AIChE J.* **55**, 2414–2421 (2009).

129 Yokozeki, A., and Shiflett, M.B., Water solubility in ionic liquids and application to absorption cycles, *Ind. Eng. Chem. Res.* **49**, 9496–9503 (2010).

130 Peng, X., Wang, W., Xue, R., and Shen, Z., Adsorption separation of CH_4/CO_2 on mesocarbon microbeads: experiment and modeling, *AIChE J.* **52**, 994–1003 (2006).

131 Camper, D., Becker, C., Koval, C., and Noble, R., Diffusion and solubility measurements in room temperature ionic liquids, *Ind. Eng. Chem. Res.* **45**, 445–450 (2006).

132 Costa Gomes, M.F., and Husson, P., Ionic liquids: promising media for gas separations, in: *Ionic liquids: from knowledge to applications*, ACS Symp. Ser., Vol. 1030, ed. N.V. Plechkova, R.D. Rogers, and K.R. Seddon, American Chemical Society, Washington, D.C. (2009), pp. 223–237.

133 Scovazzo, P., Kieft, J., Finan, D.A., Koval, C., DuBois, D., and Noble, R., Gas separations using non-hexafluorophosphate $[PF_6]^-$ anion supported ionic liquid membranes, *J. Membr. Sci.* **238**, 57–63 (2004).

134 Gan, Q., Rooney, D., and Zou, Y., Supported ionic liquid membranes in nanopore structure for gas separation and transport studies, *Desalination* **199**, 535–537 (2006).

135 Jiang, Y.-Y., Zhou, Z., Jiao, Z., Li, L., Wu, Y.-T., and Zhang, Z.-B., SO_2 gas separation using supported ionic liquid membranes, *J. Phys. Chem. B* **111**, 5058–5061 (2007).

136 Pennline, H.W., Luebke, D.R., Jones, K.L., Myers, C.R., Morsi, B.I., Heintz, Y.J., and Ilconich, J.B., Progress in carbon dioxide capture and separation research for gasification-based power generation point sources, *Fuel Proc. Technol.* **89**, 897–907 (2008).

137 Cserjési, P., Nemestóthy, N., Vass, A., Csanádi, Z.S., and Bélafi-Bakó, K., Study on gas separation by supported liquid membranes applying novel ionic liquids, *Desalination* **245**, 743–747 (2009).

138 Bara, J.E., Carlisle, T.K., Gabriel, C.J., Camper, D., Finotello, A., Gin, D.L., and Noble, R.D., Guide to CO_2 separations in imidazolium-based room-temperature ionic liquids, *Ind. Eng. Chem. Res.* **48**, 2739–2751 (2009).

139 Bara, J.E., Camper, D.E., Gin, D.L., and Noble, R.D., Room-temperature ionic liquids and composite materials: platform technologies for CO_2 capture, *Acc. Chem. Res.* **43**, 152–159 (2010).

140 Tempel, D.J., Henderson, P.B., Brzozowski, J.R., Pearlstein, R.M., and Cheng, H., High gas storage capacities for ionic liquids through chemical complexation, *J. Am. Chem. Soc.* **130**, 400–401 (2008).

141 Shiflett, M.B., Drew, D.W., Cantini, R.A., and Yokozeki, A., Carbon dioxide capture using ionic liquid 1-butyl-3-methylimidazolium acetate, *Energy Fuels* **24**, 5781–5789 (2010).

INDEX

Ionic Liquids UnCOILed: Critical Expert Overviews, First Edition. Edited by Natalia V. Plechkova and Kenneth R. Seddon.
© 2013 John Wiley & Sons, Inc. Published 2013 by John Wiley & Sons, Inc.